THERMAL DESIGN OF AEROASSISTED ORBITAL TRANSFER VEHICLES

Edited by
H. F. Nelson
University of Missouri-Rolla
Rolla, Missouri

Volume 96
PROGRESS IN
ASTRONAUTICS AND AERONAUTICS

Martin Summerfield, Series Editor-in-Chief
Princeton Combustion Research Laboratories, Inc.
Monmouth Junction, New Jersey

Technical papers selected from the AIAA 22nd Aerospace Sciences Meeting, January 1984, and the AIAA 19th Thermophysics Conference, June 1984, and subsequently revised for this volume.

Published by the American Institute of Aeronautics and Astronautics, Inc.
1633 Broadway, New York, NY 10019

American Institute of Aeronautics and Astronautics, Inc.
New York, New York

Library of Congress Cataloging in Publication Data
Main entry under title:

Thermal design of aeroassisted orbital transfer vehicles.

(Progress in astronautics and aeronautics; v. 96)
Technical papers selected from the AIAA 22nd Aerospace Sciences Meeting, January 1984, and the AIAA 19th Thermophysics Conference, June 1984, and subsequently revised for this volume.
Includes index.
1. Orbital transfer (Space flight) 2. Space vehicles—Thermodynamics 3. Space vehicles—Atmospheric entry—Addresses, essays, lectures. I. AIAA Aerospace Sciences Meeting (22nd:1984:Reno, Nev.) II. AIAA Thermophysics Conference (19th:1984:Snowmass, Colo.) III. Nelson, H. F. IV. Series.
TL507.P75 vol. 96 629.1s [629.44'1] 85-3853
[TL1075]
ISBN 0-915928-94-9

Table of Contents

Preface

The need for the solution of complex technological problems involving the generation, absorption, and transport of thermal energy has evolved into the field of science called thermophysics. It encompasses the classical macro-disciplines of the engineering sciences of materials, thermofluids, heat transfer, and electromagnetic theory along with the micro-sciences of physical optics and atomic and molecular physics.

Currently, there is a great deal of interest in using aerodynamic forces generated during a pass through the atmosphere of a planet in order to change the orbit of a space vehicle, thereby reducing the expenditure of energy as compared to all-propulsive maneuvers. The typical aeroassisted orbital transfer vehicle (AOTV) maneuver has the vehicle initially in a geosynchronous Earth orbit, so that it must undergo a large velocity decrement as it transfers to low Earth orbit in order to rendezvous with the Space Shuttle. A combination of aerodynamic drag and propulsion produces this velocity decrement more efficiently than propulsion alone. During the pass through the atmosphere, the AOTV can be banked slightly to give the lift vector a lateral direction to produce an orbit plane change as well. The aeroassist concept opens new mission opportunities, especially with regard to the initiation of a permanent Space Station.

The present volume, *Thermal Design of Aeroassisted Orbital Transfer Vehicles,* is devoted to the thermophysics of AOTVs. As its title suggests, it contains a selection of recent studies dealing with the design of AOTVs. It is truly a progress volume for it illustrates the recent advances and current thinking with regard to AOTVs. The papers within this volume were selected from the Thermophysics and Atmospheric Flight Mechanics sessions of the AIAA 22nd Aerospace Sciences Meeting in Reno, Nevada, in January 1984 and the AIAA 19th Thermophysics Conference in Snowmass, Colorado, in June 1984. They were revised and updated especially for this volume and are grouped into four chapters, each of which is a distinct and coherent unit: flowfield analysis, trajectories, thermal projection, and surface effects.

Chapter I, Flowfield Analysis, contains four papers concerned with the basic equations for the flight regimes of AOTVs and the numerical calculations of flowfields over AOTVs. In the first paper, *Lee* derives the flowfield equations from gas-kinetic principles for a four-component ionized gas consisting of neutral molecules and atoms and singly ionized ions and electrons, assuming a continuum flow. The following two papers deal with the calculation of flowfields. *Gnoffo* presents solutions that he obtained by using a coupled technique involving the compressible Navier-Stokes time-asymptotic methods for blunt-body flow and a parabolized Navier-Stokes spatial marching method for the primarily supersonic shock-layer flow over low-angle bodies. *Lombard, Venkatapathy,* and *Bardina* report on a new, computationally efficient, single level, effectively explicit algorithm for flowfield calculations. With this algorithm, one is able to make accurate computations of AOTV flowfields substantially faster and less costly than the currently available explicit or two-level time-dependent implicit methods. *Moss* and *Bird* present transitional flow regime results for hypersonic re-entry conditions typical of the nose region of the Space Shuttle Orbiter in the altitude range from 150 to 92 km. The direct simulation Monte Carlo method was used to account for translational, rotational, vibrational, and chemical nonequilibrium effects.

Chapter II, Trajectories, contains five papers which focus on the trajectories of AOTVs. Included are papers on the atmospheric phase of the trajectories, the effects of multiple passes through the atmosphere, and the influence of atmospheric uncertainties. In the first paper, *Powell, Stone,* and *Naftel* discuss a six-degree-of-freedom simulation analysis for two AOTV configurations; a low lift-to-drag ratio vehicle and a medium lift-to-drag ratio vehicle, using both predictive and adaptive guidance techniques. The performance of AOTVs is discussed by *Wilhite, Arrington,* and *McCandless.* They present a method to predict the atmospheric performance of AOTVs developed from Space Shuttle Orbiter experience and the Hypersonic Arbitrary Body program. The method is applicable throughout the AOTV flight profile from the high-altitude noncontinuum regime to the low-altitude viscous continuum regime. *Rehder* reports on results that show the effects of multiple-pass trajectories on reduction of aeroheating and sensitivity to off-normal trajectories and atmospheric conditions. He demonstrates that the maximum heat transfer rate can be greatly reduced by using multiple passes. Next, *Talay, White,* and *Naftel* discuss the effects of high-altitude viscous interactions and off-nominal atmospheric structure on AOTV weight, heating, and performance. In the last paper of this chapter,

Desautel presents an analytical method for use in preliminary design studies for characterizing the aerothermodynamic regime at perigee as a function of AOTV lift-to-drag ratio and ballistic coefficient.

Chapter III, Thermal Protection, is composed of eight papers dealing with the thermal protection of AOTVs. In the first paper, *Menees* analyzes the effect of vehicle configuration on surface heating rates and the selection of heat-shield materials. He shows that drag-brake concepts can be used advantageously over a broad range of orbital transfer missions and are highly attractive compared to all-propulsive orbital change maneuvers. *Menees* and *Park* present a preliminary design and performance analysis of a conical lifting-brake AOTV concept for a broad range of missions. The beneficial effects of finite-rate catalysis, negative lift, and multiple atmospheric passes in reducing aerothermal heating and thermal protection requirements are shown. *Scott, Ried, Maraia, Li,* and *Derry* report on a study of the aerothermodynamics and thermal protection of ellipsoidally blunted, raked-off, elliptic cone-shaped AOTVs. Sizing curves are presented which enable the designer to make preliminary estimates of the required aerobrake size. *Menees, Davies, Wilson,* and *Brown* present results that establish the approximate aerothermal heating and thermal protection requirements for return missions from geosynchronous orbit to the Shuttle Orbit. A crucial factor in the drag-brake design is shown to be the wake-flow heating environment. *Pitts* and *Murbach* study the thermal response of an AOTV to aerobrake heating. The study investigates the sensitivity of the vehicle temperature to the thermal and physical properties of the materials used to fabricate the aerobrake. *Shih* and *Gay* present heat transfer data for a low lift-to-drag ratio AOTV at Mach 10. The final two papers of this chapter study the effects of radiation heating on AOTVs. *Park* discusses a computer code that calculates the radiative properties of nonequilibrium air in low-density regimes typical of AOTV trajectories. Sample results are presented which show the departure of radiative power emission from the binary scaling law at low densities. *Sutton* reviews and compares the experimental data for air radiation from ground-based tests, the Fire II project, and Apollo 4 flights.

Chapter IV, Surface Effects, consists of five papers which focus on phenomena occurring at the surface of an AOTV. The majority of these papers involve comparisons of theory with Space Shuttle data. *Zoby, Gupta,* and *Simmonds* develop a temperature-dependent, oxygen-surface, reaction-rate coefficient for altitudes corresponding roughly to AOTV perigee from comparison with Space Shuttle data.

Gupta, Scott, and *Moss* derive the surface slip equations for low Reynolds number multicomponent binary- and single-species gas flows. *Shinn* and *Simmonds* study the viscous shock-layer heating in slip-flow regimes. They demonstrate that the heating reduction due to slip-flow effects becomes greater as the flow becomes more rarefied. *Park's* contribution deals with dissociating and ionizing non-equilibrium flows for density and velocity regimes appropriate to AOTVs. He allows the vibrational and electronic temperatures to be different from the gas temperature and considers viscous transport phenomena. The final paper of this chapter deals with the magnitude of the atom recombination coefficients needed if metal heat shields are to be effective. Low catalytic activity is desired, and tests are currently being run to identify prospective surface materials with low catalytic efficiency for both oxygen and nitrogen atom recombination.

Many individuals have provided invaluable guidance and assistance during the planning and preparation of this volume. The Editorial Committee provided great support in the review of manuscripts. I acknowledge the contribution of Dr. C. Perry Bankston, who organized the thermophysics sessions at the AIAA 22nd Aerospace Sciences Meeting, and the support of Dr. George R. Cunnington Jr. who served as Technical Program Chairman of the AIAA 19th Thermophysics Conference. Furthermore, I am grateful to Dr. Martin Summerfield, Editor-in-Chief of the *AIAA Progress in Astronautics and Aeronautics* series. Finally, appreciation must be expressed to Jeanne Godette, Managing Editor of the Progress Series, for her assistance and support in editing this volume.

H. F. Nelson
October 1984

Editorial Committee for Volume 96

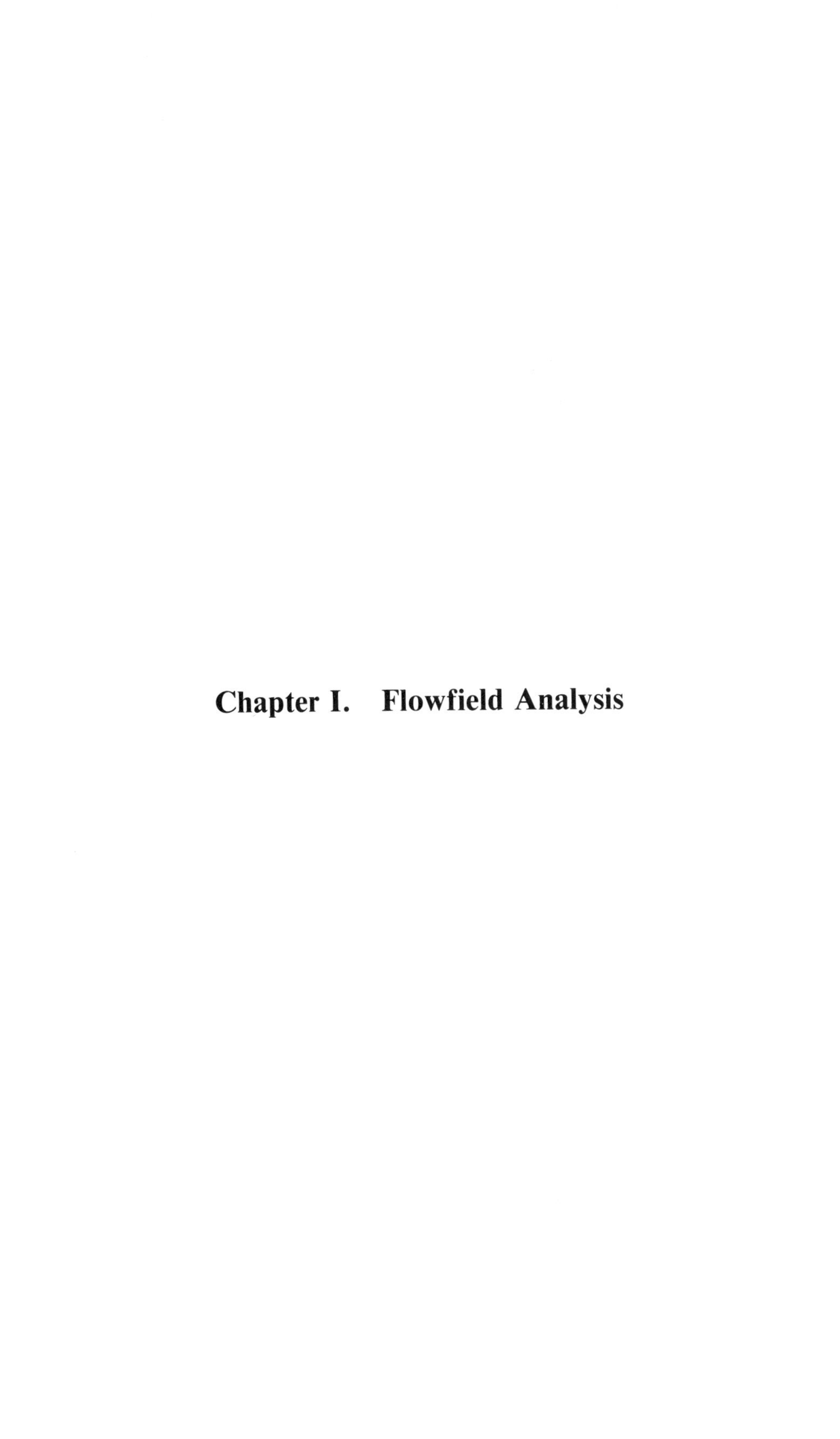

Chapter I. Flowfield Analysis

Basic Governing Equations for the Flight Regimes of Aeroassisted Orbital Transfer Vehicles

Jong-Hun Lee*

NASA Ames Research Center, Moffett Field, California

Abstract

The basic governing equations for the low-density, high-enthalpy flow regimes expected in the shock layers over the heat shields of the proposed aeroassisted orbital transfer vehicles are derived by combining and extgending existing theories. The conservation equations are derived from gas kinetic principles for a four-component ionized gas consisting of neutral molecules, neutral atoms, singly ionized ions, and electrons, assuming a continuum flow. The differences among translational-rotational, vibrational, and electron temperatures are accounted for, as well as chemical nonequilibrium and electric-charge separation. Expressions for convective and viscous fluxes, transport properties, and the terms representing interactions among various energy modes are explicitly given. The expressions for the rate of electron-vibration energy transfer, which violates the Landau-Teller conditions, are derived by solving the system of master equations accounting for the multiple-level transitions.

Nomenclature

AOTV = aeroassisted orbital transfer vehicle

$c_{pe,s}$ = specific heat of species s at constant pressure for electronic excitation

---

Presented as Paper 84-1729 at AIAA 19th Thermophysics Conference, Snowmass, Colorado, June 25-28, 1984.

*National Research Council Research Associate, Computational Chemistry and Aerothermodynamics Branch.

$c_{pf,s}$ = frozen specific heat of species s at constant pressure for heavy particles and electron
$c_{p,s}$ = specific heat of species s per unit mass at constant pressure
$c_{pv,s}$ = specific heat of molecular species at constant pressure for vibrational excitation
c_s = random thermal velocity of species s with respect to mass-averaged velocity
$c_{v,s}$ = specific heat of species s at constant volume
$\mathscr{D}_e$ = effective diffusion coefficient of electron
$\mathscr{D}_s$ = effecctive diffusion coefficient of species s
$\mathscr{D}_{sr}$ = binary diffusion coefficient for s-r colliding pair
$\mathscr{D}_s^a$ = ambipolar diffusion coefficient of ionic species s
e = overall thermodynamic energy per unit mass or electronic charge
e_s = thermodynamic energy of species s
e_s^o = chemical energy of species s
e_v = overall vibrational energy of molecules
$e_{v,s}$ = vibrational energy of species s
$e_{v,s}^*$ = equilibrium vibrational energy of species s at translational temperature T
$e_{v,s}^{**}$ = equilibrium vibrational energy of species s at electron temperature T_e
$\vec{E}$ = electric field
$E_{i,f}$ = first ionization energy
f_e = velocity distribution function of electron
f_s = velocity distribution function of species s
$g_{i,s}$ - degeneracy of electronic level i for species s
h_e = static enthalpy of electron
h_s = static enthalpy of species s
h_s^o = heat of formation of species s
$\vec{j}_e$ = conduction current of electron
k_{ij} = rate coefficient for vibrational transition from state i to j owing to e-V process
K_o = equivalent heat conductivity
K_c = rate coefficient for chemical reaction
K_{ij} = rate coefficient for vibrational transition from state i to j owing to T-V process

m_e = electron mass
m_s = mass of species s
M_s = molecular weight of species s
n_c = number density of heavy-particle collision partners
n_e = number density of electron
n_i = number density of ith excited state
n_{iE} = equilibrium number density in state i
n_I = overall number density of ions
n_M = number density of molecule
n_s = number density of species s
$\dot{n}_{e,f}$ = rate of ionization by fth electron impact ionization process
$\hat{N}$ = Avogadro's number
NEQAIR = nonequilibrium air radiation
p_e = static pressure of electron
p_s = static pressure of species s
P = static pressure of mixture
$\vec{q}$ = overall heat-flux vector
$\vec{q}_e$ = heat-flux vector of electron
$\vec{q}_s$ = heat-flux vector of species s
Q_{eV} = rate of electron thermal energy consumed for vibrational excitation by e-V process
Q_{rad} = rate of radiation energy loss
QSS = quasisteady state
R_{ij} = equilibrium transition rate from state i to j
t = time
T = translational-rotational temperature of heavy particles
T_e = electron temperature
T_v = vibrational temperature
$\vec{u}_e$ = mean velocity of electron
$\vec{u}_s$ = mean velocity of species s
$\vec{u}_o$ = mass-averaged velocity
$\vec{v}_s$ = particle velocity of species s
$\vec{V}_e$ = diffusion velocity of electron
$\vec{V}_s$ = diffusion velocity of species s
$\dot{w}_s$ = mass production rate of species s per unit volume
$\hat{\dot{w}}_s$ = molar production rate of species s
x^j = jth component of general orthogonal coordinates
X_i = normalized number density of ith excited state

y_s = molar fraction of species s
Z_s = ionic valency of species s
γ_t = total molar concentration of mixture
γ_s = molar concentration of species s per unit mass

δ^{ij} = Kronecker delta
ε = degree of charge separation
$\varepsilon_{int,s}$ = internal energy of species s
η = frozen thermal conductivity of heavy particles
η_e = frozen thermal conductivity of electrons owing to collisions between electrons and all particles
η'_e = frozen thermal conductivity of electrons owing to collisions between electrons
η_v = frozen thermal conductivity of vibrational energy owing to collisions between molecules and all particles
η'_v = frozen thermal conductivity of vibrational energy owing to collisions between molecules

$\theta_{el,s}$ = characteristic temperature of species s for electronic excitation

$\theta_{v,s}$ = characteristic temperature of molecular species s for vibration
κ = Boltzmann constant
μ = viscosity of mixture
ν = frequency of oscillator

ν^*_{er} = effective collision frequency of electrons with heavy particles

ρ = overall mass density

ρ_e = mass density of electrons
ρ_M = overall mass density of molecules
ρ_s = mass density of species s
σ_{er} = collision cross section for electron/heavy-particle interaction

τ_s = vibrational relaxation time of molecular species s for T-V process

τ_{es} = vibrational relaxation time of molecular species s for e-V process
τ^{ij} = overall viscous shear stress
ϕ_s = physical property of species s

<u>Subscripts</u>

e = electron
i,j = ith and jth vibrational states
s = species s; 1 = N, 2 = O, 3 = N_2, 4 = O_2, 5 = NO, 6 = N^+, 7 = O^+, 8 = N_2^+, 9 = O_2^+, 10 = NO^+, 11 = e^-

<u>Superscripts</u>

i,j = ith and jth components in general orthogonal coordinates

I. Introduction

Aeroassisted orbital transfer vehicles (AOTVs) have been proposed recently as a means of better implementing man's presence in space.[1-5] The primary mission of the AOTVs will be to carry transport-satellite payloads between a low Earth orbit and other Earth-centered orbits. Presently, the Space Shuttle can carry the payloads from the ground only up to the low orbit, and expendable upper-stage rockets must be used to carry them from there to other orbits. The cost of the expended upper-stage rocket and the cost of the fuel used by the Shuttle in transporting the rocket from the surface make the procedure inefficient. A space-based, reusable AOTV would eliminate both these costs. On return to the low orbit from other distant orbits, the AOTV would utilize aerodynamic drag produced over its heat shield to decrease velocity instead of using rocket fuel for retrobraking.

Due to this requirement, an AOTV must fly through the upper atmosphere of Earth, where density is low, at a superorbital velocity. In such an environment, the flow behind the bow shock wave formed over the AOTV's heat shield will tend to be chemically and thermally out of equilibrium. The flow energy will in general be sufficiently high to cause significant ionization. In the nonequilibrium region of the flow, the thermal energy of the gas will be higher than the equilibrium value because absorption of energy by dissociation does not take place to the extent dictated by the equilibrium conditions. However, since ionization rates are usually faster than dissociation rates, ionization approaches the local equilibrium value corresponding to the locally high thermal energy more closely. Hence, ionization can exceed its true equilibrium value. The combination of high thermal energy and the ionization overshoot causes enhancement of radiation. Determining the magnitude of radiation in this environment is considered to be one of the most important problems associated with AOTVs.[6]

At the peak-heating point in the flight trajectory of the AOTV, the shock-layer pressure over its heat shield is of the order of 0.01 atm.[5] At such pressures, translational, rotational, and vibrational modes of thermal energy tend to establish a Boltzmann distribution, though not necessarily at the same temperature. Rotational temperature tends to equilibrate very fast with translational temperature and, hence, can be considered to be equal to heavy-particle translational temperature T. Electron temperature T_e deviates from heavy-particle translational temperature T because of the slow rate of energy transfer between electrons and heavy particles caused by the large mass disparity between electrons and heavy particles.[7] Vibrational temperature T_v departs from both electron temperature T_e and heavy-particle translational temperature T because of the slow equilibration of vibrational energy with electron and translational energies. Both electron temperature T_e and vibrational temperature T_v play important roles in determining chemical reaction rates and radiation intensity.[6] In the low-density flow regimes under consideration, transport phenomena become important. Transport properties of a gas mixture, such as diffusion coefficient, viscosity, and thermal conductivity, must be evaluated, properly taking account of the differences among these temperatures. In addition, a charge separation occurs in the flowfields because electrons and ions drift with different diffusion velocities, thus producing an electric field $\vec{E}$ (see Ref. 7). In the presence of the electric field so produced, an electric force acts on the charged particles and changes the energy of the particles. The electric field also affects the diffusion processes of electrons and ions.

A procedure for determining radiative properties in such environments has been developed recently by Park,[8] using the concept of quasisteady-state distribution of internal states. The rates of collisional and radiative excitation and de-excitation of internal states are much faster than the rates of change of thermodynamic properties in most continuum situations.[8] As a result, the rates of transition that populate one state can be taken to be approximately equal to those that depopulate the state. This relationship, referred to as a quasisteady-state (QSS) form of the master equation,[8] yields a system of linear algebraic equations with number densities of internal states as unknowns in the left hand side and the species number densities (i.e., the sum of all internal states of the chemical species) as the given known variables in the right-hand side. By solving this system of equations, one obtains the number densities of arbitrary specified internal states as linear functions of the species number densities. Following this procedure, Park[8] developed a computer code named nonequilibrium air radiation (NEQAIR) for use in the flight regimes of the AOTVs. To utilize this code, one must specify the species number densities and the various temperatures that control the transition rates, accounting for the nonequilibrium phenomena mentioned above.

The problems of finite-rate chemical reactions, nonequilibrium ionization, and vibrational relaxation in a flow have been studied extensively, but separately, in the past.[9-12] In most studies of finite-rate chemically reacting flows, the peculiar features of nonequilibrium ionization, such as multitemperature phenomena or charge separation, have been ignored.[9] Under this assumption, a chemically relaxing flow can be solved adequately, provided chemical reaction-rate constants are known. Appleton and Bray[7] developed the conservation equations for a three-component plasma consisting of neutral atoms, single-ionized positive ions, and electrons; their equations make it possible to calculate the differences between the translational temperature T of heavy particles, the electron temperature T_e, and a charge separation. It was also shown that radiation phenomena affects the behavior of electrons. However, their treatment was confined to a monatomic gas mixture. On the other hand, there are numerous studies of vibrational relaxation in gases and its influence on the bulk thermodynamic properties; the results are well summarized in the literature.[10-12] However, those studies treated relatively low-temperature gases in which ionization is negligible.

To my knowledge, there has not been a study of the flow situation in which the above three phenomena occur simultaneously with radiative heat transfer. Consequently, there is a need to derive the basic governing equations accounting for these phenomena. It is possible today to determine the flow properties through numerical computation. Highly sophisticated numerical techniques are currently available to solve the flowfields around an entry vehicle (see, e.g., Refs. 13-15). In order to carry out numerical computation of the flowfield around an AOTV, however, one must first derive a set of conservation equations that correctly accounts for these phenomena. The purpose of the present paper is to derive a set of conservation equations that includes all significant physicochemical phenomena by combining and extending existing theories. The derivation of conservation equations is intended to be compatible with the requirements of the NEQAIR code. According to NEQAIR, the accuracy in the computed number densities of any given internal state is the same as that of the number density of the species that contains the state. Tolerating an error in the number density of the order of 1%, say, means that basic governing equations need to be accurate only to an accuracy of about 1%.

II. Assumptions

In the analysis leading to the basic governing equations, the following assumptions are made:

1) Continuum assumption: The flow is assumed to be a continuum. The AOTV, as proposed, is approximately disk-like with a diameter of about 30 m; this produces a shock standoff distance of the order of 1 m. The mean free path behind the shock wave is the order of 1 mm in the peak heating flight regimes (altitudes of about 70-90 km). Consequently, the Knudsen number, which is the ratio of the mean free path and the shock standoff distance, is much less than 1. This validates the continuum approach in the above flight regimes.

2) Assumption of a unique vibrational temperature: Vibrational energy levels of a molecule are assumed to be populated by a Boltzmann distribution corresponding to a

vibrational temperature. This assumption holds rigorously for low vibrational states.[16] High vibrational levels deviate from this assumption, but their energy content is small. Vibrational temperatures of the various molecules (N_2, O_2, NO, N_2^+, O_2^+, and NO^+) are taken to be the same because of the highly efficient vibration-vibration (V-V) energy transfer.[17,18] The harmonic oscillator model is applied to describe the molecular vibration. The harmonic oscillator model is valid rigorously only for the lower vibrational energy levels, the high vibrational states contribute negligibly to the total energy, however, and therefore can be neglected.

3) Three-temperature assumption: The heavy-particle translational temperature T, vibrational temperature T_V, and electron temperature T_e are different.[6] The rotational mode of molecules is fully equilibrated with the translational mode of heavy particles, and, therefore, rotational temperature equals the translational temperature. The electronic levels are excited by the translational mode of electrons and are in Boltzmann distribution corresponding to the electron temperature T_e. This last assumption is approximately true for the low electronic states.[8] The highly excited electronic states deviate from the assumption, but their contribution to total energy is negligibly small. The assumption does not preclude the later computation of the non-Boltzmann number densities of the high electronic states in the NEQAIR code. As mentioned earlier, the species number densities computed by the present conservation equations are to be used only in the right-hand side of the quasisteady-state master equation in the NEQAIR code. The number densities of the high electronic states will be obtained by solving the equations in the code by refining the crude assumption made in the present work.

III. Conservation Equations

The principle of conservation dictates that the various average quantities, such as mass, momentum, and energy, be conserved in a flowing gas. If one expresses the value of a conserved property averaged over a macroscopically small volume element by $\langle \phi_s \rangle$, the conservation law states

[time variation of $\langle \phi_s \rangle$]

= [difference of inflow and outflow of $\langle \phi_s \rangle$]

+ [production rate of $\langle \phi_s \rangle$ owing to sources contained]

Consequently, a general macroscopic conservation equation per unit volume for a property of species s may be written as[7]

$$\frac{\partial}{\partial t} (n_s \langle \phi_s \rangle) + \frac{\partial}{\partial x^j} (n_s \langle \phi_s v_s^j \rangle) = S(\langle \phi_s \rangle) \qquad (1)$$

where n_s is the number density of particles of type s; $\langle v_s^j \rangle$ is the average value of the jth component of particle velocity v_s^j; t is time; x^j represents the jth component of the general orthogonal coordinates; and the source term $S(\langle \phi_s \rangle)$ represents the change in $\langle \phi_s \rangle$ caused by external forces and the interactions among s-type particles and all the remaining particles, including the s-type particles themselves. In order to determine the flow quantities, the following conservation equations are necessary: 1) species mass, 2) overall mass, 3) electron momentum, 4) overall momentum, 5) electron energy, 6) vibrational energy, and 7) overall energy. The three temperatures (translation-rotation, vibration, and electron) are determined from Eqs. (5-7). The electron momentum equation (3) is necessary to predict the electric field $\vec{E}$ that is produced.

Mass Conservation Equation for Each Species

The mass conservation equation for species s is obtained by identifying the property $\langle \phi_s \rangle$ with the species mass m_s. The source term in this case is the mass production rate of species s, due to chemical reactions, per unit volume; it is denoted by $\dot{w}_s$. Then, the species mass conservation equation is given by

$$\frac{\partial \rho_s}{\partial t} + \frac{\partial}{\partial x^j} (\rho_s u_o^j + \rho_s V_s^j) = \dot{w}_s \qquad (2)$$

where $\rho_s = m_s n_s$ is the s-species mass density; u_o^j is the jth component of the mass-averaged velocity given by

$$u_o^j = \sum_s \rho_s u_s^j / \sum_s \rho_s$$

$u_s^j = \langle v_s^j \rangle$ is the jth component of the mean velocity of the species; $V_s^j = u_s^j - u_o^j$ is the diffusion velocity of the species.

Overall Mass conservation Equation

The overall mass conservation equation is obtained by summing Eq. (2) over all species and noting that the overall mass production rate

$$\sum_s \dot{w}_s$$

and the overall mass flux owing to diffusion

$$\sum_s \rho_s V_s^j$$

are zero. Thus, the overall mass conservation equation is given by

$$\frac{\partial \rho}{\partial t} + \frac{\partial}{\partial x^j} (\rho u_o^j) = 0 \tag{3}$$

where

$$\rho = \sum_s \rho_s$$

is the overall mass density.

Electron Momentum Conservation Equation

The momentum conservation equation for each species is obtained by identifying the property $\langle \phi_s \rangle$ with $m_s u_s^i$. The source term $S(m_s u_s^i)$ consists of the interaction force owing to collisions $\vec{F}_{int,s}$ and the electric field force $\vec{F}_{ele,s}$ which acts on the charged particle only. The interaction force term $\vec{F}_{int,s}$ is the summation of the terms $\vec{F}_{ela,s}$ which accounts for changes in momentum during elastic collisions, and the term $\vec{F}_{inela,s}$ which accounts for changes in momentum during inelastic collisions. Since we can assume that all species share the total momentum in proportion to their relative mass density, the contribution of inelastic terms may be neglected. Consequently, the ith momentum equation for s-species is given by

$$\frac{\partial}{\partial t}(\rho_s u_s^i) + \frac{\partial}{\partial x^j}(\rho_s u_o^i u_o^j) + \frac{\partial}{\partial x^j}(\rho_s u_o^i V_s^j + \rho_s u_o^j V_s^i)$$

$$+ \frac{\partial p_s}{\partial x^i} - \frac{\partial \tau_s^{ij}}{\partial x^j} = F^i_{ele,s} + F^i_{ela,s} \qquad (4)$$

where $p_s = \rho_s \langle c_s^2 \rangle / 3$, is defined as the average pressure over the normal components of the thermal momentum flux, and $\vec{c}_s$ is the random thermal velocity with respect to the mass-averaged velocity $\vec{u}_o$. The viscous stress $\tau_s^{ij} = -[\rho_s \langle c_s^i c_s^j \rangle - \delta^{ij} p_s]$ is defined as the negative of the thermal momentum with the pressure p_s subtracted from the normal components, and where δ^{ij} is the Kronecker delta; $\delta^{ij} = 1$ if $i = j$, $\delta^{ij} = 0$ if $i \neq j$.

The electric field force $\vec{F}_{ele,s}$ per unit volume is given (in Ref. 7) by

$$\vec{F}_{ele,s} = n_s e Z_s \vec{E} \qquad (5)$$

where e is the electronic charge; Z_s is the ionic valency, which is -1 for electrons and 1 for single-ionized positive ions; and $\vec{E}$ is the induced electric field. The electric field $\vec{E}$ is assumed to be a continuous function of space and time.

The elastic interaction force term $\vec{F}_{ela,s}$ can be obtained through the standard methods of kinetic theory by assuming that species velocity distribution functions have Maxwellian forms corresponding to the mean velocities of the species. Thus, the elastic momentum transfer term for electrons becomes (from Ref. 19)

$$\vec{F}_{ela,s} = -n_e m_e \sum_r (\vec{u}_e - \vec{u}_r) \frac{n_r}{3n_e \kappa T_e} \int \sigma_{er} c_e'^3 f_e \, d\vec{c}_e' \qquad (6)$$

where κ is the Boltzmann constant, $\vec{c}_e'$ is the random velocity of electrons with respect to the mean velocity of electrons, f_e is the velocity distribution function of electrons, and σ_{er} is the collision cross section of electrons with heavy particles. The first effective collision frequency of electrons with heavy particles ν^*_{er} can be defined as

$$\nu^*_{er} = \frac{n_r}{3n_e \kappa T_e} \int \sigma_{er} c_e'^3 f_e \, d\vec{c}_e' \qquad (7)$$

The quantity of ν^*_{er} can be evaluated when the form of the velocity distribution function f_e and the collision cross sections σ_{er} are known. The resulting expression of $\vec{F}_{ela,s}$ takes the form

$$\vec{F}_{ela,e} = -n_e m_e \sum_r \nu^*_{er} (\vec{u}_e - \vec{u}_r) \qquad (8)$$

Consequently, the electron momentum equation may be written as

$$\frac{\partial}{\partial t}(\rho_e u_e^i) + \frac{\partial}{\partial x^j}(\rho_e u_o^i u_o^j) + \frac{\partial}{\partial x^j}(\rho_e u_o^i V_e^j + \rho_e u_o^j V_e^i)$$

$$+ \frac{\partial p_e}{\partial x^i} - \frac{\partial \tau_e^{ij}}{\partial x^j} = -n_e e E^i - n_e m_e \sum_r \nu^*_{er} (u_e^i - u_r^i) \qquad (9)$$

Overall Momentum Equation

The overall momentum equation is obtained by summation over all species. Note that the terms such as

$$\sum_s \rho_s V_s^j$$

are zero by definition. In addition, the overall average collisional interaction force is zero because of conservation of momentum during collisions. Thus, the sum of Eq. (4) becomes

$$\frac{\partial}{\partial t}(\rho u_o^i) + \frac{\partial}{\partial x^j}(\rho u_o^i u_o^j) + \frac{\partial p}{\partial x^i} - \frac{\partial \tau^{ij}}{\partial x^j} = \sum_s n_s e Z_s E^i \quad (10)$$

where τ^{ij} is the overall viscous shear stress defined as

$$\tau^{ij} \equiv \sum_s \tau_s^{ij}$$

p is the total pressure defined as

$$p = \sum_s p_s$$

and n_s is the number density of charged species (electrons and ions). If the gas is electrically neutral, the right-hand side of Eq. (10) becomes zero.

Electron Energy Conservation Equation

The general energy conservation equation for species s is obtained by identifying the property $\langle \phi_s \rangle$ with the total energy of the s-species $m_s u_s^2\ 2 + \langle \varepsilon_{int,s} \rangle$, where $m_s u_s^2/2$ is the average kinetic energy of translation, and $\langle \varepsilon_{int,s} \rangle$ is the average internal energy (rotational, vibrational, electronic, and chemical) of the particles. The source term $S(\langle \phi_s \rangle)$ consists of the rate of work done by the electric field on charged particles $P_{ele,s}$, the rate of energy supplied in elastic collisions $Q_{ela,s}$, and the rate of energy transport due to inelastic collisions including radiation loss $Q_{inela,s}$.

The electric field term $P_{ele,s}$ is given (in Ref. 7) by

$$P_{ele,s} = n_s e Z_s \vec{E} \cdot \vec{u}_s \quad (11)$$

The elastic collision term $Q_{ela,s}$ can be obtained through the kinetic theory and is given (in Ref. 19) by

$$Q_{ela,e} = -\frac{m_e^2}{m_r} n_r \left(1 - \frac{T_r}{T_e}\right) \int \sigma_{er} c_e'^3 f_e \, d\vec{c}_e'$$

$$- m_e n_r (\vec{u}_e - \vec{u}_r) \cdot \vec{u}_g \int \sigma_{er} c_e' f_e \, d\vec{c}_e' \qquad (12)$$

where T_e and T_r are the electron temperature and the heavy-particle translational temperature, respectively, and $\vec{u}_g = (m_r \vec{u}_r + m_e \vec{u}_e)/(m_r + m_e)$ is the average gas velocity of the two species. Since $m_e \ll m_r$, $\vec{u}_g$ may be approximated by $\vec{u}_r$. The first integral in the right-hand side of Eq. (12) has been defined in terms of ν_{er}^* by Eq. (7). The second effective collision frequency ν_{er}^{**} can be defined as follows:

$$\nu_{er}^{**} = \frac{n_r}{n_e} \int \sigma_{er} c_e' f_e \, d\vec{c}_e' \qquad (13)$$

The second effective collision frequency ν_{er}^{**} may be approximated by the first effective collision frequency ν_{er}^* (Ref. 19). Thus, Eq. (12) becomes

$$Q_{ela,e} = 2 m_e n_e \sum_r \frac{\nu_{er}^*}{m_r} \left(\frac{3}{2} \kappa T_r - \frac{3}{2} \kappa T_e\right)$$

$$+ m_e n_e \sum_r \nu_{er}^* (\vec{u}_r - \vec{u}_e) \cdot \vec{u}_r \qquad (14)$$

The first term on the right-hand side of Eq. (14) accounts for the change in electron temperature caused by elastic encounters between electrons and heavy particles, and the second term represents the frictional heating of electrons caused by differences in the electron velocity and heavy-particle velocities.

The inelastic energy-transport processes for electrons consist of electron-impact ionization, electron-vibration energy transfer, and radiative energy loss. Thus, the inelastic collision term $Q_{inela,e}$ for electrons may be expressed as

$$Q_{inela,e} = -\sum_f \dot{n}_{e,f} E_{i,f} - Q_{eV} - Q_{rad} \tag{15}$$

The first term of the right-hand side of Eq. (15) represents the energy loss of electrons by electron-impact ionizing collisions, wherein $\dot{n}_{e,f}$ is the rate of ionization by the fth electron-impact ionization process and $E_{i,f}$ is the corresponding ionization energy.[19] The second term, Q_{eV}, expresses the electron thermal energy consumed for the excitation of molecular vibrational energy by the e-V process. The actual form of Q_{eV} will be derived later. The final term in Eq. (15), Q_{rad}, represents radiation energy loss. For air, radiation is caused mostly by free-free bremsstrahlung, free-bound ionic recombination, atomic bound-bound line transitions, and molecular bound-bound electronic-vibrational-rotational transitions. The bremsstrahlung and ionic recombination processes directly deplete electron thermal energy. The atomic and molecular bound-bound processes deplete electron thermal energy indirectly because the electronic-vibrational-rotational upper states made vacant by the processes must be repopulated by electron collisions at the expense of electron thermal energy.[8]

By using Eqs. (11-15), one obtains the electron energy equation as

$$\begin{aligned}
&\frac{\partial}{\partial t}\left[\rho_e\left(\frac{1}{2}u_o^2 + u_o^i V_e^i + e_e\right)\right] + \frac{\partial}{\partial x^j}\left[\rho_e u_o^j\left(\frac{1}{2}u_o^2 + e_e\right)\right] \\
&+ \frac{\partial q_e^j}{\partial x^j} + \frac{\partial}{\partial x^j}\left(\frac{1}{2}\rho_e u_o^2 V_e^j + \rho_e u_o^j u_o^i V_e^i\right) + \frac{\partial}{\partial x^i}(u_o^i p_e) \\
&- \frac{\partial}{\partial x^j}(u_o^i \tau_e^{ij}) = -n_e e E^i u_e^i + 2m_e n_e \sum_r \frac{\nu_{er}^*}{m_r}\left(\frac{3}{2}\kappa T_r - \frac{3}{2}\kappa T_e\right) \\
&+ m_e n_e \sum_r \nu_{er}^*(u_r^i - u_e^i)u_r^i \\
&- \sum_f \dot{n}_{e,f} E_{i,f} - Q_{eV} - Q_{rad}
\end{aligned} \tag{16}$$

where $e_e = [(3/2)\kappa T_e + \langle \varepsilon_{int,e} \rangle]/m_e$ is the total thermodynamic energy per unit mass of electron, $q_e^j = n_e \langle c_e^j \rangle [(1/2) m_e \langle c_e^2 \rangle + \langle \varepsilon_{int,e} \rangle]$ is the jth component of the heat-flux vector for electrons due to random motion, and $\langle \varepsilon_{int,e} \rangle$ is the average internal energy of electron and its value is zero.

Overall Energy Conservation Equation

The overall energy conservation equation is obtained by summation over all species. Noting that

$$\sum_s \rho_s V_s = 0$$

one obtains

$$\frac{\partial}{\partial t}\left[\rho\left(\frac{1}{2}u_o^2 + e\right)\right] + \frac{\partial}{\partial x^j}\left[\rho u_o^j\left(\frac{1}{2}u_o^2 + e\right)\right] + \frac{\partial q^j}{\partial x^j}$$

$$+ \frac{\partial}{\partial x^i}(u_o^i p) - \frac{\partial}{\partial x^j}(u_o^i \tau^{ij}) = \sum_s n_s e Z_s E^i u_s^i - Q_{rad} \qquad (17)$$

where e is the overall thermodynamic energy per unit mass and is defined by

$$e = \sum_s \rho_s e_s / \rho$$

where $e_s = [(3/2)\kappa T_s + \langle \varepsilon_{int,s} \rangle]/m_s$ is the thermodynamic energy of species s. The jth component of the overall heat-flux vector q^j is defined by

$$q^j = \sum_s q_s^j$$

where $q_s^j = n_s \langle c_s^j \rangle [(1/2) m_s \langle c_s^2 \rangle + \langle \varepsilon_{int,s} \rangle]$ is the jth component of the heat-flux vector for species s. The first term on the right-hand side in Eq. (17) expresses the work done on the charged particles by the electric field $\vec{E}$; and the second term, Q_{rad}, represents the radiative loss term described in Sec. III under "Electron Energy Conservation Equations." The contribution of the particle collisions does not appear because the sum of the collision terms is zero.

IV. Vibrational Relaxation

In order to derive the equation that governs the vibrational relaxation process, we first consider the gas to be quiescent. As mentioned in Sec. II, the number density of vibrationally excited molecules is assumed to have a Boltzmann distribution with vibrational temperature T_v. The population distribution relaxes in the event of vibrational excitation to a final equilibrium Boltzmann distribution via the continuous sequence of a Boltzmann distribution with T_v (Ref. 20). The vibrational relaxation rate equation is derived by considering the separate contributions from the translation-vibration (T-V) and the electron-vibration (e-V) energy-exchange processes.

T-V Energy-Exchange Process

Neglecting the effects of rotation-vibration and dissociation, the rate of change of the number density of vibrationally excited state n_i in the T-V process, where the subscript s is omitted, is given by

$$\frac{\partial n_i}{\partial t} = \sum_{j=0}^{\infty} (K_{ji} n_j - K_{ij} n_i) n_c \tag{18}$$

where K_{ij} is the rate coefficient for the transition from state i to j due to the T-V process and n_c is the number density of heavy-particle collision partners. The following rate equation for the T-V process can be deduced from Eq. (18) through the Landau-Teller conditions (see Ref. 10 for details)

$$\frac{\partial e_v}{\partial t} = \frac{e_v^*(T) - e_v}{\tau} \tag{19}$$

where e_v is the vibrational energy per unit mass, $e_v^*(T)$ is the equilibrium vibrational energy at translational temperature T, and τ is a relaxation time for the T-V process. The relaxation time τ is given (in Ref. 10) by

$$\tau = \frac{1}{n_c K_{10}\left(1 - e^{-\theta_v/T}\right)} \tag{20}$$

where K_{10} is the rate coefficient for transition from the first excited state to the ground state; and $\theta_v = h\nu/\kappa$ is the characteristic temperature for vibration, where h is Planck's constant and ν is a frequency of a harmonic oscillator.

e-V Energy-Exchange Process

In the case of the e-V process, it is doubtful that the Landau-Teller conditions are satisfied. In the Landau-Teller conditions: 1) only the transitions between adjacent energy levels are allowed, 2) the molecules are assumed to oscillate frequently during a collision, and 3) the trajectories of the colliding particles are unaffected by the vibrational oscillation. All these conditions are violated by an e-V encounter. Consequently, a more general method to describe the e-V process accounting for multiple-quantum transitions is necessary. This can be done following the classical treatment of Keck and Carrier.[16]

The rate of change of the number density of excited state n_i in the e-V process is given by the master equation

$$\frac{\partial n_i}{\partial t} = \sum_{j=0}^{\infty} (k_{ji}n_j - k_{ij}n_i)n_e \tag{21}$$

where k_{ij} is the rate coefficient for the transition from state i to j, due to the e-V process, and n_e is the number density of electrons. Based on the classical treatment, Eq. (21) can be reduced to an equivalent diffusion equation which describes analogously the flow of heat in a medium.[16] By solving the equivalent heat-conduction problem, one can obtain approximately the following

Landau-Teller-type rate equation for the e-V energy exchange process (see the Appendix for further details):

$$\frac{\partial e_v}{\partial t} = \frac{e_v^{**}(T_e) - e_v}{\tau_e} \tag{22}$$

where $e_v^{**}(T_e)$ is the equilibrium vibrational energy at electron temperature T_e and τ_e is the relaxation time for the e-V process. The relaxation time τ_e is given by (see Appendix)

$$\tau_e = \frac{1}{n_e(1 - e^{-\theta_v/T_e})^2 \frac{1}{2}\int k_{oj}(j)^2 \, dj} \tag{23}$$

where k_{oj} is the rate coefficient from the ground state to the upper state j. The multiple-quantum transitions are not likely to occur at low temperature, and the possibility for multiple-quantum transitions increases rapidly with temperature. Thus, there is a strong dependency of τ_e on temperature compared with τ for single-quantum transitions [see Eqs. (20) and (23)].

Vibrational Energy Conservation Equation

One can obtain the vibrational rate equation by superimposing the contributions of the T-V and e-V energy-exchange processes. The vibrational relation rate equation in terms of vibrational energy e_v for the quiescent gas becomes

$$\frac{\partial e_v}{\partial t} = \frac{e_v^{*}(T) - e_v}{\tau} + \frac{e_v^{**}(T_e) - e_v}{\tau_e} \tag{24}$$

Next, we derive the vibrational energy conservation equation for a flowing gas. The vibrational energy conservation equation per unit volume is obtained by identifying the property $\langle \phi_s \rangle$ with $m_s e_{v,s}$, where $e_{v,s}$ is the vibrational energy for molecular species s. The source term $S(m_s e_{v,s})$, which is the nonequilibrium production rate inside the volume, is the sum of the T-V and e-V relaxation rates. Note that the net energy flux inflow across the surfaces of the volume is caused by convection, diffusion, and heat conduction resulting from the vibrational temperature gradient. The heat conduction takes place as a result of energy exchange between vibrational energies, that is, through the vibration-vibration (V-V) process. Thus, one obtains the vibrational energy conservation equation per unit volume as

$$\frac{\partial}{\partial t}(\rho_s e_{v,s}) + \frac{\partial}{\partial x^j}(\rho_s e_{v,s} u_o^j)$$

$$= \frac{\partial}{\partial x^j}\left(\eta'_{v,s}\frac{\partial T_v}{\partial x^j}\right) - \frac{\partial}{\partial x^j}(\rho_s e_{v,s} V_s^j)$$

$$+ \rho_s \frac{e^*_{v,s}(T) - e_{v,s}}{\tau_s} + \rho_s \frac{e^{**}_{v,s}(T_e) - e_{v,s}}{\tau_{es}} \tag{25}$$

where $\eta'_{v,s}$ is the thermal conductivity for vibrational energy, and τ_s and τ_{es} are the T-V and e-V relaxation times for molecular species s, respectively. Note that $\eta'_{v,s}$ includes only the contribution of vibration-vibration interactions, that is, of collisions between molecules and molecules. The fourth term on the right-hand side of Eq. (25) expresses the rate of energy gain by molecular species s due to the e-V process. The overall energy gain by all molecules through the e-V processes is equivalent to the energy loss term Q_{ev} in the electron energy equation (16).

V. Simplification of Basic Governing Equations

Since the mass of an electron is very small relative to that of a heavy particle, the electron momentum and energy equations can be significantly simplified. In addition, other conservation equations, such as those for species, overall momentum, overall energy, and vibrational energy, can be rearranged in more convenient forms.

Hereafter, we will consider the mas mixture to be made up of four components--neutral molecules, neutral atoms, singly ionized positive ions, and electrons, with 11 species (N, O, N_2, O_2, NO, N^+, O^+, N_2^+, O_2^+, NO^+, and e^-) (see Ref. 6). We also specify s by denoting that $s = 1 = N$, $2 = O$, $3 = N_2$, $4 = O_2$, $5 = NO$, $6 = N^+$, $7 = O^+$, $8 = N_2^+$, $9 = O_2^+$, $10 = NO^+$, and $11 = e^-$.

Species Conservation Equation

Species conservation equation (2) can be written in terms of species molar concentration defined in the unit of moles per gram of mixture $\gamma_s = c_s/(\hat{N}m_s)$ (Ref. 21), where $c_s = \rho_s/\rho$ is the species mass fraction, $\hat{N}$ is the Avogadro's number, and m_s is the species mass. Then, Eq. (2) becomes

$$\frac{\partial}{\partial t}(\rho\gamma_s) + \frac{\partial}{\partial x^j}(\rho u_o^j \gamma_s) = -\frac{\partial}{\partial x^j}(\rho\gamma_s V_s^j) + \hat{\dot{w}}_s \tag{26}$$

where $\hat{\dot{w}}_s = \dot{w}_s/(\hat{N}m_s)$ is the molar production rate of species s.

Electron Momentum Conservation Equation

Because of the electron's small mass, we may neglect the inertial terms $\partial(\rho_e u_e^i)/\partial t$, $\partial(\rho_e u_o^i u_o^j)/\partial x^j$, and $\partial(\rho_e u_o^i V_e^j + \rho_e u_o^j V_e^i)/\partial x^j$ in the electron momentum equation (9). Since the viscous stress owing to electrons is generally negligible, the viscous term $\partial\tau_e^{ij}/\partial x^j$ is also neglected, as it was by Sutton and Sherman.[19] Thus, the electron momentum equation becomes

$$\frac{\partial p_e}{\partial x^i} = -n_e e E^i - n_e m_e \sum_{r=1}^{10} \nu_{er}^* (u_e^i - u_r^i) \tag{27}$$

If we let $u_r^i \simeq u_o^i$, Eq. (27) can be rewritten as

$$\frac{\partial p_e}{\partial x^i} = -n_e e E^i - \frac{m_e}{e} \sum_{r=1}^{10} \nu_{er}^* n_e e (u_e^i - u_o^i)$$

$$= -n_e e E^i + \frac{m_e}{e} \sum_{r=1}^{10} \nu_{er}^* j_e^i \tag{28}$$

where j_e^i is the conduction current of electrons defined as

$$j_e^i = -en_e V_e^i = -en_e(u_e^i - u_o^i) \tag{29}$$

Since there is no conduction current in the present flow-field, Eq. (28) finally becomes

$$E^i = -\frac{1}{n_e e}\frac{\partial p_e}{\partial x^i} \tag{30}$$

Physically, Eq. (30) means that the electric field E^i generated by a charge separation is proportional to the electron pressure gradient.

Overall Momentum Conservation Equation

The overall momentum conservation equation (10) can be rearranged by means of the electron momentum equation (30) as

$$\frac{\partial}{\partial t}(\rho u_o^i) + \frac{\partial}{\partial x^j}(\rho u_o^i u_o^j) = -\frac{\partial p}{\partial x^i} + \frac{\partial \tau^{ij}}{\partial x^j} + \left(1 - \frac{n_I}{n_e}\right)\frac{\partial p_e}{\partial x^i} \tag{31}$$

where

$$n_I = \sum_{s=6}^{10} n_s$$

is the overall number density of ions. As is the case for most partially ionized gas flow, it is considered that the degree of charge separation, $\varepsilon = 1 - n_e/n_I$, is very small. Therefore, we may assume that $n_e \simeq n_I$ as a numerical approximation.

Vibrational Energy Conservation Equation

If we sum over all molecular species s (s = 3, 4, 5, 8, 9, 10), Eq. (25) becomes

$$\frac{\partial}{\partial t}(\rho_M e_v) + \frac{\partial}{\partial x^j}(\rho_M e_v u_o^j)$$

$$= \frac{\partial}{\partial x^j}\left(\eta_v' \frac{\partial T_v}{\partial x^j}\right) - \frac{\partial}{\partial x^j}\left(\hat{N}\rho \sum_{s=M} m_s \gamma_s e_{v,s} V_s^j\right)$$

$$+ \rho\hat{N} \sum_{s=M} m_s \gamma_s \frac{e_{v,s}^*(T) - e_{v,s}}{\tau_s}$$

$$+ \rho\hat{N} \sum_{s=M} m_s \gamma_s \frac{e_{v,s}^{**}(T_e) - e_{v,s}}{\tau_{es}} \qquad (32)$$

where $M = 3, 4, 5, 8, 9, 10$ and

$$\rho_M = \sum_{s=M} \rho_s$$

is the overall mass density of molecules; e_v is the overall vibrational energy for the molecular species; and η_v' is the frozen thermal conductivity of vibrational energy of the molecular mixture resulting from collisions between molecules. From the harmonic oscillator model, the vibrational energy $e_{v,s}$ for species s and the equilibrium energies $e_{v,s}^*$ and $e_{v,s}^{**}$ are given by

$$e_{v,s} = \frac{\kappa}{m_s} \frac{\theta_{v,s}}{e^{\theta_{v,s}/T_v} - 1}$$

$$e_{v,s}^* = \frac{\kappa}{m_s} \frac{\theta_{v,s}}{e^{\theta_{v,s}/T} - 1} \qquad (33)$$

$$e_{v,s}^{**} = \frac{\kappa}{m_s} \frac{\theta_{v,s}}{e^{\theta_{v,s}/T_e} - 1}$$

where κ is Boltzmann's constant and $\theta_{v,s}$ is the vibrational characteristic temperature (vibrational energy gap divided by κ) for molecular species s. The overall vibrational energy e_v is given by

$$e_v = \sum_{s=M} \frac{\rho_s e_{v,s}}{\rho_s} \tag{34}$$

Because the collision partners for particular species s through the T-V process are a mixture of heavy particles including the s-species, the T-V relaxation time τ_s in Eq. (32) must be replaced by the average relaxation time $\langle \tau_s \rangle$ given by

$$\langle \tau_s \rangle = \sum_{r=1}^{10} \gamma_r \Big/ \sum_{r=1}^{10} \gamma_r / \tau_{sr} \tag{35}$$

where τ_{sr} is the relaxation time for species s with collision partner r.

Electron Energy Conservation Equation

It must be noted that the elastic energy transfer term $Q_{ela,e}$ [Eq. (14)] is deduced assuming that the species velocity distribution functions have Maxwellian forms corresponding to the mean velocities of the species. Therefore, the electron temperature T_e in Eq. (14) is defined with respect to the electron mean velocity. To be consistent with this definition, it is desirable to describe the electron energy equation in terms of the electron mean velocity $\vec{u}_e$. Thus, the alternative form of the electron energy equation is given by

$$\frac{\partial}{\partial t} \rho_e \left(\frac{1}{2} u_e^2 + \frac{3}{2} \frac{\kappa T_e}{m_e}\right) + \frac{\partial}{\partial x^j} \left[\rho_e u_e^j \left(\frac{1}{2} u_e^2 + \frac{3}{2} \frac{\kappa T_e}{m_e}\right)\right]$$

$$+ \frac{\partial q_e^j}{\partial x^j} + \frac{\partial}{\partial x^i} (u_e^i p_e) - \frac{\partial}{\partial x^j} (u_e^i \tau_e^{ij})$$

$$= -n_e e u_e^i E^i + 2 m_e n_e \sum_{r=1}^{10} \frac{\nu_{er}^*}{m_r} \left(\frac{3}{2} \kappa T - \frac{3}{2} \kappa T_e\right)$$

(continued)

$$+ m_e n_e \sum_{r=1}^{10} \nu^*_{er}(u^i_r - u^i_e)u^i_r - \sum_{f=6}^{10} \dot{n}_{e,f} E_{i,f}$$

$$- \rho\hat{N} \sum_{s=M} m_s \gamma_s \frac{e^{**}_v(T_e) - e_v}{\tau_{es}} - Q_{rad} \quad (36)$$

Letting $u^i_r \approx u^i_o$ and using the electron momentum equation (28) along with the zero conduction current condition, the first and fourth terms of the right-hand side of Eq. (36) can be rewritten as

$$-n_e e u^i_e E^i + m_e n_e \sum_{r=1}^{10} \nu^*_{er}(u^i_o - u^i_e)u^i_o$$

$$= j^i_e E^i + u^i_o \frac{\partial p_e}{\partial x^i} = u^i_o \frac{\partial p_e}{\partial x^i}$$

By neglecting the inertial term $(1/2)u^2_e$ and the viscous stress term in Eq. (36), and by means of the electron momentum equation (30), the electron energy equation becomes

$$\frac{\partial}{\partial t}\left(\frac{3}{2} n_e \kappa T_e\right) + \frac{\partial}{\partial x^j}\left(\frac{3}{2} n_e \kappa T_e u^j_e\right) + \frac{\partial q^j_e}{\partial x^j} + \frac{\partial}{\partial x^i}(u^i_e p_e)$$

$$= 2m_e n_e \sum_{r=1}^{10} \frac{\nu^*_{er}}{m_r}\left(\frac{3}{2}\kappa T - \frac{3}{2}\kappa T_e\right) + u^i_o \frac{\partial p_e}{\partial x^i}$$

$$- \sum_{f=6}^{10} \dot{n}_{e,f} E_{i,f} - \rho\hat{N} \sum_{s=M} m_s \gamma_s \frac{e^{**}_v(T_e) - e_v}{\tau_{es}} - Q_{rad} \quad (37)$$

Overall Energy Conservation Equation

The overall energy conservation equation (17) can be rearranged by means of the electron momentum equation (30). By letting $u^i_s \approx u^i_o$ for heavy particles, the first term of the right-hand side of Eq. (17) can be written as

$$\sum_s n_s e Z_s E^i u_s^i = \left(u_e^i - \frac{n_I}{n_e} u_o^i\right) \frac{\partial p_e}{\partial x^i} \tag{38}$$

Thus, the overall energy equation becomes

$$\frac{\partial}{\partial t}\left[\rho\left(\frac{1}{2} u_o^2 + e\right)\right] + \frac{\partial}{\partial x^j}\left[\rho u_o^j \left(\frac{1}{2} u_o^2 + e\right)\right]$$

$$= - \frac{\partial q^j}{\partial x^j} - \frac{\partial}{\partial x^i} (u_o^i p) + \frac{\partial}{\partial x^j} (u_o^i \tau^{ij})$$

$$+ \left(u_e^i - \frac{n_I}{n_e} u_o^i\right) \frac{\partial p_e}{\partial x^i} - Q_{rad} \tag{39}$$

VI. Various Fluxes

The various fluxes (mass, momentum, and energy) can be expressed in terms of the gradients of the physical properties and the transport properties. To a first approximation, the fluxes are derived based on a first-order kinetic theory (Chapman-Enskog linear flux theory), retaining only the first spatial derivatives of the various physical properties. The flux of mass and the flux of momentum are not appreciably affected by the internal degrees of freedom.[22] On the other hand, the flux of energy includes both the energy of translation and the energy of the internal degrees of freedom. This effect is takon into account by separating the contribution of each degree of freedom to the thermal conductivity. In addition, the forced diffusion by electric field must be considered.[23] Because electrons possess higher mobility, they try to diffuse faster than ions at first, leaving behind an excess of positive charges. The electric field produced by this charge separation has a retarding influence on electrons and an accelerating influence on ions. Thus, the electric field links together the diffusion of electrons and ions. Howe and Sheaffer[23] rigorously treated the problem of forced diffusion by electric field. However, their method involves numerical difficulties in the actual calculation because of the very small degree of charge separation. Therefore, we consider this effect approximately by introducing the ambipolar diffusion concept.[19] In deriving the actual forms of various fluxes, we neglect the contributions of thermal and pressure diffusions, which are generally small in comparison with the concentration diffusion.

Diffusion Velocity

The rigorous form of diffusion velocity given by Hirschfelder et al.[22] is usually difficult to handle in actual problems; therefore, a simplification of the expression for the diffusion velocity is necessary. Curtiss and Hirschfelder[24] give the useful relation for the diffusion velocity of species s. In the absence of thermal and pressure diffusions, the diffusion velocity of species s is given by

$$n_s m_s V_s^j = -\left[\frac{n^2 m_s\{1 - (n_s m_s/\rho)\}}{\sum_{\substack{r=1 \\ r\neq s}}^{11} n_r/\mathscr{D}_{sr}}\right]\frac{\partial}{\partial x^j}\left(\frac{n_s}{n}\right) \tag{40}$$

where

$$n = \sum_{s=1}^{11} n_s$$

is the total number density of the mixture and $\mathscr{D}_{sr}$ is the finary diffusion coefficient of the s-r pair. In terms of molar concentration γ_s, Eq. (40) is reduced to the following form:

$$V_s^j = -\frac{1}{\gamma_s M_s}\left[\frac{\gamma_t^2 M_s(1 - M_s\gamma_s)}{\sum_{\substack{r=1 \\ r\neq s}}^{11} \gamma_r/\mathscr{D}_{sr}}\right]\frac{\partial}{\partial x^j}\left(\frac{\gamma_s}{\gamma_t}\right) \tag{41}$$

where

$$\gamma_t = \sum_{s=1}^{11} \gamma_s$$

is the total molar concentration of mixture and $M_s = Nm_s$ is the molecular weight of species s.

If one defines the effective diffusion coefficient $\mathscr{D}_s$ as

$$\mathscr{D}_s = \frac{\gamma_t^2 M_s (1 - M_s \gamma_s)}{\sum_{\substack{r=1 \\ r \neq s}}^{11} \gamma_r / \mathscr{D}_{sr}} \tag{42}$$

then, the diffusion velocity V_s^j can be expressed as

$$V_s^j = - \frac{1}{\hat{N} m_s \gamma_s} \mathscr{D}_s \frac{\partial y_s}{\partial x^j} \tag{43}$$

where $y_s = \gamma_s / \gamma_t$ is the mole fraction of species s.

Ambipolar diffusion in an ionized gas occurs when an electron pressure gradient exists in a partially ionized gas, but only when there is zero current.[19] Because of the linking of the electron and ion diffusion by the electric field, an ambipolar diffusion coefficient $\mathscr{D}_I^a$ for ions has the double value of an ionic diffusion coefficient,[19] that is,

$$\mathscr{D}_I^a = 2\mathscr{D}_I$$

where $\mathscr{D}_I$ is the effective ionic diffusion coefficient in the absence of electric field.

For multicomponent ionized gas,there exist several ions with different diffusion coefficients $\mathscr{D}_I$. The resulting ambipolar diffusion coefficients are different for different species. To determine the effective diffusion coefficients for individual ions in this situation, one must solve for the charge separation rigorously.[23] In order to avoid this, however, we assume that all ions diffuse as if each was the only ion present; that is,

$$\mathscr{D}_s^a = 2\mathscr{D}_s \qquad (s = 6, 7, 8, 9, 10) \tag{44}$$

where $\mathscr{D}_s^a$ is the ambipolar diffusion coefficient of ionic species s and $\mathscr{D}_s$ is the effective diffusion coefficient in the absence of electric field. The effective diffusion coefficient of electron $\mathscr{D}_e$ can be obtained by letting $V_e^j \approx V_I^j$, where

$$V_I^j = \sum_{s=6}^{10} \rho_s V_s^j \Big/ \sum_{s=6}^{10} \rho_s$$

is the average diffusion velocity of ions. Then, $\mathscr{D}_e$ may be expressed as

$$\mathscr{D}_e = m_e \sum_{s=6}^{10} \mathscr{D}_s^a \gamma_s \Big/ \sum_{s=6}^{10} m_s \gamma_s \tag{45}$$

Overall Viscous Shear Stress

The overall viscous shear stress τ^{ij}, which appears in the overall momentum equation (31) and in the overall energy equation (39), can be expressed as[10]

$$\tau^{ij} = \mu \left(\frac{\partial u_o^i}{\partial x^j} + \frac{\partial u_o^j}{\partial x^i} \right) - \frac{2}{3} \mu \frac{\partial u_o^\ell}{\partial x^\ell} \delta^{ij} \tag{46}$$

where δ^{ij} is the Kronecker delta and μ is the viscosity of mixture that depends on compositions and temperatures.

Energy Flux

The flux of energy through a multicomponent gas is a result of temperature gradient and diffusion. In the present flow, there are three temperature gradients: translational-rotational temperature, vibrational temperature, and electron temperature gradients. By considering these contributions to energy transport, the jth component of the overall heat-flux vector q^j can be expressed as the sum of four terms as follows:

$$q^j = -\eta \frac{\partial T}{\partial x^j} - \eta_v \frac{\partial T_v}{\partial x^j} - \eta_e \frac{\partial T_e}{\partial x^j} + \sum_{s=1}^{11} \rho_s h_s V_s^j \tag{47}$$

where h_s is the enthalpy of species s; and η, η_v, and η_e are the frozen thermal conductivities of heavy particles, vibrational energy, and electrons, respectively. Note that η_v and η_e include the contributions of all collisions between particles. Since the enthalpy of electrons is $(5/2)\kappa T_e/m_e$, the jth component of the heat-flux vector q_e^j for electrons is given by

$$q_e^j = -\eta_e' \frac{\partial T_e}{\partial x^j} + \frac{5}{2} n_e \kappa T_e V_e^j \tag{48}$$

where η_e' is the frozen thermal conductivity of electrons resulting from collisions between electrons, n_e is the electron number density, and V_e^j is the jth component of the diffusion velocity of electrons.

Transport Properties

The transport properties are evaluated by extending Yos's formula,[25] which is based on the first Chapman-Enskog approximation, to the multitemperature gas mixture. This is done assuming that all distribution functions are close to Maxwellian about their inherent temperature T_s; that is, T for heavy particles and T_e for electrons. We assume additionally that the controlling temperature in $\Delta_{sr}^{(1)}$ and $\Delta_{sr}^{(2)}$ in Yos's formula is the electron temperature T_e for collisions involving electrons. This is justified because the thermal velocity of heavy particles can be neglected in comparison with the thermal velocity of electrons due to the large mass ratio of heavy particles to electrons.

Binary Diffusion Coefficient

The binary diffusion coefficient of an s-r pair of heavy particles is given by

$$\mathscr{D}_{sr} = \frac{\kappa T}{p\, \Delta_{sr}^{(1)}(T)} \tag{49}$$

where p is the pressure and $\Delta_{sr}^{(1)}(T)$ is defined as

$$\Delta_{sr}^{(1)}(T) = \frac{8}{3}\left[\frac{2m_s m_r}{\pi\kappa T(m_s + m_r)}\right]^{1/2} \pi\bar{\Omega}_{sr}^{(1,1)} \tag{50}$$

In the above expression, $\pi\bar{\Omega}_{sr}^{(\ell,\ell')}$ is the collision integral defined as the weighted average of a collision cross section of the form[25]

$$\pi\bar{\Omega}_{s,r}^{(\ell,\ell')}$$

$$= \frac{\int_0^\infty \int_0^\pi e^{-\gamma^2} \gamma^{2\ell'+3}(1 - \cos^\ell \chi) 4\pi\sigma_{sr} \sin\chi \, d\chi \, d\gamma}{\int_0^\infty \int_0^\pi e^{-\gamma^2} \gamma^{2\ell'+3}(1 - \cos^\ell \chi) \sin\chi \, d\chi \, d\gamma} \tag{51}$$

where $\sigma_{sr} = \sigma_{sr}(\chi, g)$ is the collision cross section for the collision pair s-r, χ is the scattering angle in the center of mass system, g is the relative velocity of the colliding particles, and $\gamma = [m_s m_r/2(m_s + m_r)\kappa T]^{1/2} g$ is the reduced velocity. The binary diffusion coefficient between electrons and heavy particles is given by

$$\mathscr{D}_{er} = \frac{\kappa T_e}{p\,\Delta_{er}^{(1)}(T_e)} \tag{52}$$

Viscosity

The viscosity of a mixture is given by

$$\mu = \sum_{s=1}^{10} \left\{ m_s y_s \Big/ \left[\sum_{r=1}^{10} y_r \Delta_{sr}^{(2)}(T) + y_e \Delta_{se}^{(2)}(T_e) \right] \right\}$$

$$+ m_e y_e \Big/ \left[\sum_{r=1}^{11} y_r \Delta_{er}^{(2)}(T_e) \right] \tag{53}$$

where $y_s = \gamma_s/\gamma_t$ is the mole fraction of species s

and $\Delta_{sr}^{(2)}(T)$ is defined as

$$\Delta_{sr}^{(2)}(T) = \frac{16}{5}\left[\frac{2m_s m_r}{\pi\kappa T(m_s + m_r)}\right]^{1/2} \pi\bar{\Omega}_{sr}^{(2,2)} \tag{54}$$

By using the molar concentration γ_s of species s, Eq. (53) can be rewritten as

$$\mu = \sum_{s=1}^{10}\left\{m_s\gamma_s \Big/ \left[\sum_{r=1}^{10}\gamma_r\,\Delta_{sr}^{(2)}(T) + \gamma_e\,\Delta_{se}^{(2)}(T_e)\right]\right\} + m_e\gamma_e \Big/ \left[\sum_{r=1}^{11}\gamma_r\,\Delta_{er}^{(2)}(T_e)\right] \tag{55}$$

where γ_e is the molar concentration of electrons.

<u>Thermal Conductivity</u>

The thermal conductivity consists of two parts which arise from molecular collisions and diffusion of reacting chemical species. The molecular collision terms are treated considering the separate contribution of each degree of freedom. The contribution of the diffusion of reacting species is taken into account separately in the cxpression of energy flux [Eq. (47)].

The translational thermal conductivity η_{tr} of heavy particles is given by

$$\eta_{tr} = \frac{15}{4}\kappa\sum_{s=1}^{10}\left\{y_s \Big/ \left[\sum_{r=1}^{11} a_{sr}y_r\,\Delta_{sr}^{(2)}\right]\right\}$$

$$= \frac{15}{4}\kappa\sum_{s=1}^{10}\left\{\gamma_s \Big/ \left[\sum_{r=1}^{10} a_{sr}\gamma_r\,\Delta_{sr}^{(2)}(T) + 3.54\gamma_e\,\Delta_{se}^{(2)}(T_e)\right]\right\} \tag{56}$$

where a_{sr} is defined by

$$a_{sr} = 1 + \frac{[1 - (m_s/m_r)][0.45 - 2.54(m_s/m_r)]}{[1 + (m_s/m_r)]^2} \tag{57}$$

Assuming that the rotational energy modes of molecules are fully excited, the overall rotational thermal conductivity of a mixture is given by

$$\eta_{rot} = \kappa \sum_{s=M} \left\{ y_s \Big/ \left[\sum_{r=1}^{11} y_r \, \Delta_{sr}^{(1)} \right] \right\}$$

$$= \kappa \sum_{s=M} \left\{ \gamma_s \Big/ \left[\sum_{r=1}^{10} \gamma_r \, \Delta_{sr}^{(1)}(T) + \gamma_e \, \Delta_{se}^{(1)}(T_e) \right] \right\} \tag{58}$$

where $M = 3, 4, 5, 8, 9, 10$.

Thus, the frozen thermal conductivity of a mixture of heavy particles is given by

$$\eta = \eta_{tr} + \eta_{rot}$$

$$= \frac{15}{4} \kappa \sum_{s=1}^{10} \left\{ \gamma_s \Big/ \left[\sum_{r=1}^{10} a_{sr} \gamma_r \, \Delta_{sr}^{(2)}(T) + 3.54 \gamma_e \, \Delta_{se}^{(2)}(T_e) \right] \right\}$$

$$+ \kappa \sum_{s=M} \left\{ \gamma_s \Big/ \left[\sum_{r=1}^{10} \gamma_r \, \Delta_{sr}^{(1)}(T) + \gamma_e \, \Delta_{se}^{(1)}(T_e) \right] \right\} \tag{59}$$

Since the number density of vibrationally excited molecules is assumed to have a Boltzman distribution with T_V, the vibrational thermal conductivity η_V is given by

$$\eta_V = \kappa \sum_{s=M} \left\{ \gamma_s \Big/ \left[\sum_{r=1}^{10} \gamma_r \, \Delta_{sr}^{(1)}(T) + \gamma_e \, \Delta_{se}^{(1)}(T_e) \right] \right\} \tag{60}$$

The V-V thermal conductivity η_V' in Eq. (32) can be obtained by considering the contribution of molecule-molecule collisions to the overall vibrational thermal conductivity η_V

$$\eta_V' = \eta_V \sum_{s=M} \left\{ \sum_{r=M} y_r \, \Delta_{sr}^{(2)} \Big/ \left[\sum_{r=1}^{11} y_r \, \Delta_{sr}^{(2)} \right] \right\} \quad \text{(continued)}$$

$$= \eta_v \sum_{s=M} \left\{ \sum_{r=M} \gamma_r \, \Delta_{sr}^{(2)}(T) \Big/ \left[\sum_{r=1}^{10} \gamma_r \, \Delta_{sr}^{(2)}(T) + \gamma_e \, \Delta_{se}^{(2)}(T_e) \right] \right\} \quad (61)$$

The electron thermal conductivity η_e is given by

$$\eta_e = \frac{15}{4} \kappa \left\{ y_e \Big/ \left[\sum_{r=1}^{11} a_{er} y_r \, \Delta_{er}^{(2)} \right] \right\}$$

$$= \frac{15}{4} \kappa \left\{ \gamma_e \Big/ \left[\sum_{r=1}^{11} 1.45 \, \gamma_r \, \Delta_{er}^{(2)}(T_e) \right] \right\} \quad (62)$$

The thermal conductivity of electrons η_e' due to collisions only between electrons is given by

$$\eta_e' = \eta_e \left\{ y_e \, \Delta_{ee}^{(2)} \Big/ \left[\sum_{r=1}^{11} y_r \, \Delta_{er}^{(2)} \right] \right\}$$

$$= \eta_e \left\{ \gamma_e \, \Delta_{ee}^{(2)}(T_e) \Big/ \left[\sum_{r=1}^{11} \gamma_r \, \Delta_{er}^{(2)}(T_e) \right] \right\} \quad (63)$$

Collision Cross Section

Since the accuracy of the calculated transport properties is greatly dependent on the accuracy of the collision integrals, it is desirable to use up-to-date collision cross sections. The collision cross sections for neutral-neutral interactions for air species were studied intensively, and fairly reliable data were obtained.[26,27] The Coulomb collision cross sections between charged particles, which have very large values, are relatively well known. However, ion-neutral and electron-neutral (especially electron-atom) cross sections for air species are still quite uncertain. Consequently, the estimated values for these interactions must be used provisionally at present. It would be greatly desirable to obtain reliable data for the elastic ion-neutral and electron-neutral cross sections as well as the inelastic electron-vibration and electron-impact dissociation cross sections in air species; this should be the subject of future studies.

VII. Other Physical Properties

Collision Frequency

The effective collision frequency of electrons with heavy particles ν^*_{er}, which appears in the electron energy equation (37), must be evaluated. To derive ν^*_{er}, it is assumed that the distribution function for each species has a Maxwellian form about its own temperature. The electron-neutral collision frequency ν^*_{er} is given by the expression[19]

$$\nu^*_{er} = n_r \sigma_{er} \left(\frac{8}{\pi} \frac{\kappa T_e}{m_e} \right)^{1/2} \qquad (r = 1,\ 2,\ 3,\ 4,\ 5) \qquad (64)$$

where n_r is the number density of the neutrals and σ_{er} is the cross section for electron-neutral interaction. For the case of Coulomb interaction between electrons and ions, the effective Coulomb cross section σ_{ei} is given by[25,28]

$$\sigma_{ei} = \frac{4}{3} (2\pi) \frac{e^4}{(3\kappa T_e)^2} \, \ell n \left[1 + \frac{h^2 (3\kappa T_e)^2}{e^4} \right] \qquad (65)$$

where e is the electronic charge and h is a cutoff impact parameter to prevent the divergence of the Coulomb cross section. The value of h is usually taken to be the Debye shielding length $(\kappa T_e / 4\pi n_e e^2)^{1/2}$ (Ref. 7), and thus, the electron-ion collision frequency ν^*_{er} has a form

$$\nu^*_{er} = \frac{64}{27} \left(\frac{\pi}{m_e} \right)^{1/2} n_r e^4 \frac{1}{(2\kappa T_e)^{3/2}} \, \ell n \left[\frac{9\kappa^3 T_e^3}{4\pi n_e e^6} \right]$$

$$(r = 6,\ 7,\ 8,\ 9,\ 10) \qquad (66)$$

Vibrational Relaxation Times

The vibrational relaxation times τ_s and τ_{es}, which appear in the vibrational energy equation (32), need to be evaluated. Various reports exist dealing with the T-V relaxation in N_2, O_2, and air.[29-31] Millikan and White[31] showed an empirical equation for the vibrational relaxation time in the form of $p\tau_{sr}$ (atm·s). The deduced empirical equation is

$$p\tau_{sr} = \exp[A_{sr}(T^{-1/3} - 0.015\mu_{sr}^{1/4}) - 18.42] \quad (67)$$

where A_{sr} is the constant and

$$\mu_{sr} = [M_s M_r/(M_s + M_r)]$$

is the reduced molecular weight of the pair of colliding particles s-r. The constant A_{sr} is given by

$$A_{sr} = 1.16 \times 10^{-3}\mu_{sr}^{1/2}\theta_{v,s}^{4/3}$$

where $\theta_{v,s}$ is the vibrational characteristic temperature for molecular species s. Thus, Eq. (67) becomes

$$p\tau_{sr} = \exp\left\{1.16 \times 10^{-3}\left(\frac{M_s M_r}{M_s + M_r}\right)^{1/2}\theta_{v,s}^{4/3} \times \left[T^{-1/3} - 0.015\left(\frac{M_s M_r}{M_s + M_r}\right)^{1/4}\right] - 18.42\right\} \quad (68)$$

The T-V relaxation time τ_{sr} for molecular species s with various collision partners can be calculated by Eq. (68). The average relaxation time τ_s can then be obtained from Eq. (35). It is to be noted that Eqs. (67) and (68) are valid only for temperatures up to about 8000 K. Serious deficiencies develop if the equations are applied to higher temperatures.[32] Determining vibrational relaxation times at high temperatures remains one of the most urgently needed future tasks. There are no available data for the e-V relaxation time τ_{es}. However, one can evaluate τ_{es} through Eq. (23), provided vibrational excitation rate coefficients are known. Slinker and Ali[33] calculated the excitation rate coefficients for N, O, N_2, O_2, and NO using measured collision cross sections. Since the e-V relaxation time τ_{es} is a function of electron temperature T_e and electron pressure p_e, one can write

$$p_e \tau_{es} = \kappa T_e \left[\left(1 - e^{-\theta_{v,s}/T_e} \right)^2 \frac{1}{2} \int k_{oj,s}(j)^2 \, dj \right]^{-1} \quad (69)$$

where κ is Boltzmann's constant and $k_{oj,s}$ is the rate coefficient from the ground state to the upper state j for molecular species s.

Chemical Reaction Rate Coefficients

The mass production rates of species s $\dot{w}_s$ in Eq. (2) can be evaluated from species number densities and chemical reaction rate coefficients. In order to do so for the AOTV flight regimes, chemical rate coefficient data are needed for high temperatures. Most available data are confined to the temperature ranges below 5000 K. In the multitemperature environment under consideration, the chemical reaction rate coefficients K_c are functions of three temperatures,[32] i.e., $K_c = K_c(T, T_V, T_e)$. No such data are presently available. Determining these rate coefficients in this environment is one of the most important problems to be solved.

Radiation Energy Loss

The radiation energy loss term Q_{rad} in Eqs. (37) and (39) can be predicted by a computer code developed by Park.[8] From the given nonequilibrium thermodynamic state variables, the code calculates number densities of internal states and the accompanying emission and absorption characteristics in AOTV flowfields. It should be noted that to calculate the nonequilibrium state variables through flow conservation equations, the energy loss due to radiation must be taken into account. That is, radiation transport is strongly coupled with the flow equations. Consequently, one must rely on an iterative method to solve the system of flow equations with radiation transport.

Thermodynamic Properties

Since it is assumed that 1) the rotational mode of molecules is fully equilibrated with the translational mode of heavy particles and that 2) the population densities of vibrational and electronic energy levels have Boltzmann distributions with vibrational temperature T_V and electron temperature T_e, respectively, the thermodynamic energy of species s is given by[10]

$$e_s = \frac{3}{2}\frac{\kappa}{m_s} T + \left(\frac{\kappa}{m_s}\right)\theta_{el,s} \frac{(g_{1,s}/g_{o,s})e^{-\theta_{el,s}/T_e}}{1 + (g_{1,s}/g_{o,s})e^{-\theta_{el,s}/T_e}} + e_s^o$$

(atoms: $s = 1, 2, 6, 7$)

$$e_s = \frac{5}{2}\frac{\kappa}{m_s} T + \left(\frac{\kappa}{m_s}\right)\frac{\theta_{v,s}}{e^{\theta_{v,s}/T_v} - 1}$$

$$+ \left(\frac{\kappa}{m_s}\right)\theta_{el,s} \frac{(g_{1,s}/g_{o,s})e^{-\theta_{el,s}/T_e}}{1 + (g_{1,s}/g_{o,s})e^{-\theta_{el,s}T_e}} + e_s^o \tag{70}$$

(molecules: $s = 3, 4, 5, 8, 9, 10$)

$$e_s = \frac{3}{2}\frac{\kappa}{m_e} T_e \quad \text{(electron: } s = 11\text{)}$$

where $g_{i,s}$ is the degeneracy of electronic energy level i for species s, e_s^o is the chemical energy of species s and $\theta_{v,s}$ and $\theta_{el,s}$ are the characteristic temperatures (i.e., energy levels divided by the Boltzmann constant κ) for vibrational and electronic excitations for species s, respectively. In the above expressions, we additionally assume that only the ground and first electronic energy levels are significantly populated. The overall thermodynamic energy e of a mixture is given by

$$e = \frac{\sum_{s=1}^{11} \rho_s e_s}{\rho} = \hat{N} \sum_{s=1}^{11} \gamma_s m_s e_s \tag{71}$$

The enthalpy h_s of species s is given by

$$h_s = c_{pf,s} T + c_{pv,s} T_v + c_{pe,s} T_e + h_s^o \tag{72}$$

where $c_{pf,s}$, $c_{pv,s}$, and $c_{pe,s}$ are the frozen specific heat for translation-rotation, specific heat for vibrational excitation, and specific heat for electronic excitation at constant pressure for species s, respectively, and h_s^o is the heat of formation. The specific heats at constant volume are given by

$$c_{vf,s} = \frac{3}{2} \frac{\kappa}{m_s}$$

(atoms and electron: $s = 1, 2, 6, 7, 11$)

$$c_{vf,s} = \frac{5}{2} \frac{\kappa}{m_s}$$

(molecules: $s = 3, 4, 5, 8, 9, 10$)

(73)

$$c_{vv,s} = \left(\frac{\kappa}{m_s}\right)\left(\frac{\theta_{v,s}}{T_v}\right)^2 \frac{e^{\theta_{v,s}/T_v}}{(e^{\theta_{v,s}/T_v} - 1)^2}$$

($s = 3, 4, 5, 8, 9, 10$)

$$c_{ve,s} = \left(\frac{\kappa}{m_s}\right)\left(\frac{\theta_{el,s}}{T_e}\right)^2 \frac{(g_{1,s}/g_{o,s}) e^{-\theta_{el,s}/T_e}}{[1 + (g_{1,s}/g_{o,s}) e^{-\theta_{el,s}/T_e}]^2}$$

($s = 1, 2, \ldots, 10$)

The specific heat $c_{p,s}$ at constant pressure can be obtained through the following thermodynamic relation[19]

$$c_{p,s} = c_{v,s} + \frac{\kappa}{m_s} \tag{74}$$

XIII. Final Forms of Basic Governing Equations

The final forms of the basic conservation equations for AOTV flowfields can be obtained by introducing the diffusion velocity V_s^j [Eq. (43)], shear stress τ^{ij} [Eq. (46)], and energy fluxes q^j [Eq. (47)] and q_e^j [Eq. (48)] into the simplified Eqs. (26), (31), (32), and (39). Since electron kinetic energy and electronic excitation energy of heavy particle are both related to electron temperature T_e, we describe the conservation equation in the form of the sum of electron energy and electronic excitation energy.

Species conservation equations:

$$\frac{\partial}{\partial t}(\rho\gamma_s) + \frac{\partial}{\partial x^j}(\rho u_o^j \gamma_s) = \frac{\partial}{\partial x^j}\left(\frac{\rho}{Nm_s}\mathscr{D}_s \frac{\partial y_s}{\partial x^j}\right) + \hat{\dot{w}}_s \tag{75}$$

Overall mass conservation equation:

$$\frac{\partial \rho}{\partial t} + \frac{\partial}{\partial x^j}(\rho u_o^j) = 0 \tag{76}$$

Overall momentum conservation equation:

$$\frac{\partial}{\partial t}(\rho u_o^i) + \frac{\partial}{\partial x^j}(\rho u_o^i u_o^j)$$

$$= -\frac{\partial p}{\partial x^i} + \frac{\partial}{\partial x^j}\left[\mu\left(\frac{\partial u_o^i}{\partial x^j} + \frac{\partial u_o^j}{\partial x^i}\right) - \frac{2}{3}\mu\frac{\partial u_o^\ell}{\partial x^\ell}\delta^{ij}\right] + \left(1 - \frac{n_I}{n_e}\right)\frac{\partial p_e}{\partial x^i} \tag{77}$$

Vibrational energy conservation equation:

$$\frac{\partial}{\partial t}(\rho_M e_v) + \frac{\partial}{\partial x^j}(\rho_M e_v u_o^j)$$

$$= \frac{\partial}{\partial x^j}\left(\eta_v' \frac{\partial T_v}{\partial x^j}\right) + \frac{\partial}{\partial x^j}\left(\rho\sum_{s=M} e_{v,s}\mathscr{D}_s \frac{\partial y_s}{\partial x^j}\right)$$

$$+ \rho\hat{N}\sum_{s=M} m_s\gamma_s \frac{e^*_{v,s}(T) - e_{v,s}}{\langle\tau_s\rangle} + \rho\hat{N}\sum_{s=M} m_s\gamma_s \frac{e^{**}_{v,s}(T_e) - e_{v,s}}{\tau_{es}} \tag{78}$$

Electron and electronic excitation energies conservation equation:

$$\frac{\partial}{\partial t}\left[\hat{N}\rho T_e\left(\frac{3}{2}\gamma_e\kappa + \sum_{s=1}^{10}\gamma_s m_s c_{ve,s}\right)\right]$$

$$+\frac{\partial}{\partial x^j}\left[\hat{N}\rho T_e\left(\gamma_e\kappa u_e^j + u_o^j\sum_{s=1}^{10}\gamma_s m_s c_{ve,s}\right)\right]$$

$$= \frac{\partial}{\partial x^j}\left(\eta_e\frac{\partial T_e}{\partial x^j}\right) + \frac{\partial}{\partial x^j}\left[\rho T_e\left(\frac{5}{2}\frac{\kappa}{m_e}\mathscr{D}_e\frac{\partial y_e}{\partial x^j}\right.\right.$$

$$\left.\left. + \sum_{s=1}^{10} c_{pe,s}\mathscr{D}_s\frac{\partial y_s}{\partial x^j}\right)\right] - \frac{\partial}{\partial x^i}(u_e^i p_e)$$

$$+ 2\hat{N}\rho m_e\gamma_e\sum_{r=1}^{10}\frac{\nu_{er}^*}{m_r}\left(\frac{3}{2}\kappa T - \frac{3}{2}\kappa T_e\right) + u_o^i\frac{\partial p_e}{\partial x^i}$$

$$- \sum_{f=6}^{10}\dot{n}_{e,f}E_{i,f} - \hat{N}\rho\sum_{s=M} m_s\gamma_s\frac{e_{v,s}^{**}(T_e) - e_{v,s}}{\tau_{es}} - Q_{rad} \quad (79)$$

Overall energy conservation equation:

$$\frac{\partial}{\partial t}\left[\rho\left(\frac{1}{2}u_o^2 + e\right)\right] + \frac{\partial}{\partial x^j}\left[\rho u_o^j\left(\frac{1}{2}u_o^2 + e\right)\right]$$

$$= \frac{\partial}{\partial x^j}\left(\eta\frac{\partial T}{\partial x^j} + \eta_v\frac{\partial T_v}{\partial x^j} + \eta_e\frac{\partial T_e}{\partial x^j} + \sum_{s=1}^{11}\rho h_s\mathscr{D}_s\frac{\partial y_s}{\partial x^j}\right)$$

$$- \frac{\partial}{\partial x^j}(u_o^i p) + \frac{\partial}{\partial x^j}\left[u_o^i\mu\left(\frac{\partial u_o^i}{\partial x^j} + \frac{\partial u_o^j}{\partial x^i}\right)\right.$$

$$\left. - \frac{2}{3}u_o^i\mu\frac{\partial u_o^\ell}{\partial x^\ell}\delta^{ij}\right] + \left(u_e^i - \frac{n_I}{n_e}u_o^i\right)\frac{\partial p_e}{\partial x^i} - Q_{rad} \quad (80)$$

Equations of state:

The equations of state can be obtained by assuming that each species obeys the perfect-gas law[19]

$$p_s = n_s \kappa T = \hat{N} \rho \gamma_s \kappa T$$

$$(\text{heavy particles:} \quad s = 1, 2, \ldots, 10) \tag{81}$$

and

$$p_e = n_e \kappa T_e = \hat{N} \rho \gamma_e \kappa T_e \qquad (\text{electron:} \quad s = 11) \tag{82}$$

Thus, total static pressure p is given by

$$p = \hat{N} \rho \left(\sum_{s=1}^{10} \gamma_s \kappa T + \gamma_e \kappa T_e \right) \tag{83}$$

IX. Concluding Remarks

A system of gas-dynamic conservation equations was developed for the thermal and chemical nonequilibrium, ionized gas flow expected in AOTV flight regimes. Based on the continuum flow approach, the general governing equations for a multicomponent ionized gas were derived by accounting for the effects of three temperatures (translational temperature T, vibrational temperature T_V, and electron temperature T_e), chemical nonequilibrium, transport effects, and electric charge separation. It was assumed that the rotational mode of molecules is fully equilibrated with the translational mode of heavy particles, and that the population densities of vibrational and electronic energy levels have Boltzmann distributions with vibrational temperature T_V and electron temperature T_e, respectively. The translation-vibration (T-V) and electron-vibration (e-V) energy-exchange processes were considered separately by assuming that the vibration-vibration (V-V) energy-exchange process is infinitely fast. The Landau-Teller-type rate equation for the multiple-quantum transition e-V process was obtained by extending the classical treatment of Keck and Carrier[16] for the T-V processes. The transport properties were evaluated by extending Yos's formula[25] to the multitemperature gas mixture. Finally, the system of equations, which is suitable for the numerical calculations for the AOTV flowfields, was shown.

Appendix: Vibrational Rate Equation for e-V Process

Assuming that the vibrational oscillators interchange their energies with an electron heat bath having a constant temperature T_e, the rate of change of the number density n_i is given by [see Eq. (21)]

$$\frac{\partial n_i}{\partial t} = \sum_{j=0}^{\infty} (k_{ji} n_j - k_{ij} n_i) n_e \tag{A1}$$

where k_{ij} is the rate coefficient for the transition from state i to j and n_e is the number density of electrons. The principle of detailed balancing provides the condition

$$k_{ji} n_{jE} = k_{ij} n_{iE} \tag{A2}$$

where n_{iE} is the equilibrium population in state i. Using the relation (A2) and the normalized number density $X_i = n_i/n_{iE}$, one obtains the following normalized master equation

$$\frac{n_{iE}}{n_e} \frac{\partial X_i}{\partial t} = \sum_{j=0}^{\infty} R_{ij} (X_j - X_i) \tag{A3}$$

where $R_{ij} = k_{ij} n_{iE}$ is the equilibrium transition rate.

If we consider that the vibrational energy levels are continuously distributed over energy level i (classical analogy), the summations in Eq. (A3) can be replaced by integrations

$$\frac{n_{iE}}{n_e} \frac{\partial X_i}{\partial t} = \int R_{ij} (X_j - X_i) dj \tag{A4}$$

If the right-hand side of Eq. (A4) is expanded in powers of $\partial^n X_i / \partial i^n$, and the terms of the order $\partial^2 (\Delta_2 \partial^2 X_i / \partial i^2) / \partial i^2$ are neglected, one obtains the following diffusion type equation[16]:

$$\frac{n_{iE}}{n_e}\frac{\partial X_i}{\partial t} = \frac{\partial}{\partial i}\left(\frac{\Delta_2}{2}\frac{\partial X_i}{\partial i}\right) \tag{A5}$$

where

$$\Delta_2 = \int R_{ji}(j - i)^2\, dj$$

is the second moment of the energy transfer (j,i) with respect to R_{ji}.

Equation (A5), together with the requirement that the sum of all n_i's equals the given number density of molecules n_M,

$$n_M = \sum_i n_i \approx \int n_i\, di \tag{A6}$$

gives the number density n_i, which is a function of time and energy state i. If we assume that all molecules are populated in the ground state at $t = 0$, the problem becomes equivalent to the one-dimensional heat conduction through a semi-infinite solid, and it corresponds to the heat conduction with heat impulse at $t = 0$ and variable heat conductivity.

Time variation of n_i is usually much smaller in magnitude than chemical reaction rates.[8] This situation yields the so-called quasisteady-state (QSS) approximation in solving the master equation. This, in turn, validates the assumption that the number density n_i has a Boltzmann distribution with a vibrational temperature T_v throughout the relaxation process. The spatial variation of $\Delta_2/2$, which corresponds to the equivalent heat conductivity K_o in equivalent heat conduction, is assumed to be small at $i = 0$. Consequently, one may write an approximately analogous equation (A5) with the one-dimensional heat conduction equation within a semi-infinite solid as

$$\frac{\partial T}{\partial t} = K_o \frac{\partial^2 T}{\partial x^2} \tag{A7}$$

where K_o is the constant heat conductivity. The boundary condition is $T = T_o$ (constant) at $x = 0$, and the initial temperature is zero for $x > 0$.

The solution of Eq. (A7) is well known and is given in Ref. 34 as

$$T(x,t) = T_o\left[1 - \text{erf}\,\frac{x}{2(K_o t)^{1/2}}\right] \tag{A8}$$

where erf(z) is the error function defined by

$$\text{erf}(z) = \frac{2}{\sqrt{\pi}}\int_o^z e^{-\xi^2}\,d\xi$$

Thus, the corresponding normalized number density X_i can be dictated as

$$X_i(i,t) = X_o\left[1 - \text{erf}\,\frac{i}{2(K_o t)^{1/2}}\right] \tag{A9}$$

where $X_o = (n_i/n_{iE})_{i=0}$ is the normalized number density at $i = 0$ and $K_o = (\Delta_2/2)_{i=0}$ is the corresponding heat conductivity. The value of x_o can be approximately assumed to be 1.

We are concerned here with the total vibrational energy of the system of oscillators. The total vibrational energy E_v per unit volume for harmonic oscillators is given by

$$E_v = \sum_{i=0}^{\infty} ih\nu n_{iE}X_i \simeq h\nu\int_o^{\infty} in_{iE}X_i\,di$$

where h is Planck's constant and ν is the frequency of oscillators. Thus,

$$E_v = h\nu n_M(1 - e^{-h\nu/\kappa T_e})\int_o^{\infty} i\,e^{-(h\nu/\kappa T_e)i}$$

$$\times\left[1 - \text{erf}\,\frac{i}{2(K_o t)^{1/2}}\right]di$$

Introducing the notations $\theta_v = h\nu/k$ and $a = \theta_v/T_e$, we write finally

$$E_v = h\nu n_M(1 - e^{-a}) \int_0^\infty i\, e^{-ai} \left[1 - \text{erf}\, \frac{1}{2(K_o t)^{1/2}}\right] di \tag{A10}$$

The equilibrium value of E_v is obtained by setting $t = \infty$ in Eq. (A10). Then, the equilibrium vibrational energy E_v^{**} is given by

$$E_v^{**} = h\nu n_M(1 - e^{-a}) \int_0^\infty i\, e^{-ai}\, di \tag{A11}$$

Performing the integrals by parts, one obtains

$$E_v(t) = A\left\{\frac{1}{a^2} - \frac{b}{a\sqrt{\pi}} - \left[1 - \text{erf}\left(\frac{ab}{2}\right)\right] e^{a^2b^2/4}\left(\frac{1}{a^2} - \frac{b^2}{2}\right)\right\} \tag{A12}$$

where $A = h\nu n_M(1 - e^{-a})$ and $b = 2(K_o t)^{1/2}$. Also one obtains

$$E_v^{**} = A/a^2 \tag{A13}$$

We can expand the error function erf(ab/2) in Eq. (A12) by power series.[34] For large b, that is, for large t, one obtains the following form for $E_v(t)$ by retaining the lowest term of t

$$E_v(t) = A\,\frac{1}{a^2}\left[1 - \frac{2}{a(\pi K_o t)^{1/2}} + \ldots\right] \tag{A14}$$

Also, one obtains for small t,

$$E_v(t) = A(K_o t + \ldots) \tag{A15}$$

From Eqs. (A12) to (A15) and from the feature of error

function, we realize that $E_v(t)$ increases monotonically with time t from the initial value $E_v = 0$ at $t = 0$, and finally approaches its equilibrium value E_v^{**} at $t = \infty$.

From the feature of $E_v(t)$ mentioned above, we describe the e-V relaxation process approximately by the Landau-Teller-type rate equation. In terms of vibrational energy $e_v(t)$ per unit mass, the vibrational rate equation for the e-V process can be written as

$$\frac{\partial e_v}{\partial t} = \frac{e_v^{**}(T_e) - e_v}{\tau_e} \tag{A16}$$

where τ_e is the corresponding relaxation time for the e-V energy exchange. The relaxation time τ_e is given by

$$\tau_e = (a^2 K_o)^{-1} = \frac{1}{(\theta_v/T_e)^2 K_o} \simeq \frac{1}{\left(1 - e^{-\theta_v/T_e}\right)^2 K_o} \tag{A17}$$

The equivalent heat conductivity K_o is

$$K_o = \frac{1}{2} n_e \int k_{oj}(j)^2 \, dj$$

where n_e is the number density of electrons and k_{oj} is the rate coefficient from the ground state to the upper state $j(j \geq 1)$. Thus, τ_e becomes

$$\tau_e = n_e \left(1 - e^{-\theta_v/T_e}\right)^2 \frac{1}{2} \int k_{oj}(j)^2 \, dj^{-1} \tag{A18}$$

Acknowledgments

The author wishes to express his sincere thanks to Dr. Chul Park for his excellent advice and suggestions in carrying out this study.

References

[1]Walberg, G. D., "A Review of Aeroassisted Orbit Transfer," AIAA Paper 82-1378, AIAA 9th Atmospheric Flight Mechanics Conference, San Diego, Calif., Aug. 1982.

[2]Howe, J. T., "Introductory Aerothermodynamics of Advanced Space Transportation Systems," AIAA Paper 83-0406, AIAA 21st Aerospace Sciences Meeting, Reno, Nev., Jan. 1983.

[3]Menees, G. P., "Trajectory Analysis of Radiative Heating for Planetary Missions with Aerobraking of Spacecraft," AIAA Paper 83-0407, AIAA 21st Aerospace Sciences Meeting, Reno, Nev., Jan. 1983.

[4]Menees, G. P., "Thermal-Protection Requirements for Near-Earth Aeroassisted Orbital-Transfer Vehicle Missions," AIAA Paper 83-1513, AIAA 18th Thermophysics Conference, Montreal, Canada, June 1983; published elsewhere in this volume.

[5]Menees, G. P., Park, C., and Wilson, J. F., "Design and Performance Analysis of a Conical-Aerobrake Orbital-Transfer Vehicle Concept," AIAA Paper 84-0410, AIAA 22nd Aerospace Sciences Meeting, Reno, Nev., Jan. 1984; published elsewhere in this volume.

[6]Park, C., "Radiation Enhancement by Nonequilibrium in Earth's Atmosphere," AIAA Paper 83-0410, AIAA 21st Aerospace Sciences Meeting, Reno, Nev., Jan. 1983.

[7]Appleton, J. P. and Bray, K. N. C., "The Conservation Equations for a Nonequilibrium Plasma," Journal of Fluid Mechanics, Vol. 20, Pt. 4, Dec. 1964, pp. 659-672.

[8]Park, C., "Calculation of Nonequilibrium Radiation in AOTV Flight Regimes," AIAA Paper 84-0306, AIAA 22nd Aerospace Sciences Meeting, Reno, Nev., Jan. 1984; published elsehwere in this volume.

[9]Chung, P. M., "Chemically Reacting Nonequilibrium Boundary Layers," Advances in Heat Transfer, Vol. 2, Academic Press, New York, 1965, pp. 109-270.

[10]Vincenti, W. G. and Kruger, C. H. Jr., Introduction to Physical Gas Dynamics, John Wiley and Sons, Inc., New York, 1965.

[11]Hansen, C. F., "Rate Processes in Gas Phase," NASA RP-1090, May 1983.

[12]McKenzie, R. L., "Vibration-Translation Energy Transfer in Vibrationally Excited Diatomic Molecules," NASA TR R-466, Oct. 1976.

[13]Beam, R. M. and Warming, R. F., "An Implicit Factored Scheme for the Compressible Navier-Stokes Equations," AIAA Journal, Vol. 16, No. 4, Apr. 1978, pp. 393-402.

[14]Lombard, C. K., Davy, W. C., and Green, M. J., "Forebody and Base Region Real-Gas Flow in Severe Planetary Entry by a Factored Implicit Numerical Method. Pt. 1. Computational Fluid Dynamics," AIAA Paper 80-0065, AIAA 18th Aerospace Sciences Meeting, Pasadena, Calif., Jan. 1980.

[15]Balakrishnan, A. and Davy, W. C., "Viscous Real Gas Flowfields about Three-Dimensional Configurations," AIAA Paper 83-1511, AIAA 18th Thermophysics Conference, Montreal, Canana, June 1983.

[16]Keck, J. and Carrier, G., "Diffusion Theory of Nonequilibrium Dissociation and Recombination," Journal of Chemical Physics, Vol. 43, No. 7, July 1965, pp. 2284-2298.

[17]Shuler, K. E. and Weiss, G. H., "Exactly Solvable Nonlinear Relaxation Processes: Systems of Coupled Harmonic Oscillators," Journal of Chemical Physics, Vol. 45, No. 4, Aug. 1966, pp. 1105-1110.

[18]Bray, K. N. C., "Vibrational Relaxation of Anharmonic Oscillator Molecules: Relaxation under Isothermal Conditions," Journal of Physics B (Proc. Phys. Soc.), Ser. 2, Vol. 1, July 1968, pp. 705-717.

[19]Sutton, G. W. and Sherman, A., Engineering Magnetohydrodynamics, McGraw-Hill, Inc., New York, 1965.

[20]Montroll, E. W. and Shuler, K. E., "Studies in Nonequilibrium Rate Processes. I. The Relaxation of a System of Harmonic Oscillators," Journal of Chemical Physics, Vol. 26, No. 3, Mar. 1957, pp. 454-464.

[21]Bitter, D. A. and Scullin, V. J., "General Chemical Kinetics Computer Program for Static and Flow Reactions with Application to Combustion and Shock-Tube Kinetics," NASA TN D-6586, Jan. 1972.

[22]Hirschfelder, J. O., Curtiss, C. F., and Bird, R. B., Molecular Theory of Gases and Liquids, John Wiley and Sons, Inc., New York, 1954.

[23]Howe, J. T. and Sheaffer, Y. S., "Role of Charge Separation and Pressure Diffusion in the Gascap of Entry Objects," AIAA Journal, Vol. 7, No. 10, Oct. 1969, pp. 1971-1977.

[24]Curtiss, C. F. and Hirschfelder, J. O., "Transport Properties of Multi-Component Gas Mixture," Journal of Chemical Physics, Vol. 17, No. 6, June 1949, pp. 550-555.

[25]Yos, J. M., "Transport Properties of Nitrogen, Hydrogen, Oxygen, and Air to 30,000 K," Technical Memorandum RAD TM-63-7, AVCO-RAD, Wilmington, Mass., Mar. 1963.

[26]Yun, K. S. and Mason, E. A., "Collision Integrals for the Transport Properties of Dissociating Air at High Temperatures," Physics of Fluids, Vol. 5, No. 4, Apr. 1962, pp. 380-386.

[27]Cubley, S. J. and Mason, E. A., "Atom-Molecule and Molecule-Molecule Potentials and Transport Collision Integrals for High-Temperature Air Species," Physics of Fluids, Vol. 18, No. 9, Sept. 1975, pp. 1109-1111.

[28]Petschek, H. and Byron, S., "Approach to Equilibrium Ionization behind Strong Shock Waves in Argon," Annals of Physics, Vol. 1, 1957, pp. 270-315.

[29]Blackman, V., "Vibrational Relaxation in Oxygen and Nitrogen," Journal of Fluid Mechanics, Vol. 1, Pt. 1, May 1956, pp. 61-85.

[30]Gaydon, A. G. and Hurle, I. R., "Measurement of Times of Vibrational Relaxation and Dissociation behind Shock Waves in N_2, O_2, Air, CO, CO_2 and H_2," 8th Int. Symp. on Combustion, Pasadena, Calif., Aug. 1960, pp. 309-318.

[31]Millikan, R. C. and White, D. R., "Systematics of Vibrational Relaxation," Journal of Chemical Physics, Vol. 39, No. 12, Dec. 1963, pp. 3209-3213.

[32]Park, C., "Problems of Rate Chemistry in AOTV Flight Regimes," AIAA Paper 84-1730, AIAA 19th Thermophysics Conference, Snowmass, Colo., June 1984; published elsewhere in this volume.

[33]Slinker, S. and Ali, A. W., "Electron Excitation and Ionization Rate Coefficients for N_2, O_2, NO, N, and O," NRL Memorandum Report 4756, Naval Research Laboratory, Washington, D.C., Feb. 1982.

[34]Carslaw, H. S. and Jaeger, J. C., Conduction of Heat in Solids, Oxford University Press, Oxford, 1959.

Complete Flowfields Over Low- and Wide-Angle AOTV Conceptual Configurations

Peter A. Gnoffo*
NASA Langley Research Center, Hampton, Virginia

Abstract

Flowfields over two conceptual aeroassisted orbital transfer vehicles, a moderate lift-to-drag ratio biconic, and an axisymmetric zero-lift aerobrake are investigated. The solution procedures employ a compressible Navier-Stokes time-asymptotic technique for blunt body flow and a parabolized Navier-Stokes spatial marching technique for the primarily supersonic shock layer flow over low-angle bodies. Emphasis is placed on the laminar convective heating predictions and comparisons to experimental data where available. Code robustness and solution times are discussed, and guidelines are given for the specification of various numerical parameters. The parabolized Navier-Stokes technique has a high probability of solution success for the range of applications considered herein, and very good comparisons with experimental heat-transfer data have been obtained. The blunt body Navier-Stokes code is more sensitive to parameter specification for the aerobrake calculations. Still, coarse grid solutions can be obtained quickly to get a good first cut at flowfield definition. Fair to good comparisons have been obtained between experimental heating and pressure data and coarse and fine grid solutions.

Presented as Paper 84-1695 at AIAA 19th Thermophysics Conference, Snowmass, Colorado, June 25-28, 1984.

*Aero-Space Technologist, Aerothermodynamics Branch, Space Systems Division.

Introduction

In light of the current interest in designing a vehicle for transferring payloads from low Earth orbit (LEO) to geosynchronous Earth orbit (GEO),[1,2] a study has been initiated to assess current capabilities for defining aerothermodynamic loads over the various proposed configurations. These vehicles are designed to pass through the upper atmosphere on the return leg of their mission using aerodynamic lift and drag to hold a trajectory that, on exit from the sensible atmosphere, will permit rendezvous with either a Space Shuttle or space station. The purpose of this study is to demonstrate the present capabilities of two codes for simulating the complete flowfields over conceptual aeroassisted orbital transfer vehicles (AOTV's), to validate the results where possible with experimental data, and to comment on algorithm "robustness" for these applications.

The configurations examined here are a bent biconic, representative of a moderate-lift vehicle, and an umbrella-like drag brake in front of a cylindrical afterbody, typical of low-lift configurations. These shapes are representative of the types of vehicles being considered for aeroassisted missions. The flow simulations are obtained using the Solution-adaptive, Finite-volume Algorithm (SOFIA) for solving the Navier-Stokes Equations for axisymmetric flow over complete configurations and the Parabolized Navier-Stokes code (PNS) for marching a solution down a low-angle body at angle of attack.

Configurations

The two configurations studied in this report, a low-angle biconic and a wide-angle aerobrake, are representative of the types of vehicles being considered for aeroassisted orbital transfer missions.[2] They do not necessarily represent a "best" guess at what the final configuration will look like. These particular geometries were chosen because there is an experimental data base for comparison and because the character of their flowfields is significantly different. The flowfield over the wide-angle aerobrake is characterized by a large region of subsonic flow over the forebody and a large recirculation region in the near wake. The flowfield over the low-angle biconic is characterized by a very small region of subsonic flow on the nose, supersonic flow over the cones, and circumferential separation (vortical flow) on the leeside.

The moderate-lift vehicle is a 12.84 deg/7 deg bent biconic with fore-cone axis bent 7 deg relative to the aft-cone axis. This model has been studied in several of the hypersonic facilities at the Langley Research Center.[3-5] The dimensions are shown in Fig. 1. The wind-tunnel model has a base radius of 1 in.; the full-scale configuration has a base radius of 7 ft.

The wide-angle body is an axisymmetric, 86-deg half-angle drag brake with a cylindrical afterbody that closely approximates a wind-tunnel model studied in the Langley Continuous Flow Hypersonic Tunnel.[6,7] The pertinent parameters of the model are the brake radius R_B, brake edge radius R_E, payload radius R_p, and model length L. These parameters are presented in Table 1 under the heading model A. The wind-tunnel model had flats and ridges on the front of the drag brake to simulate a ceramic fabric material stretched over radial ribs. Details of the construction of the rear of the drag brake vary between the wind-tunnel model and the model shown in Fig. 2. The rear corner of the cylindrical afterbody was boat-tailed for the numerical computation in order to simplify grid generation with the coarse grid. The experimental heat-transfer data on the brake edge were taken on a two-dimensional model with surface area equal to the surface area of the 4-in.-diam brake.

Toward the end of this study, a 7% error was found in the specification of R_E and R_p. All of the coarse grid results were recomputed for the correct configuration; however, publication deadlines precluded a rerun of the fine grid computation. Fine grid results are presented for the incorrect model configuration denoted as model B in Table 1.

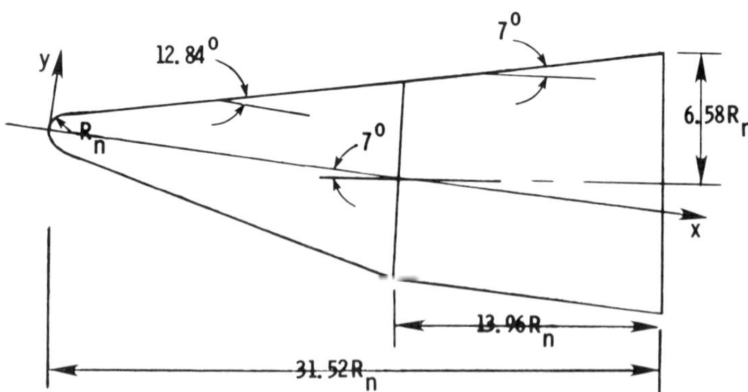

Fig. 1 Geometry of moderate-lift bent biconic.

Table 1 Brake geometry

Model	A	B
R_B	2	2
R_E	0.172	0.160
R_p	0.6	0.558
L	4.52	4.52
θ_c	86 deg	86 deg

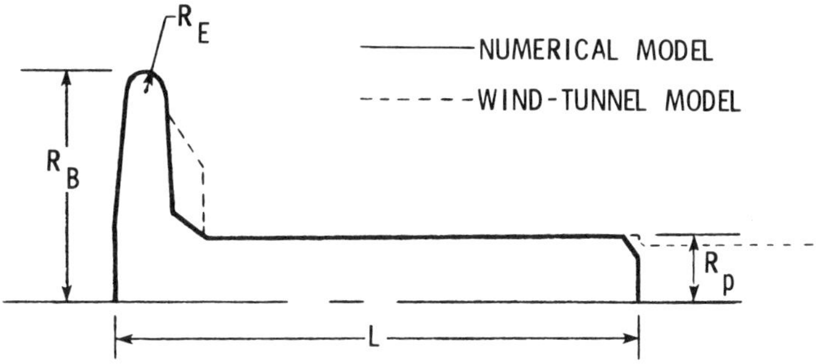

Fig. 2 Geometry of zero-lift aerobrake.

Governing Equations

The governing equations used herein are the Navier-Stokes equations for the flow over both the aerobrake and over the nose of the low-angle biconic and the parabolized Navier-Stokes equations for the flow over the conic sections of the biconic. The Navier-Stokes equations (including continuity and energy conservation) are expressed in integral conservation form as

$$\iiint_{\Omega} \bar{q}_t \, d\Omega + \iint_{\sigma} (\bar{f}\bar{i} + \bar{g}\bar{j} + \bar{h}\bar{k})\cdot\bar{n} \, d\sigma \qquad (1)$$

where $\bar{q}$, $\bar{f}$, $\bar{g}$, and $\bar{h}$ are the vector components of the Navier-Stokes equations; Ω is a control volume; σ is the surface area of the control volume, and $\bar{i}$, $\bar{j}$, $\bar{k}$, and $\bar{n}$ are unit vectors in the x, y, z, and outward normal directions, respectively.

The parabolized Navier-Stokes equations are obtained by writing the Navier-Stokes equations in a coordinate system in which one of the coordinate directions is aligned with a principal flow direction and then dropping the dissipation terms in this direction. Furthermore, the pressure gradient term in this principal direction requires special treatment in the subsonic boundary layer to maintain computational stability as the solution is marched down the body. Further details of this procedure are given in the section entitled Computational Method.

The equation of state for a perfect gas is

$$p = (\gamma - 1)\rho e \tag{2}$$

where γ, the ratio of specific heats, is a constant. At high temperatures, vibrational energy modes are excited, and dissociation of various molecule species occurs. In these situations, the ratio of specific heats is not a constant; however, the equation of state for a perfect gas can be recovered by defining

$$\gamma \equiv h/e \tag{3}$$

where γ is now a variable, and h and e are the mixture enthalpy and internal energy, respectively. This approach requires some external procedure to determine the value of this ratio at equilibrium as a function of pressure and enthalpy. A free energy minimization routine is used for this purpose, thus enabling calculations involving equilibrium flow of arbitrary gas mixtures.

The viscosity is calculated from Sutherland's law when the equilibrium option is not used. When the equilibrium option is implemented, Sutherland's law is used up to temperatures of 500 K. For $T > 500$ K, the Wilke relation for mixture viscosity is employed using curve fits for viscosities of the individual species in the mixture. The curve fit data for thermal conductivities in the present version of the code (linear fit) were not valid over a wide temperature range. Consequently, the Prandtl number was held constant at 0.72 for the $M_\infty \approx 10$ perfect gas calculation and at 0.70 for the $M_\infty \approx 30$ equilibrium air calculations.

Nondimensional convective heating is defined by

$$\dot{q}_c = \frac{\mu}{RePr} \frac{\partial h}{\partial n} \tag{4}$$

where h is the mixture enthalpy and n is the surface normal in this equation. The Lewis number is set equal to 1 for all calculations.

Computational Method

Wide-Angle Bodies: SOFIA

A solution-adaptive finite-volume algorithm (SOFIA) is used to numerically solve the laminar time-dependent Navier Stokes equations over the aerobrake shown in Fig. 1 and over the spherical nose of the biconic shown in Fig. 2. SOFIA is a fully vectorized explicit code that has options for calculating the 1) equilibrium flow of (H, He, O, N, C, e) mixtures, 2) shock capturing or discontinuous shock (Rankine Hugoniot) boundary conditions, and 3) wall slip or no-slip with or without wall blowing boundary conditions. In the finite-volume formulation, the integral form of the governing conservation laws is approximated on cells whose corners are defined by the position of grid points in physical space. The evaluation of flux through a cell wall is made using quantities from opposite ends of the wall on the predictor and corrector steps, respectively. One cycle is completed after four time steps so that the predicted/corrected quantities for a given cell are evaluated once at every corner of the cell. Details of the approximation scheme are presented in Refs. 8 and 9. The formulation is quite simple, reducing exactly to MacCormack's method[10] in a rectangular coordinate system. For illustrative purposes, one time step (one quarter cycle) is written in two dimensions as follows:

$$\left(Q_{ij}^{\overline{n+1}} - Q_{ij}^{n}\right) A_{ij}/\Delta t_{ij} =$$

$$- \left[\Delta_i (F\,\Delta_j y - G\,\Delta_j x) + \Delta_j (G\,\Delta_i x - F\,\Delta_i y)\right]_{ij}^{n} \qquad (5a)$$

$$\left(Q_{i+1,j+1}^{\overline{\overline{n+1}}} - Q_{i+1,j+1}\right) A_{ij}/\Delta t_{i+1,j+1} -$$

$$- \nabla_i (F\,\nabla_j y - G\,\nabla_j x) + \left[\nabla_j (G\,\nabla_i x - F\,\nabla_i y)\right]_{i+1,j+1}^{\overline{n+1}} \qquad (5b)$$

$$Q_{ij}^{n+1} = \left(Q_{ij}^{\overline{n+1}} + Q_{ij}^{\overline{\overline{n+1}}}\right)\Big/ 2 \qquad (5c)$$

Here Q, F, and G are the net function approximations to $\bar{q}$, $\bar{f}$, and $\bar{g}$ of the Navier-Stokes equations (including continuity and energy conservation). A_{ij} is the area of the (i,j) cell, and $(x,y)_{ij}$ are the Cartesian coordinates of the (i,j) point. The operators ∇ and Δ are backward and forward difference operators, respectively. The extension to axisymmetric flow is straightforward and is detailed in Ref. 9.

Solutions are obtained by 1) constructing the body with line segments and circular arcs; 2) defining the number of points desired on each section, using a polynomial blending function[11] to distribute the points; 3) defining the shock boundary using a "best" guess or data from a previous run; 4) generating the grid and initializing the flowfield; 5) running cases sequentially starting from "easy" $M_\infty \approx 2$, $Re_\infty \approx 10{,}000$ and working to the desired freestream condition; 6) sequentially running cases in ≈3-min blocks of CYBER 203 computer time (5000 iterations for 46x41 grid, 1500 iterations for a 91x81 grid), adjusting numerical parameters as needed until the solution converges. Steps 1-5 are only required when starting from scratch.

Calculations are made using local maximum time advancement on a relatively coarse grid (46x41). For the spherical nose calculations required for the biconic, the grid is not coarse and this procedure alone will produce high-quality converged solutions--so long as grid stretching is not excessive. Excessive stretching can distort the converged solution because of errors introduced by local maximum time step advancement or explicit smoothing. Such errors are detected by running the program with a constant time step and dropping the smoothing coefficient to $O(\Delta t)$.

The aerobrake geometry and flowfield are poorly defined using the coarse (46x41) grid. Considerable grid stretching is required to define high gradient regions near the walls and at expansion or compression corners. Consequently, convective heating calculations using local maximum time step can be as much as 25% in error in the stagnation region as compared with constant time step calculations. Pressure distributions are much less sensitive to this effect, differences generally remaining within 3%.

The fourth-order explicit smoothing is applied once per cycle--using data from the beginning of the cycle, i.e.,

$$Q^{n+4} = Q^{n+4} - \varepsilon_{ij}[(\nabla_i\{\Delta_i[\nabla_i(\Delta_i Q^n_{ij})]\} + \nabla_j\{\Delta_j[\nabla_j(\Delta_j Q^n_{ij})]\})] \tag{6}$$

In the region just behind the aerobrake, approximately bounded by the coordinate lines extending from the upper brake edge and the brake-payload corner out to one brake edge diameter into the recirculation region, the value of ε_{ij} must be kept small in order to maintain numerical stability. Typical values of ε_{ij} in this region are equal to 0.0002. Conversely, when using local maximum time step advancement of the solution, values of ε_{ij} in the free shear layer behind the brake must satisfy the relation

$$0.01 < \varepsilon_{ij} < 0.035 \tag{7}$$

in order to maintain numerical stability. For simplicity, the value of ε_{ij} in the rest of the flowfield is set equal to the value required in the shear layer. As mentioned earlier, when using constant time step advancement, the maximum value of ε_{ij} is $O(\Delta t)$.

The energy equation is rewritten in a form so that solution advancement involves discretization of $\partial\rho e/\partial t$ as a function of $\partial\bar{q}/\partial t$. Furthermore, a tendency to compute negative densities or internal energies in the very low-density region behind the brake is overcome by the following strategy. First, numerically compute $\partial\rho/\partial t$ and $\partial\rho e/\partial t$ from the finite-volume approximation to the integral conservation laws. Then, using the continuity equation as an example, compute the predicted/corrected values of ρ by dividing both sides of the continuity equation by ρ and expressing the time derivative as a derivative of $\ln\rho$. Thus,

$$\frac{1}{\rho}\frac{\partial\rho}{\partial t} = \frac{\partial \ln\rho}{\partial t} = \frac{-1}{\rho}\,\text{rhs}$$

$$\ln\rho^{\overline{n+1}} - \ln\rho^{n} = \frac{-\Delta t}{\rho}\,\text{rhs} \tag{8}$$

$$\rho^{\overline{n+1}} = \rho^{n}\exp\,(-\Delta t\ \text{rhs}/\rho)$$

where rhs is right side of the equation. Evaluating exponentials at every mesh point for every iteration is computationally expensive. However, the exponential can be expanded in series form to

$$\rho^{\overline{n+1}} = \rho^{n}\left[1 - \frac{\Delta t}{\rho^{n}}\,\text{RHS} + \frac{1}{2}\left(\frac{\Delta t}{\rho^{n}}\,\text{rhs}\right)^{2}\right] + 0\left[\frac{(\Delta t\ \text{RHS})^{3}}{6\rho^{2}}\right] \tag{9}$$

When the higher-order terms are neglected, the first bracketed term is always positive and, consequently, $\overline{\rho^{n+1}}$ is always positive. Clearly, this procedure cannot guarantee convergence, because if the rhs begins to grow large (diverge) the higher-order terms are significant and ρ will diverge. However, when the solution is converging, the rhs goes to zero, the higher-order terms are small, and ρ is always positive. Furthermore, the procedure involves minimal computational effort over the standard scheme.

Grid Generation and Adaption

In addition to the one-dimensional grid adaption scheme described in Ref. 8, the variational approach of Ref. 12 has also been employed for grid generation and solution adaption. As in the previous case, grids are updated at user-specified intervals. Routines have been developed that interpolate data from the old grid to the new grid. The partial differential equations of the variational approach are relaxed using a fully vectorized four-color successive over-relaxation scheme.

The present mode of operation is to generate the grid using the variational approach with zero weighting for orthogonality and volume control while controlling mesh spacing near the wall with a user-specified forcing function on arc length. This process gives a smoothly varying grid over the domain. The fine tuning of the grid is still accomplished using one-dimensional adaption. The volume control of the variational approach tends to bow the η coordinate lines between the forebody and the shock out in the direction of the wake, where cell volumes are, on the average, significantly larger than in the forebody region. Arc length control in the variational approach at present seems to solve this problem; however, it is not yet clear how to specify the forcing functions on arc length to accurately resolve all regions of the flow. These problems are currently being studied.

Low-Angle Bodies: PNS

The parabolized-Navier-Stokes (PNS) code of Ref. 13, with modifications to allow equilibrium flow calculations as described in Ref. 14, is used to march a solution down the bent biconic body shown in Fig. 2. The starting solution is interpolated from the flowfield over the spherical nose as calculated by program SOFIA. The PNS code is a factored-implicit, noniterative marching algorithm. The governing equations are written in an axis-normal coordinate system, and the axis of the fore cone is used as the principal axis. Both second-order implicit and fourth-order explicit smoothing are used to enhance computational stability. Only a fraction, ω, of the pressure gradient term in the principal direction is computed implicitly through the subsonic boundary layer. The term ω must satisfy the inequality

$$\omega < \min \{\gamma M_x^2/[1 + (\gamma - 1) M_x^2], 1\} \tag{10}$$

where M_x is the Mach number in the principal direction.

The fraction $(1-\omega)$ of the pressure gradient term should contain some numerical mechanism for treating wave propagation traveling upstream through the subsonic boundary layer. Such treatment is impossible using a single-sweep marching code as employed herein. Consequently, this $(1-\omega)$ fraction of the pressure gradient term is set equal to zero in the present application. Algorithms whereby this term is computed using explicitly defined backward differences from either the exact point within the boundary layer or at some supersonic flow point immediately above the boundary layer have caused catastrophic instabilities for a fore-cone angle of attack of 27 deg studied herein.

Combinations of large factors of implicit and explicit smoothing with large coordinate stretching at the surface can significantly influence (≈10%) heat-transfer calculations and sometimes cause the solution to diverge. Experience with the PNS code has shown that as coordinate stretching near the wall is increased ($\beta \to 1.$), the explicit and implicit smoothing coefficients must be decreased and the marching step size may need to be decreased. The implicit smoothing parameter is the most critical one to adjust. Once these parameters are brought within a reasonable range, heat-transfer calculations are relatively insensitive to further changes in smoothing, stretching, or step size. For the problems presented herein, this range has been approximately bounded as follows:

$$0.005 < \varepsilon_E < 0.02 \quad 0.003 < \varepsilon_I < 0.015 \qquad (11)$$
$$1.01 < \beta < 1.02 \quad 0.003 < \sigma < 0.005$$

where ε_E and ε_I are the explicit and implicit smoothing coefficients, respectively; β is the grid stretching parameter; and σ is the step size control parameter,

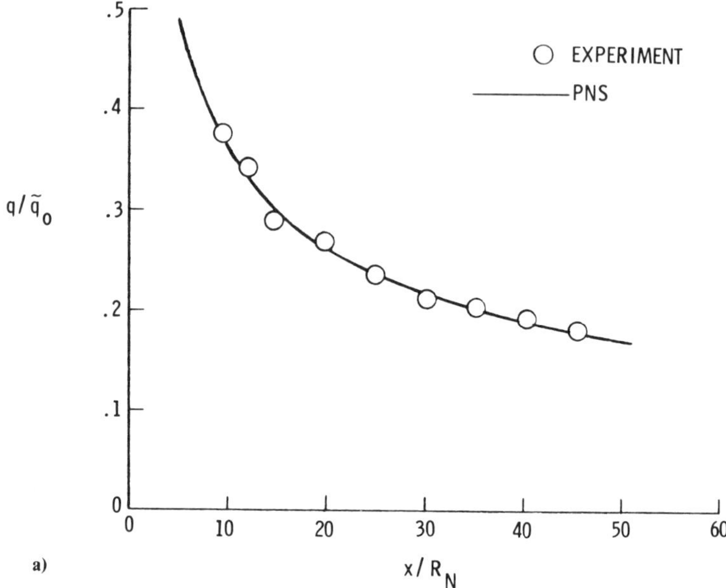

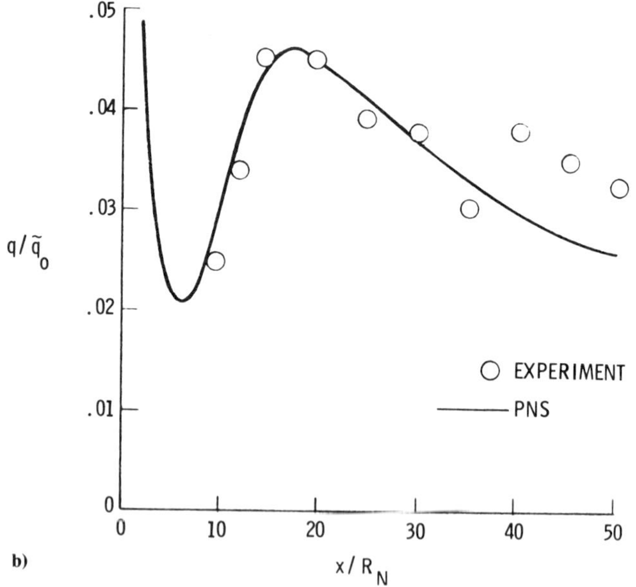

Fig. 3 Heating distributions over cone at M_∞ = 10.6, Re_∞ = 46500, and α = 20 deg. a) Windward, b) leeside.

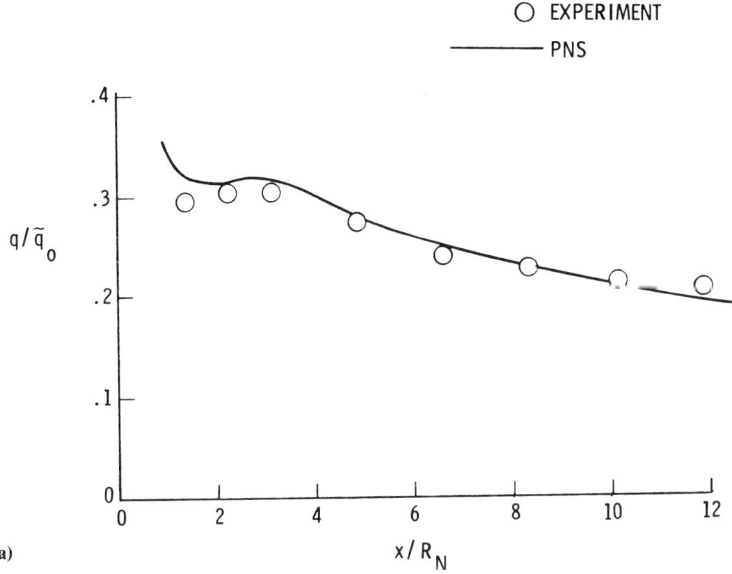

a)

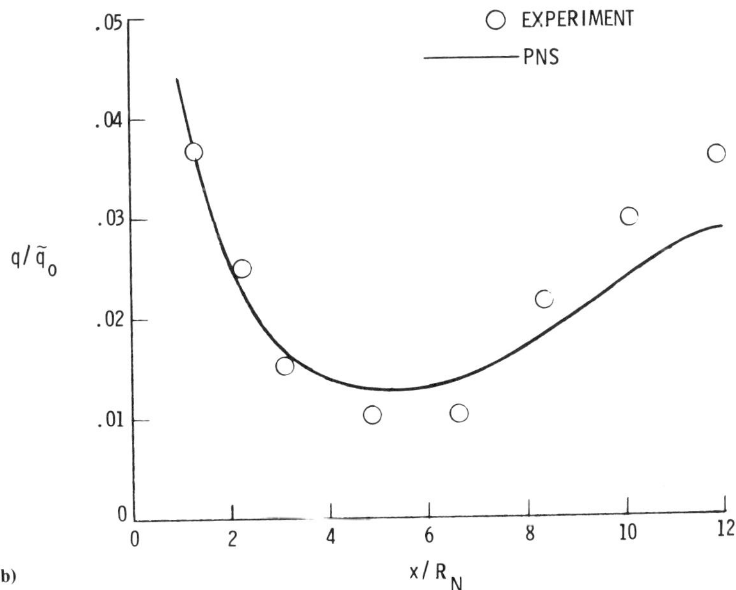

b)

Fig. 4 Heating distributions over cone at $M_\infty = 10.6$, $Re_\infty = 136400$, and $\alpha = 20$ deg. a) Windward, b) leeside.

i.e., $\Delta x/R_N = \sigma x/R_N$. More complete details of the PNS code and parameter definition appear in Refs. 13 and 14.

Results and Discussion

Low-Angle Bodies

Several comparisons have been made between heat-transfer predictions made using the PNS code and experimental data over low-angle cones and biconics. A few of these comparisons are presented herein.

The first case, taken from Ref. 15, is for a spherically blunted 15-deg cone at 20-deg angle of attack. The nose radius is 0.375 in., and the freestream Mach number and Reynolds number based on nose radius are 10.6 and 46,500, respectively. Windward and leeward heating rates are shown in Fig. 3. The comparisons on the windward side are excellent. The comparisons on the leeside are excellent up to $x/R_N = 35$ at which point the experimental data show a transitional jump in heating, and the PNS code underpredicts by 20%. (In these figures, x is measured from the vertex of the equivalent sharp cone, and $q_o = 0.2156$ MW/m^2 as used in Ref. 15.)

The second case is identical to the first except that the nose radius is 1.1 in., and thus the freestream Reynolds number based on nose radius is 136,400. This model had more gages in the region immediately behind the sphere-cone junction. Results of these comparisons are presented in Fig. 4. There is a slight (7% maximum) difference between experiment and calculation within three nose radii of the sphere-cone junction on the windward side. Further downstream, the comparisons are very good. The leeside comparisons are quite good near the junction but gradually fall (≈20%) below the experimental data. It appears there may be some transition in the experimental data between seven and eight nose radii, although this jump is not as sharp as in the previous case.

The third case is taken from a comprehensive series of tests (currently unpublished) on the biconic (Fig. 1) with a base radius equal to 1 in. (nose radius equal to 0.152 in.) taken in the Langley Continuous Flow Hypersonic Tunnel by C. G. Miller. Test data were taken at two different Reynolds numbers, and the results show clearly the effect of Reynolds number on leeside heating. Freestream conditions are the Mach and Reynolds number based on nose radius equal to 9.86 and 6980,

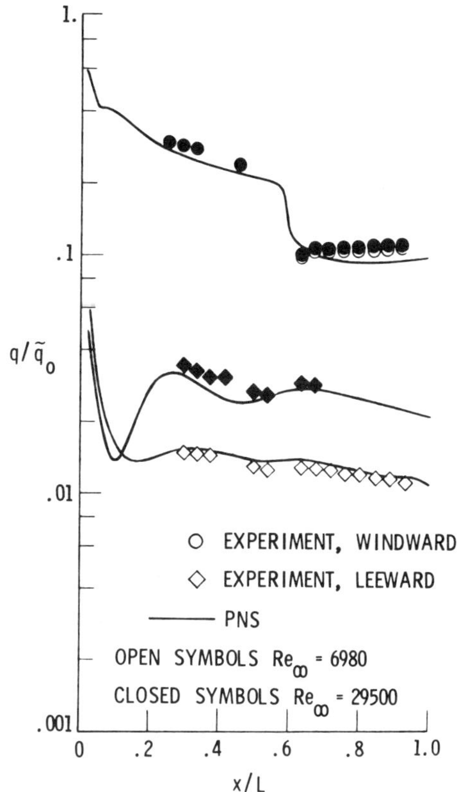

Fig. 5 Reynolds number effect on heat transfer for biconic at M_∞ = 10 and α = 20 deg.

respectively, for the low Reynolds number case, and 10.16 and 29,500, respectively, for the high Reynolds number case. Comparison of PNS code results with experiment are shown in Fig. 5, where q_0 is a reference heating rate to a sphere and equals 0.326 MW/m^2 for the low Reynolds case and 0.682 MW/m^2 for the high Reynolds number case. No Reynolds number effect is present on the windward side, where differences between prediction and experiment are within 10%. A significant increase in heating is observed on the leeside with increasing Reynolds number over the range of the test data. The PNS code follows these changes quite closely with differences generally within 15%.

Heat-transfer data taken in high-temperature air in an expansion tube were compared with the PNS code using the equilibrium air option in Refs. 5 and 14. (The experimental data in Ref. 4 should be multiplied by a factor of 1.13 to account for an error in a material property discovered after the release of that paper.) As detailed in Ref. 5, the air was not fully in chemical

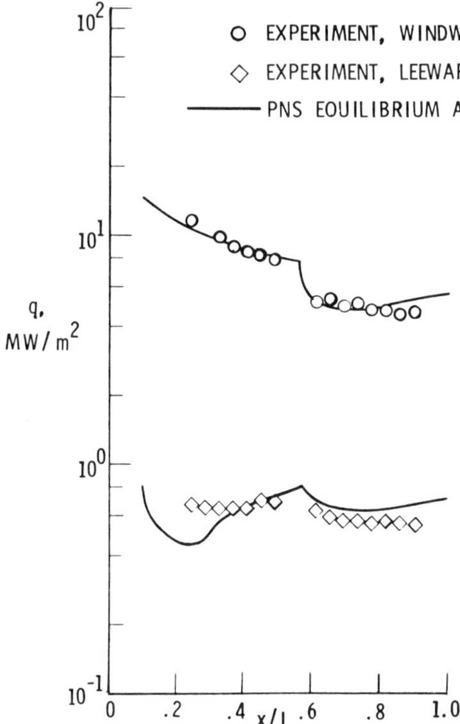

Fig. 6 Heat transfer over straight biconic in high-temperature air at α = 20 deg.

equilibrium; however, the quality of comparison between theory and experiment is very good on the windward side (within 10%) and somewhat poorer on the leeside (within 25%) as seen in Fig. 6. Freestream conditions for this case are p_∞ = 2182 N/m^2, ρ_∞ = 4.726x10^{-3} kg/m^2, T_∞ = 1604 K, V_∞ = 5326 m/s, and angle of attack equal to 20 deg. The wall temperature for this case was 300 K, and no experimental evidence of wall catalytic effects was observed in the experiment.

The trajectory of a full-scale (base radius equals 7 ft) biconic on a GEO-to-LEO aerocapture mission has been calculated using the POST code of Ref. 16. It is assumed that the biconic will maintain a 20-deg angle of attack and use roll control for energy management. Freestream conditions at eight points along the trajectory are presented in Table 2 along with the computed aerodynamic coefficients relative to the aft-cone axis and center of pressure location measured from the nose along the fore-cone axis. The boundary condition on wall temperature was set using the following equation:

$$Tw_{i+1} = 0.5[Tw_i + (q_i/\varepsilon\sigma)^{1/4}]$$

where q_i is the laminar convective heating rate; i is the index on axial location; and ε and σ are the emissivity (equal to 0.89) and the Stefan-Boltzmann constant, respectively. This specification very closely approximates a radiative equilibrium boundary condition without significant recoding of the implicit body boundary condition. A summary of the results of this trajectory analysis is presented in Figs. 7-9.

The equilibrium, stagnation point, laminar, convective heating pulse is presented in Fig. 7. Maximum heating occurs at approximately 150 s into the entry (t = 0 at 120 km) at 73 km. (A constraint on the trajectory was imposed that attempted to hold peak heating at a constant level for a longer time at a higher altitude as compared with passing deeper into the atmosphere for a shorter time with higher peak heating levels.)

Figure 8 shows the windward side and leeside symmetry plane results for laminar convective heating. The trajectory point at 150 s is at peak stagnation-point heating. The heating distribution on the windward side at t = 98 is substantially the same as the distribution at t = 150 but at approximately a 10% lower level. Much later in the trajectory, at t = 358, the heating distribution on the windward side aft cone is significantly flatter than the distributions near peak heating. This trend is characteristic of a γ effect seen in earlier investigations in which both the pressure and heating

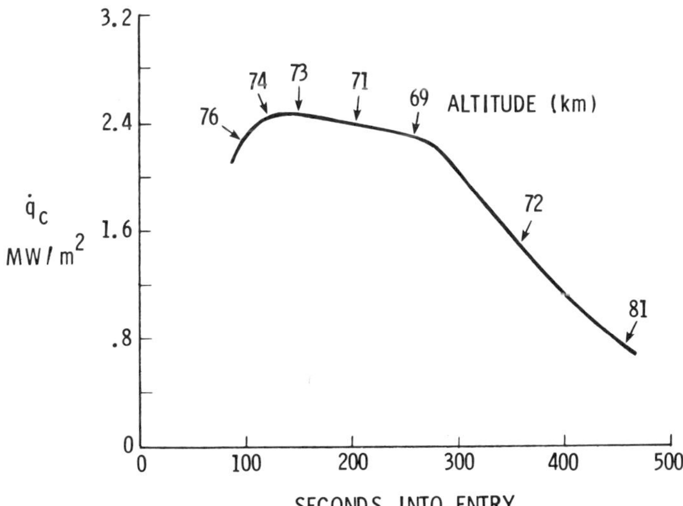

Fig. 7 Equilibrium stagnation-point laminar convective heating for full-scale moderate-lift biconic during aeroassisted maneuver.

Table 2 Trajectory data for biconic

Time	98	120	150	198	210	258	358	458
ρ_∞, kg/m^3	3.762-5	4.954-5	5.769-5	7.271-5	7.712-5	9.881-5	6.686-5	1.670-5
T_∞, K	196.42	203.72	207.88	214.35	216.04	223.25	211.99	180.65
p_∞, N/m^2	2.1212	2.897	3.443	4.474	4.782	6.332	4.069	0.8663
V_∞, km/s	9.786	9.674	9.495	9.17	9.086	8.695	7.959	7.654
M_∞	34.83	33.81	32.85	31.24	30.83	29.02	27.26	28.40
Altitude, km	75.95	74.08	73.02	71.36	70.93	69.08	71.96	80.96
Re_{∞, R_N}	9093	11,474	12,891	15,290	15,963	19,044	12,319	3396
C_A aft-cone axis	0.3080	0.2990	0.2947	0.2874	0.2873	0.2818	0.2901	0.3427
C_N aft-cone axis	0.6593	0.6589	0.6589	0.6604	0.6592	0.6592	0.6583	0.6634
X_{cp}/L fore-cone axis	0.5411	0.5402	0.5396	0.5388	0.5385	0.5371	0.5359	0.5390

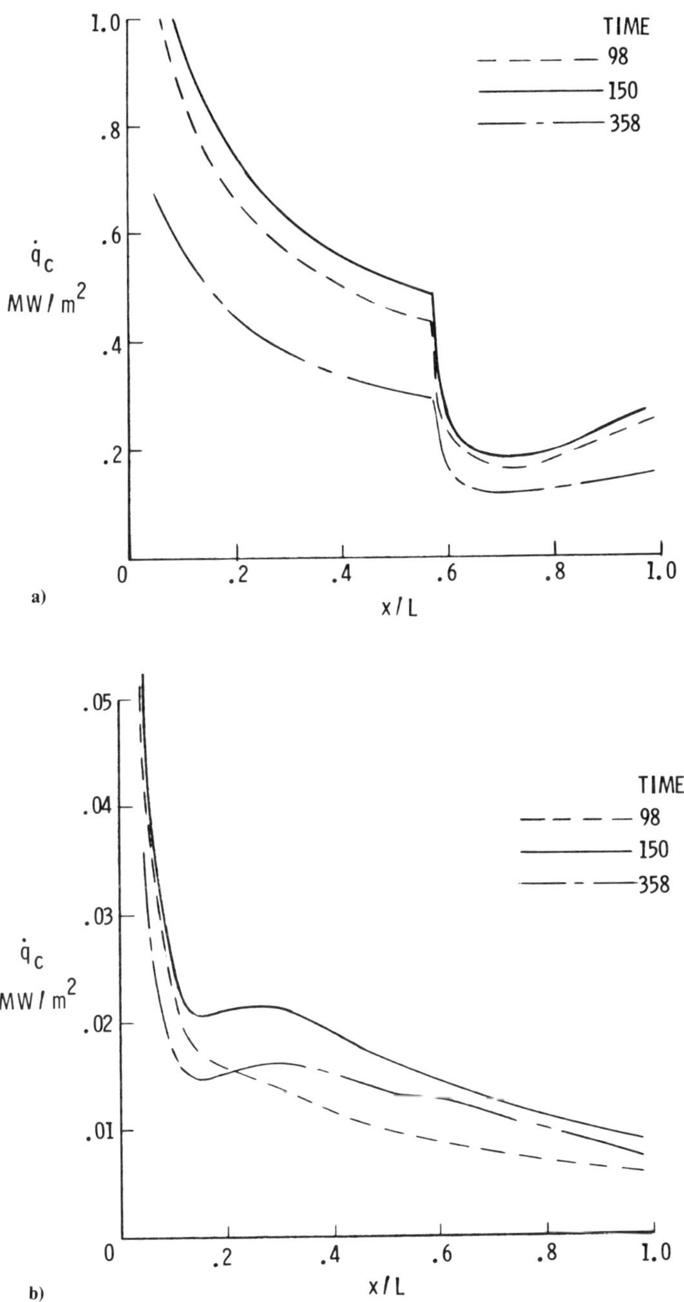

Fig. 8 Laminar convective heating on full-scale moderate-lift biconic during aeroassisted maneuvers. a) Windward, b) eeside.

distributions start to increase after the expansion to the aft cone as the local values of γ in the shock layer decrease. The different distributions on the leeside show the Reynolds number effect on the leeside heating as controlled by the leeside vortex. The freestream Reynolds numbers based on nose radius are 9100, 12,900, and 12,300 for t = 98, 150, and 358, respectively. Figure 8b shows the same trend in leeside heating levels with the Reynolds number as presented in Fig. 5.

A typical circumferential heating distribution is presented in Fig. 9. This distribution is located at a point approximately 2.5 nose radii upstream of the cone-cone junction. The flow separates (zero shear in circumferential direction) at 158 deg around the body. The minimum heating point occurs at the next circumferential grid point at 163 deg around the body.

All of the flowfields generated in this study with the PNS code used the numerical parameters bounded by Eq. (11) in the Computational Method section. A solution

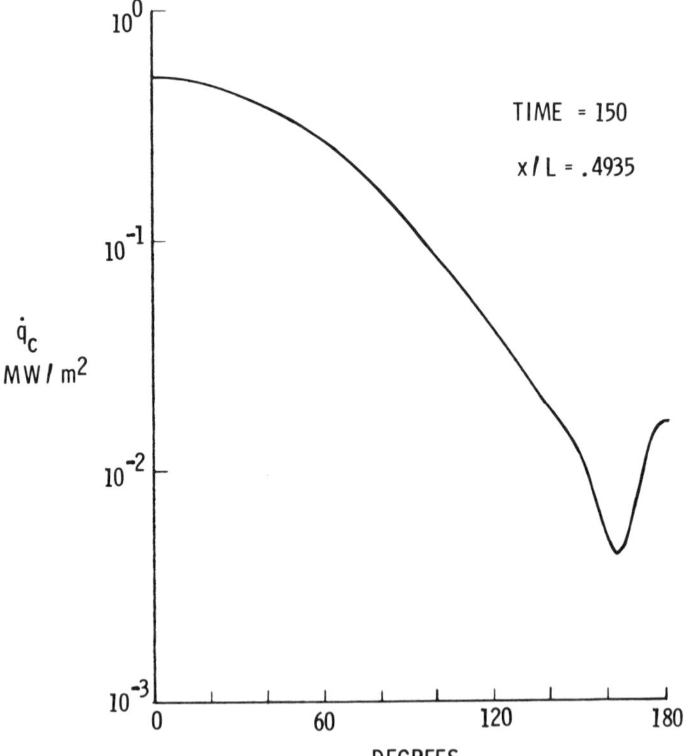

Fig. 9 Circumferential heating distribution at peak heating.

for a flowfield in chemical equilibrium takes less than 40 min of CYBER 203 computer time. The starting solutions are run sequentially from one trajectory point to the next, and each of these solutions takes 6-10 min of CYBER 203 computer time. The PNS code can be described as robust for the applications considered in this study.

The major question remaining is the accuracy of the results for the given trajectory. Given the successful comparisons that have been made with wind-tunnel data, the following two points are considered the primary causes of concern.

1) Assumption of chemical equilibrium: The concern here is that there is no capability of computing nonequilibrium chemistry with catalytic/noncatalytic surface effects that have been shown to be significant (≈40% decrease) on the Space Shuttle.[17] As shown in Ref. 18, these catalytic effects for flows in chemical nonequilibrium are enhanced over the velocity range characteristic of an AOTV trajectory. For example, at 73 km with a velocity of 9.3 km/s, there is approximately a factor of 3 decrease in stagnation heating to a 2.5-m nose radius hyperboloid in going from equilibrium chemistry to nonequilibrium chemistry with a noncatalytic wall. However, the magnitude of the peak heating rates in the nose region on the windward side may preclude the use of any ceramic thermal protection system (TPS) possessing noncatalytic surface effects. In this case, for the example mentioned above, Ref. 18 shows only a 20% decrease in heating going from an equilibrium flow to a nonequilibrium flow with a fully catalytic wall.

2) Approximate treatment of pressure gradient through the subsonic portion of the boundary layer: In a recent paper, Rakich[19] has shown that for large values of viscous interaction parameter $\bar{\chi} > 1$, where

$$\bar{\chi} = M_\infty^3 (C_\infty / R_{ex_\infty})^{1/2}$$

and

$$C_\infty = T_\infty \mu_w / T_w \mu_\infty$$

the approximate treatment of the pressure gradient term through the subsonic boundary layer can lead to errors of 10% in the evaluation of skin friction, C_f, for flow over a flat plate. These errors are measured relative to an iterated PNS solution in which the pressure gradient term can be approximated correctly (using forward differences). The viscous interaction parameter at peak heating on the fore cone is on the order of 10. There is no capability to perform such iterations in the present code, so potential errors on heat-transfer predictions cannot be quantified accurately. The expansion tube tests discussed earlier had a value of viscous interaction parameter approximately one-half the value of the peak heating case. If all of the errors in that calculation on the windward side are due to viscous interaction (probably a poor assumption), then the neglect of the pressure gradient term results in approximately a 10% decrease in the predicted heating levels. Before closing this discussion, it should be noted that about ten global iterations were required to recover this 10% enhancement of skin friction in Ref. 19. The cost of computing one trajectory point would, therefore, be increased by a factor of 10 unless some action could be taken to save the results outside of the boundary layer.

Wide-Angle Body

A comprehensive series of tests was performed on the aerobrake configuration shown in Fig. 2 in the Langley Continuous Flow Hypersonic Tunnel.[6,7] Nominal free-stream conditions used in that test and studied herein are $M_\infty = 9.82$, the stagnation pressure $P_S = 280$ psia, and the stagnation temperature $T_S = 1810°R$. The calculated Reynolds number for these conditions based on brake radius is Re = 79,248. The coarse grid used to define the flowfield has 46 points around the body and down the wake centerline and 41 points between the body and the

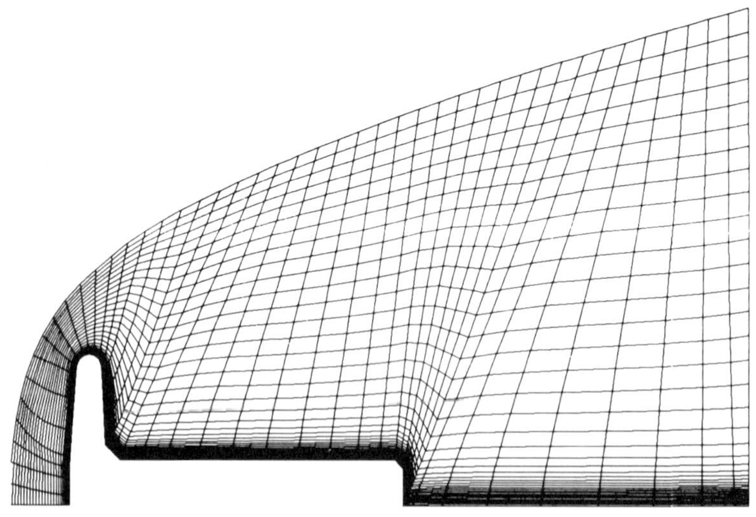

Fig. 10 Coarse grid distribution over aerobrake.

shock (see Fig. 10). The fine grid doubles the number of intervals in both coordinate directions.

Pressure and heating distributions on the brake face are presented in Figs. 11 and 12, respectively. Experimental and predicted values of C_p agree very well. Coarse and fine grid values of pressure are in excellent agreement and plot as one line in Fig. 11. In Fig. 12, the heat-transfer coefficients are nondimensionalized by h_{ref}, the stagnation-point heat-transfer coefficient to a sphere of radius 2 in. The experimental values were obtained using a phase-change paint technique. Predictions are approximately 20% below experiment except at the stagnation point, where the coarse grid results agree with experiment, and near the brake edge, where the disparity increases.

The differences between the coarse and fine grid results in the stagnation region are caused by the very coarse grid along the body (only eight points define the brake face up to the edge radius junction) and have been observed in earlier applications of SOFIA. The experimental stagnation-point value appears to be high when compared to the theory and data of Zoby and Sullivan in Ref. 20. Given an effective nose radius R_N, edge radius R_E, and body (or brake) radius R_B, one can

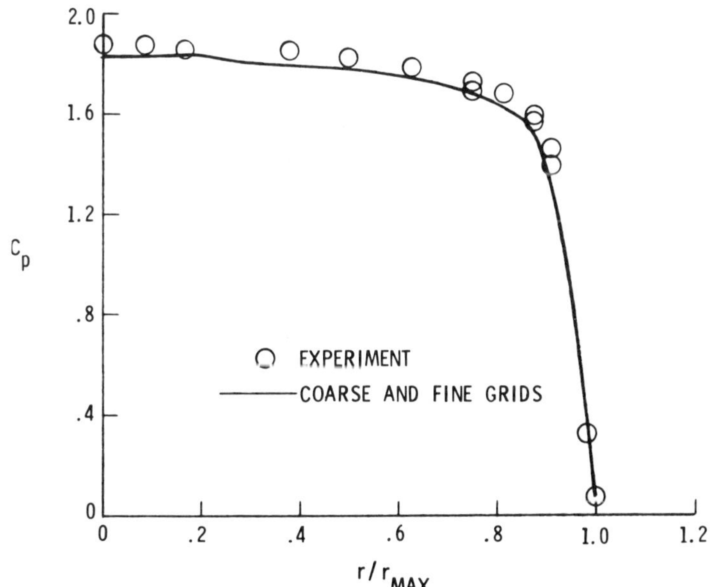

Fig. 11 Brakeface pressure distributions on aerobrake at M_∞ = 9.82, Re_∞ = 79248.

compute h_s/h_{ref}. For the present application, one can compute an effective radius for the aerobrake from the circle that intersects the stagnation point of the body (α = 0 deg) and the junction of the brake edge with the 86-deg conical surface. This results in values of R_N = 18.11 in. and h_s/h_{ref} = 0.575, the value shown in Fig. 12. Alternately, one can take the experimental value of stagnation-point heating and compute a value of R_N = 4.59 in. that gives an effective body shape that is a poor approximation to the original body (see Fig. 13). The fine grid results for h_s/h_{ref} do agree well with the square root of the experimentally determined velocity gradient ratio $[(du_e/dx)_s/(du_e/dx)_{ref}]$.

The coarse grid results completely miss the sharp rise in heating occurring immediately ahead of the brake edge, although agreeing well with thermocouple data at the front edge of a two-dimensional rectangular-faced model with brake area equal to that of the aerobrake. The fine grid results show a steep increase in heating near the brake edge, but the heating is not as large as the experimental data. It is suspected that still more grid refinement along the body near the brake edge is needed to define this peak. In this regard, it is inter-

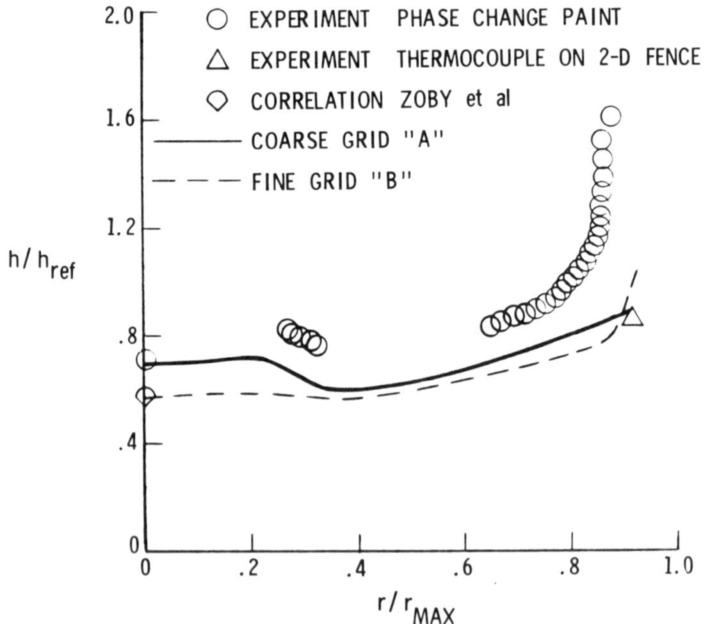

Fig. 12 Heating distributions on face of aerobrake.

esting to note that, even if there was a grid adaption capability to move points along the body in response to gradients in heating along the body, the coarse grid results completely smooth out the gradients. Thus, the solution adaption process is defeated if the original grid does not resolve at least some part of the various high-gradient regions.

Heat-transfer data over the rectangular face with edge radii of R_E = 0.172 in. and 0.062 in. are presented in Fig. 14. Results from coarse and fine grid calculations are in fair agreement with the experimental data over this rectangular approximation to the axisymmetric model.

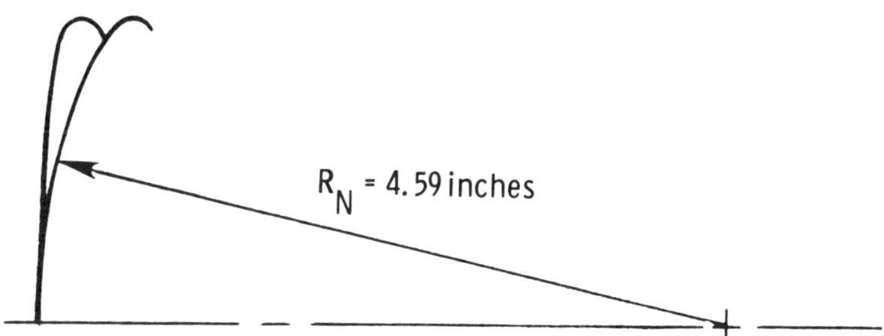

Fig. 13 Effective body corresponding to experimental stagnation-point heating.

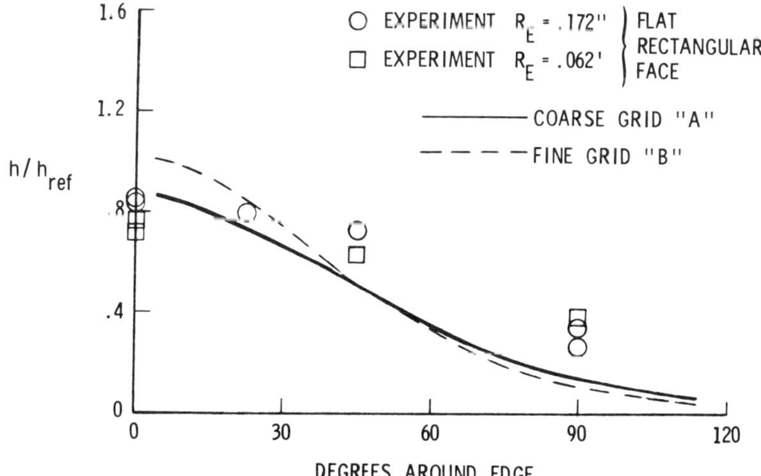

Fig. 14 Heat-transfer distribution over brake edge.

Pressure and heat-transfer data over the cylindrical payload behind the brake are presented in Figs. 15 and 16. The differences between the geometries of the wind-tunnel model and its numerical counterpart are presented in Fig. 2. Both computational bodies A and B were ramped in the back corner to facilitate coarse grid generation. The wind tunnel has a 1-in.-diam sting that would extend through the outflow boundary if modeled numerically. The computational bodies are 4.52 in. long. (This scale model of a conceptual flight vehicle was studied to see if the additional geometric complexity would pose any significant problems in computing a flowfield. In general, it was found that boattailing the afterbody solved some coarse grid generation problems, and no significant stability problems were encountered in this region.) Except for the pressure port located 0.65 in. upstream of the brake backface (x = 1.395), the computed trends in pressure and heating distributions are very good. Quantitative errors caused by differences in geometry cannot be measured here.

The coarse grid pressure coefficients are approximately 23% higher than the fine grid results on the payload. (Recall that the fine grid results are on a payload with a radius of 93% of coarse grid payload radius.) Experimental data from the two rearward pressure ports are bounded by the coarse and fine grid results. Experimental and computational data at the most forward port appear to be influenced by differences in geometry.

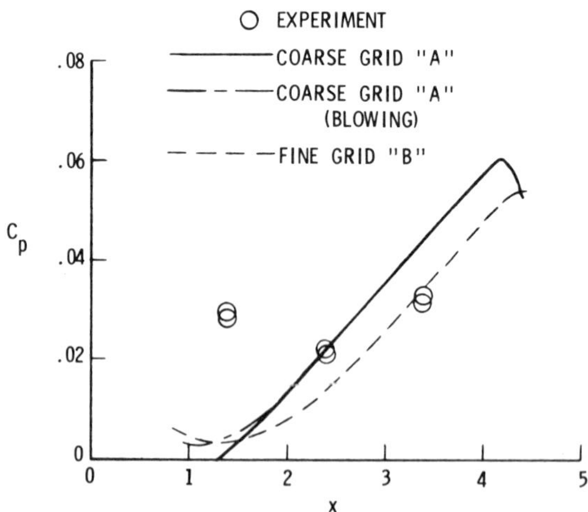

Fig. 15 Pressure distribution over payload of aerobrake.

In Fig. 16, the coarse and fine grid approximations to heating are in good agreement over most of the payload surface. The fine grid heating near the rear of the payload is about 22% lower than the coarse grid results. The dividing streamline passes just over the rear edge of the payload in the fine grid calculation (see Fig. 17), whereas the coarse grid prediction shows dividing streamline impingement on the rear of the payload. Coarse grid numerical viscosity appears to have delayed separation. While physical distances between the two predictions of dividing streamline location are small, the oblique angle of the streamline causes large differences in reattachment location. Preliminary phase II data of Ref. 6 indicate that the fine grid results present an accurate picture of the reattachment point. The difference in dividing streamline behavior accounts for the different peak payload heating values. Fine and coarse grid results are about 60% higher than experimental data at these very low heating rates.

The blowing boundary condition of program SOFIA was tested to study the effects of blowing air through the ramped fillet at the brake-payload junction on computational stability and payload pressures and heating. The blowing rate, nondimensionalized by ρ_∞/V_∞, is $\dot{m} = 0.001$ ($\rho_\infty V_\infty \dot{m}$ Area $= 5.3\times10^{-6}$ kg/s). No stability problems were encountered in this test. The small blowing rate has little effect on pressure levels over most of the payload (Fig. 15). However, mass injection from the fillet reduces coarse grid heating values by 20%.

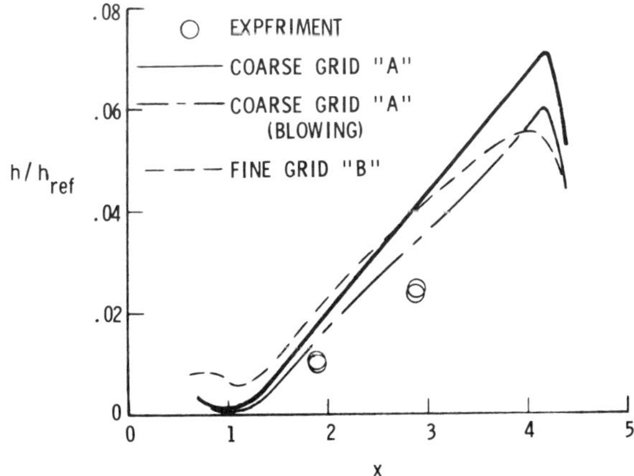

Fig. 16 Heating distribution over payload of aerobrake.

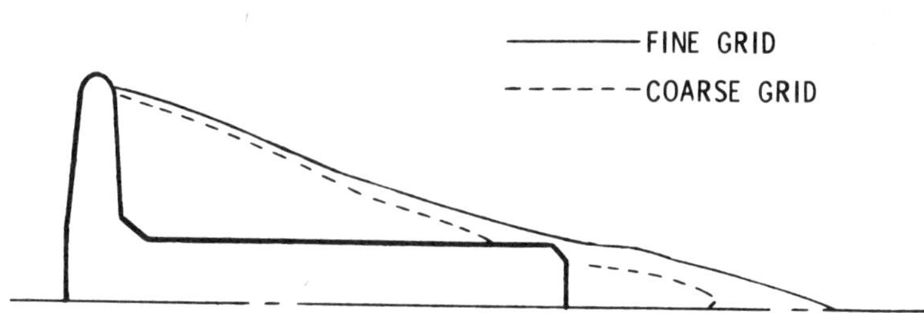

Fig. 17 Dividing streamline location from fine and coarse grid.

A qualitative picture of the entire flowfield is presented in Fig. 18 from the fine grid calculations. Pressure, internal energy, and Mach number contour plots illustrate most of the high-gradient regions in the flowfield.

Total computer time for the coarse grid blowing results, which was started from a coarse grid no-blowing solution, was approximately 18 min on the CYBER 203. The fine grid results are estimated at about 36 min starting from a coarse grid result. The estimated time for a coarse grid solution started from scratch is approximately 40 min. The actual computer time for this entire study is on the order of 7 h, which includes studying effects of various numerical parameters and grid configurations, as well as time taken in running cases that eventually became unstable because of a sensitivity to $\varepsilon_{i,j}$ and/or the use of local maximum time step.

No trajectory was run on a full-scale aerobrake using program SOFIA. All of the comments made about the PNS code with regard to restrictions to equilibrium chemistry apply as well to this program for a flight application. Program SOFIA is also restricted to axisymmetric flows (zero angle of attack), so that a true trajectory analysis would not be possible.

In contrast to the PNS code, program SOFIA cannot be considered robust for the present application. While the coarse grid results can be computed relatively quickly, they are more sensitive (in terms of stability) to parameter specification because of the large amount of grid stretching necessarily present. Fine grid computations can be affected similarly if the grid is overstretched. Program SOFIA can be used to calculate the flowfield for axisymmetric flow over aerobrakes, but the success of the computation or the quality of the converged solution can be compromised by improper numerical parameter specification and grid distribution.

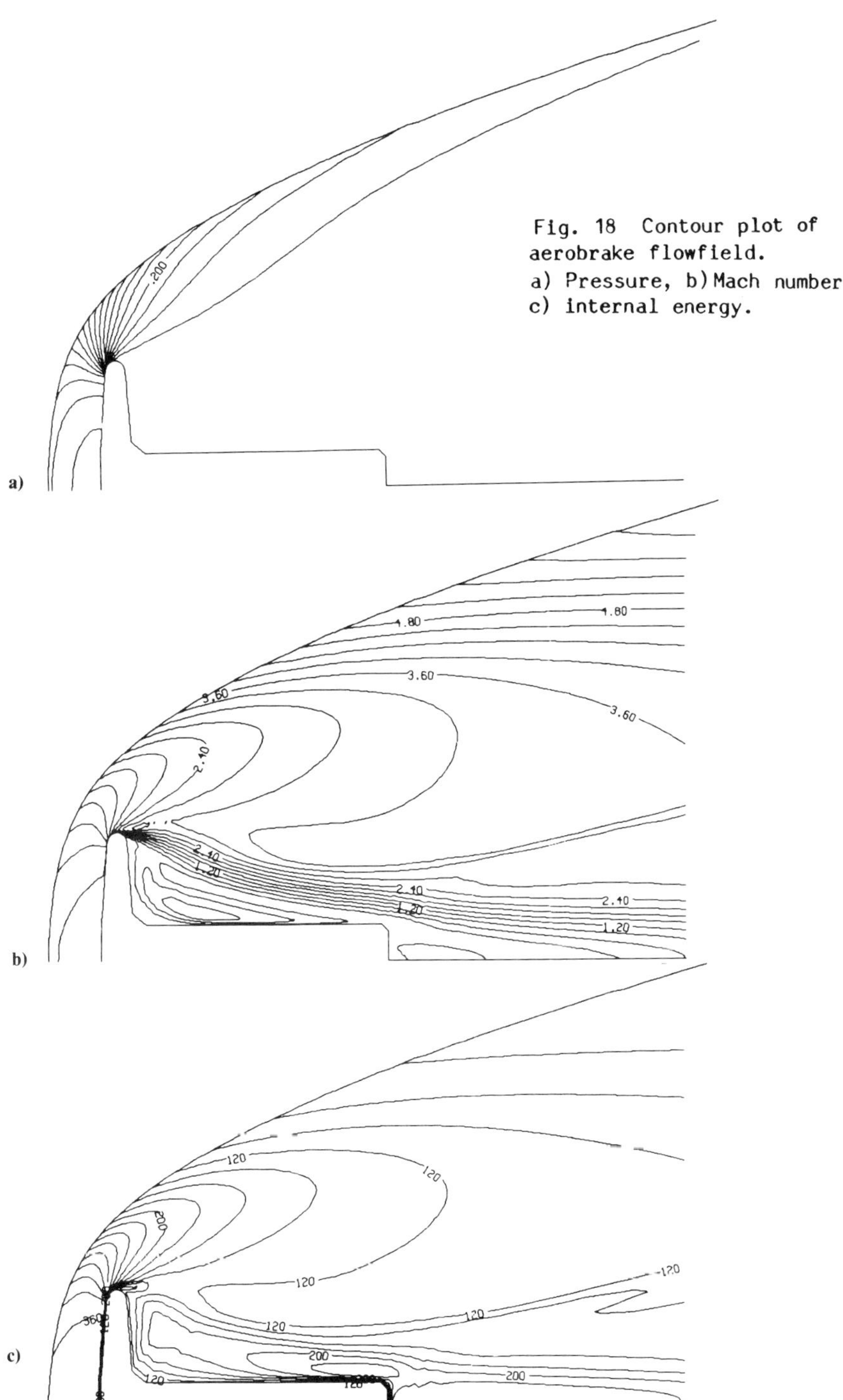

Fig. 18 Contour plot of aerobrake flowfield.
a) Pressure, b) Mach number,
c) internal energy.

Concluding Remarks

The parabolized Navier-Stokes algorithm was used to define the flowfield over a 12.84 deg/7 deg bent biconic at a 20-deg angle of attack (fore cone at 27-deg angle of attack). The program yields very good comparisons to laminar convective heat-transfer data. There are several numerical parameters that require adjustment, but the computational procedure appears to be robust for AOTV applications studied here, provided the parameter specification remains within the bounds defined in this report. On the basis of considerable experience with this code, it is believed that application of the method to similar geometries for AOTV missions would provide no significant problems as long as the shock layer flow remains substantially supersonic. The principal shortcoming of the code for AOTV applications is its restriction to flows in chemical equilibrium.

Program SOFIA was used to define the flowfield over an 86-deg half-angle umbrella-like aerobrake at 0-deg angle of attack. Coarse and fine grids were used to generate solutions under wind-tunnel conditions. The heating and pressure distributions were in excellent agreement with the trends in experimental data everywhere except just before the brake edge, where coarse grid predictions smoothed out a sudden rise in heating, and just behind the brake face on the payload, where differences in geometry cause different surface pressure distributions. There are differences in heating levels of about 20% between prediction and experiment on the brake face. The experimental data for stagnation-point heating are about 20% higher than the method of Zoby and Sullivan[20] and the present fine grid prediction. The very low pressure and heating levels from prediction and experiment on the brake edge and payload are in fair agreement. Coarse grid predictions show a delayed separation from the brake edge as compared with the fine grid results. The coarse grid dividing streamline impinges on the rear upper surface of the payload, while the fine grid results show the dividing streamline passing just over the back corner of the payload. The fine grid results are generally in better agreement with experimental data, though the coarse grid results give a quick first cut at the solution.

Program SOFIA cannot be considered robust for the present applications. Severe stability problems were most often encountered on or just behind the brake edge in the high-gradient expansion region. The algorithm modifications described in the report did much to alleviate such problems. However, the sensitivity of the program to grids and numerical smoothing can sometimes defeat rapid convergence when using local maximum time step solution advancement. The principal shortcomings of this code for AOTV applications are the restrictions to flows at zero angle of attack in chemical equilibrium.

Acknowledgments

The author would like to thank C. G. Miller for providing the unpublished experimental biconic data and R. W. Powell for providing the trajectory data.

References

[1]Walberg, G. D., "Aeroassisted Orbit Transfer--Window Opens on Missions," Astronautics and Aeronautics, Vol. 21, Nov. 1983, pp. 36-43.

[2]Walberg, G. D., "The Next Generation of Reentry Vehicles--NASA's View," AGARD Lecture Series on Hypersonic Aerothermodynamics, von Karman Institute for Fluid Dynamics, Feb. 6-10, 1984.

[3]Miller, C. G. and Gnoffo, P. A., "Pressure Distributions and Shock Shapes for 12.84/7° On-Axis and Bent-Nose Biconics in Air at Mach 6," NASA TM 83222, Dec. 1981.

[4]Miller, C. G., Blackstock, T. A., Helms, V. T., and Midden, R. E., "An Experimental Investigation of Control Surface Effectiveness and Real-Gas Simulation for Biconics," AIAA Paper 83-0213, AIAA 21st Aerospace Sciences Meeting, Reno, Nevada, Jan. 1983.

[5]Miller, C. G., Micol, J. R., and Gnoffo, P. A., "Laminar Heat-Transfer Distributions on Biconics at Incidence in Hypersonic-Hypervelocity Flows," NASA TP 2213, Dec. 1984.

[6]Hair, L. M., Engel, C. D., Shih, P., Bithell, R., and Bowman, M., "Aerothermal Test Data of Low L/D Aerobraking Orbital Transfer Vehicle at Mach 10," Remtech Inc., Huntsville, Alabama, RTR 069-1, April 1983.

[7]Hair, L. M., Engel, C. D., and Sulyma, P. R., "Low L/D Aerobrake Test at Mach 10," AIAA Paper 83-1509, AIAA 18th Thermophysics Conference, Montreal, Canada, June 1983.

[8]Gnoffo, P. A., "A Solution-Adaptive Finite-Volume Algorithm with Application to Problems in Planetary Entry," Ph.D. Dissertation, Princeton University, N. J., May 1983.

[9]Gnoffo, P. A., "A Finite-Volume, Adaptive Grid Algorithm Applied to Planetary Entry Flowfields," AIAA Journal, Vol. 21, Sept. 1983, pp. 1249-1254.

[10]MacCormack, R. W., "The Effect of Viscosity in Hypervelocity Impact Cratering," AIAA Paper 69-354, AIAA Hypervelocity Impact Conference, Cincinnatti, Ohio, April 30-May 2, 1969.

[11]Chaussee, D. S. and Steger, J. L., "Three Dimensional Viscous Flow Field Program, Pt. 2: A Curvilinear Grid and Body Generation Program for Generalized Configurations (Interim Report)," AFWAL-TM-81-64-FIMG, 1981.

[12]Brackbill, J. V. and Saltzman, J. S., "Adaptive Zoning for Singular Problems in Two Dimensions," Journal of Computational Physics, Vol. 46, June 1982, pp. 342-368.

[13]Vigneron, Y. C., Rakich, J. V., and Tannehill, J. C., "Calculation of Supersonic Viscous Flow Over Delta Wings with Sharp Subsonic Leading Edges," AIAA Paper 78-1137, AIAA 11th Fluid and Plasma Dynamics Conference, Seattle, Washington, July 1978.

[14]Gnoffo, P. A., "Hypersonic Flows Over Biconics Using a Variable-Effective-Gamma, Parabolized-Navier-Stokes Code," AIAA Paper 83-1666, AIAA 16th Fluid and Plasma Dynamic Conference, Danvers, Massachusetts, July 1983.

[15]Cleary, J. W., "Effects of Angle of Attack and Bluntness on Laminar Heating-Rate Distributions of a 15° Cone at a Mach Number of 10.6," NASA TN D-5450, Oct. 1969.

[16]Bauer, G. L., Cornick, D. E., Habeger, A. R., Petersen, F. M., and Stevenson, R., "Program to Optimize Simulated Trajectories (POST)--Volume II--Utilization Manual," NASA CR-132690, April 1975.

[17]Shinn, J. L., Moss, J. N., and Simmonds, A. L., "Viscous Shock-Layer Heating Analysis for the Shuttle Windward Symmetry Plane with Surface Finite Catalytic Recombination Rates," AIAA Paper 82-0842, 3rd AIAA/ASME Joint Thermophysics, Fluids, Plasma & Heat Transfer Conference, St. Louis, Missouri, June 1982.

[18]Shinn, J. L. and Jones, J. J., "Chemical Nonequilibrium Effects on Flowfields for Aeroassist Orbital Transfer Vehicles," AIAA Paper 83-0214, Jan. 1983.

[19]Rakich, J. V., "Iterative PNS Method for Attached Flows with Upstream Influence," AIAA Paper 83-1955, AIAA 6th Computational Fluid Dynamics Conference, Danvers, Massachusetts, July 1983.

[20]Zoby, E. V. and Sullivan, E. M., "Effects of Corner Radius on Stagnation-Point Velocity Gradients on Blunt Axisymmetric Bodies," NASA TM-1067, March 1965.

AOTV Bluff Body Flow—Relaxation Algorithm

C. K. Lombard,* J. Bardina,† and E. Venkatapathy‡
PEDA Corporation, Palo Alto, California

Abstract

We present a new, computationally efficient single-level effectively explicit/implicit algorithm for gasdynamics. The method meets all the requirements for unconditionally stable global iteration over flows with mixed supersonic and subsonic zones including bluff body flow and boundary-layer flows with strong interaction and streamwise separation. For hyperbolic (supersonic flow) regions the method is automatically equivalent to contemporary space marching methods. For elliptic (subsonic flow) regions, rapid convergence is facilitated by alternating direction solution sweeps, which bring both sets of eigenvectors and the influence of both boundaries of a coordinate line equally into play. Point by point updating of the data with local iteration on the solution procedure at each spatial step as the sweeps progress not only renders the method single level in storage but, also, improves nonlinear accuracy to accelerate convergence by an order of magnitude over related two-level linearized implicit methods. The method derives robust stability from the combination of an eigenvector split upwind difference method, conservative supra-characteristics method (CSCM), with diagonally dominant alternating direction implicit (DDADI) approximate factorization and computed characteristic boundary approximations. The properties and performance of the technique are demonstrated in a variety of quasi-one-dimensional nozzle flows including completely subsonic or supersonic or mixed

---

Presented as Paper 84-1699 at AIAA 19th Thermophysics Conference, Snowmass, Colorado, June 25-27, 1984.

*President.

†Research Scientist II.

‡Research Scientist I.

subsonic/supersonic with sonic points and shocks. The technique is applied as a method of lines in two hypersonic blunt body flow problems: a classical sphere cylinder problem previously studied experimentally and computationally, and the coupled forebody and base flow of a model drag brake aeroassisted orbital transfer vehicle (AOTV). The early results support the belief that the new algorithm has the potential to make accurate computations of AOTV flowfields substantially faster and less costly than currently available explicit or two-level time-dependent implicit methods.

I. Introduction

Preliminary system studies[1–3] and assessment of the physics[4,5] of the flight regime have led to two classes of energy efficient AOTV concepts, with high and low L/D. These re-entry space vehicles which are to fly hypersonically in the continuum upper reaches of the atmosphere are subject to convective and nonequilibrium radiative heating. Convective heating will be a problem on the sharp leading edges of the slender aeromaneuvering vehicle and on the shoulder termination of the bluff drag brake (aerobraking) AOTV. Nonequilibrium radiative heating poses a potential problem for the payload in the wake of the drag brake as well as over the forebody.

The flight regime is such that experimentation is severely limited in the information that is available or can be obtained, and reliance must be placed on computational techniques to simulate testing of design concepts. This state of affairs is not completely new since computational analysis was also required to play a significant role in the design of the Galileo probe to Jupiter[6–8]. The AOTV problem is made significantly more computationally demanding than that facing the analysts for Galileo in that the reacting gas chemistry is nonequilibrium[5] rather than equilibrium in the most important flight regime. However, equilibrium reactive gas analyses may have a useful role to play in qualitative and semiquantitative analyses, and are being applied[9,10] in parabolized Navier-Stokes (PNS) and time-dependent procedures for the compressible Navier-Stokes equations.

Park[11] has recently made a preliminary exploration of the nonequilibirum shock-layer computational problem in a simulation of one-dimensional shock tube flow with a time-dependent implicit method. An important feature of Park's analysis is the incorporation (also see Lee[12]) of a multitemperature model for the reacting gas. Park's early results were obtained in the order of 10,000 iterations of 1-s CPU time each on a CRAY XMP computer. Such times imply that significant advances in numerical technique will need to be made in order for multidimensional, and particularly three-dimensional flowfield simulations to serve more than a benchmark role.

Recent work with a conservative flux difference eigenvector split implicit upwind scheme CSCM[13–16] has led to a new relaxation algorithm[17–19] for the compressible Navier-Stokes equations. Unlike the precursor two-level linearized implicit time-dependent method, but like the efficient computation PNS procedures,[20,21] the new method CSCM-S is single level in storage and marches in space through the flow domain. In supersonic flow where all eigenvalues in a streamaligned coordinate have the same sign, the PNS and CSCM-S methods are equivalent. Unlike the classical PNS methods, however, the new eigenvector split upwind method is (linearly) unconditionally stable and (nonlinearly) robust for subsonic flow, boundary-layer flow with streamwise separation, or massively separated flows. In such elliptic flow regions, i.e., having eigenvalues of mixed sign and both upstream and downstream domains of dependence, the split upwind method can be likened to stable marching of each scalar characteristic wave system in the direction of its associated eigenvalue (simple wave velocity). In elliptic regions, the CSCM-S method iteratively marches back and forth in alternating direction solution sweeps between computational boundaries.

A substantial variety of results for internal flows in one and two space dimensions and axisymmetric blunt body flow demonstrate that the new method is robust and finds convergence accelerated between one and two orders of magnitude over two-level linearized time-dependent methods.

In this paper we will first sketch the CSCM implicit method for the inviscid terms of the axisymmetric flow equations. Then, for quasi-one-dimensional nozzle flow we will give the associated bidiagonal solution algorithm based on diagonally dominant approximate factorization, and present the single-level marching scheme that derives therefrom. Some results of supersonic, subsonic, and transonic quasi-one-dimensional nozzle flows that illustrate the properties of the technique are given. These are followed by a single-level method of lines formulation for axisymmetric flow. Lastly, based on that algorithm, we present early results for axisymmetric bluff body flow including complete forebody and base flow for a free flight range model drag brake AOTV. Other results for two-dimensional and axisymmetric internal flows can be found in Ref. 18.

II. Conservative Difference Formulation and CSCM

A finite-difference formulation of the compressible Navier-Stokes equations for two-dimensional axisymmetric flow can be expressed for general difference operators in general curvilinear coordinates in the form

$$\delta_\tau q + \overline{\xi}_x \delta_\xi F + \frac{\overline{\xi}_y}{\overline{y}} \delta_\xi (yG) + \overline{\eta}_x \delta_\eta F + \frac{\overline{\eta}_y}{\overline{y}} \delta_\eta (yG) = 0 \tag{1}$$

Here, F and G are the combined viscid and inviscid conservative flux vectors for the x and y Cartesian coordinate directions. The viscous terms are second-order central differenced and treated in the well-known thin layer approximation in a manner very similar to Steger[22] for two-dimensional flow or Lombard et al.[8] for two-dimensional axisymetric flow, and will not be described further here. The inviscid fluxes and the conservative dependent variables are given in terms of the usual primitive variables by

$$\begin{array}{ccc} \underline{F} & \underline{yG} & \underline{q} \\ \rho u & y\rho v & \rho \\ \rho u^2 + p & y\rho v u & \rho u \\ \rho u v & y\rho v^2 + \overline{y}p & \rho v \\ u(\epsilon + p^2) & v(\epsilon + p) & \epsilon \end{array} \tag{2}$$

where the volumetric total energy $\epsilon = p/(\gamma - 1) + 1/2\rho(u^2 + v^2)$. The metric coefficients of Eq. (1) will also be given finite-difference approximations symbolized by the formulas $\xi_x = \delta_\eta y/\widetilde{J}$, $\xi_y = -\delta_\xi x/\widetilde{J}$, $\eta_x = -\delta_\xi y/\widetilde{J}$, $\eta_y = \delta_\eta y/\widetilde{J}$ with $\widetilde{J} = \delta_\xi x \delta_\eta y - \delta_\xi y \delta_\eta x$.

In the following development of the CSCM splitting for the system of Eq. (1) with Eq. (2), we shall express relations Eq. (2) in terms of the volumetric internal energy P rather than the pressure p through the substitution $p = (\gamma - 1)P$. This makes certain transformations simpler and is natural for intended real-gas applications, of which perfect gas is a special case.

The basis for the CSCM flux difference splitting is the sequence of discrete difference transformations that mimic the differential transformations between the conservative and nonconservative representations of the PDE, on the one hand, and between the nonconservative and characteristic representations, on the other. Specializing to one-dimension for purposes of discussion, we can write differentially

$$\partial_x F = A\partial_x q \equiv MT\Lambda T^{-1} M^{-1} \partial_x q \tag{3a}$$
$$= MT\Lambda T^{-1} \partial_x \widetilde{q} \equiv MA' \partial_x \widetilde{q} \tag{3b}$$
$$= MT\Lambda \partial_x \widetilde{\widetilde{q}} \tag{3c}$$

Here, A is the Jacobian matrix of the transformation of the flux vector by the conservative variables q ; and $\widetilde{q}$ and $\widetilde{\widetilde{q}}$ are the associated primitive and characteristic variables defined, respectively, using the matrices M^{-1} and T^{-1} of the diagonalizing similarity transformation of A. The diagonal matrix Λ contains, as elements, the real eigenvalues u, $u + c$, and $u - c$ where c is the sound speed. Indeed, under the transformations of Eq. (3), the conservative, nonconservative (or primitive), and characteristic representations of the elementary PDE's

$$\partial_\tau q + \partial_x F = 0 \tag{4a}$$

$$\partial_\tau \widetilde{q} + A' \partial_x \widetilde{q} = 0 \tag{4b}$$

$$\partial_\tau \widetilde{\widetilde{q}} + \Lambda \partial_x \widetilde{\widetilde{q}} = 0 \tag{4c}$$

are all equivalent. The goal of the CSCM derivation is to arrive at a matrix formulation for discrete differences having transformation properties similar to Eq. (3) and satisfying

$$\Delta_x F \equiv \widetilde{A} \Delta_x q \tag{5}$$

These conditions satisfy Roe's[23] property U to guarantee accurate shock capturing capability.

In axisymmetric flow, where for any curvilinear coordinate direction ς the contravariant (cell face normal) velocity component is $\overline{W} = \overline{\varsigma_x u} + (\overline{\varsigma_y/y})\overline{yv}$, the eigenvalues of the locally one-dimensional splitting are $\overline{W}$, $\overline{W} + \overline{C}$, and $\overline{W} - \overline{C}$ with $\overline{C} = (\overline{\varsigma_x^2} + \overline{\varsigma_y^2})^{1/2}\overline{c}$. Then, the splitting of the Jacobian matrix $\hat{A}_\varsigma$ which satisfies in conformity with Eq. (1)

$$\Delta_\varsigma \hat{F} = \overline{\varsigma_x} \Delta_\varsigma F + \frac{\overline{\varsigma_y}}{\overline{y}} \Delta_\varsigma y G \equiv \hat{A}_\varsigma \Delta_\varsigma q \tag{6}$$

can be expressed[15] as

$$A_\varsigma^{\pm} = (\overline{M}\overline{T} D^{\pm} \overline{T}^{-1} \overline{M}^{-1}) \hat{A}_\varsigma \tag{7a}$$

Here $\overline{M}$ and $\overline{T}$ are the interval average matrices of the similarity transform diagonalizing $\hat{A}_\varsigma$, and $D^{\pm}$ are the diagonal unit vector truth functions for the signs of the eigenvalues according to

$$D^{\pm} = \frac{1}{2}(\overline{\Lambda} \pm |\overline{\Lambda}|)/|\overline{\Lambda}| \quad , D^{+} + D^{-} = I \tag{7b}$$

Equations (7) with Eq. (6) effectively define the splitting of the conservative flux difference into pieces ΔF^{+} and ΔF^{-} associated with characteristic waves propagating with positive and negative eigenvalue signs.

III. CSCM-S One-Dimensional Formulation

By analogy with the well-known result for the scalar wave equation, an unconditionally stable implicit finite-difference equation for the gasdynamic system at the jth grid point can be written from the adjacent upwind flux difference split pieces ΔF_{j-1}^{+} and ΔF_{j}^{-} as

$$(I + A^{+} \nabla + A^{-} \Delta) \delta q_j = -A^{+} \Delta q)_{j-1}^{n} - A^{-} \Delta q)_{j}^{n} \tag{8}$$

Here, ∇ and Δ are backward and forward spatial difference operators. In the notation the interval averaged matrices between node points j and $j+1$ are labeled j. Thus, in the equation for the jth grid point, $A^{+} \Delta_\xi q)_{j-1}$ represents stable characteristic spatial differencing backward for positive eigenvalue contributions, and $A^{-} \Delta_\xi q)_j$ forward for negative ones. The right-hand side of Eq. (8) is written for the first-order method. Higher-order methods in space are given with results in Refs. 13 and 16, and a related method is proposed herein.

With $\delta q = q^{n+1} - q^{n}$, Eq. (8) defines a two-level linearized coupled block tridiagonal matrix implicit scheme that can be solved by a block elimination procedure. In Ref. 16, a new diagonally dominant bidiagonal approximately factored alternating sweep solution procedure (DDADI) for Eq. (8) is presented that is shown to be very robust and is effectively explicit, i.e., requires only a decoupled sequence of local block matrix inversions rather than the solution of the coupled set.

$$D\delta \overset{*}{q}_J = RHS_j + A^+ \delta \overset{*}{q}_{j-1} \tag{9a}$$

$$D\delta q_j = D\delta \overset{*}{q}_j - A^- \delta q_{j+1} \tag{9b}$$

In Eqs. (9), $D = I + A^+ - A^-$ is the central block of the unfactored tridiagonal system. For the linear problem, Eq. (9a) is equivalent to the single-level space marching procedure

$$D\delta \overset{*}{q}_j = A^+ \overset{*}{q}_{j-1} - A^+ q_j^n - A^- \Delta q)_j^n \tag{10a}$$

Nonlinearity enters in the single-level space marching form (10a) in that with data updating at each step of the forward sweep, the matrices A^+ are averaged between $\overset{*}{q}_{j-1}$ and q_j^n, rather than homogeneously at the old iteration level n. Similarly, a companion backward space marching sweep that is symmetric to Eq. (10a) and is intimately related to the backward sweep of equation (9b) with Eq. (9a) is

$$D\delta q_j = -A^+ \Delta \overset{*}{q})_{j-1} + A^- \overset{*}{q}_j - A^- q_{j+1}^{n+1} \tag{10b}$$

The CSCM-S method given by Eqs. (10) is von Neumann unconditionally stable for the scalar wave equation. References 16 and 18 show heuristically the significance of retaining both sets of eigenvalues in the diagonal block D to scale the contributions to the right-hand side (RHS) of the difference equations, rendering separately both the forward and backward sweeps highly stable regardless of eigenvalue sign. Consequently as the local Courant number (CFL) is taken large, the robust method becomes a very effective (symmetric Gauss-Seidel) relaxation scheme for the steady equations, a fact which substantially contributes to the extremely fast performance that will be demonstrated.

At a right computational boundary on the forward sweep, we solve the characteristic boundary point approximation [14,16]

$$(\tilde{\tilde{A}}^+ + A^+)\delta q_N = A^+ \overset{*}{q}_{N-1} - A^+ q_N^n \tag{11a}$$

$q_N^{n+1} = \overset{*}{q}_N$ and at a left, on the backward sweep,

$$(\widetilde{\widetilde{A}}^{-} - A^{-})\delta q_1 = A^{-} q_1^{n} - A^{-} q_2^{n+1} \tag{11b}$$

Analysis of a model system with upwind differenced scalar equations and coupled boundary conditions of the type above
linearized bidiagonal scheme[16] by Oliger
also strongly supports the numerically confirmed robust stability of the present nonlinear method for gasdynamics. A final factor in the rapid convergence to steady state that the scheme affords is the use of local iteration on the solution procedure at each marching step. The iteration serves to make the eigenvectors in the coefficient matrices consistent with the advanced state and, thus, provides improved accuracy for the nonlinear system. It appears cost effective to do this inner iteration everywhere, i.e., in both subsonic and supersonic regions, as the use of two inner iterations has been found to reduce the number of global iteration steps to convergence by a factor of three to four.

Quasi-One-Dimensional Flow Results

In the following we will give results from some numerical experiments with quasi-one-dimensional nozzle flows to illustrate the properties and performance of the technique. From the computed solutions, we show here plots of density. The agreement shown between the computations and the exact solutions for density extends to all the variables.

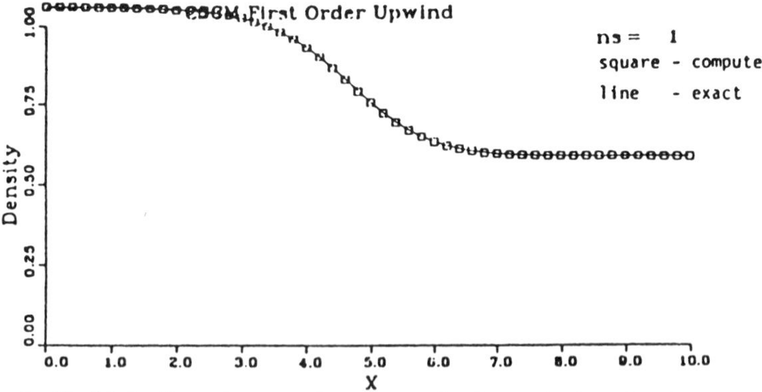

Fig. 1 Shubin's diverging nozzle supersonic flow solution developed in one forward sweep from supersonic initial data.

First, we present results for quasi-one-dimensional supersonic flow with no shock in Shubin's diverging nozzle. In purely supersonic zones, the experience with the present method is that the solution can be marched accurately in one global iteration, as ought to be the case. Figure 1 shows the exact solution (in solid line) and the computed result from the first forward sweep. With subsequent sweeps, the error (the difference between the exact and the computed solution) reduces to machine accuracy in less than three global iterations. In fact, by increasing (from two) the number of inner iterations on the solution procedure at each space marching step, one can guarantee convergence to prescribed accuracy in one forward sweep. This is also true of contemporary locally linearized unsplit PNS methods in supersonic flow.

With the globally iterative nonlinear space marching formulation, early experience in two quasi-one-dimensional nozzle problems with mixed supersonic/subsonic zones is that solutions are obtained in roughly an order of magnitude fewer iteration steps than had been required with the previously fast pseudo time-dependent technique and block tridiagonal solving.

The two nozzle problems, which are described and solved by Yee et al.[25] and also solved with the CSCM time-dependent technique in Refs. 13 and 14 are Shubin's diverging nozzle flow and Blottner's converging/diverging nozzle flow. Both problems involve unmatched overpressures at the outflow which result in internal shock terminated supersonic zones and subsonic outflow. For the experiments involving flow of mixed type, the same initial data given by Yee et al.[25] — a linear interpolation between inflow and outflow values for effectively exact solutions of the problems — is used that was used previously with the time-dependent approach.

For flows of mixed type, in Figs. 2 and 3, respectively, results are shown for successive forward and backward sweeps for five global iteration steps with Shubin's and Blottner's nozzle flows. In both cases, the exact solution as given by Yee et al.[25] is shown by a solid line, and the present computational results solved on a 51-point mesh is indicated by boxes. Blottner's nozzle flow is shown converged after 10 global iteration steps. There is substantial evidence in results not shown that with further work, the number of global iterations required to compute flows such as Blottner's can be reduced by a factor of two to about five. In these early experiments, no provision was made for inhibiting the formation of expansion shocks, a subject treated in Ref. 18 and applied in the present two-dimensional code.

In Fig. 4, we show a subscritical, i.e., completely subsonic, flow solution computed in only two global iteration steps for the Blottner

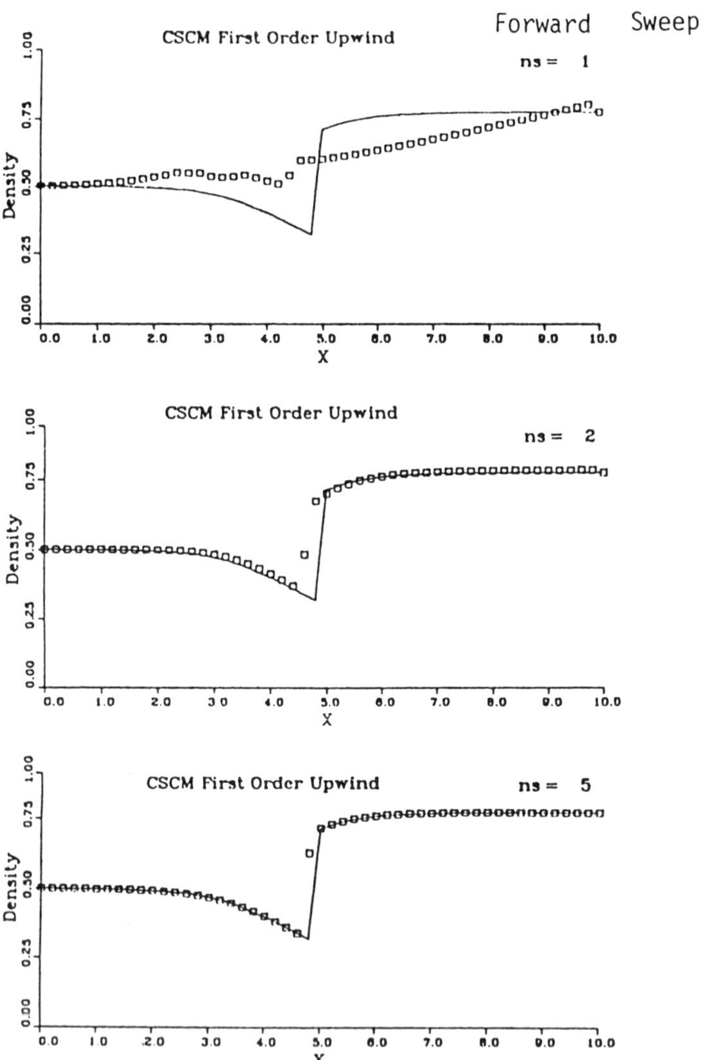

Fig. 2 Shubin's diverging nozzle flow solution developed in alternating forward and backward sweeps, one each per global iteration step for five steps: square – computed results; line – exact solution.

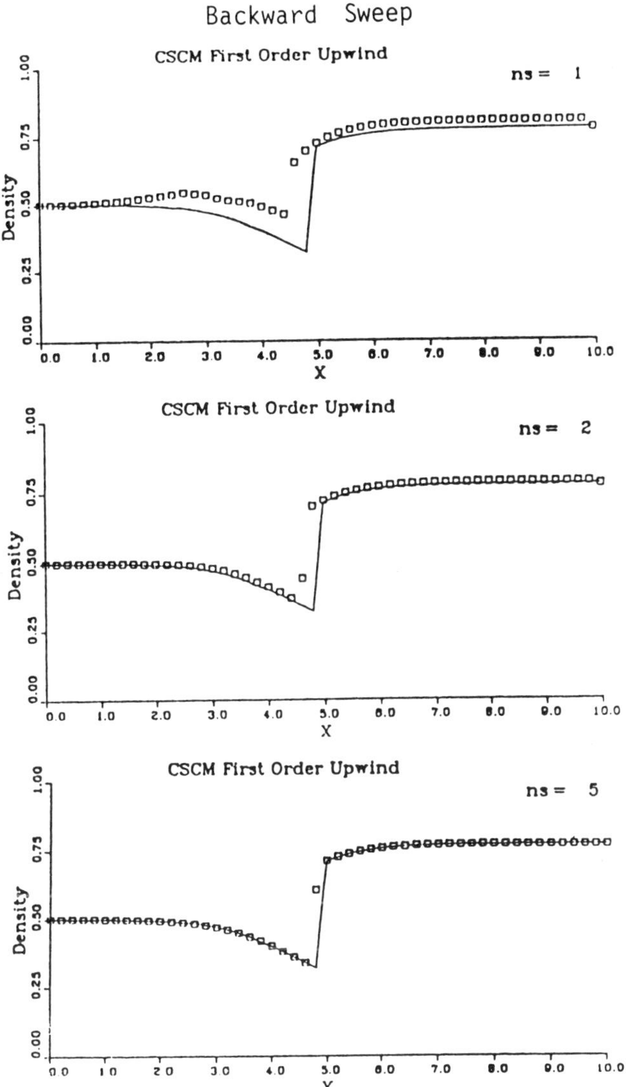

Fig. 2 (continued) Shubin's diverging nozzle flow solution developed in alternating forward and backward sweeps, one each per global iteration step for five steps: square - computed results; line - exact solution.

nozzle geometry with different inflow conditions. Here, the exact analytical solution derived by Venkatapathy is shown by a solid line and our computed results by boxes.

The alternating direction sweeps in our method have been derived directly out of theory for solving the implicit set of difference

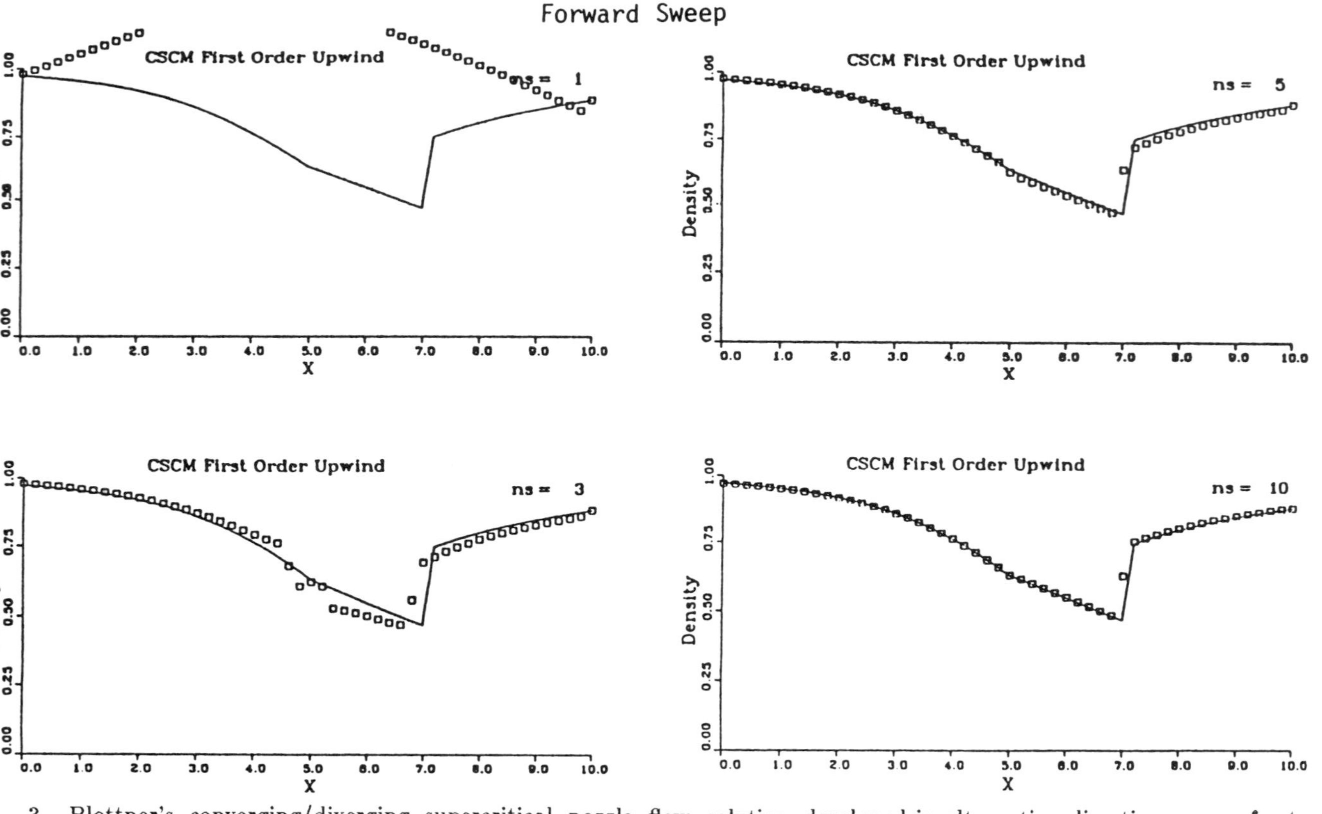

Fig. 3 Blottner's converging/diverging supercritical nozzle flow solution developed in alternating direction sweeps for ten global iteration steps: square – computed results; line – exact solution.

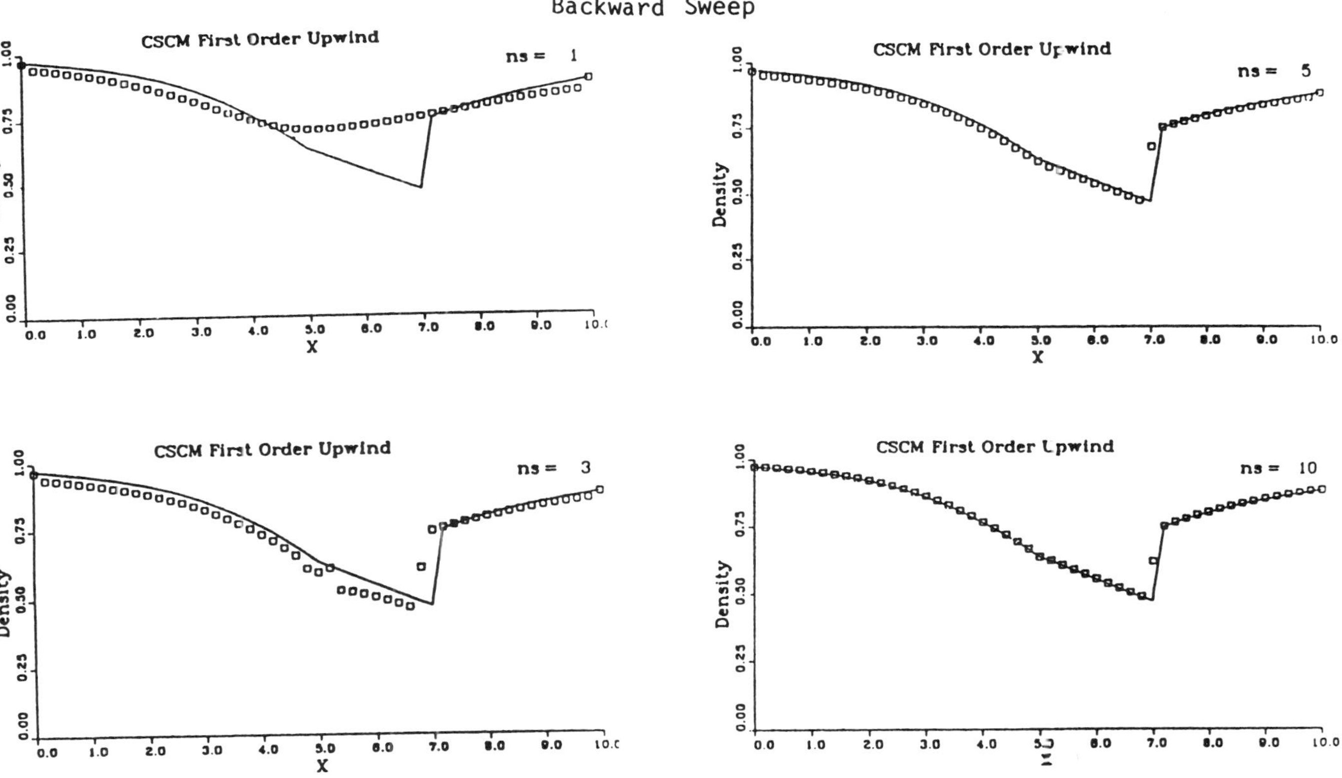

Fig. 3 (continued) Blottner's converging/diverging supercritical nozzle flow solution developed in alternating direction sweeps for ten global iteration steps: square - computed results; line - exact solution.

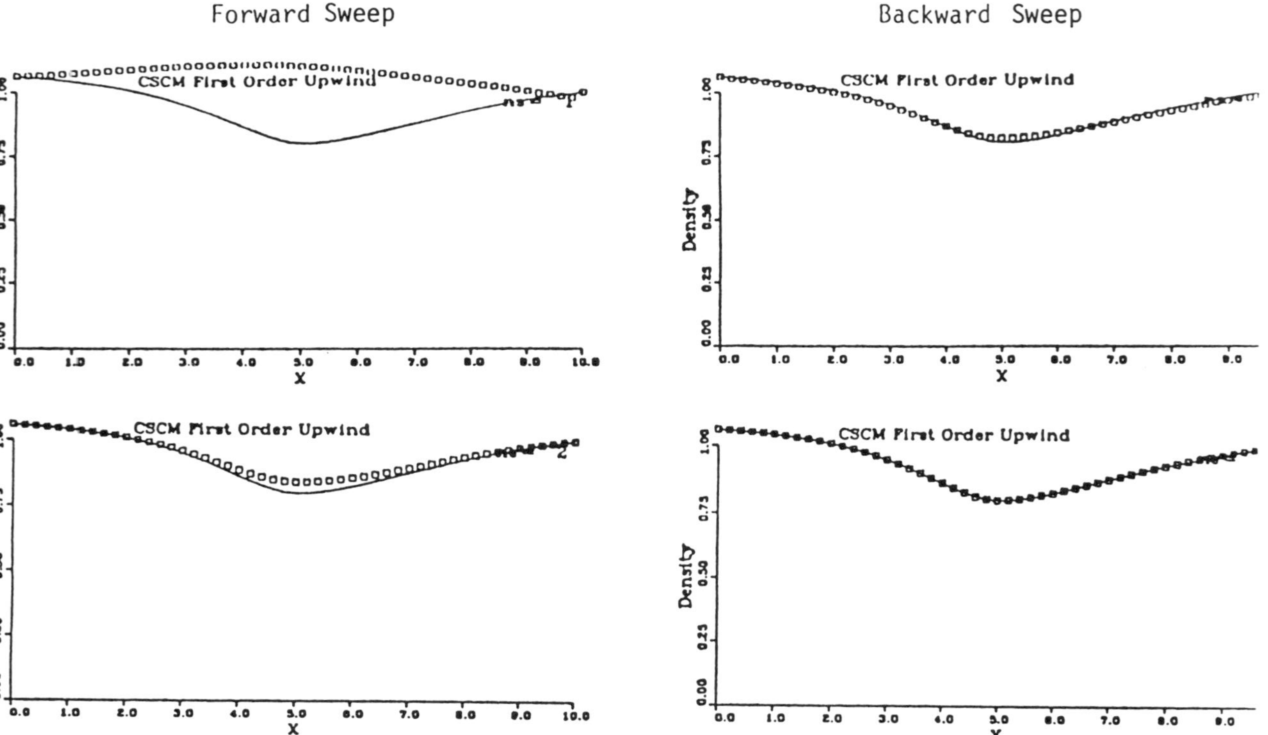

Fig. 4 Subcritical solution to Blottner's converging/diverging nozzle developed with alternating sweeps in two global iteration steps: square – computed results; line – exact solution.

equations. However, with a moments thought, one can see mechanistically, numerically speaking, that omitting the backward sweep from the pair and globally iterating only with the forward sweep equation (10a), will result in permitting the influence of a subsonic outflow boundary (or interior disturbance) to propagate upstream only one grid point per global iteration. In such a case, which relates to other global iteration methods found in the literature and which sweeps only in the main flow direction, the rate of convergence is greatly inhibited relative to symmetric sweeping by a factor of order roughly the number of grid points in the subsonic zone. Mathematically, this inhibition is the result of the failure to include the effect of the eigenvectors governing upstream influence in the implicit process, but to treat these waves explicitly with effective CFL unity.

In Fig. 5, we illustrate the progress of the transient solution to the subcritical nozzle problem after 15 forward sweeps, with the backward sweeps Eq. (10b) omitted. One can clearly see that the wave influence of the outflow boundary has progressed only 15 mesh points forward of the outflow boundary. In Ref. 14, it is argued with further evidence from the foregoing computational experiment, that the characteristic time in number of iterations to relaxing the solution with unidirectional sweep methods is the number of points in the subsonic zone. For this problem with 50 grid points, the solution is found to converge to roughly the same rms error

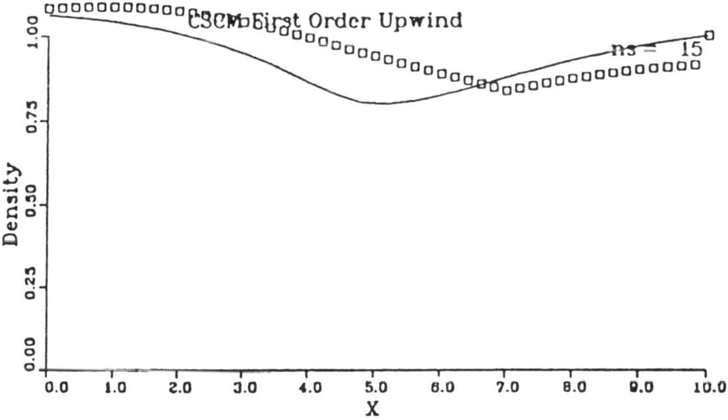

Fig. 5 Solution to subcritical nozzle problem after 15 global iterations with forward (streamwise) marching sweeps only. The upstream influence has propagated only 15 mesh points showing, that the method without backward sweeps is explicit in the upstream waves.

after three characteristic times (150 steps) as the solution obtained with the symmetric alternating sweep sequence after only 3 global iterations.

Blottner's supercritical nozzle problem, which involves subsonic inflow accelerating through a sonic point to a supersonic zone terminated by a shock to subsonic outflow, is the most computationally demanding of the test cases and indicates the capability for the method to compute simply and consistently over the subsonic forebody and base regions of blunt bodies in supersonic flow. Thus, the need for separate time-dependent codes is obviated by this new method.

Finally, in Figs. 6 and 7, we present the convergence history for the present nonlinear scheme and the linearized time-dependent scheme for completely subsonic and supersonic nozzle flows. The x-axis shows the number of iterations each scheme requires to reduce

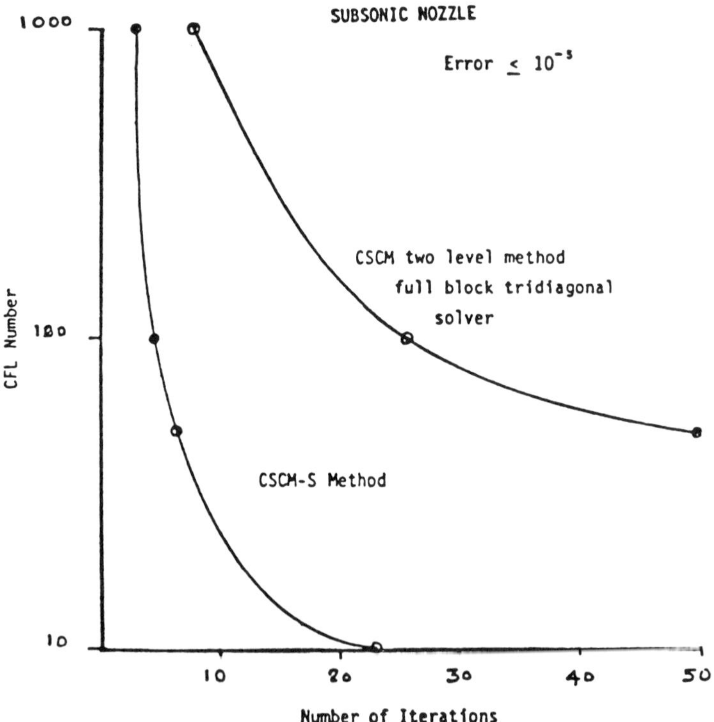

Fig. 6 Convergence history comparison between the full block tridiagonal CSCM solver and the CSCM-S method for the subcritical nozzle. The solutions have reached an rms error less than 1×10^{-5}.

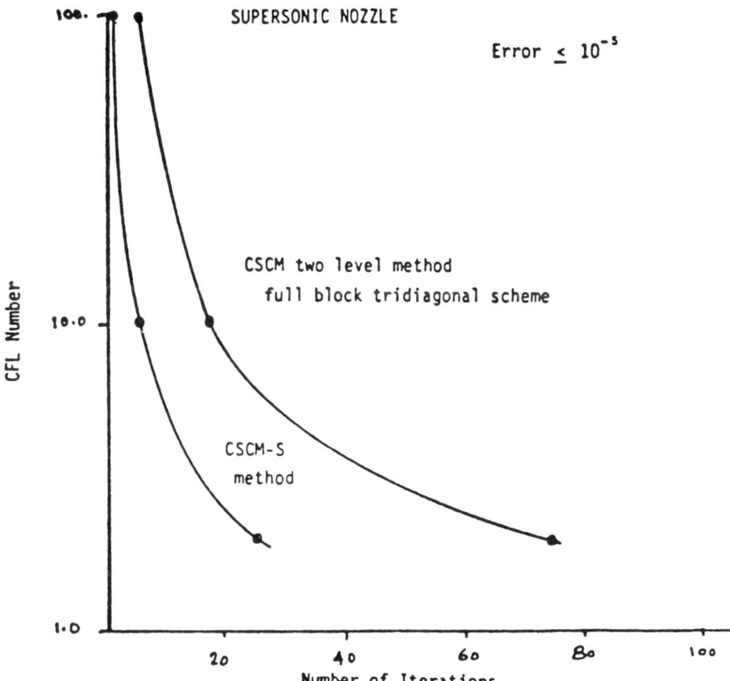

Fig. 7 Convergence history comparison between the full block tridiagonal CSCM solver and the CSCM-S method for the supersonic nozzle. The solutions have reached an rms error less than 1×10^{-5}.

the exact error to five orders of magnitude for various Courant numbers. It is evident that the present scheme converges extremely fast at all CFL numbers compared with the method based on the linearized block tridiagonal solver, and that the lower the CFL number at which the respective methods operate, the more favorable is the relative rate of convergence of the nonlinear single-level scheme.

IV. CSCM-S Two-Dimensional Formulation

For two-dimensional inviscid flow, the implicit block matrix equations can be approximately factored in a number of different ways,[18] each leading to a scheme a number of which conveniently vectorize for highly efficient computation on modern super-computers. A computationally slower but comparatively simple scheme to test for robustness and rate of convergence in upwind implicit time-dependent codes, and the one for which we have obtained results to date, is the method of lines. Then assuming a marching coordinate ξ, inviscid terms

$$(B^+\nabla_\eta + B^-\Delta_\eta)\delta q \tag{12a}$$

$$-B^+\Delta_\eta q)_{k-1} - B^-\Delta_\eta q)_k \tag{12b}$$

are added to the left- and right-hand sides, respectively, of both the forward and backward sweep Eqs.(10a) and (10b). For viscous flow, second centrally differenced, viscous terms in the Baldwin-Lomax thin layer approximation[26] are also added in the η-direction as is conventionally practiced.[22] Along each η coordinate line, one can solve the resulting coupled equations with a block tridiagonal procedure as has been done here. Alternatively, a further DDADI bidiagonal approximate factorization could be employed in the η-direction and solved either linearly as in Ref. 16 or non-linearly as here in the ξ-direction. As shown in the quasi-two-dimensional numerical experiments of Ref. 16, DDADI bidiagonal approximate factorization is stable for viscous as well as inviscid terms.

Second-Order Methods

Here we make some comments on higher-order spatial differencing for the right-hand side of the equations. In previous papers, we have advanced the three point upwind difference operator as a second-order method. Experience with that method alone has shown it to be accurate but not very robust for the breadth of flow conditions to be generally encountered. Also, Li[27] has suggested the possibility of lowering truncation error and improving heat transfer through the use of a central difference scheme in the boundary layer. While Li considered a switch, from one method to the other, we propose a linear combination of the central and three point upwind. The scheme that uses 2/3 of the central method and 1/3 of the upwind is the well-known third-order biased upwind method. However, the scheme that results from the use of half central and upwind differencing is very interesting.

$$\begin{aligned} &-\frac{1}{4}A^+\Delta q_j - A^+\Delta q_{j-1} + \frac{1}{4}A^+\Delta q)_{j-2} \\ &-\frac{1}{4}A^-\Delta q_{j-1} - A^-\Delta q_j + \frac{1}{4}A^-\Delta q)_{j+1} \end{aligned} \tag{13}$$

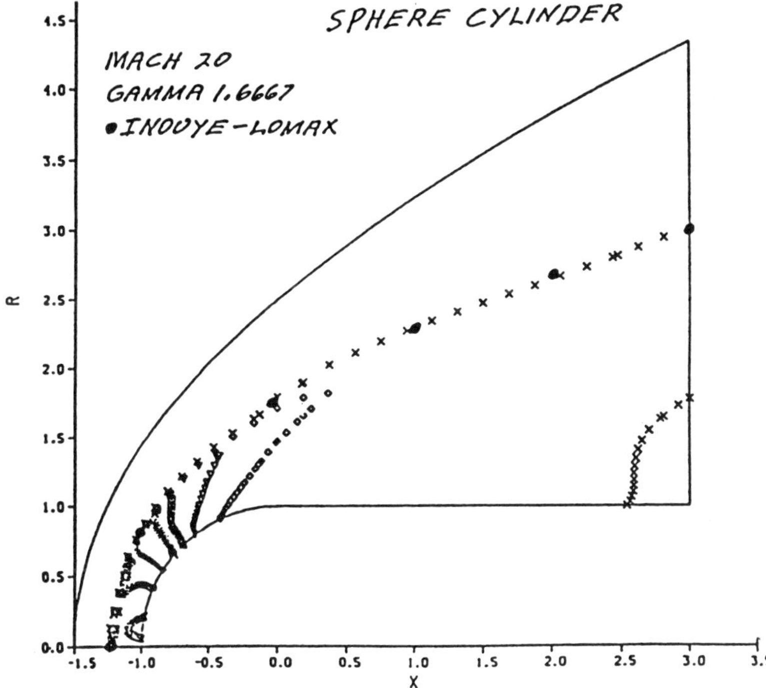

Fig. 8 Pressure contours for the Mach 20 blunt body problem solved in 20 global iterations with first-order CSCM-S. Inouye-Lomax data also fit experiment for shock shape.

This scheme would appear to have two virtues. The first is, linearly, the truncation error $O(h^2)$, reduced by a factor of 4 over either difference operator alone. The second is that the terms with one-quarter weights on either side of the first-order upwind contributions (with weight one) are balanced. Since both sets of these terms with one-quarter weight are of themselves destabilizing, through the contribution of a wrong eigenvalue sign, balancing them with minimum weight for either can, on the average, be expected to minimize their adverse impact on robustness in the real case of variable coefficients. For this second-order scheme, the first-order upwind contributions to the diagonal are highly stabilizing in the sense of Leonard.[28]

Axisymmetric Flow Results

In Fig. 8, we show pressure contours for our solution of a Mach 20 inviscid perfect gas flow of helium $\gamma = 1.667$ over a

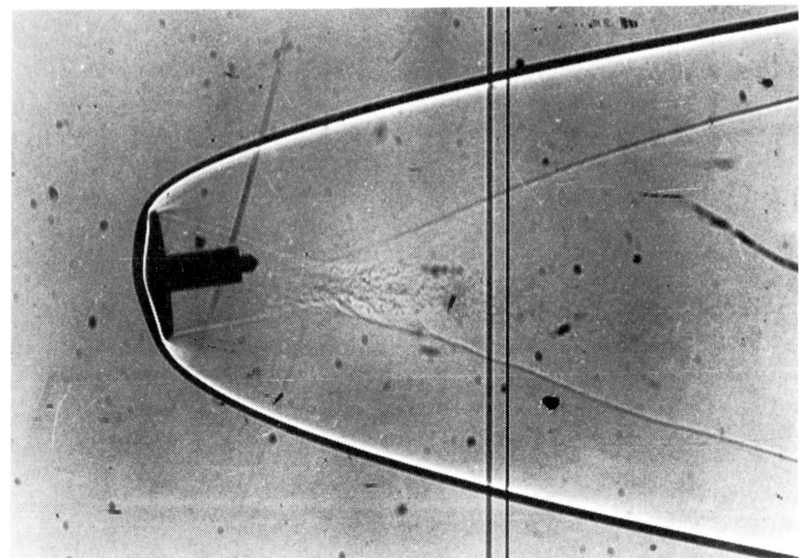

Fig. 9 Schlieren photo of Mach 13 flight of AOTV model in ballistic range at NASA Ames Research Center (courtesy of Intrieri).

sphere cylinder. Also indicated by bold dots are the experimentally and computationally determined shock location given by Inouye and Lomax.[29] The location of the bow shock captured in our calculation finds excellent agreement with the referenced work.

Though the bow shock could also have been fitted, we believe an accurate shock capturing capability may be particularly important for nonequilibrium flow problems such as will be encountered by the AOTV to which the technique is intended to be applied. There, the number density function for the more mobile electrons is thought to have a greatly smeared gradient relative to and, indeed, extending upstream of the shock in the neutral gas. Thus, it may be of interest to try to resolve this distribution and the associated reaction chemistry in the vicinity of the gasdynamic shock.

In Fig. 9, we show a Schlieren photo obtained by Intrieri[30] during a Mach 13 flight of a model AOTV in air in the ballistic range at NASA Ames Research Center. We show in Fig. 10 the simple algebraically generated single quadrilateral block grid in 72×56 mesh points we have used for our present early calculations of the coupled forebody and base flow about a model AOTV. In Fig. 11, we show pressure contours obtained with the CSCM-S method at Mach 20 for

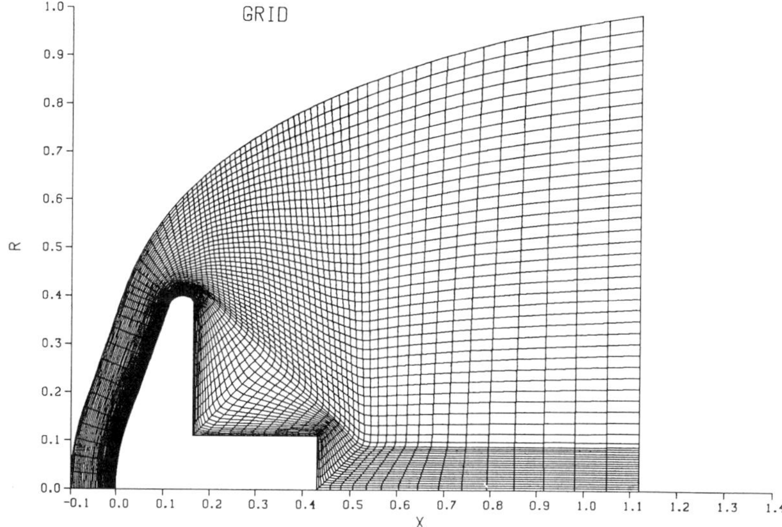

Fig. 10 Computational mesh in 72×56 points for model AOTV flow simulation.

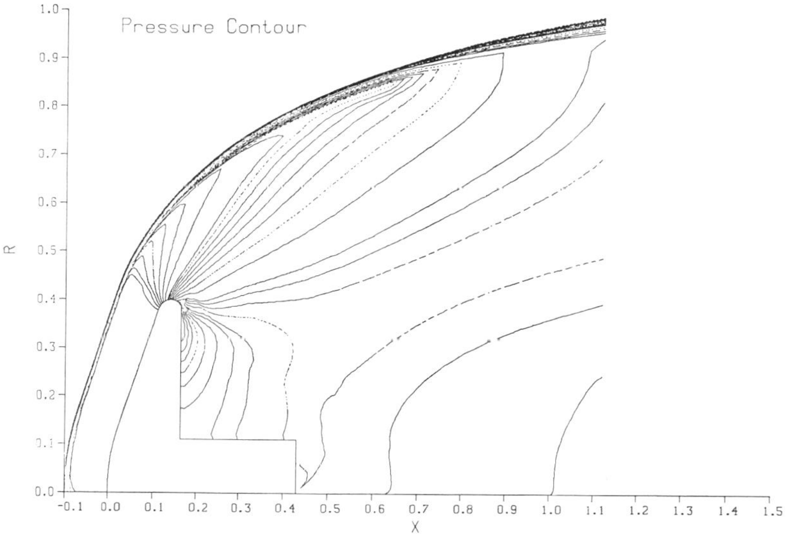

Fig. 11 Pressure contour for Mach 20 model AOTV flowfield simulation in ideal gas at gamma 1.4 and Re 10^5; computation on 72×56 mesh.

perfect gas $\gamma = 1.4$ and laminar Re 10^5 in the thin layer viscous approximation. The treatment of the viscous terms computed on a flow aligned mesh in that approximation is certainly as accurate as we can resolve the gradients in the streamwise direction, know the Reynold's number, or model the turbulence in the case of transitional flow.

We note that the shock standoff distances that can be observed in the experimental results of Fig. 9 and the computation of Fig. 11 are quite comparable. This is as it should be since the shock layer in chemical equilibrium (experimental conditions) at Mach 13 is not much thinner[31] than for perfect gas, and the perfect gas shock layer of our computation at Mach 20 is somewhat thinner than at Mach 13.

Also evident in the Schlieren photo are the lip shock over the shoulder expansion in the model flow and the shear layer or dividing streamline separating the inner recirculating base flows and the outer wake. These features are evident in the Mach contours (Fig. 12) obtained from the present computation.

Both the pressure and Mach contours show favorable qualitative agreement with results obtained by Gnoffo[10] for a similar model at Mach 9.82. The numerical treatment of the internal and external

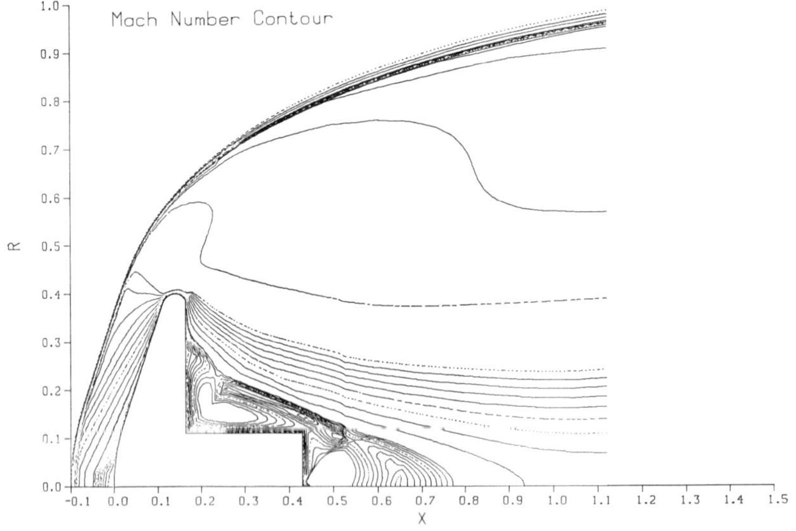

Fig. 12 Mach number contour for solution corresponding to Fig. 11.

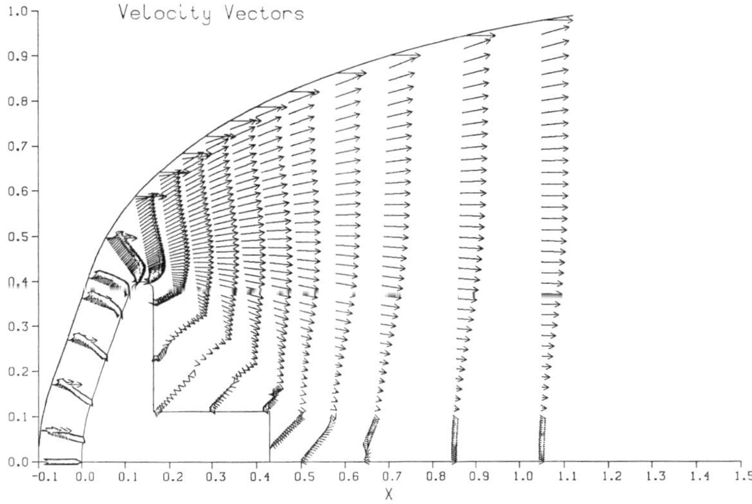

Fig. 13 Velocity vector plot for solution corresponding to Fig. 11.

corners on the payload portion of the models is somewhat sharper in the present work than for the case of Gnoffo who rounded the corners. In Fig. 13, we show a velocity vector plot for the flowfield. The plot exhibits the slow recirculating flow in the base region and the classic viscous wake profile in the merger with the outer flow. The calculation was not carried sufficiently far downstream to clearly exhibit the wake recompression shock. The outflow boundary is stably computed both along the boundary and to the boundary using characteristic boundary-point procedures as described in Ref. 18.

In Fig. 14, we show results in Mach number contours for the solution on a twice refined mesh, i.e., half node spacing in each coordinate direction. From comparison of Figs. 12 and 14 for the two meshes, it appears that the separation and reattachment points are not substantially mesh sensitive nor are other details of the contour lines. Thus, it is believed that the solution on the coarser mesh is not overly subject to numerical viscosity. However, the details of the solution in the base region for this complex geometry are not sufficiently well explored to rule out some mesh dependency, particularly where lines turn abruptly as in the interior corners. This issue will be further explored in the future through solving on alternative (cartesian) topology meshes admitted by a version of the code with a more flexible data structure.[32]

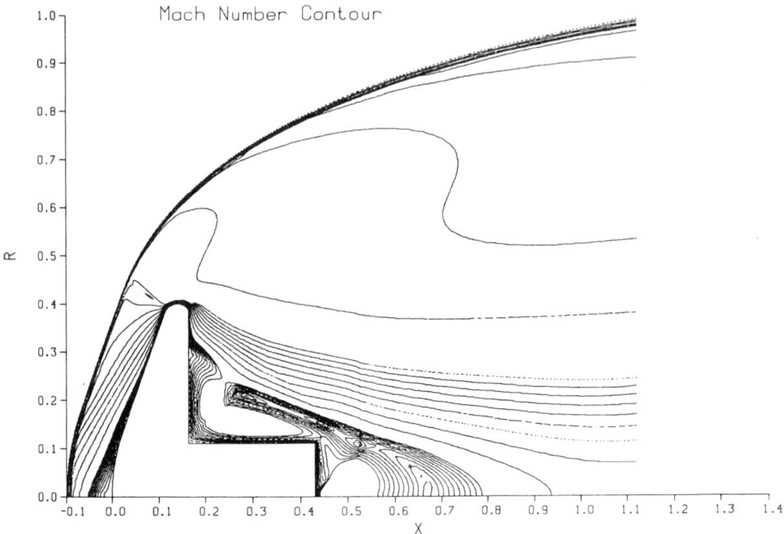

Fig. 14 Mach number contour on refined mesh on half the spacing for Fig. 12.

The present results were run with adiabatic wall. At this writing early results are being obtained for constant temperature wall. The temperature contours appear reasonable. Heat transfer results from these developing computations will be presented in a later report.[33]

Discussion

Gnoffo's results[10] were obtained with an explicit method in the order of 20,000 iterations along with the present results in order of a few hundred. Given the added computation of the implicit procedure, there is believed to be at least an order of magnitude computational advantage in the single-level implicit procedure over explicit methods. (The limitations of explicit schemes is well recognized by Gnoffo who is also working on an implicit upwind method.) The two factors that largely account for the rapid convergence of the present method are the use of physically consisent upwind differenced convection and locally iterated space marching of the solution. The two together serve to communicate data between the interior and the boundaries rapidly and accurately taking local nonlinear dependence of the convection into account. The realizable computational speed of vectorized variations[18] of the single-level algorithm will become clear after substantially more work.

Finally, we remark that some type of solution adaptive gridding is definitely to be desired to accurately and efficiently capture the details of the flow structure in such complex wake flows. Candidate schemes are the mesh redistribution approaches of Gnoffo,[10] Saltzman and Brackbill,[34] and Nakahashi and Deiwert[35]. Another approach being explored by the present authors is adaptive refinement on either joint patched composite mesh[8] or disjoint patched meshes,[36] the latter concept pioneered by Berger et al.[37] We note that the single-level algorithm is particularly convenient for obtaining implicit communication among patches at patch boundaries. Indeed, the method on such grid systems appears ideal for computing asynchronously on multiprocessor machines.

Acknowledgments

This work was partially supported by NASA Ames Research Center under Contract NAS2-11920, by AFOSR under Contract F49620-83-C-0084 and by ARO under Contract DAAG29-84-C-0002.

References

[1]Walberg, C.D., "A Review of Aeroassisted Orbital Transfer," AIAA 82-0137, 1982, (in addition, a synopsis appears in Astronautics and Aeronautics, Nov. 1983, pp. 36-43).

[2]Menees, G.P., "Thermal-Protection Requirements for Near-Earth Aeroassisted Orbital-Transfer Vehicle Missions," AIAA 83-1513, AIAA 18th Thermophysics Conference, Montreal, Canada, June 1983; published elsewhere in this volume.

[3]Davies, C.B. and Park, C., "Aerodynamic Characteristics of Generalized Bent Biconic Bodies for Aero-Assisted, Orbital-Transfer Vehicles," AIAA 83-1512, June 1983.

[4]Howe, J.T., "Introductory Aerothermodynamics of Advanced Space Transportation Systems," AIAA 83-0406, Jan. 1983.

[5]Park, C., "Calculation of Nonequilibrium Radiation in AOTV Flight Regimes," AIAA 84-0306, AIAA 22nd Aerospace Sciences Meeting, Reno, Nev., Jan. 1984; published elsewhere in this volume.

[6]Moss, J.N., "Advancements in Aerothermodynamics in Support of the Galileo Probe," Thirteenth International Symposium on Space Technology and Science, Tokyo, Japan, June 28 - July 2, 1982.

[7]Green, M.J. and Davy, W.C., "Galileo Probe Forebody Thermal Protection Thermophysics of Atmospheric Entry," Progress in Astronautics and Aeronautic Vol.82, edited by T. E. Horton, AIAA 1982.

[8]Lombard, C.K., Davy, W.C., and Green, M.J., "Forebody and Base Region Real-Gas Flow in Severe Planetary Entry by a Factored Implicit Numerical Method – Part I (Computational Fluid Dynamics)," AIAA 80-0065, Jan. 1980.

[9]Scott, C. D., Reid, R. C., Maraia, R. J., Li, C.-P. and Derry, S. M., "An AOTV Aeroheating and Thermal Protection Study," AIAA 84-1710, AIAA 19th Thermophysics Conference, Snowmass, Colorado, June 1984; published elsewhere in this volume.

[10]Gnoffo, P.A., "Complete Flowfields Over Low and Wide Angle AOTV Conceptual Configurations," AIAA 84-1695, AIAA 19th Thermophysics Conference Snowmass, Colorado, June 1984; published elsewhere in this volume.

[11]Park, C., "Problems of Rate Chemistry in AOTV Flight Regimes," AIAA 84-1730, AIAA 19th Thermophysics Conference, Snowmass, Colorado, June 1984; published elsewhere in this volume.

[12]Lee, J.H., "Basic Governing Equations for AOTV Flight Regimes," AIAA 84-1729, AIAA 19th Thermophysics Conference, Snowmass, Colorado, June 1984; published elsewhere in this volume.

[13]Lombard, C.K., Oliger, J., and Yang, J.Y., "A Natural Conservative Flux Difference Splitting for the Hyperbolic Systems of Gasdynamics," AIAA Paper 82-0976, July 1982.

[14]Lombard, C.K., Oliger, J., Yang, J.Y., and Davy, W.C., "Conservative Supra-Characteristics Method for Splitting the Hyperbolic Systems of Gasdynamics with Computed Boundaries for Real and Perfect Gases," AIAA Paper 82-0837, July 1982.

[15]Lombard, C.K., Oliger, J., and Yang, J.Y., "A Natural Conservative Flux Difference Splitting for the Hyperbolic Systems of Gasdynamics," Lecture Notes in Physics, Vol. 170, edited by E. Krause, June 1982, pp. 364-370.

[16]Lombard, C.K., Bardina, J., Venkatapathy, E., and Oliger, J., "Multi-Dimensional Formulation of CSCM - An Upwind Flux Difference Eigenvector Split Method for the Compressible Navier-Stokes Equations," AIAA-83-1895, AIAA 6th Computational Fluid Dynamics Conference, Danvers, Mass., July 1983.

[17]Lombard, C.K., and Venkatapathy, E., "Universal Single Level Implicit Algorithm for Gasdynamics," NASA Contractor Report 166531, Jan. 1984.

[18]Lombard, C.K., Venkatapathy, E., and Bardina, J., "Universal Single Level Implicit Algorithm for Gasdynamics," AIAA 84-1533, AIAA 17th Fluid

Dynamics, Plasma Dynamics and Lasers Conference Snowmass, Colorado, June 1984.

[19] Venkatapathy, E. and Lombard, C.K., "Universal Single Level Implicit Algorithm for Gasdynamics," Presented at the Ninth International Conference on Numerical Methods in Fluid Dynamics, Saclay, France, June 1984.

[20] Vigneron, Y.C., Rakich, J.V., and Tannehill, J.C., "Calculation of Supersonic Viscous Flow over Delta Wings with Sharp Subsonic Leading Edges," NASA TM-78500, 1978.

[21] Schiff, L.B. and Steger, J.L., "Numerical Simulation of Steady Supersonic Viscous Flow," AIAA-79-0130, Jan. 1979.

[22] Steger, J.L., "Implicit Finite Difference Simulation of Flow About Arbitrary Geometries with Application to Airfoils," AIAA Paper 77-665, 1977.

[23] Roe, P.L., "The Use of the Riemann Problem in Finite-Difference Schemes," Seventh International Conference on Numerical Methods in Fluid Dynamics Lecture Note in Physics, Vol. 141, 1981, pp. 354-359.

[24] Oliger, J. and Lombard, C.K., "Boundary Approximations for Alternating Sweep Implicit Upwind Methods for Hyperbolic Systems," SIAM Fall Meeting, Norfolk, Virginia, 1983.

[25] Yee, H.C., Beam, R.M., and Warming, R.F., "Stable Boundary Approximations for a Class of Implicit Schemes for the One-Dimensional Inviscid Equations of Gas Dynamics," AIAA Paper 81-1009-CP, June 1981.

[26] Baldwin, B.S. and Lomax, H., "Thin Layer Approximation and Algebraic Model for Separated Turbulent Flows," AIAA 73-257, 1973.

[27] Li, C.P., "On A Finite-Difference Method for Solving Unsteady Viscous Flow Problems," AIAA Paper 83-0560, Jan. 1983.

[28] Leonard, B.P., "The QUICK Algorithm: A Uniformly Third-Order Finite-Difference Method for Highly Convective Flows," Computer Methods in Fluids, edited by K. Morgan, C. Taylor, and C.A. Brebbia, Pentech Press, London, Ontario, Canada, 1980, pp. 159-195.

[29] Inouye, M. and Lomax, H., "Comparison of Experimental and Numerical Results for the Flow of a Perfect Gas About Blunt- Nosed Bodies," NASA TND-1426, 1962.

[30] Intrieri, P.F., recent unpublished experiment.

[31] Balakrishnan, A. and Davy, W.C., "Viscous Real Gas Flowfields about Three Dimensional Configurations," AIAA-83-1511, June 1983.

[32] Venkatapathy, E., Bardina J., and Lombard, C.K., "Flow Over Steps and Cavities with the CSCM Upwind Method," Abstract presented for the AIAA 18th Fluid Dynamics, Plasma Dynamics and Lasers Conference, Cincinnati, Ohio, July 1985.

[33]Bardina, J., Venkatapathy, E., and Lombard, C.K., "Two Dimensional and Axisymmetric Heat Transfer Results with the Single Level CSCM-S Upwind Implicit Algorithm," Abstract presented for the AIAA 20th Thermophysics Conference, Williamsburg, Virginia, June 1985.

[34]Saltzman, J. and Brackbill, J., "Applications and Generalizations of Variational Methods for Generating Adaptive Meshes," Numerical Grid Generation, edited by J. F. Thompson, North-Holland, 1982.

[35]Nakahashi, K. and Deiwert, G. S., "A Practical Adaptive-Grid Method for Complex Fluid-Flow Problems," presented at the Ninth International Conference on Numerical Methods in Fluid Dynamics, Saclay, France, June 1984.

[36]Lombard, C.K. and Venkatapathy, E., "Boundary Treatment for CSCM Solved on Disjoint Patched Meshes," Abstract presented for the AIAA 7th Computational Fluid Dynamics Conference, Cincinnati, Ohio, July 1985.

[37]Berger, M., Gropp, W., and Oliger, J., "Grid Generation for Time Dependent Problems: Criteria and Methods," Numerical Grid Generation Techniques, NASA CP-2166, 1980, pp. 181-188.

Direct Simulation of Transitional Flow for Hypersonic Re-entry Conditions

James N. Moss*
NASA Langley Research Center, Hampton, Virginia
and
Graeme A. Bird†
University of Sydney, New South Wales, Australia

Abstract

This paper presents results of flowfield calculations for typical hypersonic reentry conditions encountered by the nose region of the Space Shuttle orbiter. Most of the transitional flow regime is covered by the altitude range of 150 to 92 km. Calculations were made with the direct simulation Monte Carlo (DSMC) method that accounts for translational, rotational, vibrational, and chemical nonequilibrium effects. Comparison of the DSMC heating results with both Shuttle flight data and continuum predictions showed good agreement at the lowest altitude considered. However, as the altitude increased, the continuum predictions, which did not include slip effects, departed rapidly from the DSMC results by overpredicting both heating and drag. The results demonstrate the effects of rarefaction on the shock and the shock layer, along with the extent of the slip and temperature jump at the surface. Also, the sensitivity of the flow structure to the gas-surface interaction model, thermal accommodation, and surface catalysis are studied.

Nomenclature

A_b = base area
C_D = drag coefficient = $2F_x/\rho_\infty U_\infty^2 A_b$

Presented as Paper 84-0223 at AIAA 22nd Aerospace Sciences Meeting, Reno, Nev., Jan. 9-12, 1984. This paper is declared a work of the U.S. Government and therefore is in the public domain.

*Research Leader, Aerothermodynamics Branch, Space Systems Division.

†Professor, Department of Aeronautical Engineering.

C_f = friction coefficient = $2\tau/\rho_\infty U_\infty$
C_H = heat-transfer coefficient = $2q/\rho_\infty U_\infty^3$
C_i = mass fraction of species i = ρ_i/ρ
$\tilde{C}_o$ = mass fraction of element oxygen
d = nominal molecular diameter
DSMC = direct simulation Monte Carlo
F = force
Kn = Knudsen number = λ/ℓ
K_r^2 = Cheng's parameter = $\rho_\infty R_N/\mu_\infty u_\infty C^* = \sqrt{\pi}\, R_N/2S_\infty\lambda_\infty C^*$
ℓ = characteristic dimension
M = molecular weight of mixture
OTV = orbital transfer vehicle
R_N = nose radius
R = universal gas constant = 8.3143 J/mol·K
Re_∞ = Reynolds number = $\rho_\infty U_\infty R_N/\mu_\infty$
S_∞ = speed ratio = $U_\infty\sqrt{M/2RT_\infty}$
STS-2 = Space Transportation System, Shuttle flight 2
T = thermodynamic temperature
T_{ov} = overall kinetic temperature
T^* = $(T_{o,\infty} + T_w)/2$
u = velocity component tangent to body surface
U_∞ = freestream velocity
VSL = viscous shock layer
X_i = mole fraction of species i
x = coordinate measured along the body centerline
x/L = nondimensionalized Space Shuttle orbital axial length
η = coordinate normal to body surface
θ = hyperboloid asymptotic half-angle
λ_∞ = freestream mean-free path
λ_1 = mean-free path adjacent to body surface
μ = viscosity
μ^* = viscosity evaluated at T^*
ρ = density
σ = total collision cross section
τ = shear stress

Subscripts

i = ith species
o = total values
w = wall values
∞ = freestream values

Introduction

Currently, there is a renewed interest in the aerothermodynamics of external flows about vehicles at very high altitudes. Two factors contributing to the enhanced interest are the development of the U.S. Space Transportation System (the Space Shuttle) and the studies underway to identify ways of enhancing the capability of the Space Transportation System by developing orbital transfer vehicles (OTVs) that utilize aeroassist technology[1] to achieve a specified low Earth orbit after returning from a high Earth orbit. The Space Shuttle orbiter is now making routine flights and provides a means of conducting experiments and of making measurements throughout the flow spectrum--encompassing the free molecule through continuum flow regimes. For the aeroassisted portion of an OTV mission, the vehicle will be at very high altitudes[2-3] (> 75 km) while traveling at speeds of about 10 to 7.5 km/s. Consequently, a portion of the atmospheric encounter of an OTV will occur in the transitional flow regime; the extent of the transitional flow encounter will depend on the aerodynamic characteristics of the vehicle. Hence, knowledge of the drag and heating experienced during the atmospheric encounter is critical in assessing the aerodynamic performance and thermal protection requirements for such a vehicle. In the transitional flow regime, both the drag and heating are very sensitive to the degree of rarefaction.

The transition flow regime, bounded by continuum and free-molecule flow, has always posed difficulties for designers who generally resort to empirical correlations based on sparse experimental data. Ground-based experimental data are not available for transitional flow conditions for atmospheric encounters at speeds of 7.5 km/s or greater. Under these conditions the speed ratio is of the order of 20, the surface temperature is of the order of 0.01 of the total temperature, and for altitudes greater than 100 km a significant amount of atomic oxygen exists in the freestream. Therefore, it is most desirable to have a computational method that can provide data for comparison with the measurements from the Shuttle.

Computationally there appears to be no alternative to the particle approach when the Knudsen number is of order unity and higher.[4] The only closed equation that is then applicable is the Boltzmann equation. Solutions of the Boltzmann equation are readily obtained in the free-molecule limit as the Knudsen number tends to infinity, but very serious analytical difficulties are encountered at Knudsen numbers characteristic of transitional flow. As a consequence of the need for an alternative approach, Monte Carlo procedures (any method that employs random numbers) have been developed. Of the various techniques for the Monte Carlo simulation of gas flows, the direct simulation Monte Carlo (DSMC) method is the one that is most readily applied to complex problems.[5] The DSMC method has been developed by Bird[6-7] over the past 20 years, during which time the applications of the method have advanced from idealized or generally artificial test cases to problems of specific engineering situations.[5,8-9]

The application of Monte Carlo simulation methods has been aimed primarily at the transition regime that is characterized by Knudsen numbers that are above the upper limit for the validity of the Navier-Stokes equations but below the level at which the flow falls into the collisionless flow or free-molecule regime[4] (see Fig. 1). The current paper focuses on the transitional regime, with special emphasis on obtaining solutions at or near conditions for which continuum solutions remain valid. The freestream conditions are those representative of the reentry conditions experienced by the Space Shuttle Orbiter during re-entry and encompass an altitude range of 150 to 92 km. Comparisons of the DSMC predictions are made with Shuttle flight data for heat-transfer rates. Furthermore, comparisons with continuum predictions[10] are made for the lower-altitude conditions for heat transfer, drag, and flowfield structure. Also, the impact of variations in surface catalytic activity and surface thermal accommodation on heating is demonstrated. The calculations were performed with sufficient resolution to isolate the extent of the temperature jump and velocity slip at the surface. The results represent the first transitional flow predictions for Earth re-entry in which nonequilibrium effects due to translation, rotation, vibration, and chemistry are modeled in the simulation. Finally, the current study has direct implications for the studies that are being conducted to define the

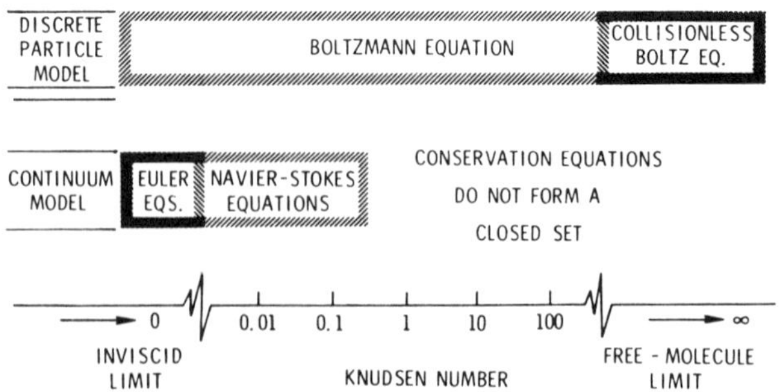

Fig. 1 The Knudsen number limits on the conventional mathematical models of neutral gas flows.

technology requirements for aeroassisted orbital transfer vehicles.

Analysis

Direct Simulation Monte Carlo Method

Any method that employs random numbers may be described as a Monte Carlo method, and there are a number of methods for which the terms Monte Carlo and simulation may be attached. In Ref. 4, Bird categorizes the various Monte Carlo simulation methods into four basic groups and outlines the features of each group.

In the DSMC method, the intermolecular collisions are dealt with on a probabilistic rather than a deterministic basis. The time parameter in the simulation may be identified with physical time in the real flow, and all calculations are unsteady. When the boundary conditions are such that the flow is steady, then the solution is the asymptotic limit of the unsteady flow. The computation is always started from an initial state that permits an exact specification, such as a vacuum or uniform equilibrium flow. Consequently, the method does not require an initial approximation to the flowfield and does not involve any iterative procedures. A computational cell network is required only in physical space rather than phase space. Furthermore, advantage may be taken of flow symmetries to reduce the dimensions of the cell network and the number of position coordinates that need be stored for each molecule, but the collisions are always treated as three-dimensional phenomena. The boundary conditions are specified in terms of the behavior of the individual molecules rather than the distribution function. All procedures may be specified in such a manner that the computational time is directly proportional to the number of simulated molecules.

References 4 and 6 review the effects of molecular complexity on the results from previous simulations in which the molecular models for monatomic gases range from the hard sphere to those including long-range attractive forces. The effects appear to be fully explained through the variation of collision cross section with relative speed or temperature. Variations in angular scattering law do not appear to have a significant influence on any flow for which comparative data are available. Consequently, the variable hard sphere (VHS) model (Ref. 6) is now recommended for the simulation of monatomic gases in an engineering context. Basically, the VHS model incorporates the essential features of the more complex models while retaining the computational simplicity of the hard sphere model. The VHS model has a well-defined diameter and follows the classical hard sphere scattering law, but the diameter is an inverse power law function of the relative collision energy between the colliding molecules, i.e.,

$$\sigma \equiv \pi d^2 \propto (1/2\ M_r\ C_r^{\ 2})^{-\omega} \tag{1}$$

where M_r is the reduced mass and C_r is the collision relative speed. The power law ω is directly related to the temperature exponent of the coefficient of viscosity (see Ref. 6).

For diatomic and polyatomic gases, the establishment of adequate models has proved to be difficult, and the difficulties associated with the physical models have led to the development of phenomenological models of which the Larsen-Borgnakke[11] model is recommended for engineering studies. The model satisfies the principle of detailed balance for translational velocities, so that, when equilibrium is achieved, the translational distribution conforms to the Maxwellian, and the translational and internal temperatures are equal. An outline of the Larsen-Borgnakke phenomenological model in a form that is compatible with a gas mixture of VHS molecules is given in Ref. 6.

The classical collision theory for chemical reaction rates is essentially a phenomenological approach based on steric factors or reactive cross sections and can be readily incorporated into direct simulation methods. The expressions for the steric factors are developed in Ref. 7 for hard spheres and inverse power law models, while the results for VHS models are given in Ref. 6.

The VHS model along with the compatible Larsen-Borgnakke model and reactive cross sections are the molecular models used in the present Monte Carlo simulations.

Viscous Shock-Layer Method

The continuum results presented for comparison purposes were obtained by solving the steady viscous shock-layer (VSL) equations where most of the results have been reported in Ref. 10. The VSL method solves a set of equations that is uniformly valid throughout the shock layer. The VSL equations are obtained from the steady-state, Navier-Stokes equations by keeping terms up to second order in the inverse square root of a Reynolds number, i.e.,

$$\varepsilon = \sqrt{\mu_{ref}/\rho_\infty U_\infty R_N} \tag{2}$$

where μ_{ref} is the reference viscosity evaluated at a reference temperature ($T_{ref} = U_\infty^2/C_{p_\infty}$, where C_{p_∞} is the freestream specific heat). Consequently, one set of equations is solved for both the inviscid and viscous regions (see Ref. 10 for the governing equations). The equations are solved as described by Davis[12] as an initial-value, boundary-value problem using an implicity, finite-difference, numerical procedure. Global iteration is used to relax an initial thin shock-layer approximation and other approximations concerning the initial shock shape.

Relevant to the present study are the results of Ref. 13, where perfect gas VSL and Navier-Stokes solutions were compared for blunt body hypersonic flow with appropriate slip and jump boundary conditions. The comparison showed that the VSL method is sufficiently accurate for freestream Reynolds numbers of the order of 1000. However, substantial difference in drag and heating (both quantities low in comparison with the Navier-Stokes results at 15 and 25%, respectively) occur in the stagnation region for Reynolds numbers of the order of 100.

Conditions for Calculations

The freestream conditions and vehicle parameters are representative of the Shuttle orbiter nose during re-entry. They are summarized in Tables 1-3. Unless stated otherwise, all results presented are for a noncatalytic wall.

Table 1 Freestream conditions and results

Case	Altitude, km	Density, kg/m^3	U_∞, km/s	T_∞, K	Mole Fractions X_{O_2}	X_{N_2}	X_O	$\bar{M}$, g/mol	λ_∞, m
1	92.35	2.184×10^{-6}	7.50	180	0.217	0.783	0	28.96	0.036
2	99.49	5.906×10^{-7}	7.50	190	0.217	0.783	0	28.96	0.134
3	104.93	2.457×10^{-7}	7.47	223	0.153	0.782	0.065	27.85	0.309
4	109.75	1.146×10^{-7}	7.47	249	0.123	0.771	0.106	27.23	0.649
5	115.00	4.380×10^{-8}	7.50	304	0.098	0.754	0.148	26.63	1.659
6	122.50	1.790×10^{-8}	7.50	401	0.080	0.723	0.197	25.98	3.961
7	130.00	8.230×10^{-9}	7.50	500	0.071	0.691	0.238	25.44	8.438
8	150.00	2.140×10^{-9}	7.50	733	0.055	0.615	0.330	24.27	30.964

Recent continuum calculations have been performed for the second Shuttle flight (STS-2), and the results of these calculations have been compared with flight measurements of the heat-transfer rates along the windward centerline of the Shuttle. The results reported by Shinn[10] for an altitude range of 92 to 48 km showed that the nonequilibrium VSL predictions gave generally good agreement with the flight results. The results of those calculations showed that the nonequilibrium chemistry had a very pronounced effect on the predicted heating for a significant portion of the heat pulse, particularly the higher-altitude portion of the heat pulse.

In order to make meaningful comparisons with existing continuum calculations, the present DSMC calculations use the same chemical kinetics model as Ref. 10 (the species O, O_2, N, N_2, and NO with 34 chemical reactions). However, as noted earlier, the rate constants have been converted to reaction cross sections. Also, most calculations are for the windward centerline of the Shuttle, as modeled in Ref. 10. An "equivalent axisymmetric body" concept was used in that study to model the windward centerline of the Shuttle at a given angle of attack with an appropriate axisymmetric body at zero angle of attack. The axisymmetric body is a hyperboloid with nose radii R_N and asymptotic body half-angles θ as given in Table 2. Some calculations were made for a three-dimensional shape for comparison with those from the computationally simpler axisymmetric model.

The freestream conditions for cases 1 and 2 correspond to STS-2, while those for cases 3 and 4 are for STS-3. The trajectory parameters, such as velocity and angle of attack, were determined from a trajectory reconstruction process, as described in Ref. 14. Free-

Table 2 Body[a] parameters and wall temperature

Case	R_N, m	θ, deg	Wall temperature range, k From	To
1	1.296	41.15	1043	800
2	1.362	42.50	800	610
3	1.362	42.50	560	453
4	1.362	42.50	420	374
5	1.362	42.50	300	300
6	1.362	42.50	300	300
7	1.362	42.50	300	300
8	1.362	42.50	300	300

[a]Hyperboloid with asymptotic half-angle θ and nose radius R_N.

stream density and temperature were determined by a procedure[15] that combined atmospheric modeling and meteorological data taken close to the time of Shuttle flight. Since measured meteorological data were not available above 90 km, the Jacchia-Roberts model[16] was used in this region, with the boundary values chosen to match the meteorological profiles below 90 km.[15] The uncertainty in freestream density above 90 km that results from this procedure is expected to increase with altitude, but its magnitude cannot be determined. The freestream conditions for cases 5 through 8 are representative of those encountered by the Shuttle during re-entry. For these cases, the freestream density, temperature, and composition is that given by Jacchia[16] for an exospheric temperature of 1200 K. Note that the composition was adjusted to that for three freestream species (O_2, N_2, and O). The freestream gas composition for cases 3 and 4 are also adjusted values from Jacchia, while those for cases 1 and 2 are the same as used in Ref. 10.

The wall temperature distributions used for cases 1 through 4 correspond to values that were measured with thermocouples mounted in the Shuttle tile coating at various locations along the windward centerline. Furthermore, the flight heating rates were calculated from the thermocouple data by using a one-dimensional, transient heat-transfer analysis.[17] For cases 5 through 8, the wall temperature was assumed to be constant at 300 K.

Several freestream parameters that are often used to characterize a flowfield are included in Table 3, and their definitions are given in the Nomenclature.

A plane view of the computational domain is depicted in Fig. 2. This consists of one or more blocks, of which

Table 3 Various parameters and results

Case	Re_∞	Kn_∞	K_r^2	S_∞	q, kW/m^2	C_H	C_D
1	1751.	0.028	7.667	23.3	81.3	0.176	1.24
2	475.	0.098	2.210	22.7	49.9	0.405	1.27
3	172.	0.227	1.739	20.5	30.5	0.595	1.36
4	73.	0.476	0.508	19.2	16.6	0.687	1.52
5	24.	1.219	0.213	17.2	7.7	0.838	1.69
6	8.	2.909	0.097	14.8	3.3	0.874	1.82
7	3.1	6.196	0.048	13.1	1.6	0.927	1.89
8	0.6	22.734	0.014	10.6	0.4	0.964	1.94

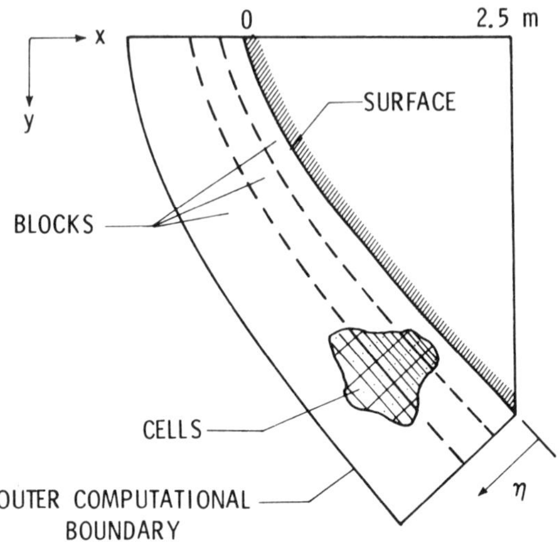

Fig. 2 Computational domain.

only one block resides in core memory at a given time. Each block may consist of one or more arbitrary regions within which the time step and the weighting factor that relate the number of computational molecules to the number of physical molecules are constants. The smallest unit of physical space is the cell, which provides a convenient reference for the sampling of the macroscopic gas properties. The dimensions of the cells must be such that the change in flow properties across each cell is small. Time is advanced in discrete steps of magnitude such that the time step is small in comparison with the mean collision time.

Most calculations were performed using three blocks with two regions per block and 275 cells per block. The number of computational molecules per cell was always greater than six, with the total number of molecules being approximately 16,000 for the three blocks. The computations were performed on a Perkin Elmer 3220 mini-computer with one megabyte of core memory. Using three computational blocks, no more than 60% of the core memory of the machine was utilized while performing the calculations. The computer run times ranged from 50 to 250 hours, depending on the freestream density.

Results and Discussion

While the results cover most of the transitional flow regime, special attention has been paid the 92.35-km case. This is because it overlaps with the apparent range of validity of the VSL continuum calculations.

Early in the study, attempts were made to isolate the sensitivity of the calculated surface heating rates to various parameters for an altitude of 92.35 km. Among those parameters considered were the effect of changing the gas phase reaction rate constants (increasing the preexponential constant by a factor of 2 relative to the nominal data set); varying the body length to investigate the possiblity of any "end of body" effects; checking for possible three-dimensional effects by using an elliptic paraboloid at angle of attack, which gives a much better representation of the Shuttle surface curvature in the transverse plane; varying the wall temperature with respect to the measured Shuttle values, varying the computational cell dimension in the direction normal to the body surface; and changing the molecular constants that describe the collision cross section of the gas species. Of these parametric variations, only the last two significantly influenced the heat transfer at the surface.

When the molecular constants [collision cross section at a reference temperature and the exponential term in Eq. (1)] were adjusted to produce a larger collision cross section, the surface heating decreased. Future studies will be made in which the molecular constants for the chemical species will be further adjusted such that the transport properties of the individual species are more closely modeled. Such a fine tuning of the molecular constants was not done in the present study, but such a process could produce changes in predicted heating of the order of 10% with respect to the results presented herein.

As the cell dimension normal to the surface is reduced from that of a coarse grid to one that resolves the macroscopic gradients, the predicted heating decreases because the particles (atoms and molecules) are able to adjust to local conditions rather than arriving at the surface from larger distances and higher energies than is appropriate.

For the 92.35 km-case, numerous calculations were made using different computational cell sizes. The results showed that the surface pressure was relatively insensitive, while the heating and skin friction was sensitive to the cell dimension normal to the surface. The final normal cell distribution used in this study was such that approximately five computational cells were within a mean-free path of the surface, and for cells removed from the surface the thickness was less than a local mean-free path.

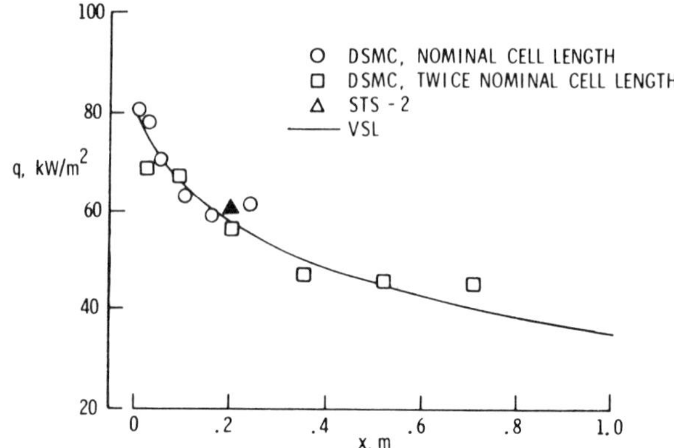

Fig. 3 Comparison of predicted and measured heating (Altitude = 92.35 km).

To achieve the aforementioned normal cell resolution, the computational requirements for case 1 was quite long, so a much shorter calculation was performed for this case, where 50 cells were used normal to the body and a much shorter body simulated such that only six cells in the direction along the body were used. Only one block with six regions was used for these computations, and the results compared favorably (Fig. 3) with the STS-2 heating results and the VSL prediction. (The drag results for case 1 as included in Table 3 is that obtained with a coarser grid and with the computational domain as previously described for the nominal configuration.) By varying the length of the computational cells in the direction tangent to the body (changing the length by a factor of 2), little effect on predicted heating is evident (Fig. 3).

As for the structure of the flowfield, Fig. 4 presents information along the stagnation streamline for the 92-km case. The DSMC results show that, although the shock thickness is of the same order as the shock layer, the maximum temperatures for the VSL and DSMC solutions differ by only about 6%. [Note that two different temperatures are being compared in Fig. 4a. For the VSL calculations, local thermodynamic equilibrium is assumed, and, therefore, only one temperature describes the translational and internal energy states. This is not necessarily the case for the DSMC calculation where thermodynamic nonequilibrium effects are modeled. The temperature shown for the DSMC solution is an overall kinetic temperature, T_{ov}, defined as the weighted mean of the translational and internal temperature (see Ref. 7).] The overall temperature begins to rise appreciably at a distance of about 0.4 m from the body, while the VSL calculation, using a discontinuous shock, has a total shock layer thickness of 0.1 m. When the individual temperature components are examined for the DSMC solution (Fig. 4b), the rotational and vibrational temperatures are seen to lag far behind the translational temperatures. The difference between the translational and internal temperature modes increases with altitude, since the collision rate decreases. In the region adjacent to the surface, the slope of the temperature profile resulting from the two computational methods (Fig. 4c) is slightly different, and the temperature jump predicted is 350 K.

The density rise will always lag behind the temperature rise as evidenced by the results in Figs. 4a and 4d. In fact, if one assumes that the center of the shock for the DSMC calculation is at the location where the density equals the mean of the freestream and post-shock continuum values, then the continuum and DSMC shock locations are in good agreement.

The chemical composition profiles along the stagnation streamline resulting from the DSMC and VSL solutions are presented in Figs. 4e and 4d for O_2 and N_2, and N, respectively. The profile resulting from the two solutions has the same general shape; however, the DSMC results show a significant influence of the thick shock wave on the chemical composition within the shock layer. A significant number of chemical reactions occur in the shock wave, producing atomic mass fraction in excess of 20% at the shock location, as given by the continuum solution. By including shock slip boundary conditions in the VSL solution, the chemical composition profile would compare more favorably with the DSMC results within the shock layer, as is demonstrated by the recent study of Shinn and Simmonds.[18]

The effect of increasing the altitude is to create a more rarefied situation in which the shock layer and shock wave merge (Figs. 5a and 5b). As the altitude increases, the extent of the flowfield disturbance continues to increase, as evidenced by the results shown in Fig. 5c. Shown are the stagnation locations for the maximum value of overall temperature and the location where the density is six times the freestream density (perfect gas condition) as a function of freestream density. Since the collision rate is proportional to the square of the density, the rate of chemical reactions

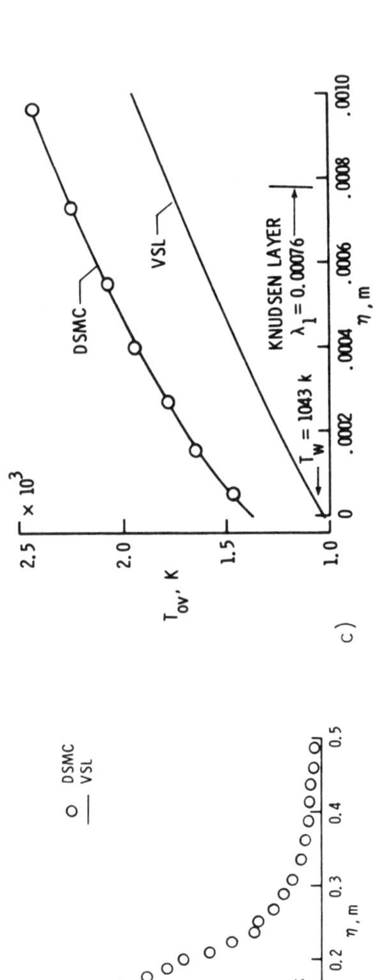

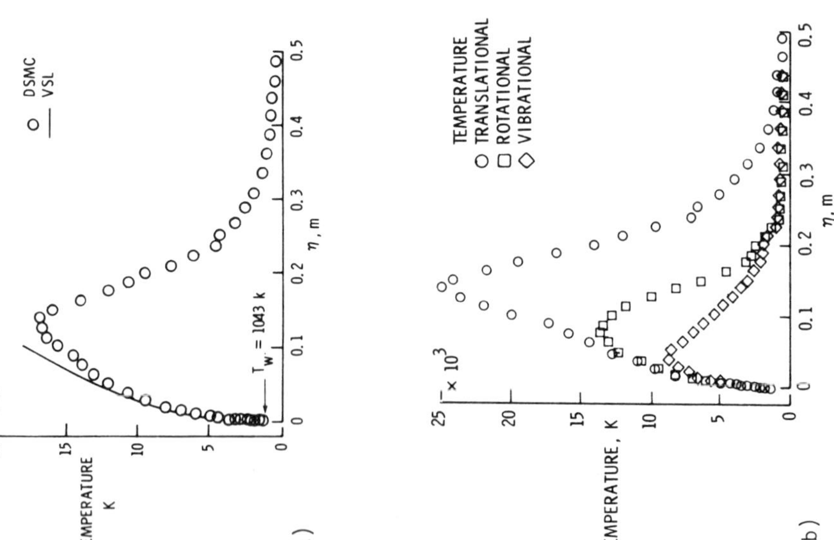

Fig. 4 Flowfield structure along stagnation streamline (Altitude = 92.35 km); a) comparison of temperature; b) extent of thermodynamic nonequilibrium; c) comparison of temperature profiles adjacent to surface.

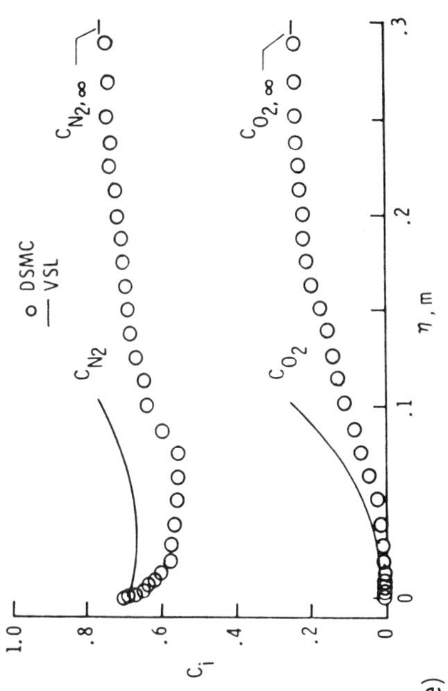

e)

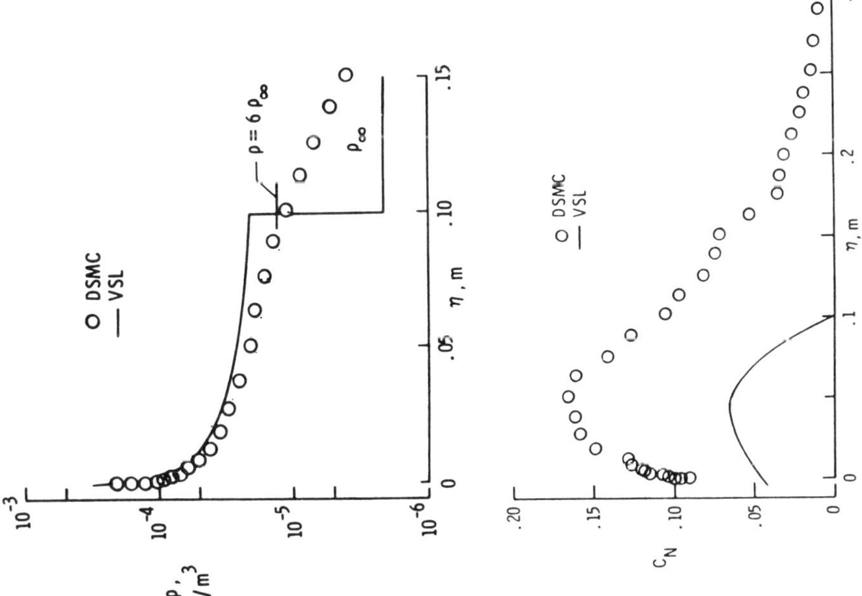

Fig. 4 Flowfield structure along stagnation streamline (Altitude = 92.35 km); d) comparison of density profiles; e) comparison of species mass fraction profiles for N_2 and O_2; and f) comparison of species mass fraction profiles for atomic nitrogen.

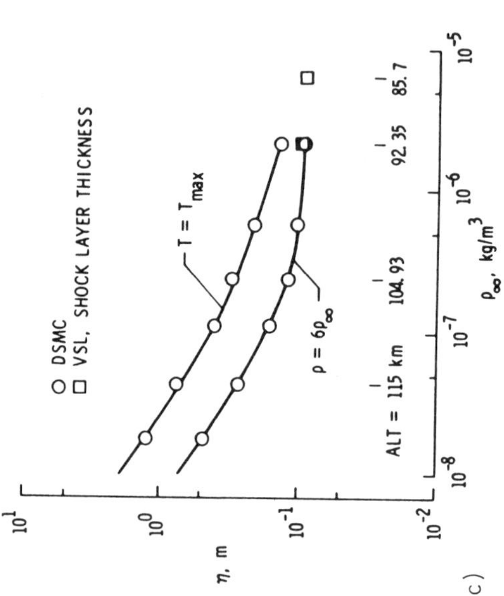

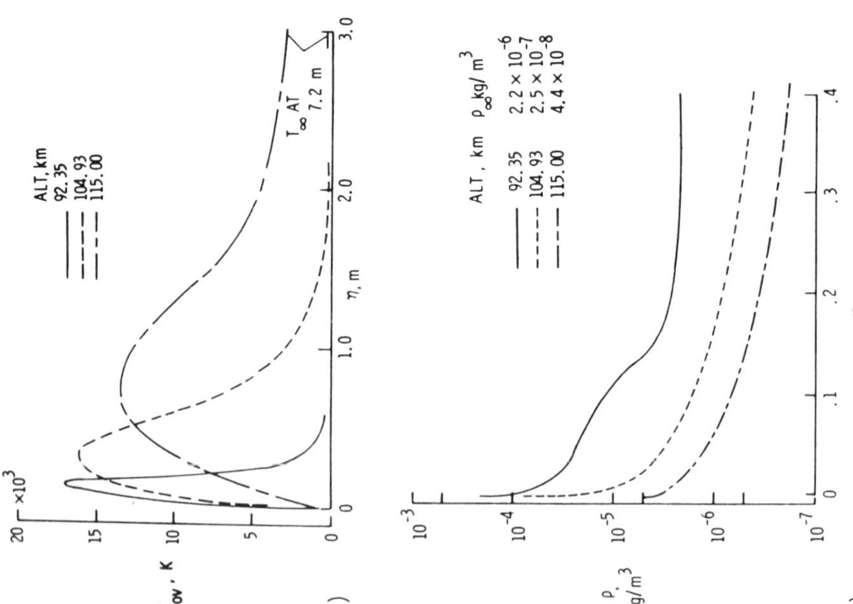

Fig. 5 Effect of rarefaction on flowfield structure (stagnation streamline); a) overall kinetic temperature profiles; b) density profiles; and c) location of stagnation point quantities.

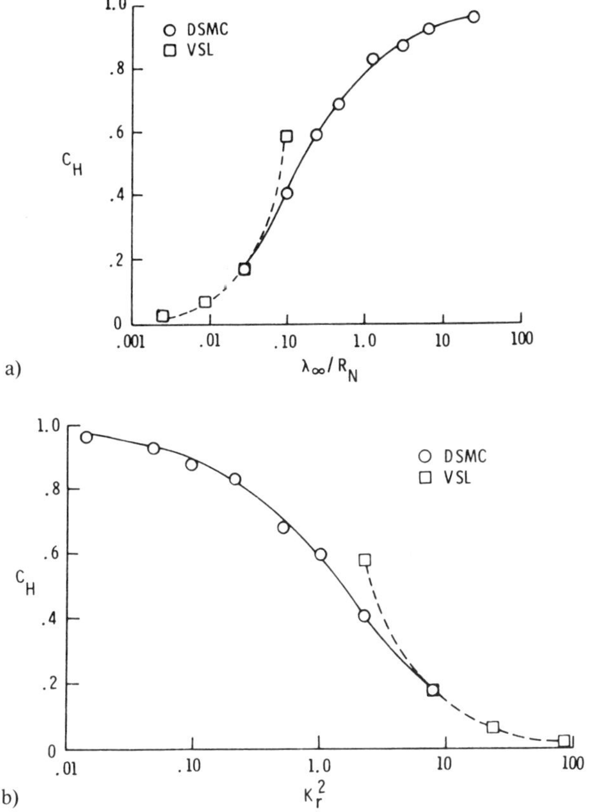

Fig. 6 Comparison of predicted stagnation point heating; a) heat-transfer coefficient versus Knudsen number; and b) heat-transfer coefficient versus Cheng's rarefaction parameter.

decreases with increasing altitude, and the present calculations show little chemical activity occurring above 105 km.

Effect of Rarefaction on Surface Quantities

Quantities of particular interest in the transitional flow regime are the heating and drag experienced by a vehicle. Figures 6a and 6b present the stagnation point heating results expressed in terms of the heat-transfer coefficient as a function of the free-stream Knudsen number and Cheng's rarefaction parameter, respectively. The predicted results show the qualitative behavior expected in that the heat-transfer coefficient increases rapidly with increasing rarefaction and

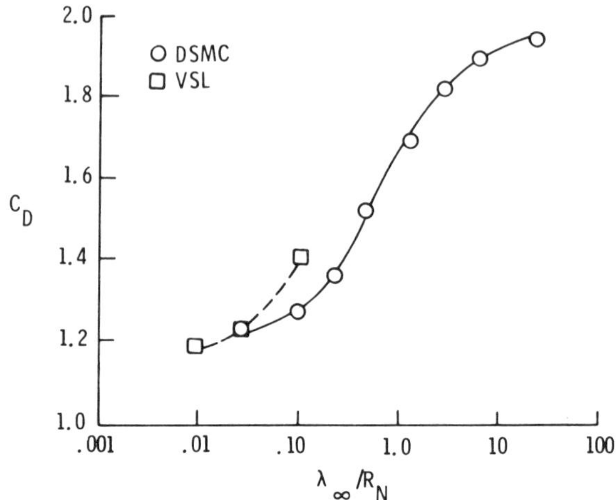

Fig. 7 Effect of rarefaction on predicted drag.

approaches a value of 1 as the free-molecule limit is approached. Also shown in Fig. 6 is a comparison of the DSMC results with those obtained with a VSL method. The VSL results presented are for altitudes of 78.9, 85.7, 92.4, and 99.5 km. The comparison shows that the VSL results, without slip and jump boundary conditions, begin to depart rapidly from the DSMC results for λ_∞/R_N of about 0.03, or K_r^2 of about 7. Both the DSMC and VSL methods are in good agreement with the flight measured results at 92.35 km.

Comparisons such as presented in Fig. 6 provide information on where a continuum method such as the VSL method fails to provide correct surface quantities. The effect of including slip and jump boundary conditions in the VSL analysis is discussed in a concurrent study by Shinn and Simmonds.[18]

The comparison of the drag prediction (Fig. 7) shows trends similar to those discussed for the heat-transfer comparisons. As was the case for the heat-transfer coefficient, the drag coefficient experiences a significant increase in value with increasing rarefaction. The change in drag is due primarily to the increase in skin friction rather than a change in pressure coefficient. The extent of the change in skin friction distribution with increasing altitude is presented in Fig. 8a. The corresponding heat-transfer distributions are given in Fig. 8b.

As the density of a flow is reduced from that of continuum conditions, the conditions of temperature

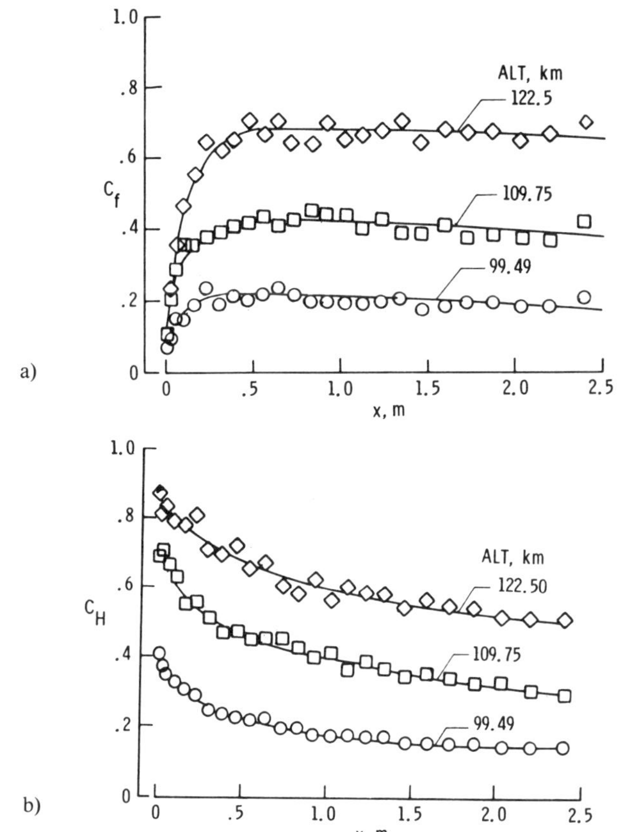

Fig. 8 Effect of rarefaction on surface distributions; skin friction coefficient; and heat-transfer coefficient.

continuity adjacent to the surface and zero-surface velocity are no longer satisfied. This occurs because the state of the molecules adjacent to the surface is affected not only by the surface but also by the flow conditions at a distance of the order of a mean-free path from the surface. Consequently, as the flow becomes more rarefied, the spatial region that influences the state of the gas adjacent to the surface increases and gives rise to significant velocity slip and temperature jump effects. In Figs. 9 and 10, the calculated temperature jump and velocity slip are presented as a function of the Knudsen number. Figure 9 presents the stagnation temperature jump expressed as a fraction of the specified wall temperature, and the values range from 0.33 to 4.64 for the Knudsen number range considered. The velocity slip

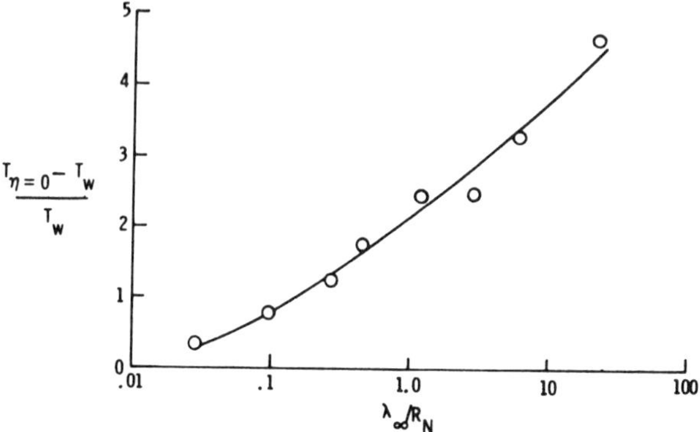

Fig. 9 Effect of rarefaction on stagnation point temperature jump.

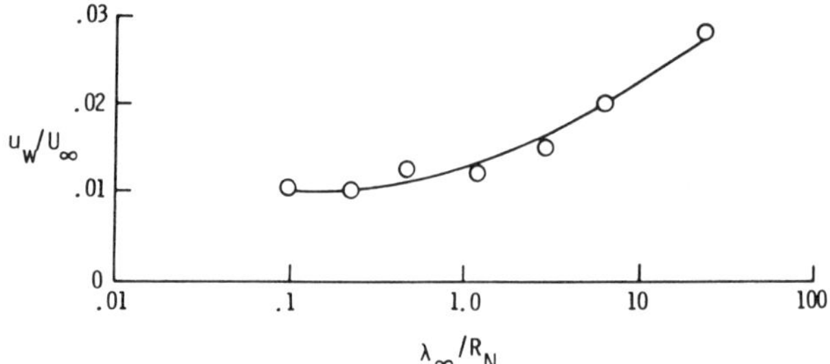

Fig. 10 Effect of rarefaction on velocity slip (X = 1.5 m; $U_\infty \simeq$ 7.5 km/s).

shown in Fig. 10 is for an axial location of 1.5 m, and the values when expressed as a fraction of the freestream velocity range from about 0.010 to 0.028.

Comparison with Flight Data

The DSMC heating results show good agreement with the corresponding flight results for the 92.35-km case. However, as the altitude is increased, the agreement between calculated and flight deduced heating rates becomes progressively worse. Figure 11 shows a comparison of these results between 92 and 110 km--that is, the early portion of the orbiter heating pulse. The heat-transfer rates are presented as a function of altitude in Fig. 11a while the same results when expressed in terms of heat-transfer coefficient and freestream Knudsen number are shown in Fig.11b. These results are for an x/L location of 0.025, where the value of x is the

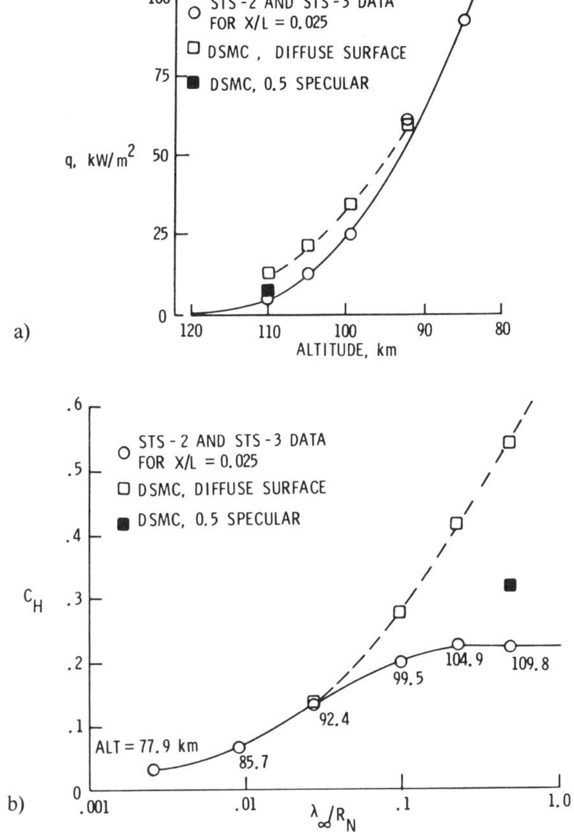

Fig. 11 Comparison of flight data with DSMC predictions; a) heating rate versus altitude; and b) heat-transfer coefficient versus Knudsen number.

distance measured from the nose of the orbiter, and L is the distance from the orbiter nose to the hinge line (L = 32.9 m). The orbiter x/L location of 0.025 corresponds to the hyperboloid x location of approximately 0.2 m.

When both the flight and calculated heating values are expressed in terms of the heat-transfer coefficient and compared in terms of the freestream Knudsen number, the differences become very apparent, as shown in Fig. 11b. The flight heat-transfer coefficient values increase with increasing Knudsen number up to an altitude of 105 km and then remain constant at a value of 0.22, rather than having an increasing trend with increasing Knudsen number as does the DSMC results. The trend of the measured C_H values is in conflict with the free-molecule limit with complete accommodation as well as with the DSMC results.

The reason for the flight C_H trend with increasing Knudsen number is not known at present. However, two areas of concern are obvious: one being the accuracy of the flight heating values at low heating conditions (Fig. 11a), and the second being that some event may occur in flight that has not been included in the Monte Carlo simulation. The accuracy of the flight heating data is not addressed in the present study, as they have been discussed previously.[17] Therefore, if it is assumed that the thermocouple measurements and the heating rates deduced from those measurements are reasonably accurate, then the question arises of the cause of the apparent low heating values for Knudsen numbers of the order of 0.1 and greater. Two events that would produce lower heating values are mass addition to the flowfield and the lack of full thermal accommodation. If there is any outgassing as the orbiter encounters the onset of the heating pulse, then this would reduce the heating. (It should be noted that a fuel dump occurs prior to entry with the fuel for the reaction control surfaces in the nose region being ejected forward of the orbiter.) The effect of mass addition was not considered in the present calculations; however, the effect of the surface reflection model was examined.

Effect of Surface Reflection Model

For case 4, an altitude of 110 km, a calculation was made assuming that half of the particles that interact with the surface do so in a diffuse manner with full thermal accommodation, and half interact in a specular manner (elastic collisions) with no thermal accommodation. The net result is a thermal accommodation of 0.5. The results of this calculation are compared with those from the diffuse calculation in Figs. 11 and 12. The stagnation point heating is 60% of that for full accommodation. Similar reductions in heating are evident at other locations downstream of the stagnation point (Fig. 12 a). Even with the assumption of 50% specular reflection, the calculated heating is high in comparison with the STS-3 flight results for x/L locations of 0.025, 0.098, and 0.140. Therefore, it is concluded that the flight heating results cannot be explained in terms of surface reflection model alone, since the thermal accommodation coefficient required to achieve agreement with flight data would be much less than 0.5.

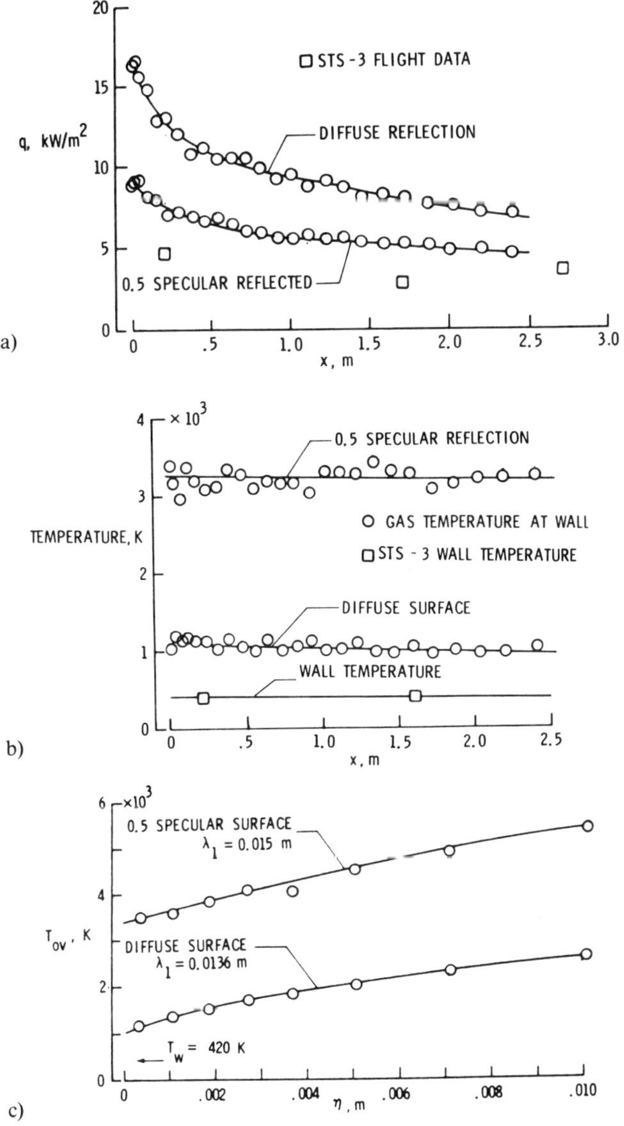

Fig. 12 Effect of surface reflection model (Altitude =109.95 km); a) heat-transfer distribution; b) temperature jump; and c) temperature profile (stagnation point).

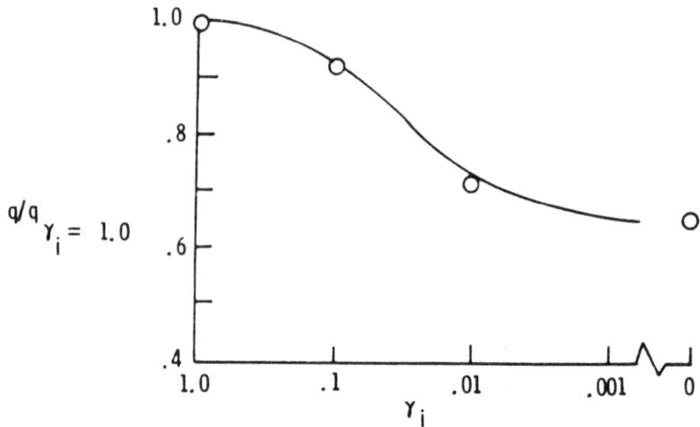

Fig. 13 Effect of surface catalysis on heating [Altitude = 92.35 km (stagnation point)].

Altering the gas-surface interaction model produces a substantial change in the state of the gas adjacent to the surface. The impact on the overall temperature adjacent to the surface is to produce a temperature jump that is about 4.5 times that for the diffuse surface with full accommodation (Fig. 12b). However, the slope of the temperature profiles is very similar, as is shown for the stagnation point in Fig. 12c.

Catalytic Wall Effect

In general, two conditions must be met before wall catalytic activity has an influence on surface heating. First, the energy of the flowfield must be such as to create dissociated flow, and second, some degree of chemical nonequilibrium must persist. Such conditions exist for much of the Shuttle entry heating pulse and will also exist for an aeroassist orbital transfer vehicle during atmospheric encounter.

In the present study, the catalytic wall effect was examined for only the 92.35-km case. The results of these calculations are shown in Fig. 13, where the stagnation point heating is presented as a function of the surface recombination probability γ_i. In the present calculations, the recombination probability was assumed to be the same for both atomic oxygen and nitrogen even though different values could be specified for the respective species. For the noncatalytic surface, the heating is 65% of the fully catalytic value, which indicates the advantage of having a surface that is basically noncatalytic.

Values of the recombination probability appropriate for the Shuttle tiles have been measured by Scott[19] for the atomic species oxygen and nitrogen. For the 92.35-km case, where the stagnation wall temperature is 1043 K, the γ_i values using Scott's results would be 0.007 and 0.0004 for nitrogen and oxygen, respectively. Consequently, the results for the noncatalytic surface are very nearly appropriate to the Shuttle tiles.

Concluding Remarks

Results obtained with the DSMC method for the transitional flow regime encountered by the Shuttle during re-entry demonstrate the practical application of the method to such problems. The solutions obtained show good agreement at the lowest altitude case (92.35 km) with both flight and VSL heating results. With increasing altitude, the agreement between flight and calculated heating becomes progressively poorer. While the calculated heat-transfer coefficient asymptotically approaches a value of 1.0 for large Knudsen numbers, the flight results never reach a value much greater than 0.22. If an energy accommodation coefficient less than 1.0 is assumed, the agreement is improved, but the discrepancy cannot be totally resolved with realistic values of accommodation coefficient. Significant temperature jump and velocity slip effects are present in all cases. Chemical reactions within the flowfield are evident up to an altitude of about 105 km, whereas at higher altitudes the gas composition is basically that of the freestream. Thermodynamic nonequilibrium effects are evident for all conditions considered, extending all the way to the surface at the higher altitudes. The results for heating and drag suggest that continuum calculations must be modified to account for slip and temperature jump effects at freestream Knudsen numbers of approximately 0.03.

Future studies should focus on the sensitivity of heating, drag, and flowfield structure to variations in molecular constants. Also, it would be of interest to extend the present calculations to lower altitudes for comparison with the results from continuum calculations. The molecular constants in the DSMC method should be further refined so that the two numerical methods use, as nearly as possible, the same transport properties.

Acknowledgment

This work was performed under the NASA Langley Floyd L. Thompson Fellowship at the University of Sydney.

References

[1]Walberg, D. G., "Aeroassisted Orbit Transfer - Window Opens on Missions," Astronautics and Aeronautics, Vol. 11, Nov. 1983, pp. 36-43.

[2]Shinn, J. L. and Jones, J. J., "Chemical Nonequilibrium Effects on Flowfields for Aeroassist Orbital Transfer Vehicles" AIAA Paper 83-0214, AIAA 21st Aerospace Sciences Meeting, Reno Nev., Jan. 1983.

[3]Howe, J. T., "Introductory Aerothermodynamics of Advanced Space Transportation Systems," AIAA Paper 83-0406, AIAA 21st Aerospace Sciences Meeting, Reno, Nev., Jan. 1983.

[4]Bird, G. A., "Monte Carlo Simulation of Gas Flows," Annual Reviews of Fluid Mechanics, Vol. 10, edited by M. D. Van Dyke, J. V. Wehausen, and J. L. Lumley, Annual Reviews Inc., Palo Alto, Calif., 1979, p. 11.

[5]Bird, G. A., "Simulation of Multidimensional and Chemically Reacting Flows," Rarefied Gas Dynamics, Vol. 1, edited by R. Campargue, CEA, Paris, 1979, pp. 365-388.

[6]Bird, G. A., "Monte-Carlo Simulation in an Engineering Context," AIAA Progress in Astronautics and Aeronautics: Rarefied Gas Dynamics, Vol. 74, Part 1, edited by S. S. Fisher, AIAA, New York, 1981, pp. 239-255.

[7]Bird, G. A., Molecular Gas Dynamics, Clarendon Press, Oxford, England, 1976.

[8]Hueser, J. E. and Brock, F. J., "Shuttle Flowfield Analysis Using the Direct Simulation Monte Carlo Technique," paper presented at the USAF/NASA International Spacecraft Contamination Conference, USAF Academy, Colorado Springs, Colo., March 1978.

[9]Chrusciel, G. T. and Pool, L A., "Knudsen Layer Characteristics for a Highly Cooled Blunt Body in Hypersonic Rarefied Flow," AIAA Paper 83-1424, AIAA 18th Thermophysics Conference, Montreal, Canada, June 1983.

[10]Shinn, J. L., Moss, J. N., and Simmonds, A. L., "Viscous Shock-Layer Heating Analysis for the Shuttle Windward Plane with Surface Finite Catalytic Recombination Rates," AIAA Paper 84-0842, AIAA/ASME 3rd Joint Thermophysics, Fluids, Plasma, and Heat Transfer Conference, St. Louis, Mo., June 1982.

[11]Borgnakke, C. and Larsen, P. S., "Statistical Collision Model for Monte Carlo Simulation of Polyatomic Gas Mixtures," Journal of Computational Physics, Vol. 18, Aug. 1975, pp. 405-420.

[12]Davis, R. T., "Numerical Solution of the Hypersonic Viscous Shock-Layer Equations," AIAA Journal, Vol. 8, May 1970, pp. 843-351.

[13]Anderson, E. C. and Moss, J. N., "Numerical Solution of the Steady-State Navier-Stokes Equations for Hypersonic Flow About Blunt Axisymmetric Bodies," NASA TM X-71977, June 1974.

[14]Compton, H. R., Findlay, J. T., Kelly, G. M., and Heck, M. L., "Shuttle (STS-1) Entry Trajectory Reconstruction," AIAA Paper 81-2459, AIAA Flight Testing Conference, Las Vegas, Nev., Nov. 1981.

[15]Price, J. M., "Atmospheric Definition for Shuttle Aerothermodynamic Investigations," Journal of Spacecraft and Rockets, Vol. 20, March-April 1983, pp. 133-140.

[16]Jacchia, L. G., "Thermospheric Temperature, Density, and Composition: New Models," Research in Space Science, Smithsonian Astrophysical Observatory, Special Report 375, Cambridge, Mass., March 1977.

[17]Throckmorton, D. A., "Benchmark Aeroheating Data from the First Flights of the Space Shuttle Orbiter," AIAA Paper 82-0003, AIAA 20th Aerospace Sciences Meeting, Orlando, Fla., Jan. 1982.

[18]Shinn, J. L. and Simmonds, A. L., "Comparison of Viscous-Shock-Layer Heating Analysis with Shuttle Flight Data in Slip Flow Regime," AIAA Paper 84-0226, AIAA 22nd Aerospace Sciences Meeting, Reno, Nev., Jan. 1984.

[19]Scott, C. D., "Catalytic Recombination of Oxygen and Nitrogen in High Temperature Reusable Surface Insulation," Progress in Astronautics and Aeronautics: Aerothermodynamics and Planetary Entry, edited by A. L. Crosbie, Vol. 77, AIAA, New York, 1981, pp. 192-212.

Chapter II. Trajectories

Performance Evaluation of the Atmospheric Phase of Two Aeromaneuvering Orbital Transfer Vehicles

Richard W. Powell,* Howard W. Stone,* and J. Chris Naftel*
NASA Langley Research Center, Hampton, Virginia

Abstract

Studies are underway to design reusable orbital transfer vehicles that would be used to transfer payloads from low Earth orbit to higher orbits and return. One promising concept is to use an atmospheric pass on the return leg to reduce the amount of fuel for the mission. This paper discusses a 6-degree-of-freedom simulation analysis for two configurations: a low lift-to-drag ratio (≈0.15) configuration and a medium lift-to-drag ratio (≈0.60) configuration using both a predictive guidance technique and an adaptive guidance technique. Both guidance schemes were evaluated using the 1962 standard atmosphere and three atmospheres that had been derived from three entries of the Space Shuttle. The predictive technique requires less reaction control system activity for both configurations, but because of the limited number of updates and because each update used the 1962 standard atmosphere, the adaptive technique produces more accurate exit conditions.

Nomenclature

b	=	reference span, ft
c	=	Chapman-Reubesin viscosity coefficient
$\bar{c}$	=	reference chord, ft
C_D	=	drag coefficient=drag/qS
C_L	=	lift coefficient=lift/qS
C_ℓ	=	rolling moment coefficient=rolling moment/q Sb

Presented as Paper 84-0405 at AIAA 22nd Aerospace Sciences Meeting, Jan. 9-12, 1984, Reno, Nev.

*Aerospace Engineer, Space Systems Division.

C_{ℓ_β} = $\partial C_\ell/\partial\beta$, deg^{-1}
C_m = pitching moment coefficient=pitching moment/q $S\bar{c}$
C_N = normal force coefficient=normal force/q S
C_n = yawing moment coefficient=yawing moment/q Sb
C_{n_β} = $\partial C_n/\partial\beta$, deg^{-1}
C_Y = side force coefficient=side force/qS
I_x = moment of inertia about body roll axis, slug-ft^2
I_y = moment of inertia about body pitch axes, slug-ft^2
I_z = moment of inertia about body yaw axis, slug-ft^2
L/D = lift-to-drag ratio=C_L/C_D
M = Mach number
$M_{x_{RCS}}$ = rolling moment generated by RCS, ft-lb
$M_{y_{RCS}}$ = pitching moment generated by RCS, ft-lb
$M_{z_{RCS}}$ = yawing moment generated by RCS, ft-lb
OTV = orbital transfer vehicle
q = dynamic pressure, psf
RCS = reaction control system
Re = Reynolds number
r_s = yaw rate about the stability axis, deg/s
S = reference area, ft^2
U_{x_c} = command to activate roll RCS
U_{y_c} = command to activate pitch RCS
U_{z_c} = command to activate yaw RCS
$\bar{v}$ = viscous correlation parameter = $M\sqrt{c}/\sqrt{Re}$
V_{rel} = relative velocity, ft/s
α = angle of attack, deg
α_c = commanded angle of attack, deg
α_{error} = $\alpha_c - \alpha$, deg
β = sideslip angle, deg
γ_{rel} = relative flight-path angle, deg
ρ = atmospheric density, slug/ft^3
ρ_{62} = atmospheric density of 1962 standard atmosphere, slug/ft^3
ϕ = bank angle, deg
ϕ_c = commanded bank angle, deg
ϕ_{error} = $\phi_c - \phi$, deg
σ_{rel} = relative heading angle, deg

Introduction

In evaluating the needs of future space transportation systems, it becomes apparent that projections of large amounts of traffic to geosynchronous and other high Earth orbits justify the development of reusable orbital transfer vehicles (OTVs). These vehicles would transfer payloads from a Space Shuttle or a low Earth orbiting space station to the high Earth orbits and then return. Many analyses (see Ref. 1) have indicated that performance advantages can be obtained using aerodynamic forces generated through an atmospheric pass to achieve the necessary orbital changes (both apogee and inclination) on the return leg as compared with all-propulsive orbital changes.

This paper will discuss the preliminary evaluation of two guidance techniques for the atmospheric portion of the flight using 6-degree-of-freedom simulations. Both techniques used roll-angle modulation without angle-of-attack modulation to control the apogee and inclination at exit. The first technique is adaptive and the second is predictive. These techniques were applied to two configurations, one with a low L/D (≈ 0.15) and one with a moderate L/D (≈ 0.6). To properly evaluate the guidance techniques, the atmospheric portion of the flight was simulated with 6 degrees of freedom through four atmospheres. These atmospheres were the 1962 standard and three atmospheres derived from entries of the Space Shuttle.

Configuration Definition

The configurations used in this analysis were selected to evaluate the aerodynamic performance impact

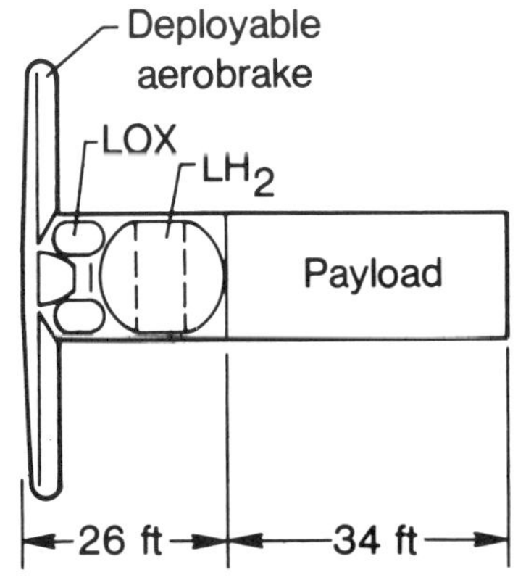

Fig. 1 Sketch of low L/D configuration.

Table 1 Physical characteristics at atmospheric entry interface

	Low L/D configuration	Moderate L/D configuration
Weight, lbs	9883	8320
S, ft^2	1963.5	168.95
b, ft	50.	58.7
$\bar{c}$, ft	50.	58.7
$I_x, slug\text{-}ft^2$	2402	887
$I_y, slug\text{-}ft^2$	23,768	43,268
$I_z, slug\text{-}ft^2$	24,093	43,361
$M_{x_{RCS}}$, ft-lb	900	350
$M_{y_{RCS}}$, ft-lb	10,000	10,000
$M_{z_{RCS}}$, ft-lb	4000	7,200

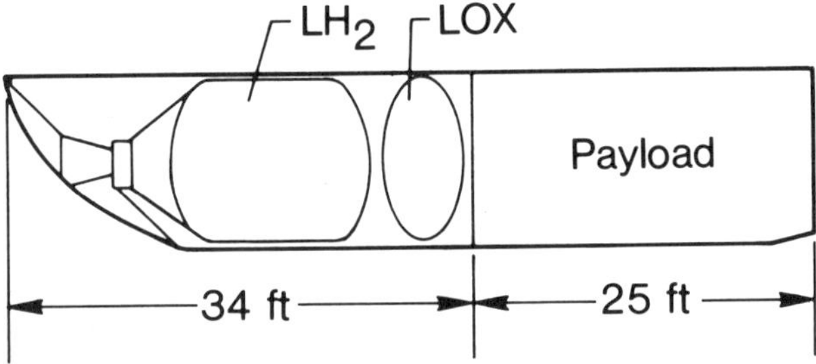

Fig. 2 Sketch of moderate L/D configuration.

on guidance techniques by examining a low L/D configuration and a moderate L/D configuration.

The low L/D configuration is depicted in Fig. 1. This configuration, known as a lifting brake, is characterized by a large-diameter (50 ft) heat shield that generates lift by flying at a nonzero angle of attack. This vehicle was designed to fly at an angle of attack of -15 deg.

The moderate L/D configuration is depicted in Fig. 2. This configuration has a 15-ft-diam cylindrical afterbody, and a nose shaped to provide the required aerodynamic performance. Table 1 lists the characteristics of the configurations used in this study. These characteristics were obtained from Ref. 2 for the low L/D configuration and Ref. 3 for the moderate L/D configuration.

The aerodynamics of these configurations shown in Figs. 3 and 4 are taken from Ref. 4. The derivation of the aerodynamics is also described in Ref. 4 and includes high-altitude effects.

Mission Description

For this analysis, the following nominal mission was assumed. The orbital transfer vehicle deorbits from geosynchronous orbit, enters the atmosphere, and then exits the atmosphere in a phasing orbit at a 28.5-deg. inclination with an apogee of 300 n.mi. The 300-n.mi. apogee was chosen to allow for phasing between the orbital transfer vehicle and the Space Shuttle.

The transfer orbits that were used in this analysis were determined as follows. A 3-degree-of-freedom analysis was used to determine the combination of transfer orbit perigee and atmospheric bank angle history that would result in the maximum atmospheric plane change. Once the maximum atmospheric inclination change capability was known, the transfer orbit used in the analysis could be determined. This transfer orbit was selected so that only 60% of the inclination change capability in the atmosphere would be required to reach the target inclination of 28.5 deg. This allowed a margin to account for the effects on inclination change capability of atmospheric density variations, errors in the aerodynamic predictions, and errors at the atmospheric interface. Table 2 shows the entry interface condition for the two configurations.

Atmospheric Profiles

To aid in the evaluation of the two configurations, four atmospheres were modeled. The first is the 1962 standard. The other three were derived from the second, fourth, and sixth entries of the Space Shuttle (STS-2, STS-4, and STS-6), as described in Ref. 5. These atmospheres were derived using the accelerations as measured onboard the orbiter and assuming that the normal force

Table 2 Atmospheric interface conditions

	Low L/D configuration	Moderate L/D configuration
Altitude, ft	400,000	400,000
V_{rel}, ft/s	32,443.7	32,391.6
γ_{rel}, deg	-4.21	-5.45
σ_{rel}, deg	117.9	113.1
Apogee, n.mi.	19,323	19,323
Perigee, n.mi.	45.3	31.6
Inclination, deg	27	22.5

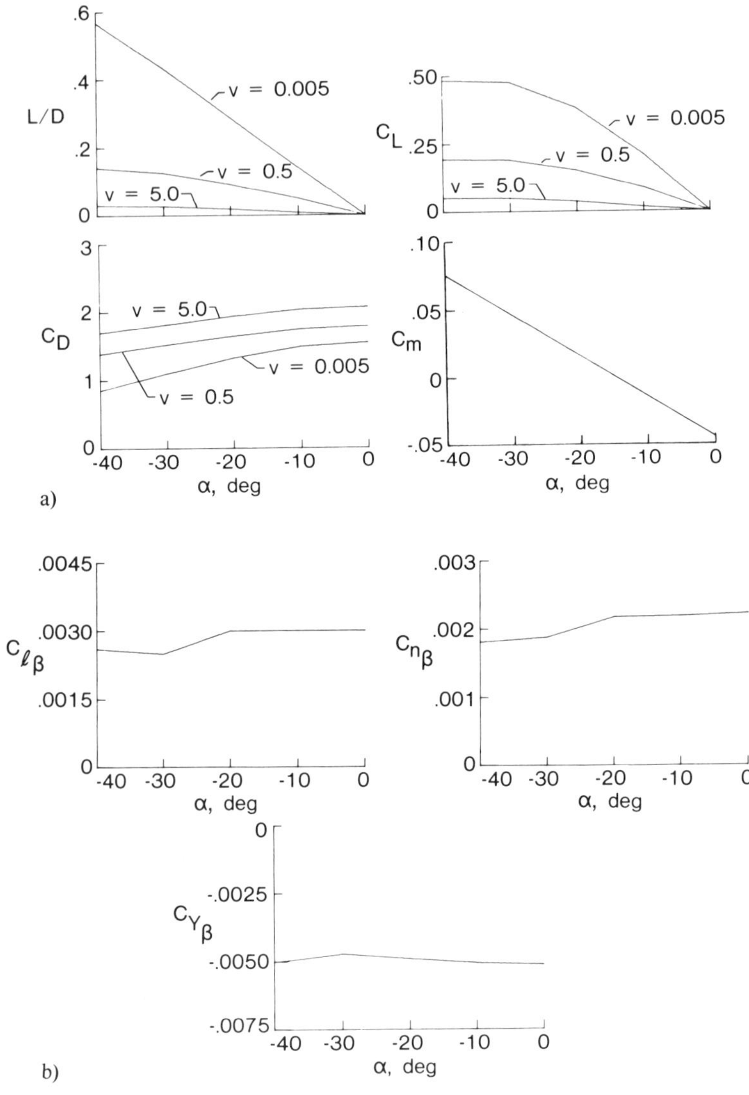

Fig. 3 Aerodynamics for low L/D configuration: a) longitudinal; b) lateral directional.

coefficient is known. This technique was used for altitudes less than 320,000 ft. For the STS-2 and STS-4 entries, the density above 320,000 ft was that which had been derived using remote soundings. For STS-6 the density profile above 320,000 ft was derived using data from additional accelerometers designed for the low dynamic pressure regime. These accelerometers are part of the

High Resolution Accelerometer Package that first flew on STS-6 (Ref. 6). The density profile for each of these atmospheres as compared with the 1962 standard atmosphere is shown in Fig. 5.

Flight Control System

A flight control system was designed for both configurations assuming that all controlling moments were provided by a reaction control system (RCS). The control

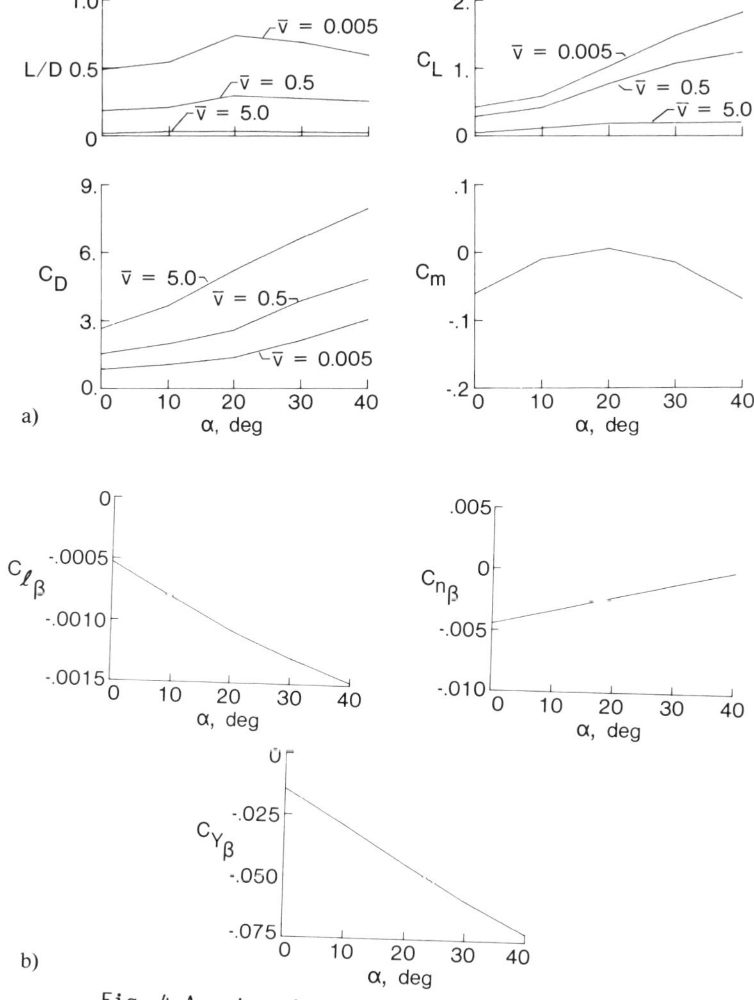

Fig. 4 Aerodynamics for moderate L/D configuration: a) longitudinal; b) lateral directional.

system was designed using a computer program known as Interactive Digikon. This program was developed under contract[7] with the Honeywell Corporation and allows for the interactive design and evaluation of control systems. This program can analyze either continuous or digital control systems. For this analysis, the design was developed as a continuous system and converted to a discrete equivalent within the computer program. Interactive Digikon was also used to size the RCS thrusters. Fig. 6 shows the flight control system for these configurations. The pitch axis circuit and the roll axis circuit are identical for both configurations, but the β circuit requires a different lead filter and gain for each configuration.

Guidance Techniques

Two guidance techniques were evaluated to determine their capability to accommodate atmospheric variations: an adaptive technique and a predictive technique. The adaptive guidance technique was the drag reference closed loop guidance algorithm developed by Oliver Hill at NASA Johnson Space Center for aeroassisted orbital transfer vehicles.[8] The controller equation used in the algorithm is a linear feedback equation with variable coefficients to control the vertical L/D. The required vehicle roll angle is calculated from the ratio of the desired vertical L/D to the vehicle L/D capability. The controller equation has three terms: the reference L/D, the drag acceleration error, and the altitude rate error. During the early portion of the atmospheric flight, the reference drag acceleration value used in the calculation of the reference parameters is the equilibrium glide value (altitude acceleration = 0) for the current velocity and altitude rate. The L/D reference value is obtained from the equations of motion by assuming that the rate of change of drag acceleration is constant. The altitude rate reference value is calculated based on the desired rate of change of drag.

At some preset velocity, usually just after pullout, the exit phase is begun. In this phase, the terms in the controller equations become functions of linear extrapolations of the reference drag and altitude rate based on velocity from the phase transition point to the exit conditions. When the sensed acceleration reaches a small value, the drag error term in the controller equation is dropped.

The inclination angle change is controlled by rolling from one side to the other within an envelope around the

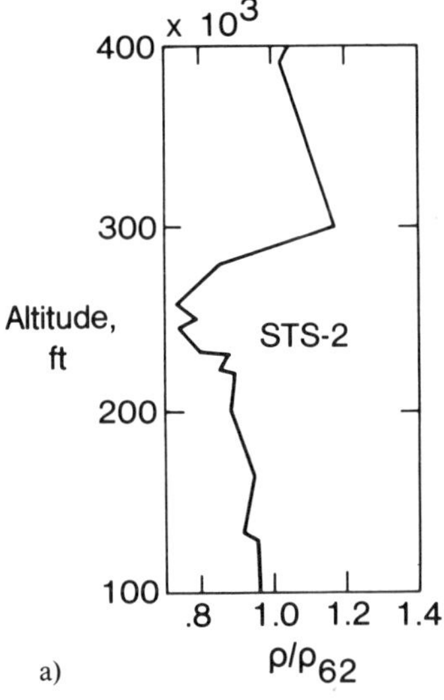

a)

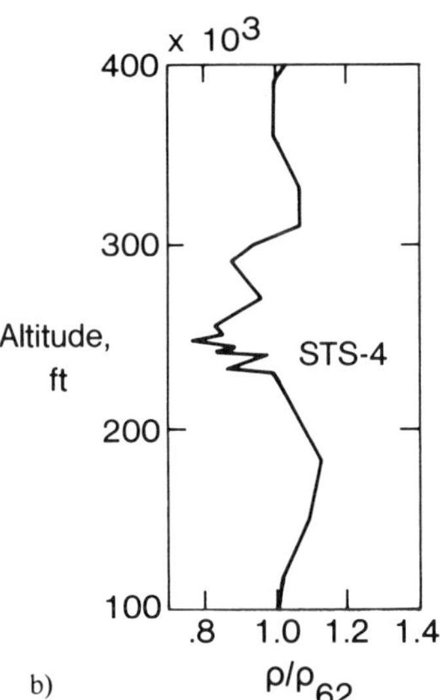

b)

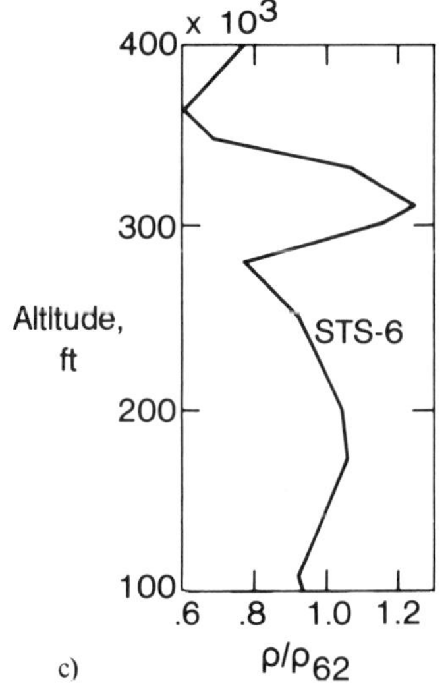

c)

Fig. 5 Atmospheric density profiles: a) STS-2 atmosphere; b) STS-4 atmosphere; c) STS-6 atmosphere.

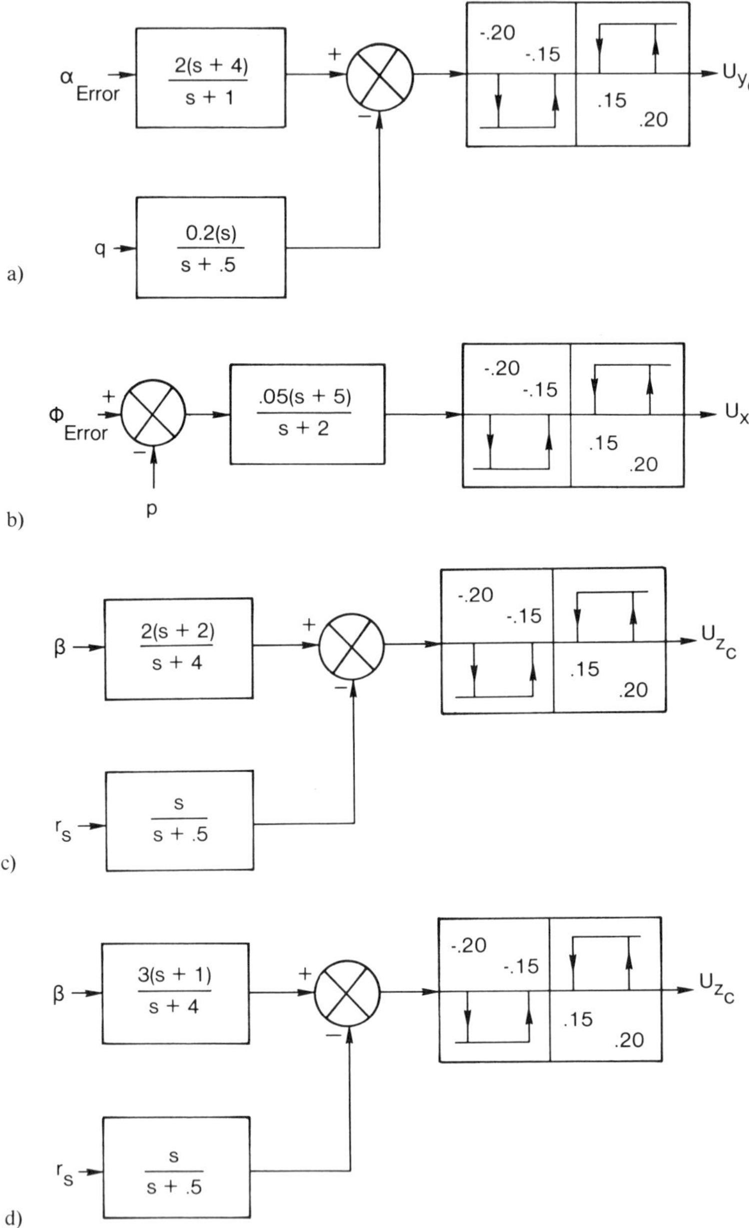

Fig. 6 Flight control system: a) pitch circuit; b) roll circuit; c) slipslide circuit for low L/D configuration; d) Sideslip circuit for moderate L/D configuration.

target inclination angle. At some chosen velocity during the exit portion of the atmospheric pass, the envelope steps down to a narrow boundary to assure convergence on the desired inclination angle at the end of the atmospheric pass.

The predictive technique was developed as follows. A nominal "optimum" trajectory was determined using a 6-degree-of-freedom simulation program. This simulation used the initial condition shown in Table 2 and determined the bank angle time history required to reach the proper exit conditions for an apogee of 300 n.mi. and an inclination of 28.5 deg for a 1962 standard atmosphere. This bank angle history is slightly different from the 3-degree-of-freedom trajectory described earlier, because the 3-degree-of-freedom simulation does not model vehicle rotational dynamics. The vehicle was commanded to follow this bank angle history through the Shuttle-derived atmospheres described earlier. When the current state condition differs significantly from the nominal "optimum," a new "optimum" is calculated from that point, and the commanded bank angle history is modified. All "optimum" trajectories, including the update trajectories, assumed a 1962 standard atmosphere. Since this study was designed to establish feasibility, the number of updates was limited to two.

Discussion of Results

Predictive Guidance Technique

This technique commands the orbital transfer vehicle to fly the bank angle profile determined for the 1962

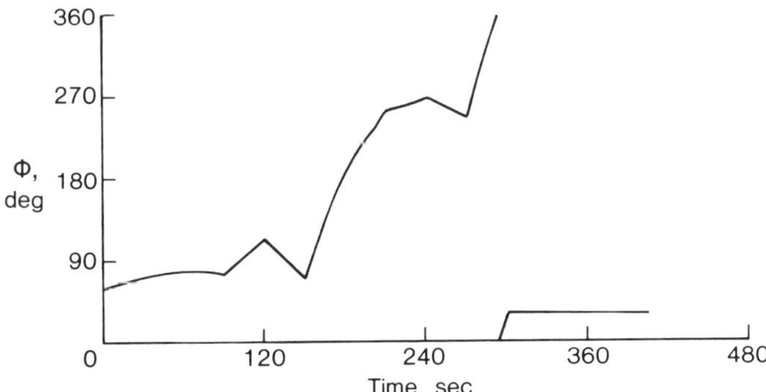

Fig. 7 Bank angle profile for low L/D configuration using predictive guidance and nominal atmosphere.

standard atmosphere until an update is required. When an updated bank angle profile is required, it is determined using the 1962 standard atmosphere.

The bank angle profiles for the low L/D configuration are shown in Figs. 7-10 for the nominal 1962 standard atmosphere and the off-nominal atmospheres derived from the three Space Shuttle entries (STS-2, STS-4, STS-6). Table 3 shows the results from these simulations.

The maximum difference between the nominal and off-nominal atmosphere trajectories at the minimum, or pull-out altitude was 132 ft out of 275,000 ft; and the inertial velocities at pullout were: 30,632 ft/s for the nominal atmosphere, 30,905 ft/s for the STS-2 atmosphere, 30,866 ft/s for the STS-4 atmosphere, and 31,076 ft/s for the STS-6 atmosphere. Since the differences between the trajectories were small through pullout, the first trajectory update was delayed until 120 s after entry, or approximately 10 s after pullout. At this point in the trajectory, the off-nominal atmosphere cases had consistently higher velocities and, therefore, dissipated less

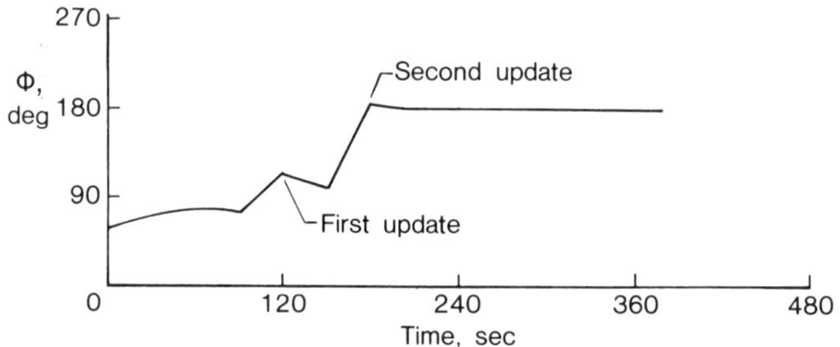

Fig. 8 Bank angle profile for low L/D configuration using predictive guidance and STS-2 atmosphere.

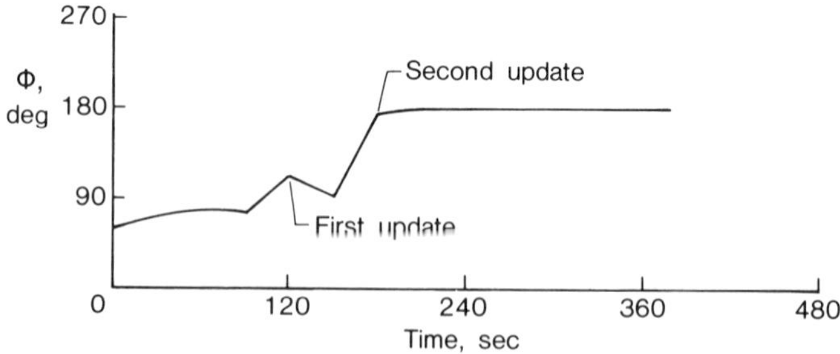

Fig. 9 Bank angle profile for low L/D configuration using predictive guidance and STS-4 atmosphere.

energy on the entry portion of the trajectory than for the nominal atmosphere case.

At the first update, a new bank angle profile was determined for the exit portion of the flight that would allow the vehicle to reach the proper exit conditions if it were subject to the 1962 standard atmosphere. A second update occurred at 180 s. At this time, the predictive analysis determined that the targeted exit conditions could not be reached for any of the off-nominal atmosphere trajectories even if the orbital transfer vehicle maintained a roll angle of 180 deg. Table 3 shows the results of the simulations after both updates. The maximum error of 746.4 n.mi. in exit apogee occurred for the STS-6 atmosphere.

The bank angle profiles for the moderate L/D configurations are shown in Figs. 11-14 for the nominal 1962 standard atmosphere and the off-nominal atmospheres. Table 4 shows the results from these simulations.

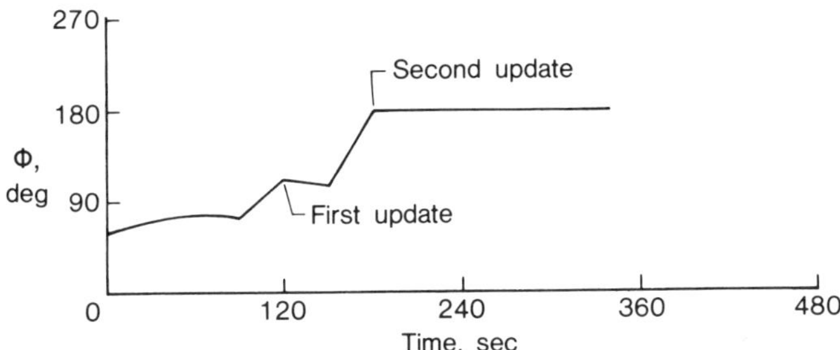

Fig. 10 Bank angle profile for low L/D configuration using predictive guidance and STS-6 atmosphere.

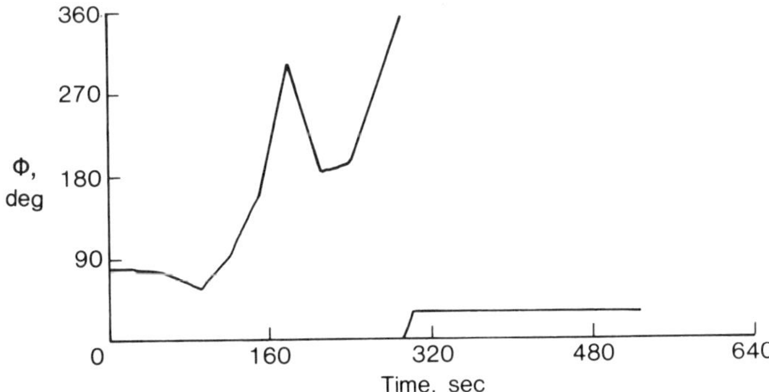

Fig. 11 Bank angle profile for moderate L/D configuration using predictive guidance and nominal atmosphere.

Table 3 Comparison of atmospheric passes for low moderate L/D configuration

	Predictive guidance atmosphere				Adaptive guidance atmosphere			
	Nominal	STS-2	STS-4	STS-6	Nominal	STS-2	STS-4	STS-6
Exit apogee, n.mi.	296.8	619.1	593.3	1046.4	305.3	423.5	287.9	456.7
Exit perigee, n.mi.	36.4	45.4	44.8	46.2	11.84	13.64	9.48	24.22
Exit inclination, deg	28.6	28.5	28.7	28.4	28.5	28.46	28.48	28.47
Minimum altitude, ft	275,504	275,372	275,441	275,393	273,433	269,201	273,081	270,617
Maximum dynamic pressure, psf	8.1	6.8	7.7	6.6	8.8	8.7	8.7	8.4
Maximum sensed acceleration, g	2.4	2.0	2.3	2.0	2.6	2.5	2.5	2.4
Maximum stagnation heating rates (1-ft-radius sphere), BTU/ft^2-s	77.3	71.3	75.7	70.1	77.1	74.4	78.0	75.0
Total stagnation heat load, BTU/ft^2	10,488	10,445	10,281	9781	9743	9587	9860	9732
On-time for roll RCS thrusters, s	8.56	6.24	9.92	9.84	52.84	44.96	56.44	67.24
On-time for yaw RCS thrusters, s	16.08	8.56	10.88	10.96	39.04	34.56	44.72	48.48
Total atmospheric pass time, s	406	380	380	340	325	301	331	320

The maximum difference between the nominal and off-nominal atmospheres at the pullout altitude was 2,312 ft out of 220,000 ft; and the inertial velocities at pullout were 30,790 ft/s for the nominal atmosphere, 30,925 ft/s for the STS-2 atmosphere, 30,837 ft/s for the STS-4 atmosphere, and 30,839 ft/s for the STS-6 atmosphere. Since, once again, the differences between the trajectories were small through pullout, the first trajectory update was delayed until after pullout. If the orbital transfer vehicle were to encounter a STS-2 atmosphere, its apogee at exit would be 386 n.mi. with no updates. Therefore, only one update, at 240 s, was used for this case. For the other two atmospheres, the first update occurred at 150 s after entry, or approximately 40 s after pullout. At the first update, a new bank angle profile was determined for the exit portion of the flight

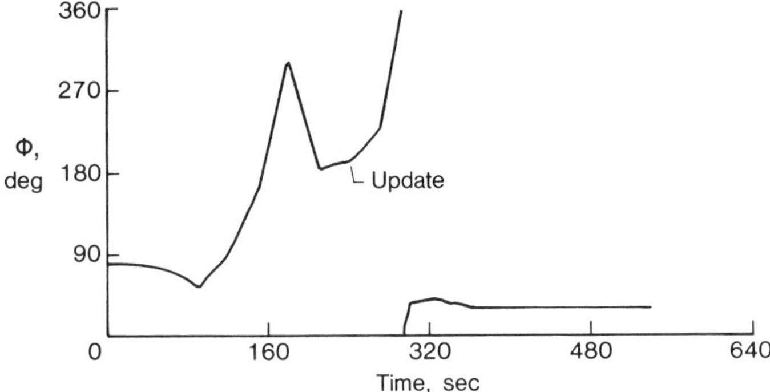

Fig. 12 Bank angle profile for moderate L/D configuration using predictive guidance and STS-2 atmosphere.

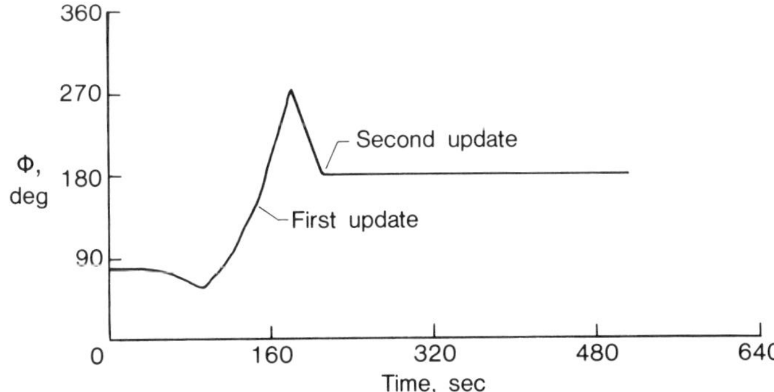

Fig. 13 Bank angle profile for moderate L/D configuration using predictive guidance and STS-4 atmosphere.

that would allow the vehicle to reach the proper exit conditions if it were subject to the 1962 standard atmosphere. The second update for these two off-nominal atmospheres occurred at 210 s. At this time, the predictive analysis determined that the proper exit conditions could not be reached for these two off-nominal atmospheres even if the vehicle maintained a roll angle of 180 deg. Table 4 shows the results of the simulations after the updates.

These results suggest that for predictive guidance to work effectively, a larger number of updates will be required, and the atmospheric model used in the predictions must be updated. This update could be done by comparing the orbital energy at the beginning of the update to the orbital energy of the "optimum" and using this comparison to determine an effective ρ/ρ_{62} multiplier for the atmospheric density. The inherent assump-

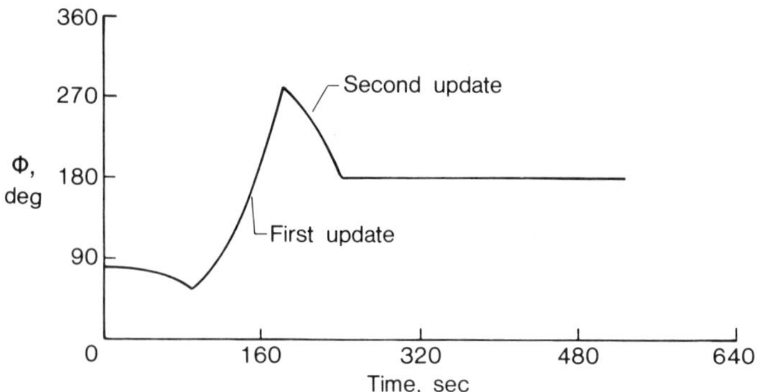

Fig. 14 Bank angle profile for moderate L/D configuration using predictive guidance and STS-6 atmosphere.

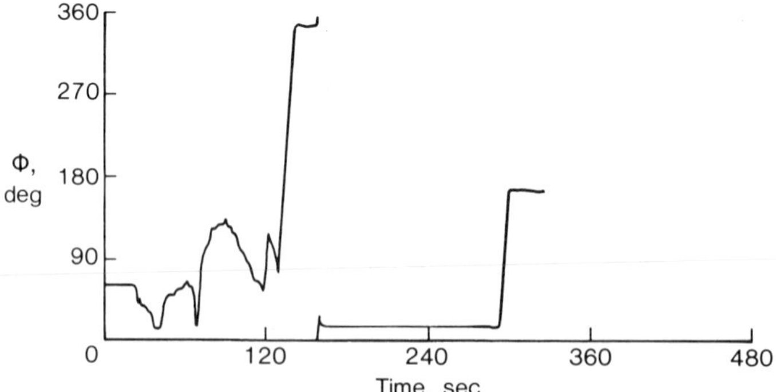

Fig. 15 Bank angle profile for low L/D configuration using adaptive guidance and nominal atmosphere.

Table 4 Comparison of atmospheric passes for moderate L/D configuration

	Predictive guidance atmosphere				Adaptive guidance atmosphere			
	Nominal	STS-2	STS-4	STS-6	Nominal	STS-2	STS-4	STS-6
Exit apogee, n.mi.	302.3	347.1	583.2	446.4	301.6	315.4	295.8	307.8
Exit perigee, n.mi.	35.6	34.9	46.1	45.5	42.36	37.2	41.53	35.1
Exit inclination, deg	28.4	27.9	28.7	28.4	28.47	28.54	28.47	28.49
Minimum altitude, ft	223,473	221,161	222,617	222,910	235,432	231,754	233,172	234,206
Maximum dynamic pressure, psf	96.8	92.7	101.2	97.9	62.8	59.0	61.3	63.6
Maximum sensed acceleration, g	3.9	3.7	4.1	3.9	2.5	2.4	2.5	2.6
Maximum stagnation heating rates (1-ft-radius sphere), BTU/ft^2-s	267.7	263.4	274.9	269.9	228.4	219.6	229.0	229.0
Total stagnation heat load, BTU/ft^2	30,834	32,133	29,296	30,062	36,215	36,184	36,442	35,823
On-time for roll RCS thrusters, s	17.36	12.48	9.92	9.16	100.36	76.68	103.7	85.16
On-time for yaw RCS thrusters, s	32.72	25.44	21.36	18.24	122.88	96.08	127.7	105.44
Total atmospheric pass time, s	530	536	500	530	649	600	653	598

tion in this method is that the atmosphere from pullout to exit is the same as entry to pullout.

Adaptive Guidance Technique

This technique commanded a constant bank angle until the sensed acceleration reached 0.05 g (see Fig. 15). Then the vehicle was commanded to roll to place the lift vector up to keep the vehicle from plunging too deeply in the atmosphere and keep the maximum acceleration to about 2.5 g (see Tables 3 and 4). Approximately 100 s into the entry, the pullout occurred, and the bank angle was modulated to achieve the equilibrium glide condition. Just after pullout,the velocity decreased very rapidly, and the linear exit phase was entered.

The point (velocity) at which the exit phase was begun was varied, but for the nominal and three off-nominal atmospheres used in this preliminary study, the best orbital parameters at exit were obtained by initiating the exit phase very soon after pullout. Several other parameters in the guidance algorithm were also varied in this study so that the proper exit conditions would be reached with only a few bank reversals for the nominal atmosphere.

The guidance algorithm parameters used for the nominal atmosphere were then used for the off-nominal atmospheres. Table 3 shows that the exit apogee for the nominal atmosphere was close to the desired 300 n.mi. value, but the STS-2 and STS-6 off-nominal atmospheres resulted in significantly higher exit apogees for the low L/D configuration. Figures 15-18 show the bank angle time histories for the low L/D configuration for the nominal 1962 standard atmosphere and the off-nominal atmospheres. The bank angle time histories show that both of these trajectories had a significant lift vector

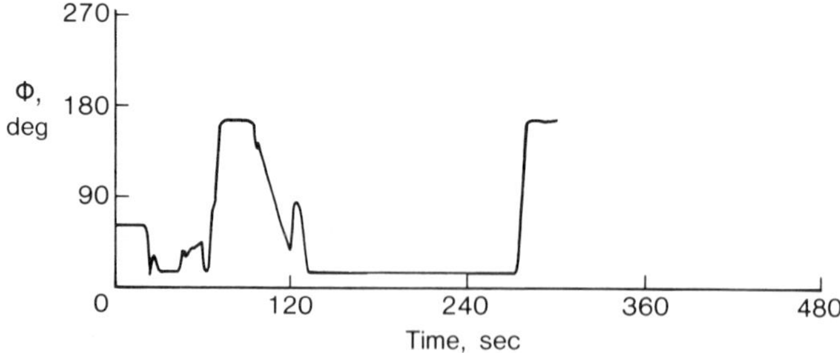

Fig. 16 Bank angle profile for low L/D configuration using adaptive guidance and STS-2 atmosphere.

up (ϕ = 15 deg) flight time during the exit phase. This was because the sensed drag error was weighted more heavily than the altitude rate error term in this portion of the trajectory; thus, the command was to climb higher in the atmosphere to reduce drag even though the final apogee will be too high. In spite of the differences in the atmospheres, the entry phase was very similar for all the trajectories for the low L/D configuration (see Table 3). The maximum acceleration, dynamic pressure, and minimum altitude were only slightly different for the four atmospheres.

The linear exit approach worked well for the moderate L/D configuration. The maximum apogee error at exit was 15.4 n.mi., and the maximum inclination error was 0.04 deg. (see Table 4). The bank angle profiles for nominal and off-nominal atmospheres are shown in Figs. 19-22.

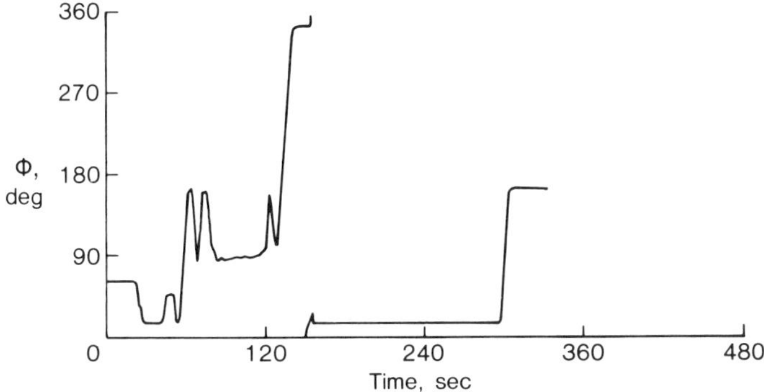

Fig. 17 Bank angle profile for low L/D configuration using adaptive guidance and STS-4 atmosphere.

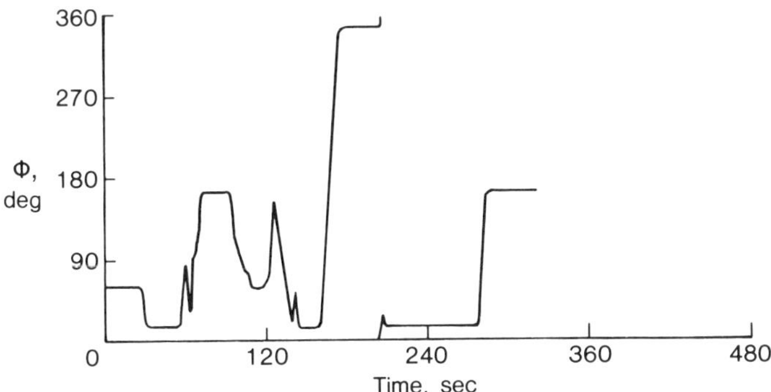

Fig. 18 Bank angle profile for low L/D configuration using adaptive guidance and STS-6 atmosphere.

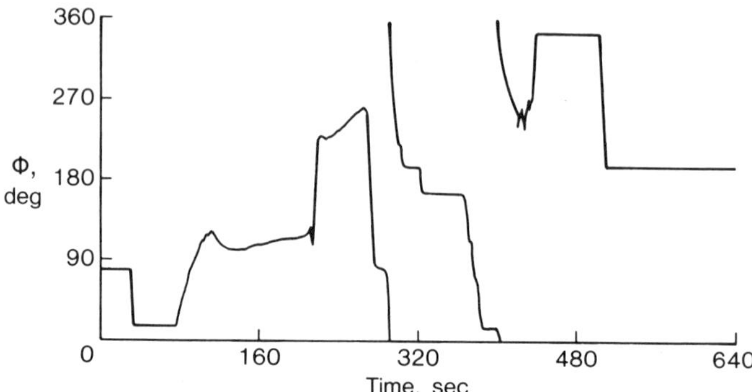

Fig. 19 Bank angle profile for moderate L/D configuration using adaptive guidance and nominal atmosphere.

The adaptive guidance algorithm does a good job of obtaining the desired exit inclination in all cases by using the multiple bank reversals. However, this use of multiple reversals does require much more RCS thruster on-time than using the predictive technique.

Comparison and Recommendations

Both guidance techniques fail to attain the desired orbital parameters for the low L/D configuration. Although further perfection of the adaptive technique for this configuration might improve the results somewhat, both techniques seemed to lack capability in the exit portion of the trajectory for a configuration that has very limited trajectory control authority. The linear exit phase approach of the adaptive technique was adequate for the moderate L/D configuration.

At pullout, the minimum altitudes are rather uniform for the various atmospheres. The maximum dynamic pressures and maximum stagnation heating rate to a 1-ft-radius sphere were also uniform and consistent for the two guidance techniques. The predictive technique should be modified to better limit maximum sensed acceleration for the moderate L/D configurations. The acceleration limiting of the adaptive technique did result in a longer flight time duration and significantly higher total heat loads for the moderate L/D configuration. Thus, a trade study of maximum sensed acceleration level vs total heat load is needed.

The adaptive technique used in this study was designed to minimize the number of bank reversals

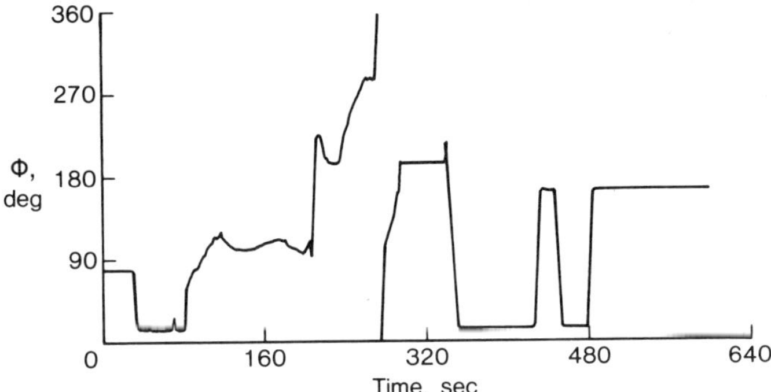

Fig. 20 Bank angle profile for moderate L/D configuration using adaptive guidance and STS-2 atmosphere.

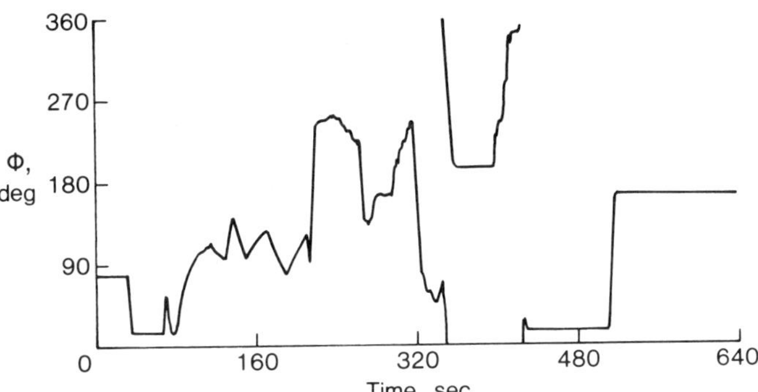

Fig. 21 Bank angle profile for moderate L/D configuration using adaptive guidance and STS-4 atmosphere.

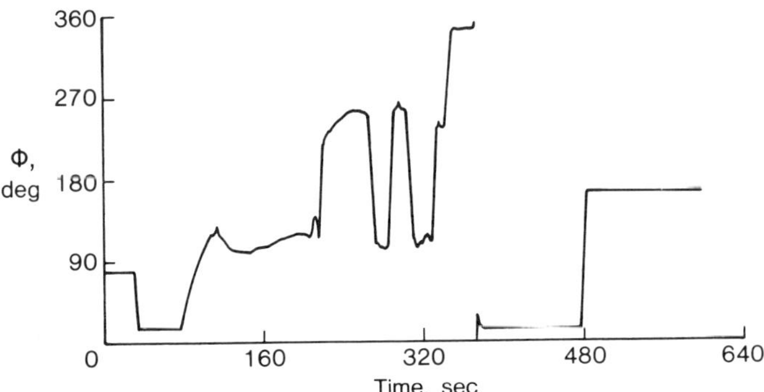

Fig. 22 Bank angle profile for moderate L/D configuration using adaptive guidance and STS-6 atmosphere.

required for inclination control, but, nevertheless, the amount of maneuvering was significantly greater than the predictive technique and resulted in relatively large on-times for the RCS thrusters.

References

[1]Walberg, G. D., "A Review of Aeroassisted Orbit Transfer," AIAA Paper 82-1378, AIAA 9th Atmospheric Flight Mechanics Conference, San Diego, California, Aug. 9-11, 1982.

[2]NASA Contract-NAS 8-33533; Orbital Transfer Vehicle (OTV) Concept Definition Study, Feb. 1981.

[3]NASA Contract-NAS 8-31452; Applications Study of Aero-Maneuvering Orbit-to-Orbit Shuttle (AMOOS), Jan. 1976.

[4]Wilhite, A. W., Arrington, J. P., and McCandless, R. S., "Performance Aerodynamics of Aeroassisted Orbital Transfer Vehicles," AIAA Paper 84-0406, AIAA 22nd Aerospace Sciences Meeting, Reno, Nevada, Jan. 1984.

[5]Findlay, J. T., Kelly, G. M., McConnell, J. G., and Compton, H. R., "Shuttle 'Challenger' Aerodynamic Performance from Flight Data Comparisons with Predicted Values and "Columbia" Experience," AIAA Paper 84-0485, AIAA 22nd Aerospace Sciences Meeting, Reno, Nevada, Jan. 1984.

[6]Blanchard, R. C. and Rutherford, J. F., "The Shuttle Orbiter High Resolution Accelerometer Package (HiRAP): Preliminary Flight Results," AIAA Paper 84-0490, AIAA 22nd Aerospace Sciences Meeting, Reno, Nevada, Jan. 1984.

[7]Mahesh, J. K, Konar, A. F., and Ward, M. D., "Interactive Flight Control System Analysis Program" NASA CR-172352, June 1984.

[8]Hill, Oliver, "An Adaptive Guidance Logic for an Aeroassisted Orbital Transfer Vehicle," AIAA Paper 83-357, ASA/AIAA Astrodynamics Specialist Conference, Lake Placid, New York, Aug. 1983.

Performance Aerodynamics of Aeroassisted Orbital Transfer Vehicles

Alan W. Wilhite,* J. P. Arrington,† and R. S. McCandless*
NASA Langley Research Center, Hampton, Virginia

Abstract

A method for predicting the performance aerodynamics of aeroassisted orbital transfer vehicles was developed based on techniques that were used in the aerodynamic data book of the Space Shuttle Orbiter and theories from the Hypersonic Arbitrary Body Program. The method spans the entire flight profile of the aeroassisted orbital transfer vehicles from the extreme high-altitude noncontinuum regime to the highly viscous continuum regime. Results from this method are compared with flight data from the Shuttle Orbiter, Apollo Capsule, and Viking Aeroshell. Finally, performance aerodynamics are estimated for three aero-assisted orbital transfer vehicles that range from low to high lift-to-drag ratio configurations.

Nomenclature

AOTV = aeroassisted orbital transfer vehicle
C'_∞ = Chapman-Rubesin viscosity coefficient
C_A = axial coefficient=axial force/$q_\infty S$
C_D = drag coefficient=drag force/$q_\infty S$
C_L = lift coefficient=lift force/$q_\infty S$
C_N = normal coefficient=normal force/$q_\infty S$
FMF = free molecular flow
HAB = Hypersonic Arbitrary Body
K = modified Newtonian coefficient
K_n = Knudsen number=$\lambda/1$
L/D = lift-to-drag ratio=C_L/C_D

Presented as Paper 84-0405 at AIAA 22nd Aerospace Sciences Meeting, Reno, Nevada, January 9-12, 1984.

*Aerospace Engineer, Space Systems Division.
†Head, Vehicle Analysis Branch, Space Systems Division.

l = reference length
M_∞ = freestream Mach number
q_∞ = freestream dynamic pressure
R = gas constant
R_∞ = freestream Reynolds number
S = reference area
T_w = wall temperature
T_∞ = freestream temperature
T' = reference temperature
V_∞ = freestream velocity
$\overline{V}'_\infty$ = viscous correlation parameter=$M_\infty\sqrt{C'_\infty/R_\infty}$
α = angle of attack
γ = ratio of specific heats
λ = mean free path
μ_∞ = freestream viscosity
σ = energy accommodation coefficient

Introduction

The Space Transportation System is now an operational system for delivering payloads to low Earth orbit. Several of these payloads and many of the future payloads must be transferred to higher energy orbits that require an orbital transportation system. Today, this function has been performed by expendable stages that use either solid or liquid propellants.

In order to support the deployment of a large number of satellites in geosynchronous orbit in an economical manner and to ultimately provide manned service, orbit transfer stages must be reusable and be able to return to low Earth orbit. Numerous studies have shown that the payload can be substantially increased if the atmosphere is used to reduce orbital energy with drag and to change inclination with lift rather than using all propulsive systems.[1] These aeroassisted orbital transfer vehicle (AOTV) concepts have ranged from low L/D configurations (Gemini/Apollo/Viking performance) to high L/D configurations (greater than the Shuttle Orbiter).

Because the AOTV flight maneuvers are performed at high altitude, there is special concern about the high-altitude aerodynamics of AOTVs. Several studies have unrealistically assumed that maximum L/D is constant throughout the flight profile. Actually, the maximum L/D degrades with increasing altitude, as shown by Maslen.[2]

Maslen estimated the variations in maximum L/D by using approximate methods to estimate lift and drag as a function of bluntness and viscosity. The degradation in maximum L/D was overpredicted with this method by as much as a factor of 3 as compared with flight data from the Shuttle Orbiter.[1]

Because of these predictions, there has been concern about the degradation of aerodynamic performance in the flight environment.[3] The purpose of this paper is to develop a method of estimating performance of AOTV configurations through their flight envelopes from noncontinuum, extreme altitude conditions through the continuum in the viscous-interaction regime.

AOTV Configurations

Three AOTV configurations from previous studies have been chosen to cover the range of expected L/D performance for aerodynamic analysis.

The first AOTV configuration (Fig. 1) is called an aerobrake.[4] It consists of a payload, propulsion, and miscellaneous subsystems that are packaged in a cylindrical structure. A deployable aerobrake, like a large umbrella, is used for deceleration and inclination change by utilizing drag and lift of the aerobrake at low angles of attack. Stable trim is maintained by an offset center-of-gravity location or by tilting the aerobrake relative to the cylindrical body. The aerobrake is considered a low L/D concept with an experimental L/D of 0.25 at 15 deg angle of attack.

The second AOTV configuration (Fig. 2) is called the Aeromaneuvering Orbit-to-Orbit Shuttle (AMOOS).[5] Its geometry is an ellipsoidal cylinder with a raked-off nose and an aft mounted trim flap (not shown). The nose was designed for stagnation heating with an ablative heat

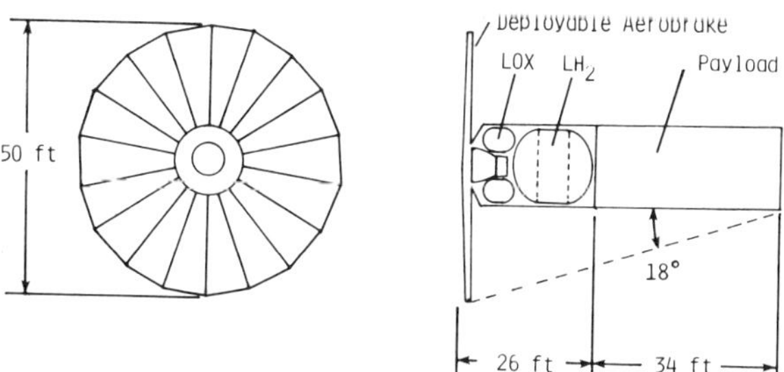

Fig. 1 Low L/D configuration.

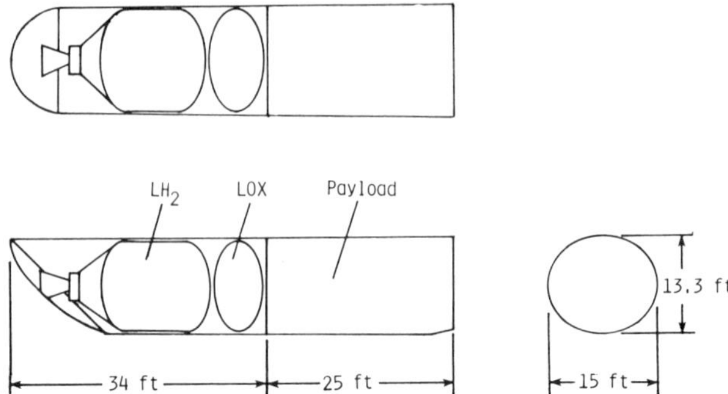

Fig. 2 Moderate L/D configuration.

shield. The AMOOS is considered a moderate L/D concept with an experimental L/D of 0.6 at 35 deg angle of attack.

The third AOTV configuration (Fig. 3), which has an estimated L/D of 2.18 at 11 deg angle of attack, was designed for a high L/D mission (large inclination change requirements). For high L/D capability, the liquid oxygen is stored in two separate tanks to provide a tapered nose, and inflated chines are used to continue this taper along the body. A large deployable flap is needed to trim the vehicle at low angles of attack for maximum L/D performance.

For each of these configurations, aerodynamics have been estimated through their flight regime based on the following analysis.

Aerodynamic Analysis

In order to predict the performance aerodynamics of AOTVs, the flight regime must be established. As shown in Fig. 4, the flight regime of a moderate L/D configuration[7] is similar to the high-altitude re-entry of the Space Shuttle Orbiter[8] and the Apollo Capsule.[9] The AOTV enters the atmosphere, maneuvers for deceleration and plane change, and exits the atmosphere for orbit circularization and launch vehicle rendezvous. Although the AOTV trajectory is similar to that of the Apollo Capsule, the techniques used to correlate and predict the aerodynamics of the Space Shuttle Orbiter have been adopted because the aerodynamic data base has been established by more wind tunnel hours and theoretical analyses than any other aerospace vehicle. In addition, high-quality flight data exist through the entire trajectory, especially in the high-altitude regimes.

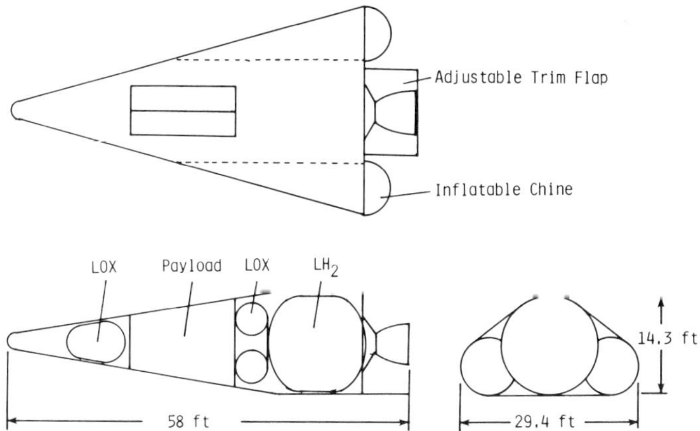

Fig. 3 High L/D configuration.

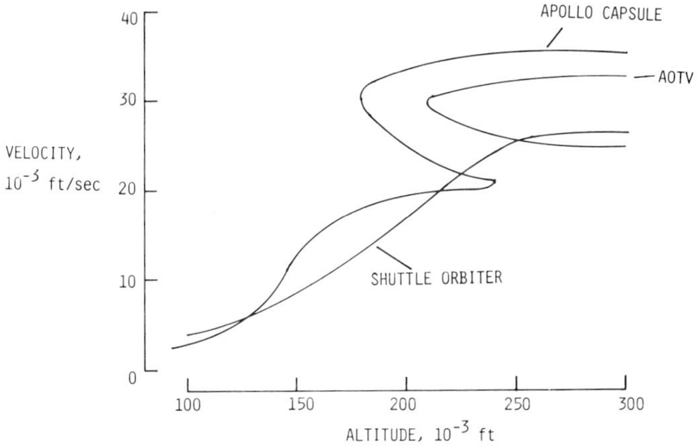

Fig. 4 Re-entry profile comparisons.

As shown in Fig. 5, the trajectory of the orbiter is divided into four separate regimes: the continuum hypersonic and viscous-interaction regimes, and the noncontinuum transition and free molecule regimes. The flow in the hypersonic regime can be characterized by a thin viscous layer near the body and an inviscid flow beyond the viscous layer that can be analyzed separately. In this flight regime, the aerodynamics of the orbiter were established by wind tunnel tests because the similarity parameters, M_∞ and R_∞, could be closely matched.

The viscous-interaction region occurs at high Mach numbers and low Reynolds numbers, where the mutual interaction of the boundary layer and inviscid flowfield

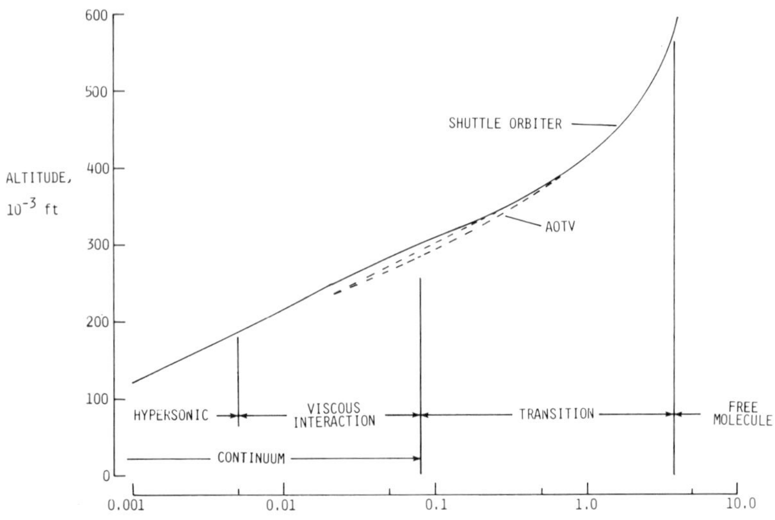

Fig. 5 Flight regimes of the Shuttle Orbiter and AOTV.

have a large effect on skin friction and surface pressures. In the Shuttle aerodynamic data book, wind tunnel tests and analytical results were correlated and extrapolated because the similarity parameters could not be matched exactly in ground facilities (Fig. 6). The parameter used for the viscous correlation was $\overline{V}'_\infty$ (Ref. 10). It is used to correct freestream conditions to the conditions behind the strong shock that exists in this regime. As shown in Fig. 7, this parameter correlates the calculated axial force on a 15-deg cone throughout the continuum regime better than other correlation parameters that have been used in the past ($0.005 < \overline{V}'_\infty < 0.08$). Reference 10 reviews the use of the parameter $\overline{V}'_\infty$ for correlating forces and moments on the Shuttle Orbiter.

The parameter $\overline{V}'_\infty$ is given by

$$\overline{V}'_\infty = M_\infty \sqrt{C'_\infty / R_\infty}$$

by the T-prime method[11,12]:

$$C'_\infty = (\mu'/\mu_\infty)\,(T_\infty/T')$$

and

$$\frac{T'}{T_\infty} = 0.468 + 0.532\,\frac{Tw}{T_\infty} + 0.195\left(\frac{\gamma-1}{2}\right)M_\infty^2$$

using Monaghan's constants[13] and a Prandl number of one. By using Keye's three-constant modified formula[14] for the

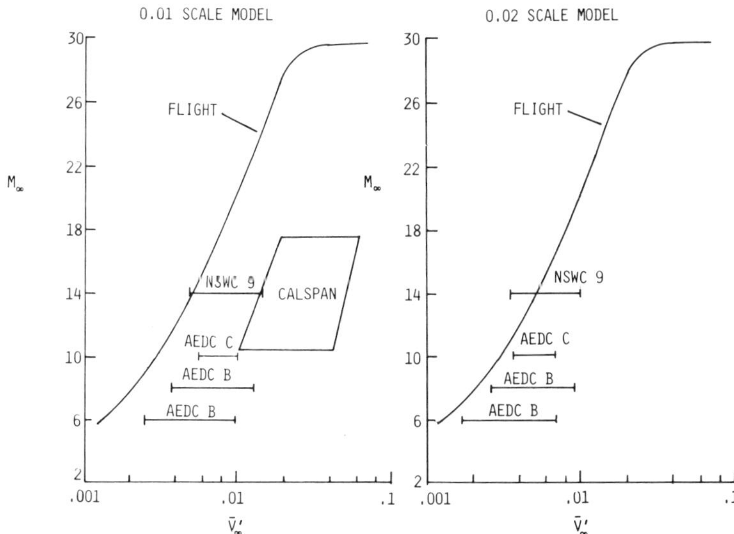

Fig. 6 Shuttle Orbiter flight profile and ground-test facility capabilities.

viscosity of air, C'_∞ can be expressed as

$$C'_\infty = \left(\frac{T'}{T_\infty}\right)^{0.5} \frac{T_\infty + (122.1)\mathrm{x}10^{-5/T_\infty}}{T' + (122.1)\mathrm{x}10^{-5/T'}}$$

where temperature is in degrees Kelvin. In the orbiter aerodynamic data book, T_w and γ were assumed to be 2459° R (1366 K) and 1.15°, respectively. For altitudes above 200,000 ft, γ can range from

$$1.10 < \gamma < 1.17$$

depending on the character of the shock (Fig. 8) based on equilibrium real-gas solutions for normal and oblique shocks at Shuttle Orbiter entry trajectory conditions.[15,16] The value of γ varies with the level of energy absorbed by the air molecules that range from complete molecular ionization at extremely high temperatures, through the dissociation of the nitrogen and oxygen molecules, to simple molecular vibration at lower temperatures.[17]

No aerodynamic test data were available for the orbiter in the free molecular flow (FMF) regime where the mean free path of the air molecules is much larger than the reference

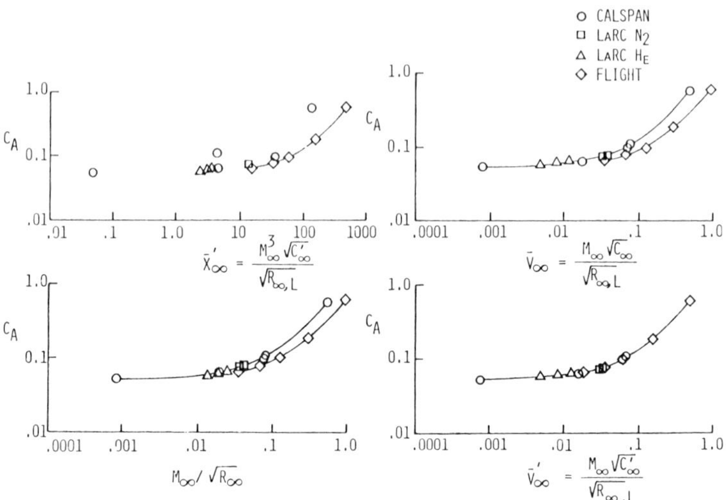

Fig. 7 Correlation of estimated Shuttle Orbiter axial-force coefficient based on a 15-deg cone axial force.

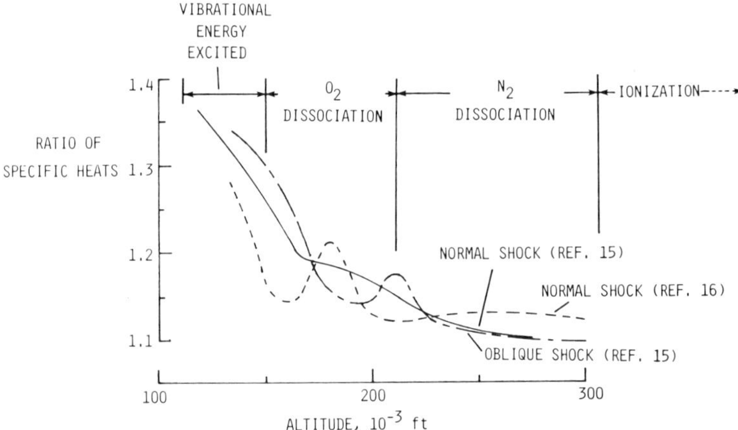

Fig. 8 Effect of high-altitude re-entry on the ratio of specific heats.

length of the vehicle (a factor of 10). The FMF aerodynamic characteristics were predicted by the Hypersonic Arbitrary Body (HAB) Program[18] in the orbiter data book. Because the mean aerodynamic chord was used for the reference length (39.56 ft), the transition regime starts at 272,000 ft (K_n = 0.001), and the free molecular regime starts at 582,000 ft (K_n = 10.0). The following flight conditions were assumed in the data book:

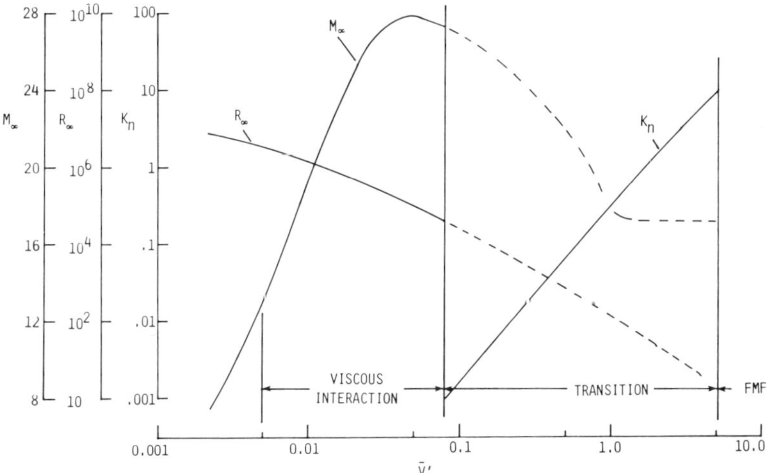

Fig. 9 Variation of similarity parameters for the Shuttle Orbiter re-entry.

$$V_\infty/\sqrt{2RT_w} = 9.03$$

and

$$T_w/T_\infty = 0.25$$

The orbital aerodynamics were based on completely diffused reflection; thus an accommodation factor, σ, of 1 for both normal and tangential directions was used.

Because there were no test data or analytical data available in the transition regime for the orbiter, aerodynamic data were predicted by interpolating between the viscous-interaction regime and the free molecular regime using the Lockheed bridging formula[19]:

$$C_T = C_C + (C_F - C_C) \sin^2 [\pi(A + B \log K_n)]$$

where C_T is the transition coefficient data;
C_C is the viscous data at $\overline{V}'_\infty = 0.08$ and $K_n = 0.001$;
C_F is the free molecular data at $\overline{V}'_\infty = 5.3$ and $K_n = 10.0$;
$A = 3/8$ and $B=1/8$.

As shown in Fig. 9, $\log K_n$ is nearly linear with the $\log \overline{V}'_\infty$. With this linearity, the results for the present aerodynamics are conveniently presented as a function of one parameter $\overline{V}'_\infty$ instead of a combination of $\overline{V}'_\infty$ and K_n

or altitude as in the orbiter aerodynamic data book. Thus, the transition regime begins at a value of $\overline{V}'_\infty$ that corresponds to $K_n = 0.001$ and ends at a $\overline{V}'_\infty$ that corresponds to $K_n = 10.0$.

In the noncontinuum regime, the parameters M_∞, R_∞, and $\overline{V}'_\infty$ have little physical meaning but can still be computed using the 1962 standard atmosphere and the velocity of the vehicle. To preserve the linearity between $\overline{V}'_\infty$ and K_n, the temperature above 400,000 ft is held constant instead of increasing as given in the standard atmosphere.

For the aerodynamic analysis of the three AOTV configurations, the Aerodynamic Preliminary Analysis System (APAS) that incorporates a modified version of the HAB Program was used.[20] The orbiter and moderate and high L/D configurations were analyzed using modified Newtonian theory with skin friction. The skin friction theory models the viscous-interaction regime by using an iterative scheme that adjusts the boundary layer and the boundary conditions for the inviscid flow due to their mutual interactions. The value for the Newtonian constant K was determined by

$$K = 2\ (\gamma+1)\ (\gamma+7)/\ (\gamma+3)^2$$

as given in Ref. 21.

For the low L/D AOTV configurations, the modified Newtonian and viscous blunt body theories were used. The viscous blunt body theory accounts for a loss of lift (as much as 40%) because of viscous shear on the base of the vehicle in the opposite direction of lift (Fig. 10). This theory was developed to explain the error on predicted splashdown positions in the early flights of the Gemini Capsule.[22] This theory is compared to flight data from the Apollo Capsule and the Viking Aeroshell.

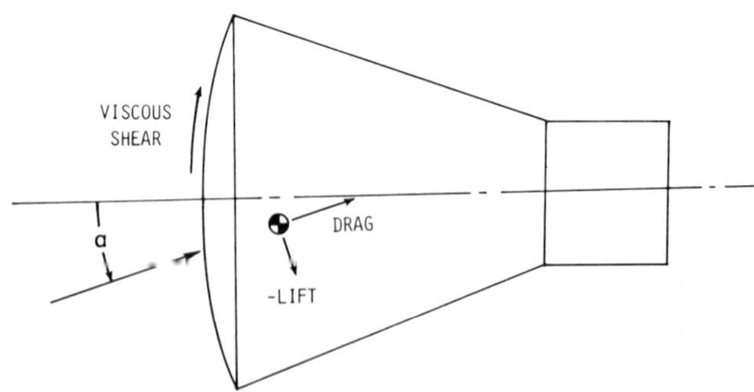

Fig. 10 Aerodynamics forces on the Gemini Capsule.

Analytical/Flight Comparisons

The results of applying the theories in the HAB Program to the methods in the Shuttle aerodynamic data book are compared with flight results for the Shuttle Orbiter, Apollo Capsule, and Viking Aeroshell. These vehicles are representative of high and low L/D configurations.

For the Shuttle Orbiter, the measured data are from flight 6, which used the inertial measurement units (IMU) for a major portion of the re-entry (from 300,000 ft to landing)[23] and the high resolution accelerometer package (HIRAP)[24] for the orbital and rarefied flow regimes where external forces are extremely small.

The L/D data from the flight are compared with the aerodynamic data book and theoretical estimates at 40 deg angle of attack (Fig. 11). The data book altitude where the viscous-interaction and transition regimes merge was rounded off from an altitude of 272,000 ft (K_n = 0.001) to 300,000 ft. As shown in Fig. 11, the data book values were faired through the assumed correct altitude (272,000 ft at $\overline{V}'_\infty$ = 0.08), thus providing a good comparison. The HAB comparison is also quite good in this regime. This figure shows a substantial reduction in L/D (from 1.08 to 0.90) in just the viscous-interaction regime ($0.005 < \overline{V}'_\infty < 0.08$), and a substantial reduction from the continuum regime through the noncontinuum regime. The L/D error using the inviscid Newtonian analysis is approximately 20% at $\overline{V}'_\infty$ = 0.08.

In the free molecular regime, the flight L/D was higher than that estimated in the data book by the free molecular theory in HAB with an energy accommodation coefficient σ of 1.0. With σ=0.94, the HAB analysis matches flight data. In Ref. 24, the flight data matched the FMF theory of Hulbert and Sherman[25] for the theoretical value of σ=1.0.

The Lockheed bridging formula for interpolating between the theories at the viscous-interaction and FMF regimes compared quite well with the flight data. The L/D curve was generated by bridging the normal- and axial-coefficient curves and then transforming to the wind axes for lift and drag rather than bridging the L/D data directly. Bridging L/D data directly results in erroneous values.

At maximum L/D (near 20 deg angle of attack), a larger reduction in L/D occurs than that at flight conditions (Fig. 12). In the viscous-interaction regime, this reduction is from 1.85 down to 1.14. The reduction from the inviscid Newtonian analysis to Newtonian with skin friction is approximately 43%. Thus, for the aerodynamics

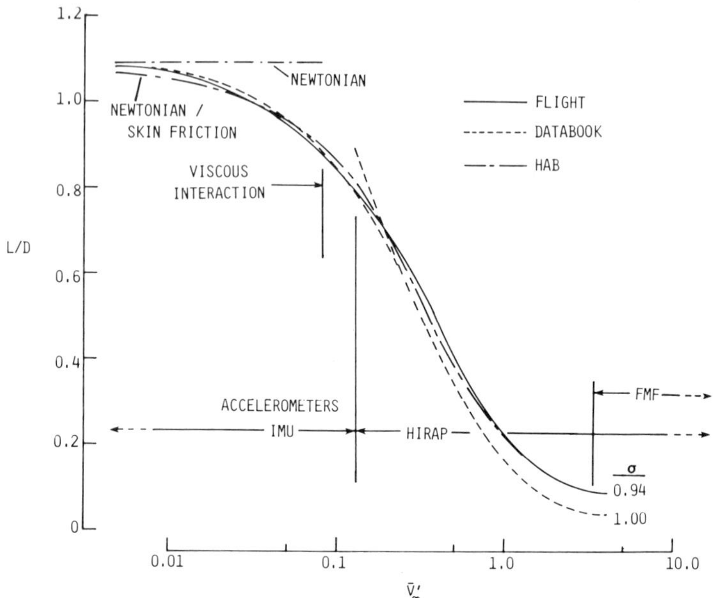

Fig. 11 Flight and prediction comparisons for the Shuttle Orbiter.

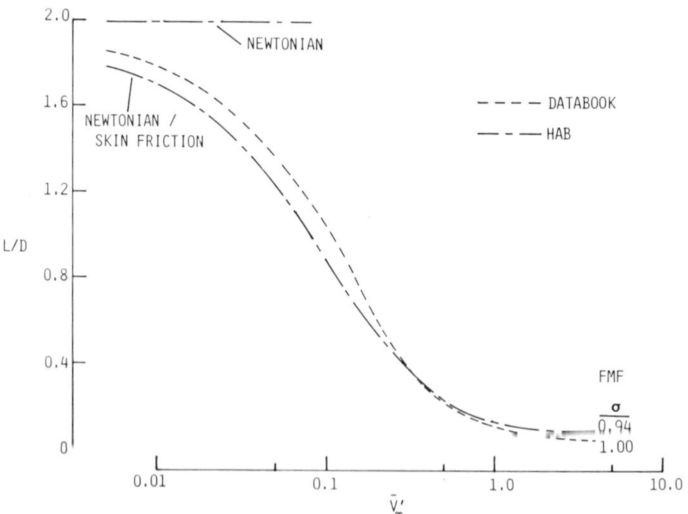

Fig. 12 Comparison of the Shuttle Orbiter data book to HAB predictions for maximum L/D.

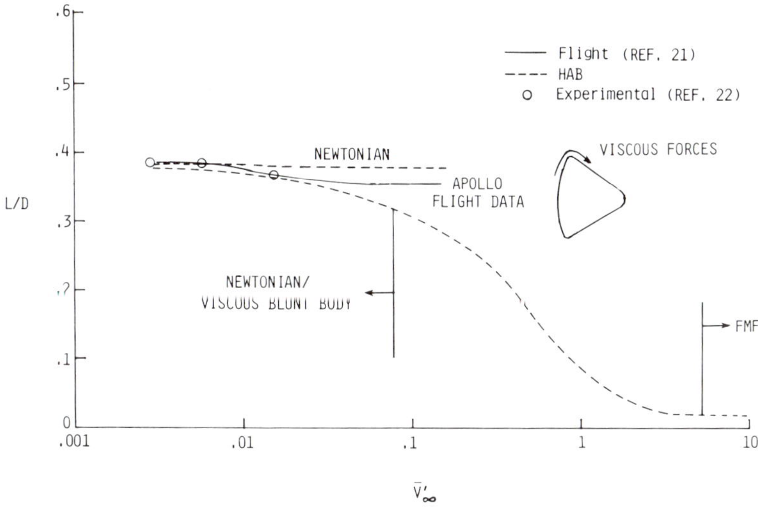

Fig. 13 Comparison of HAB estimates with Apollo flight and wind tunnel data.

of AOTVs, the viscous effects must be considered, because the majority of their flight trajectory is in the viscous-interaction regime. It should be noted that the HAB data have the same trends as the data book data in the viscous-interaction regime, but the predicted value of L/D was approximately 4% less than the data book value.

For low L/D configurations, the Newtonian/viscous blunt body theory is compared with the Apollo Capsule flight data[9] (Fig. 13). From a $\overline{V}'_\infty$ of 0.005 to 0.01, the data agree reasonably well. The reduction in L/D in this range is caused by a 2-deg reduction in angle of attack. In the viscous regime ($\overline{V}'_\infty \approx 0.08$), the flight data remain constant, but the theoretical data decrease because of the viscous effects. For the Gemini Capsule, on which the the blunt body viscous effects were modeled, the flow separates at the base edge. For the Apollo Capsule at these low Reynolds numbers, the flow remains attached, and the viscous shear is recovered on the afterbody.[26] Thus, for blunt bodies with a faired afterbody similar to the Apollo capsule, the viscous effects on lift are negligible.

A second source of flight data on a low L/D configuration is the Viking Aeroshell. As shown in Fig. 14, drag decreases in the viscous-interaction regime with increasing altitude at an altitude of 70,000 ft. On the other hand, the viscous blunt body theory shows an increase in drag that is expected because Reynolds number is decreasing. The decrease in drag was discussed in Ref. 27, but no definitive reason was given.

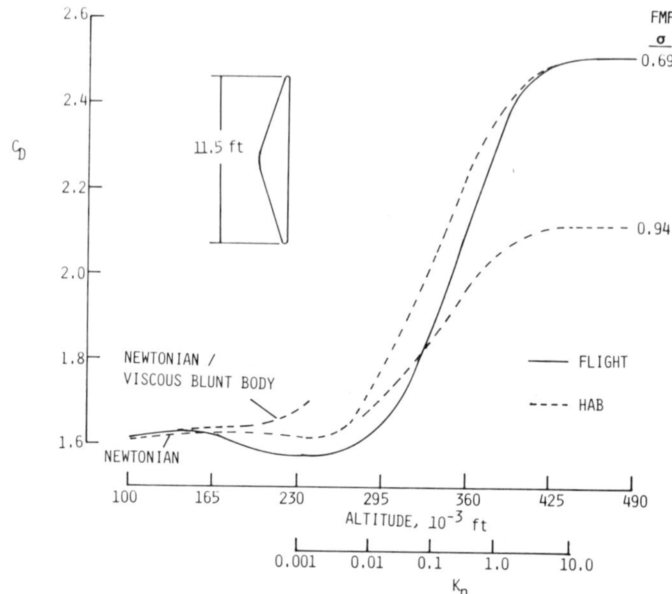

Fig. 14 Comparison of HAB estimates to Viking Aeroshell flight data.

The Lockheed bridging formula was used with the Newtonian data and the flight FMF data. As with the Shuttle flight data, the bridging formula gives excellent trends and a reasonable estimate. As with any empirical formula, the flight data could have been matched by adjusting the constants in the bridging formula.

To match the flight FMF regime, a σ of 0.69 was used with the HAB theory. In Ref. 25, a σ of only 0.98 was needed with the Hulbert and Sherman theory[25] to match the flight results. Thus, with the Viking Aeroshell and Shuttle Orbiter data comparisons, the Hulbert and Sherman theory provides a much better basis for predicting aerodynamics of various shapes in the FMF regime.

AOTV Performance Aerodynamics

The performance aerodynamics of the three AOTV configurations are presented in Figs. 15-17. For each of these configurations, there are significant changes in all the aerodynamic coefficients through the speed regimes.

Based on previous studies, AOTVs fly near maximum L/D for best performance. For the low L/D configuration, maximum L/D occurs above 40 deg angle of attack, but, as shown geometrically in Fig. 1, the flow should impact the payload cylinder at approximately 18 deg. In heating tests of the configuration, the payload canister has heating that is 78, 64, 56, and 45% of stagnation at angles of attack of 30, 20, 15, and 10 deg, respectively.[28] In these tests, the separated flow off the aerobrake turned toward the base of the vehicle and reattached on the cylindrical afterbody and caused high heating areas. Thus, for these vehicles, the attitude should be restricted to low angles of attack

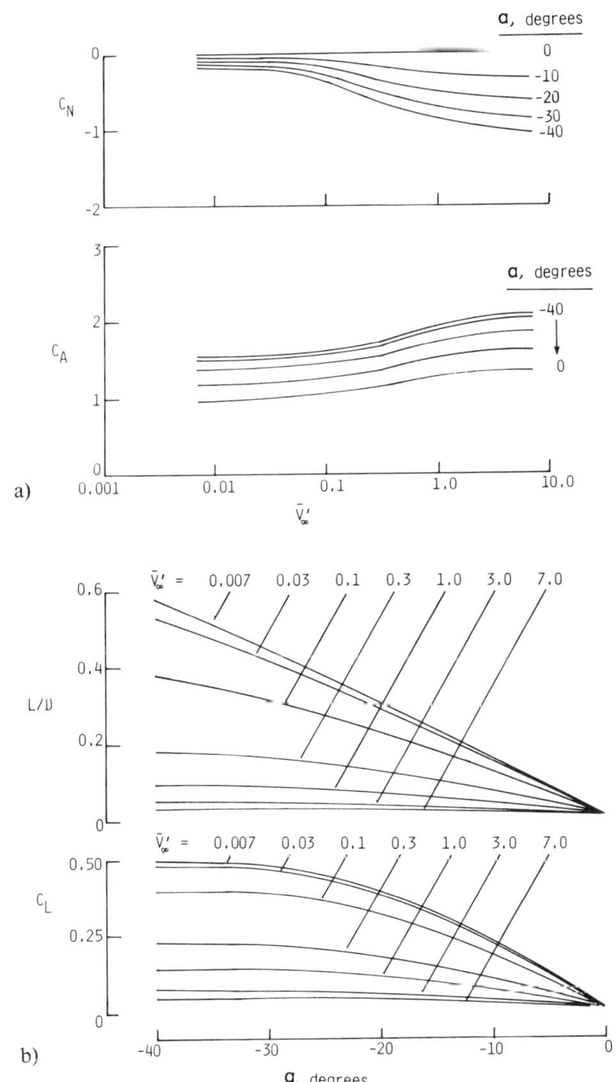

Fig. 15 Performance aerodynamics of the low L/D configuration.

(approximately 15 deg), giving a maximum L/D of 0.22 instead of an L/D that approaches the moderate L/D configuration. The moderate L/D configuration has a maximum L/D of 0.65 at 20 deg angle of attack, and the high L/D configuration has a maximum L/D of 1.71 at 10 deg angle of attack.

The L/D values along the flight profile for the three configurations are presented in Fig. 18. As expected, the high L/D configuration loses a much greater percentage of its performance in the transition and FMF regimes than the other configurations. The L/Ds for the high, moderate, and low L/D configurations were reduced by 95, 88, and 64%,

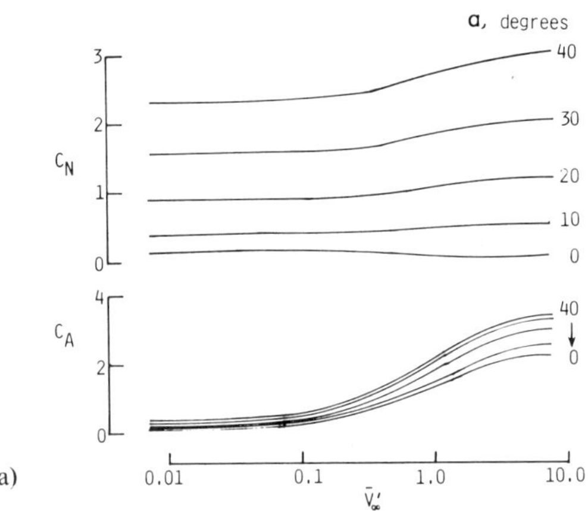

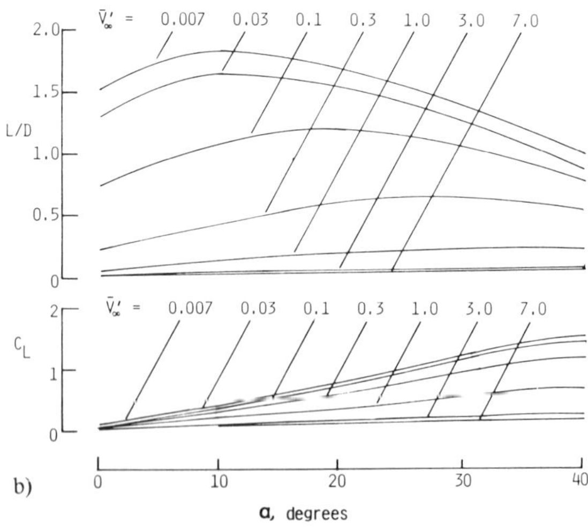

Fig. 16 Performance aerodynamics of the moderate L/D configuration.

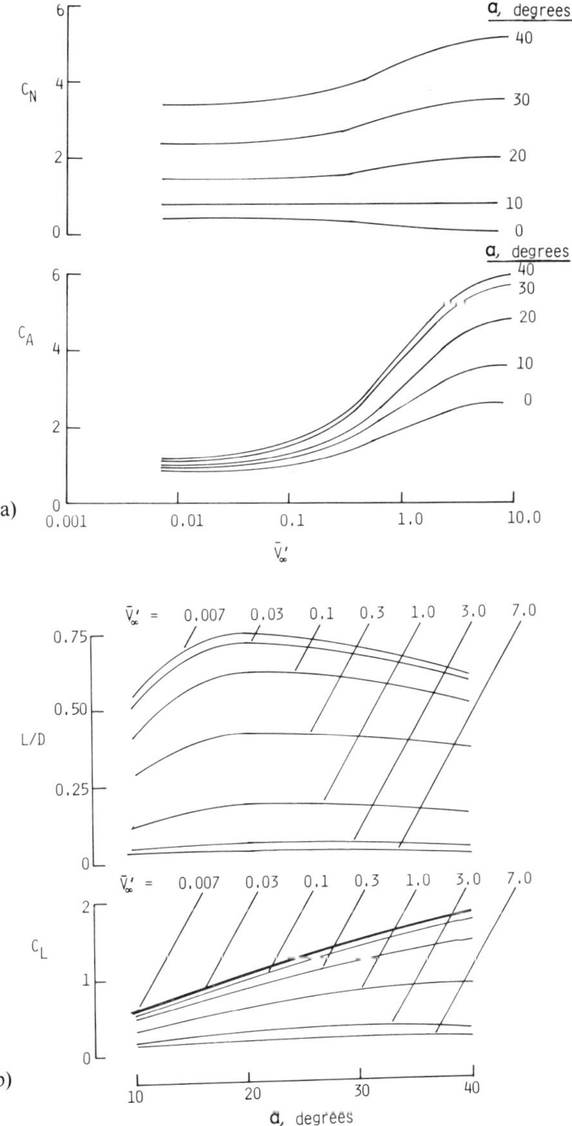

Fig. 17 Performance aerodynamics of the high L/D configuration.

respectively. Reference 7 evaluated the trajectory performance of these three configurations using the present aerodynamics. The performance results showed little impact on the trajectory using the transition and FMF aerodynamics because much of the trajectory was performed in the viscous-interaction regime. Thus, the severe degradation in aerodynamic performance in the noncontinuum regime does

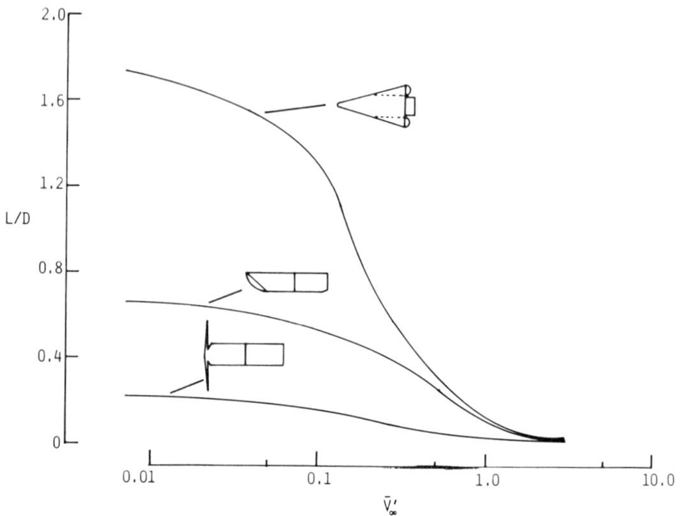

Fig. 18 Aerodynamic comparison of the three AOTV configurations.

not significantly impact the overall performance of the vehicles.

Conclusions

A method was developed to estimate the performance aerodynamics of AOTV configurations through the flight regime based on Space Shuttle experience and theories from the Hypersonic Arbitrary Body Program. The theoretical results showed excellent agreement for flight data comparisons for the Shuttle Orbiter throughout the speed regime. Correct trends were estimated for maximum L/D as compared with the Shuttle Orbiter aerodynamic data book, but the absolute values were underestimated by approximately 4%.

For the Apollo Capsule (blunt base with a faired afterbody), the modified Newtonian theory without blunt body viscous effects correlated well with flight data. The same results apply to the Viking Aeroshell.

The Lockheed bridging formula for interpolating the aerodynamics between the viscous-interaction regime and the free molecular flow regime agreed well with both the Shuttle Orbiter and Viking Aeroshell flight data.

The Hulbert and Sherman free molecular flow theory seems to predict the aerodynamics of various shapes much better than the theory in the Hypersonic Arbitrary Body Program.

Finally, the L/Ds of the high, moderate, and low configurations were reduced by 95, 88, and 64%, respectively through the flight envelopes from the continuum hypersonic regime to the noncontinuum free molecular flow regime.

References

[1]Walberg, G. D., "A Review of Aeroassisted Orbit Transfer," AIAA Paper 82-1378, 9th Atmospheric Flight Mechanics Conference, San Diego, Calif., Aug. 9-11, 1982.

[2]Maslen, S. H., "Synergetic Turns With Variable Aerodyanmics," Journal of Spacecraft and Rockets, Vol. 4, Nov. 1967, pp. 1475-1482.

[3]Cruz, M. I., "Trajectory Optimization and Closed Loop Guidance of Aeroassisted Orbital Transfer," AAS Paper 83-413, Astrodynamics Specialists Conference, Lake Placid, N.Y., Aug, 22-25, 1983.

[4]Heald, D. A. "Is a Versatile Orbit Transfer Stage Feasible: Orbit Transfer Concepts, Potential Missions and Evolution," AIAA Paper 79-08666, Conference on Advanced Technology for Future Space Systems, Hampton, Va., May 1979.

[5]White, J., "Applications Study of Aeromaneuvering Orbit-to Orbit Shuttle," LMSC-HREC TR D496705, Lockheed, Huntsville, Al., Contract NAS8-31452, Jan. 1976.

[6]Andrews, D., "Technology Identification for Aero-Configured Orbital Transfer Vehicles," Boeing Aerospace, Seattle, Wa., Contract F33615-82-C-3014, April 1983.

[7]Talay, T. A., White, N. H., and Naftel, J. C., "Impact of Atmospheric Uncertainties and Real Gas Effects on the Performance of Aeroassisted Orbital Transfer Vehicles," AIAA Paper 84-0408, 22nd Aerospace Sciences Meeting, Reno, Nev., Jan. 9-12, 1984.

[8]Aerodynamic Design Data Book, Vol. 1 Orbiter Vehicle, STS-1, Rockwell International, Downey, Calif., SD72-SH-0060, Vol. 1M, 1980.

[9]Hillje, E. R., "Entry Aerodynamics at Lunar Return Conditions Obtained from the Flight of Apollo 4 (AS-501)," NASA TND-3599, Oct. 1969.

[10]Woods, W. C., Arrington, J. P., and Hamilton, H. H. II, "A Review of Preflight Estimates of Real-Gas Effects on Space Shuttle Aerodynamic Characteristics," NASA CP-2283, March 1983.

[11]Chapman, D. R. and Rubesin, M. W., "Temperature and Velocity Profiles in the Compressible Laminar Boundary

Layer with Arbitrary Distribution of Surface Temperature," Fluid Mechanics Session, 17th Annual Meeting of the IAS, New York, N.Y., Jan. 24-27, 1949.

[12]Bertram, M. H., "Hypersonic Laminar Viscous Interaction Effects on the Aerodynamics of Two-Dimensional Wedge and Triangular Planform Wings," NASA TN D-3523, Aug. 1966.

[13]Monaghan, R. J., "On the Behavior of Boundary Layers at Supersonic Speeds," Fifth International Aeronautical Conference, Los Angeles, Calif., June 20-23, 1955, pp. 277-315.

[14]Keyes, F. G., "The Heat Conductivity, Viscosity, Specific Heat, and Prandtl Number for Thirteen Gases," Project SQUID, Technical Report 37, MIT, Cambridge, Mass., April 1, 1952.

[15]Hunt and Souders, "Normal and Oblique Shock Flow Parameters in Equilibrium Air Including Attached Shock Solutions for Surfaces at Angles of Attack, Sweep, and Dihedral," NASA SP-3093, July 1975.

[16]Trimpi, R. L. and Jones, R. A., "A Method of Solution with Tabulated Results for the Attached Oblique Shock-Wave System for Surfaces of Various Angles of Attack, Sweep, and Dihedral in an Equilibrium Real Gas Including the Atmosphere," NASA TR R-63, October 1959.

[17]Kuethe, A. M. and Schetzer, I. D., Foundations of Aerodynamics, John Wiley and Sons, New York, 1959, p. 385.

[18]Gentry, A. E., "The Mark IV Supersonic-Hypersonic Arbitrary Body Program," AFFDL-TR-73-159, Nov. 1973.

[19]Warr, J., "Orbital Aerodynamic Computer Program to Calculate Force and Moment Coefficients on Complex Vehicle Configurations," LMSC/HREC D162498, TM 54-20-275, Lockheed, Houston, Tx., Aug. 1970.

[20]Divan, P., "Aerodynamic Preliminary Analysis System II Part II, User's Manual," NASA CR 165628, April 1981.

[21]Love, E. S., Woods, W. C., and Rainey, R. W., "Some Topics in Hypersonic Body Shaping," AIAA Paper 69-181, 7th Aerospace Sciences Meeting, New York, N.Y., Jan. 20-22, 1969.

[22]Goldberg, L., "Forces and Moments on the Front Face of a Blunt Lifting Re-entry Vehicle," AIAA Paper 66-464, 4th Aerospace Sciences Meeting, Los Angeles, Calif., June 27-29, 1966.

[23]Findlay, J. T., Kelley, G. M., Heck, M. L.; and McConnel, J. G., "Results From the First Entry Flight of the NASA Space Shuttle Challenger, STS-6," AMA Report 83-9, Contract NAS1-16087, Analytical Mechanics Assoc., Hampton, Va., June 1983.

[24]Blanchard, R. C. and Rutherford, J. F., "The Shuttle Orbiter High Resolution Accelerometer Package Experiment: Preliminary Flight Results," AIAA Paper 84-0490, 22nd Aerospace Sciences Meeting, Reno, Nev., Jan. 9-12, 1984.

[25]Hulbut, F. C. and Sherman, F. S., "Application of the Nocilla Wall Reflection Model to Free-Molecule Kinetic Theory," Phys. Fluids, Vol. 11, March 1968, pp. 486-496.

[26]Harris, J. E., "Longitudinal Aerodynamic Characteristics of the Apollo Command Module at a Mach Number of 20 and Data Comparisons Over a Wide Mach Number-Reynolds Number Range," NASA TM X-1395, June 1967.

[27]Blanchard, R. C. and Walberg, G. D., "Determination of the Hypersonic Continuum/Rarefied-Flow Drag Coefficient of the Viking Lander Capsule 1 Aeroshell From Flight Data," NASA TP 1793, Dec. 1980.

[28]Hair, L. M. and Engel, C. D., "Aerothermal Test Data of Low L/D Aerobraking Orbital Transfer Vehicle at Mach 10," Remtech Report 069-1, Contract NAS8-34590, NASA LaRC Test 117, Remtech Inc., Huntsville, Ala., April 1983.

Multiple Pass Trajectories for an Aeroassisted Orbital Transfer Vehicle

John J. Rehder*
NASA Langley Research Center, Hampton, Virginia

Abstract

This paper investigates the use of multiple pass trajectories for aeroassisted orbital transfer vehicles (AOTVs) as a means of reducing the severe aeroheating environment and the extreme sensitivity to off-nominal trajectory and atmospheric conditions. One-, two-, and three-pass trajectories were calculated for AOTVs with ballistic coefficients ranging from 5 to 150 psf. Compared to the single-pass case, the maximum heat rate was reduced by 30% for two passes and 45% for three passes, while the maximum acceleration was reduced by 40% and 55%, respectively. The sensitivity of the trajectories to variations in atmospheric and orbital parameters was not significantly reduced.

Nomenclature

a_{max} = maximum acceleration, g
AOTV = aeroassisted orbital transfer vehicle
GEO = geosynchronous orbit
L/D = lift-to-drag ratio
LEO = low Earth orbit
POST = Program to Optimize Simulated Trajectories
W/C_DA = ballistic coefficient, psf
$\dot{q}_{max}$ = maximum reference convective stagnation point heat rate to a 1-ft-radius sphere, $BTU/ft^2 \cdot s$
n = number of atmospheric passes

Presented as Paper 84-0407 at AIAA Aerospace Sciences Meeting, Reno, Nev., Jan. 9-12, 1984.

*Aerospace Engineer, Space Systems Division.

Introduction

In the future, there will be a need to deliver large payloads to high orbits, such as geosynchronous (GEO). Although the current Space Transportation System can deliver 5000 lb to GEO, using an expendable upper stage, more payload and reusability are desirable in an orbital transfer vehicle (OTV). One way to achieve both of these goals is to use aerodynamic drag produced during a grazing pass through the atmosphere rather than a rocket engine to slow the vehicle upon return from high orbit, thereby reducing the propellant required to transport the payload.

Although the amount of round trip payloads to GEO can be doubled with aerobraking,[1] there are penalties associated with this technique. During the atmospheric pass, the vehicle experiences severe aeroheating, requiring the additional weight and complexity of a thermal protection system. Also, the trajectory is extremely sensitive to off-nominal conditions, such as variations in atmospheric density and navigational errors. This sensitivity presents a challenge to control system designers.

One possible means of alleviating these problems is to use more than one pass through the atmosphere, as illustrated in Fig. 1. The vehicle does not penetrate as deeply into the atmosphere during the first pass, and the apogee of the orbit does not decrease all the way down to low

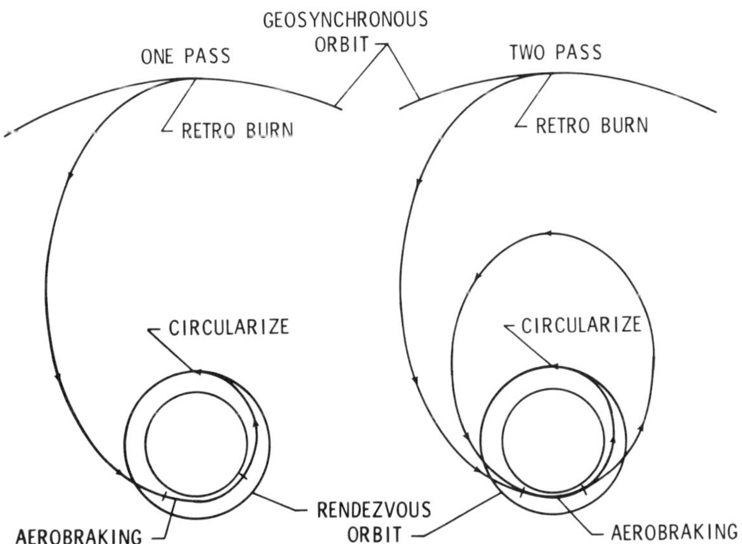

Fig. 1 Comparison of AOTV trajectories.

Earth orbit (LEO). The AOTV continues in this medium-altitude orbit until it re-enters the atmosphere for another pass. Although the figure shows a two-pass trajectory, any number of passes can be made as the apogee gradually is reduced down to LEO. During any given pass, the severity of the aerodynamic heating should be less than that experienced for the single-pass case. Also, having more than one pass gives an opportunity to correct errors in the trajectory due to off-nominal conditions.

Other studies have looked at multiple pass schemes for AOTVs. In one, where a drag-brake kit was added to an OTV, the vehicle weight was minimized for 30 passes.[2] The total flight time of the return from GEO was 6 days, which may be unacceptable, particularly for manned missions. Also, that many passes through the Van Allen radiation belts would be a concern. Another study determined that using more than one pass did not reduce the heat rate enough to save a significant amount of weight.[3] This conclusion was affected by the use of a nonoptimum multiple pass strategy and the decision to use an ablative heat shield.

Method

Depending on the assumptions one makes, such as plane-change requirements and predictability of the atmosphere, an almost limitless variety of AOTV configurations can be conceived. These range from umbrella-shaped drag brakes and inflatable ballutes to lifting bodies.[4] In this paper, the aerodynamic characteristics of the vehicle were

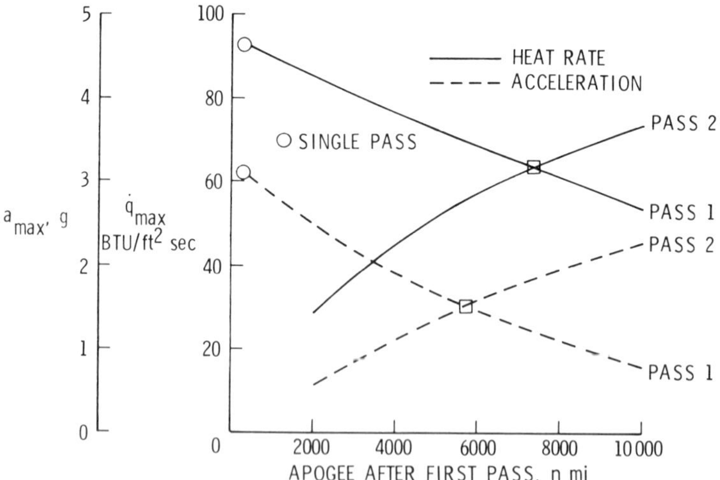

Fig. 2 Results of two-pass trajectories; $W/C_DA = 5$ psf.

varied parametrically to avoid tying the results to any particular configurations. The lift-to-drag ratio (L/D) and the ballistic coefficient W/C_DA were varied from 0 to 0.8 and from 5 to 150 psf, respectively.

The trajectories were calculated using the Program to Optimize Simulated Trajectories (POST).[5] A round trip mission from LEO to GEO was assumed with an initial vehicle weight of 68,000 lb. The number of atmospheric passes ranged from one to three. Before each pass, an engine burn was performed at the apogee of the orbit to give the target perigee for the pass that would yield the desired apogee after the pass. The length of the burn was selected by POST and depended on the desired apogee and the aerodynamic characteristics of the vehicle. Except for the initial burn at GEO, the burns were very small (< 1 ft/sec) and could be thought of as midcourse corrections. The apogee after the final pass was targeted to be 300 n.mi., where an engine burn was used to circularize the orbit. During each pass, the vehicle flew at a constant bank angle of 90 deg to maximize aerodynamic plane change and at a constant angle of attack. The stagnation point covective heat rate during the trajectory was calculated using Chapman's method.

Results

In the single-pass case, for a given W/C_DA, varying L/D had no effect on the main parameter of interest,

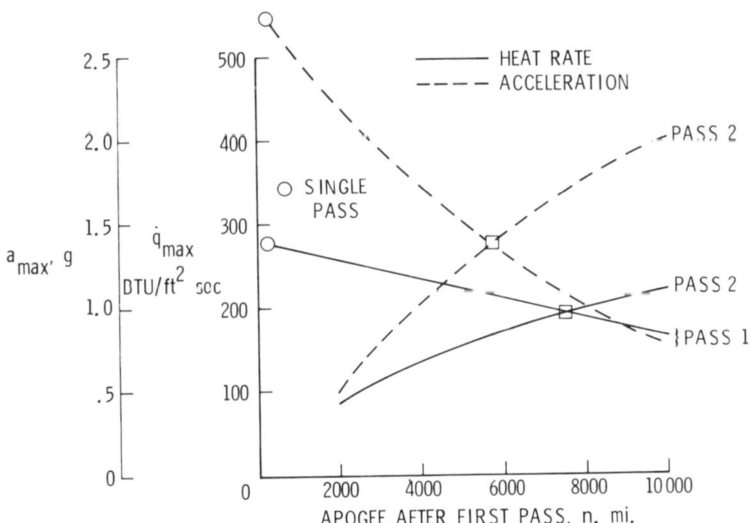

Fig. 3 Results of two-pass trajectories; $W/C_DA = 50$ psf.

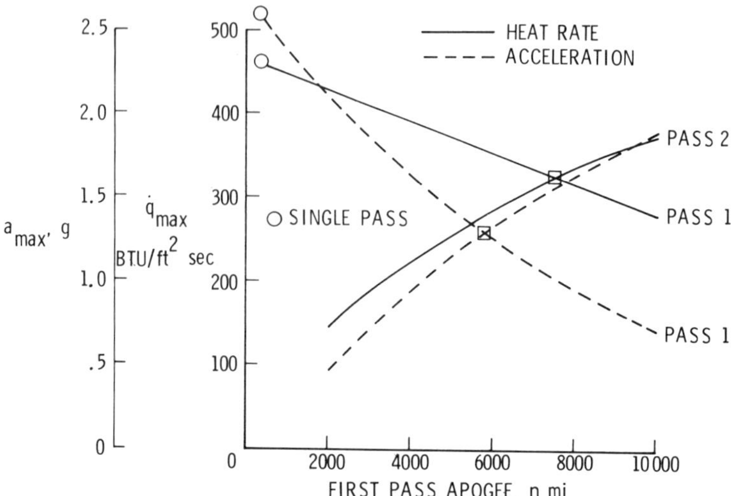

Fig. 4 Results of two-pass trajectories; W/C_DA = 150 psf.

stagnation point heat rate. Therefore, to reduce the number of trajectories to be calculated, L/D was kept constant at 0.8 for the remainder of the study. In the two-pass case, one of the parameters to be varied was the apogee after the first pass. Figure 2 shows the maximum heat rate and acceleration encountered during each of the passes as the apogee was varied for W/C_DA of 5 psf. As expected, the heat rate for the first pass decreased as the apogee after the pass increased, since less energy is dissipated. That leaves more energy to be dissipated, with a corresponding increase in heat rate, during the second pass. The minimum value for $\dot{q}_{max}$ occurs when the heat rates for the two passes are equal. As shown on the figure, this crossover point is at an apogee of about 7500 n.mi. The heat rate is about 30% less than that for the single-pass case, which also appears on the figure. The maximum acceleration for the minimum heat rate trajectory was about 40% less than that for the single pass.

Figures 3 and 4 have the results for W/C_DA's of 50 and 150 psf, respectively. The result is the same as shown in the previous figure. That is, the optimum value for apogee after the first pass is 7500 n.mi., and the maximum heat rate is reduced by about 30%. The actual value of $\dot{q}_{max}$ decreases sharply with decreasing W/C_DA. Since the optimum value of the intermediate apogee remains constant, the distribution of velocity losses, or ΔV's between the two passes, apparently remains constant and independent of the aerodynamic characteristics of the vehicle.

In the three-pass case, the apogees after the first and the second pass are variables. For a fixed value for apogee after the first pass, the second-pass apogee can be varied in the same manner as the two-pass case. The results for a first-pass apogee of 10,000 n.mi. are shown in Fig. 5 with data from the two-pass case shown for reference. The maximum heat rate was reduced 30% from the two-pass case and the acceleration was reduced about 40%. Figs. 6 and 7 show the results for first-pass apogees of 12,500 and 15,000 n.mi. The curves are similar to the 10,000-n.mi. case, but there are no comparable two-pass data. The data from Figs. 5-7 were then plotted for the first-pass apogee. Figure 8 shows the heat rate and the optimum second-pass apogee. The crossover in heat rate occurred at an apogee of 10,600 n.mi., and the corresponding second-pass apogee was 5050 n.mi. Although data are presented only for W/C_DA of 50 psf, the same combination of apogee altitudes was found for the other values of W/C_DA, an echo of the results from the two-pass analysis.

Figure 9 is a summary of all the optimum trajectories calculated. The data generated for up to three passes are represented by the equation

$$q_{max} = 41.97(W/C_DA)^{0.485}(n)^{-0.556}$$

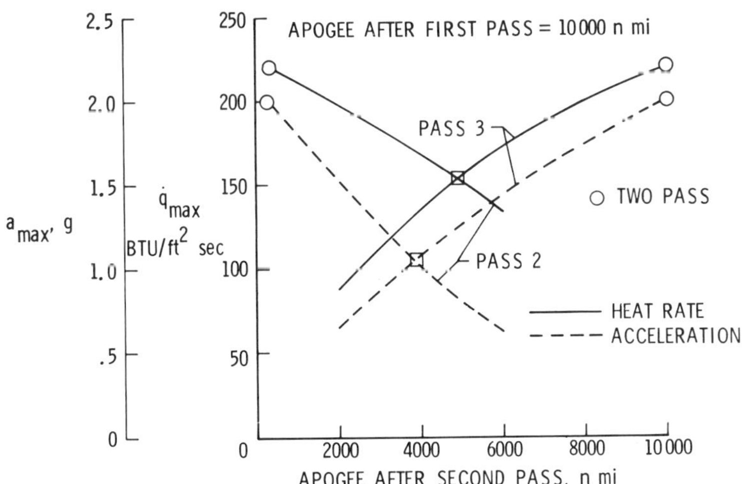

Fig. 5 Results of three-pass trajectories; W/C_DA = 50 psf; first-pass apogee = 10,000 n.mi.

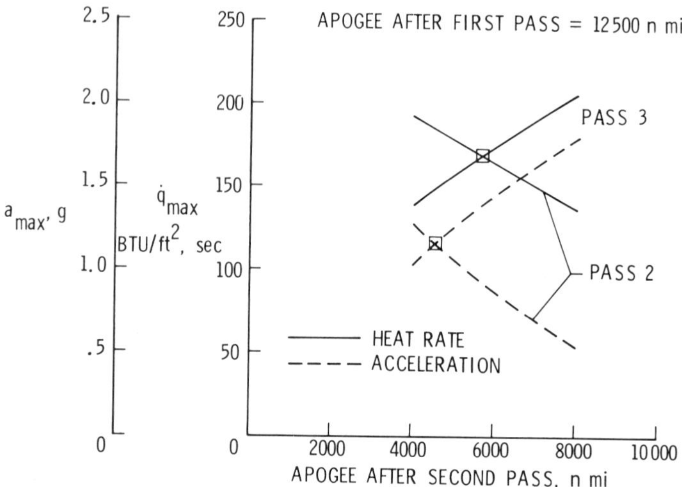

Fig. 6 Results of three-pass trajectories; W/C_DA = 50 psf; first-pass apogee = 12,500 n.mi.

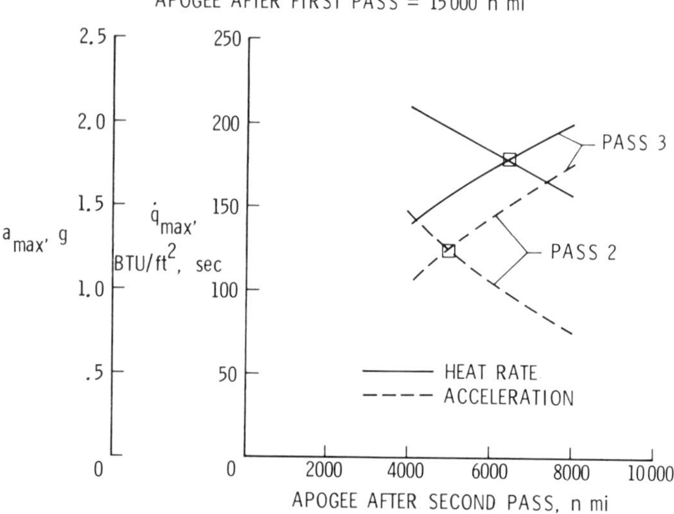

Fig. 7 Results of three-pass trajectories; W/C_DA = 50 psf, first-pass apogee = 15,000 n.mi.

This equation was used to generate results for up to five passes. The heat rate is seen to be dependent on W/C_DA and the number of passes, though there is little gain in going beyond five passes. Figure 10 is the same data plotted to emphasize the dependence of heat rate on

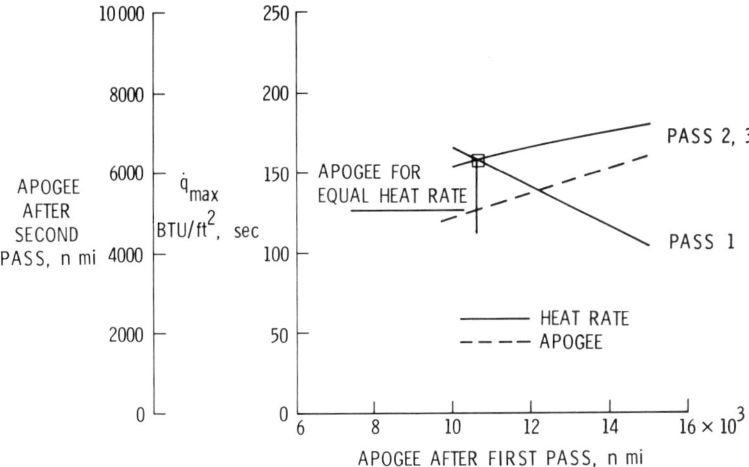

Fig. 8 Optimum apogee determination for three-pass AOTV.

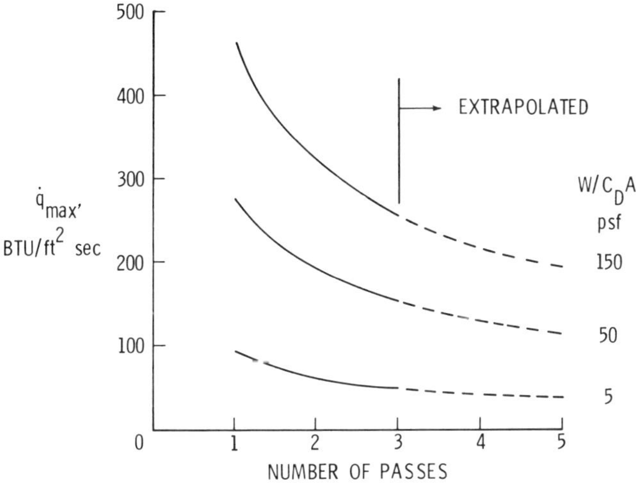

Fig. 9 The effect of the number of passes on heat rate.

W/C_DA. These data can be viewed in several ways: By using more than one pass, the maximum heat rate may be reduced enough to allow the use of a lighter or more durable material for the thermal protection system. Another view is that for a given heat rate, using more passes allows a higher value of W/C_DA, which usually means a smaller vehicle.

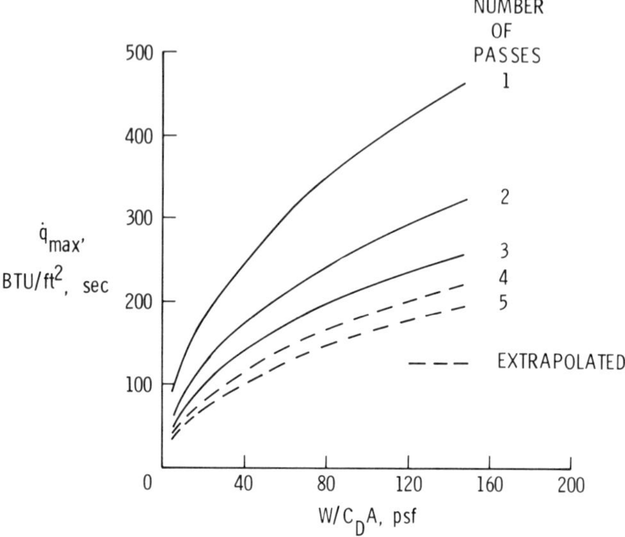

Fig. 10 The effect of the ballistic coefficient on heat rate.

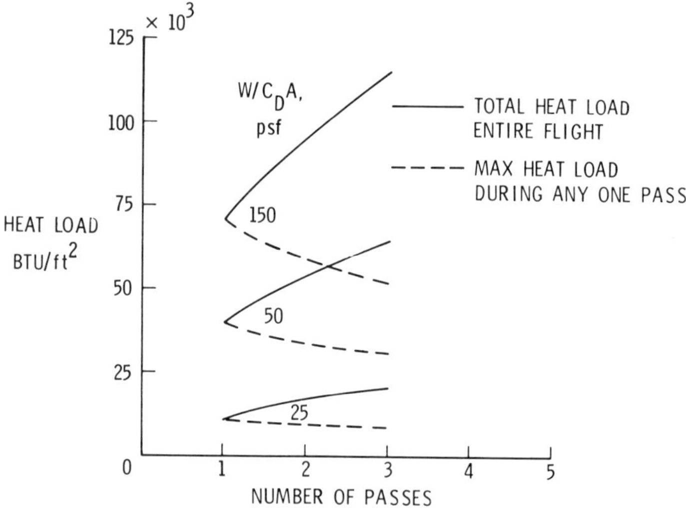

Fig. 11 The effect of the number of passes on heat load.

The integrated heat load is shown in Fig. 11. The solid lines are the total heat load for the entire trajectory (the sum of the heat load for all the passes). The dashed lines are the maximum heat load for any one pass. The total heat load for the entire flight increases sharply with the number of passes, which would present a problem if

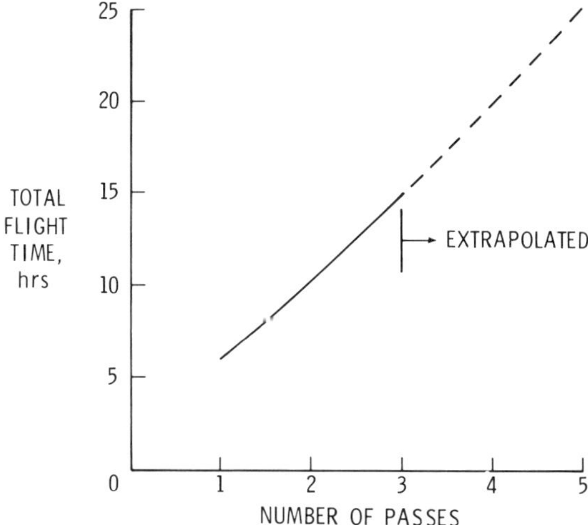

Fig. 12 The effect of the number of passes on flight time.

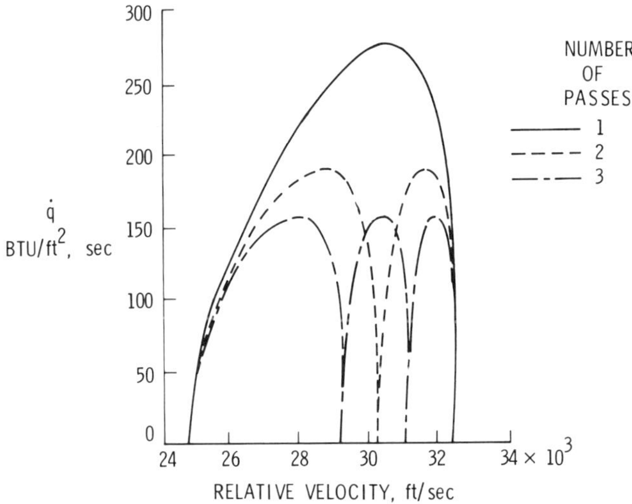

Fig. 13 Heat rate history during the atmospheric passes.

the vehicle is unable to dissipate heat during the coast period of several hours between passes. If the heat can be dissipated, then there is some benefit in going to more passes since the heat load during any one pass decreases with the number of passes.

The total flight time is dependent only on the number of passes, as shown in Fig. 12. The increase from 6 h for the single-pass case to 15 h for the three-pass case can be

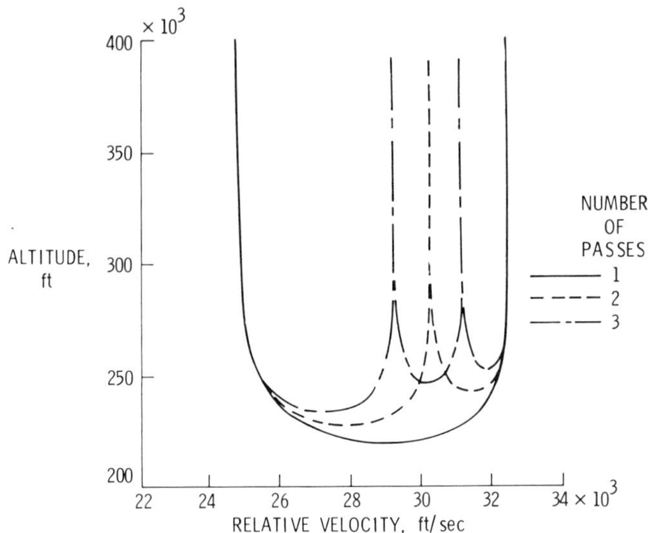

Fig. 14 Altitude history during the atmospheric passes.

considered to be only a slight addition to missions that typically last several days.

Figures 13 and 14 are time histories of heat rate and altitude for the optimum trajectories for W/C_DA of 50 psf. In both cases, the single-pass trajectory essentially forms an envelope for the others. It can be seen in Fig. 13 that the maximum heat rate is the same for all the passes of a given trajectory. Figure 14 shows that for multiple pass trajectories the perigee decreases and the velocity loss increases for each succeeding pass.

Summary

Using a detailed three-dimensional trajectory analysis, the maximum heat rate for an AOTV trajectory was reduced substantially by using a multiple pass strategy. The amount of the reduction is 30% for two passes and 45% for three passes. This result holds true for a wide range of vehicle ballistic coefficients. The integrated heat load and acceleration that occur during the passes also decrease, and the flight time increases only slightly.

References

[1]Walberg, Gerald D., "A Review of Aeroassisted Orbit Transfer," AIAA Paper 82-1378, AIAA 9th Atmospheric Flight Mechanics Conference, San Diego, CA., Aug. 1982.

[2]"Space Tug Aerobraking Study," NASA CR-12405, The Boeing Company, April 1972.

[3]White, John, "Feasibility and Tradeoff Study of An Aeromaneuvering Orbit-to-Orbit Shuttle (AMOOS)," Lockheed Missiles and Space Company, Huntsville, ALA, "Final Report LMSC-HREC TR D390272, July 1974.

[4]Wilhite, Alan W., "Integrated Design of an Aeroassisted Orbital Transfer Vehicle," AIAA Paper 84-0406, AIAA 22nd Aerospace Sciences Meeting, Reno, Nevada, January 1984.

[5]Brauer, G. L., Cornick, D. E., and Stevenson, R., "Capabilities and Applications of the Program to Optimize Simulated Trajectories (POST)," NASA CR-2770, Feb. 1977.

Impact of Atmospheric Uncertainties and Viscous Interaction Effects on the Performance of Aeroassisted Orbital Transfer Vehicles

Theodore A. Talay,* Nancy H. White,* and J. Chris Naftel*
NASA Langley Research Center, Hampton, Virginia

Abstract

Simulations of aerobraking trajectories of aeroassisted orbital transfer vehicles (AOTVs) returning from geosynchronous orbit were analyzed to examine the effects of high-altitude viscous interactions and off-nominal atmospheres on AOTV return weight, heating, and loads performance. Viscous interaction effects encountered at high altitudes had few detrimental effects on the return weight capabilities for AOTVs representing a range of lift/drag ratios. Most of the AOTV return weight increase over an all-propulsive OTV occurred for a low lift/drag ratio. Smaller increases in return weight were observed for higher lift/drag ratios, at the expense of significantly higher heating and aerodynamic loads. Off-nominal atmospheres based on Shuttle-derived data and multipliers on a U.S. standard atmosphere were considered. AOTVs intended for entry under standard atmospheric conditions either deorbited during the pass through the off-nominal atmospheres or missed the target phasing orbit by wide margins. The AOTVs could successfully negotiate these atmospheres when new bank angle histories were implemented with little loss and sometimes with a gain in return weight.

Nomenclature

a_{max} = maximum acceleration, g
AOTV = aeroassisted orbital transfer vehicle

Presented as Paper 84-0408 at AIAA 22nd Aerospace Sciences Meeting, Reno, Nev., Jan. 9-12, 1984.

*Aerospace Engineer, Space Systems Division.

c'_∞ = parameter that accounts for viscosity and temperature changes across a shock
GEO = geosynchronous orbit
L/D = lift-to-drag ratio
LEO = low Earth orbit
M_∞ = freestream Mach number
$q_{\infty,max}$ = maximum dynamic pressure, lb/ft^2
$\dot{q}_{max}$ = maximum convective stagnation point heat rate referenced to a 1-ft-radius sphere, BTU/ft^2-s
R_n = Reynolds number
TPS = thermal protection system
$\overline{V}'_\infty$ = viscous interaction parameter = $M_\infty\sqrt{c'_\infty / R_n}$
W/C_DA = ballistic coefficient, lb/ft^2

Introduction

For payloads to reach high Earth orbits, which are beyond the performance capability of the Space Shuttle, upper stages are required. Presently, these upper stages are expendable. The current outlook for future space transportation predicts the placement of greater numbers of large payloads in high orbits, including geosynchronous orbit (GEO). Advanced upper-stage concepts are under study that could significantly increase the payload weight while lowering operational costs of delivery to GEO. A key feature of these orbital transfer vehicles (OTVs) is their reusability, which, while attractive from an economic viewpoint, also represents an enabling technology for the return of relatively large payloads from GEO.[1]

The OTV, on its return from GEO, can use a propulsive maneuver to decelerate into a low Earth orbit (LEO). For the aeroassisted orbital transfer vehicle (AOTV) concept, this maneuver is accomplished by aerobraking. The vehicle passes through the Earth's upper atmosphere one or more times, dissipating orbital energy and lowering the orbit apogee. For an AOTV with an aerodynamic lift capability, the orbit inclination can also be changed. This last maneuver is similar to a synergetic plane change described for low Earth orbital maneuvers.[2]

The payoff for an AOTV comes in the form of greatly reduced propellant requirements compared to the all-propulsive OTV. For a Shuttle-based system, this savings translates into a more than doubling of the payload delivery and return capability.[3] The aeroassist technique, however, introduces a new set of technological demands and penalties.

Aerodynamic heating requires that the AOTV have a thermal protection system (TPS). Aerodynamic and deceleration loads necessitate an adequate load-bearing structure. Weight increases associated with these systems detract from the total possible payload increases brought about by atmospheric braking. Furthermore, the atmospheric flight trajectory and resulting final orbit of an AOTV are very sensitive to the initial entry conditions, placing demands on the guidance, navigation, and control systems. For the atmospheric pass, density variations from a standard must be accommodated by adequate aerodynamic control. Although certain density variations are of a predictable nature, unsteady variations caused, for example, by "gravity waves" have led to a behavior that has been described as "potholes in the sky" as observed on some Space Shuttle entries.[4] The vehicle aerodynamics also depend on the flow regime encountered. The AOTV is expected to traverse all the possible regimes from free molecular flow to hypersonic continuum flow. Recent evidence based on Shuttle orbiter flight data indicates that the impact of the variations in vehicle aerodynamics brought about by these high-altitude viscous effects may have been overestimated.[3]

In this paper, the performance of several classes of AOTVs with widely varying lift/drag capabilities is examined in the presence of uncertainties in atmospheric density and where high-altitude viscous effects exist for the vehicle. The measures of performance include the maximum AOTV weight (including payload) returned to the Shuttle following the AOTV mission, the heating rates, total heat loads, dynamic pressures, and accelerations encountered by the vehicle. In addition to presenting the effects of the various flow regimes on the performance of AOTV concepts, this paper also compares the performance of several AOTVs for the same mission and demonstrates the impact of off-nominal atmospheres on their performance.

Mission Description

In this analysis, the AOTV baseline mission was a round trip to GEO with a maximum return weight to low Earth orbit. An AOTV and its payload, weighing a combined 66,000 lb, is launched from the Kennedy Space Center as a single Shuttle payload into a 160-n.mi. circular orbit inclined at 28.5 deg. The AOTV delivers its payload propulsively to GEO (19,323 n.mi.) inclined at 0 deg, and then returns with a maximum payload to a 300-n.mi. phasing orbit, using a single-pass atmospheric braking maneuver. Following phasing, the AOTV transfers propulsively to a 160-n.mi.

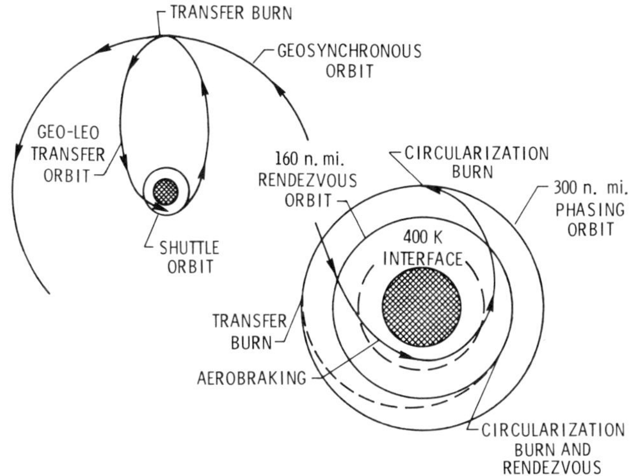

Fig. 1 AOTV GEO baseline mission.

orbit for subsequent rendezvous and retrieval by the Shuttle (Fig. 1). Multiple-pass aerobraking maneuvers are discussed in Ref. 5. The timing of the GEO to LEO transfer propulsive maneuver to place the orbit perigee within the Earth's atmosphere and affect a portion of the plane change back to 28.5 deg is a function of the L/D capability of the vehicle under consideration.

AOTV Descriptions

Three classes of AOTVs were examined in the analysis. The primary distinguishing feature of these AOTVs is the various ranges of L/D capability. The general configurations of these vehicles are depicted in Figs. 2-4. The low L/D vehicle, shown in Fig. 2, uses a deployable aero brake at the upstream end. The moderate L/D vehicle, shown in Fig. 3, has basically a cylindrical shape, but the nose contouring produces the desired L/D. The high L/D vehicle, depicted in Fig. 4, has a conical shape with inflatable chines and body flap producing the high L/D desired. Further configuration details concerning these vehicles are available in Ref. 6. For this performance analysis, all configurations were packaged to a maximum allowable 66,000 lb for the Space Shuttle payload bay. No consideration was given to the breakdown of this weight with regard to vehicle structure (including TPS) and actual delivered and returned payload. Rather, performance was measured by the maximum weight fraction of the 66,000 lb actually returned

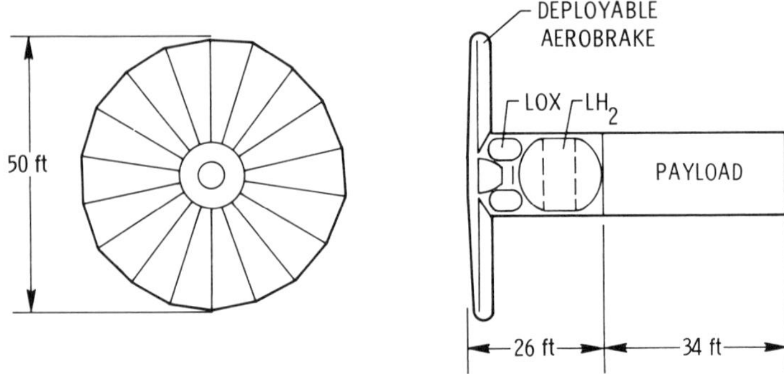

Fig. 2 AOTV with low L/D capability (Ref. 6).

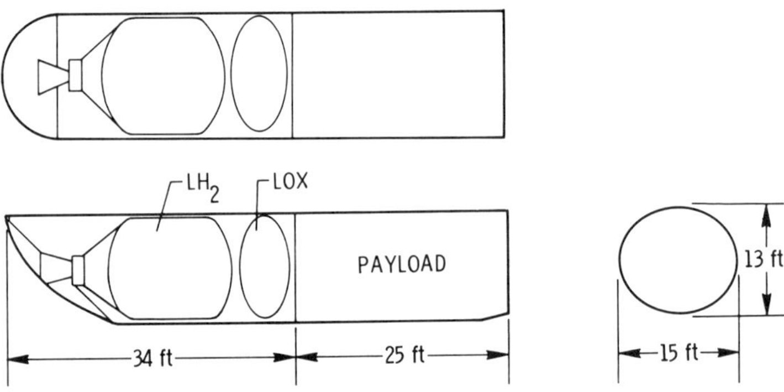

Fig. 3 AOTV with moderate L/D capability (Ref. 6).

to the Shuttle following completion of the mission. More detailed analyses, beyond the scope of the present work, would be required to apportion these final weights in terms of payload and vehicle system hardware.

Aerodynamic characteristics of the three configurations are presented in Ref. 6. These characteristics were derived using Shuttle and Apollo flight data and hypersonic computational methods. Figures 5-7 show the L/D ratios, derived in Ref. 6, for the three configurations versus angle of attack for various values of the viscous interaction parameter, $\bar{V}'_\infty$, defined by

$$\bar{V}'_\infty = M_\infty \sqrt{c'_\infty / R_n}$$

Thus, $\bar{V}'_\infty$ is dependent on freestream Mach number M_∞; Reynolds number R_n; and c'_∞, a term that accounts for

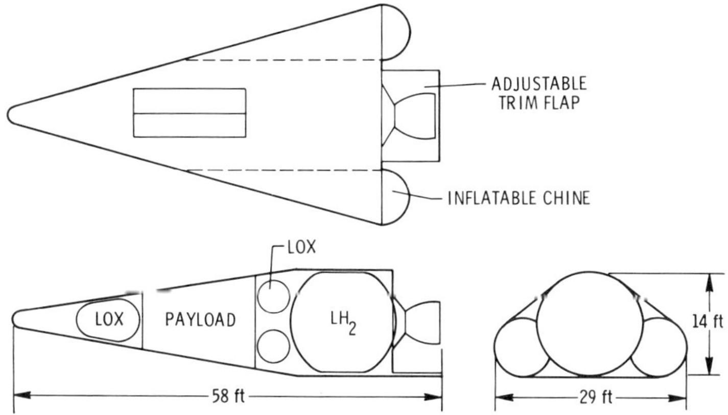

Fig. 4 AOTV with high L/D capability (Ref. 6).

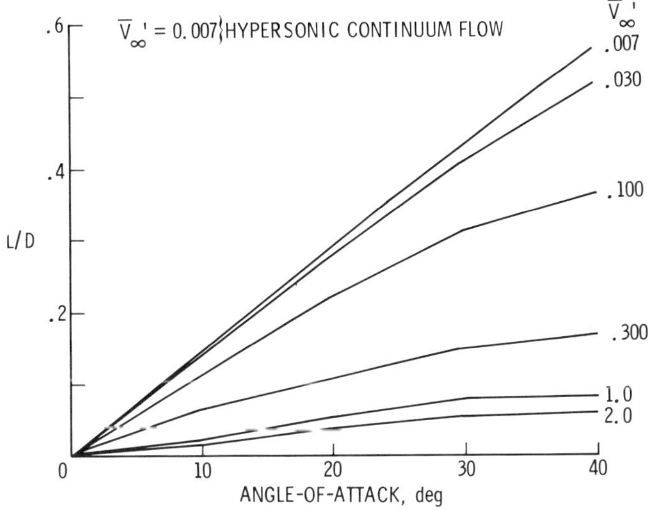

Fig. 5 Lift/drag of low L/D vehicle.

viscosity and temperature changes across the strong shock wave that envelops an entering AOTV. It also accounts for changes in the specific heat ratio under high heating. Detailed discussions of the methods used to obtain the aerodynamics of the study vehicles under viscous interaction conditions are found in Ref. 6.

Readily evident in Figs. 5-7 are the marked reductions in the L/D ratio for the different AOTV configurations for increasing values of $\overline{V}'_\infty$. Increases in $\overline{V}'_\infty$ occur for AOTV flight at high altitudes where differing flow regimes

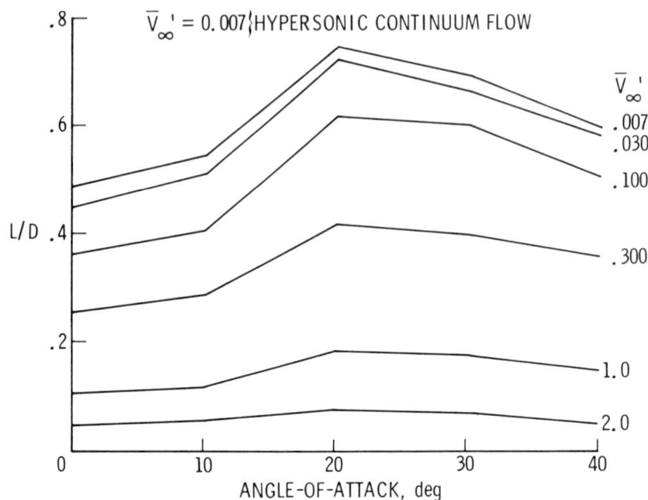

Fig. 6 Lift/drag of moderate L/D vehicle.

ranging from free molecular and transitional to slip and continuum flow are encountered. In the ensuing discussions, the vehicles whose aerodynamics are depicted in Figs. 5-7 are referred to as the low, moderate, and high L/D vehicles, respectively. These vehicles were examined under conditions of hypersonic continuum flow ($\bar{V}'_\infty$ = 0.007) where the vehicle aerodynamics remained constant at fixed angles of attack, and for variable $\bar{V}'_\infty$ conditions where the aerodynamics varied, even at fixed angles of attack. Under hypersonic continuum flow conditions, $(L/D)_{max}$ values are 0.755 and 1.878 for the moderate and high L/D vehicles, respectively. For the low L/D vehicle, at 15 deg angle of attack, the L/D is 0.215 for continuum flow.

Analysis Technique

Trajectories for the AOTVs were calculated using the three-dimensional version of the Program to Optimize Simulated Trajectories (POST).[7] In the first part of this analysis, a 1962 U.S. standard atmosphere was used for the aerobraking studies. An all-propulsive maneuver (chemical propulsion specific impulse of 456 s and engine thrust of 15000 lb) at GEO targets the AOTV to a return trajectory to Earth where the transfer orbit perigee lies within the Earth's atmosphere. Some of the 28.5-deg plane change (to reacquire the Shuttle inclination) is accomplished during this event, with the remainder obtained

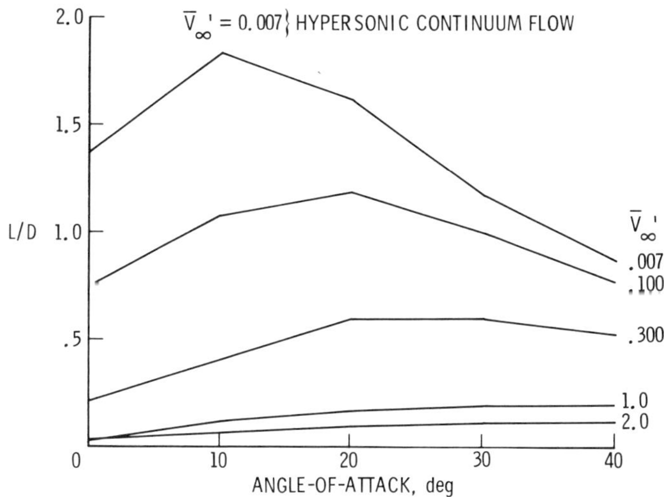

Fig. 7 Lift/drag of high L/D vehicle.

from the aeromaneuvering capability of the AOTVs. Upon reaching the atmospheric interface at 400,000 ft, the AOTV flies the aerodynamic portion of its trajectory at constant angle of attack but with variable bank angle, rolling the lift vector about the velocity vector. Rotating the lift vector modulates the drag via altitude control but also turns the vehicle through the remaining inclination change. The atmospheric phase is assumed to end as the AOTV exits through 400,000 ft. Sufficient energy is dissipated during aerobraking for the orbit apogee to be reduced to 300 n.mi., the assumed phasing orbit for subsequent Shuttle rendezvous, while achieving a 28.5-deg inclination with the same ascending node longitude as the Shuttle. Thereafter, a three-burn all-propulsive sequence in LEO circularizes the vehicle at 300 n.mi. for phasing, places the AOTV in a transfer orbit to 160 n.mi., and finally circularizes in the orbit of the Shuttle for rendezvous. POST is allowed to seek an initial propulsion event time, burn duration, and thrust angle at GEO as well as bank angle history through the atmosphere to ensure a maximum weight returned to the Shuttle.

Using the described method, three-dimensional POST trajectories were calculated for all three vehicle configurations for a wide range of fixed angles of attack and under the assumptions of a hypersonic continuum ($\overline{V}_\infty'$ constant), and then including the effects of viscous interactions encountered in high-altitude flow regimes ($\overline{V}_\infty'$ variable). Results of interest included the maximum

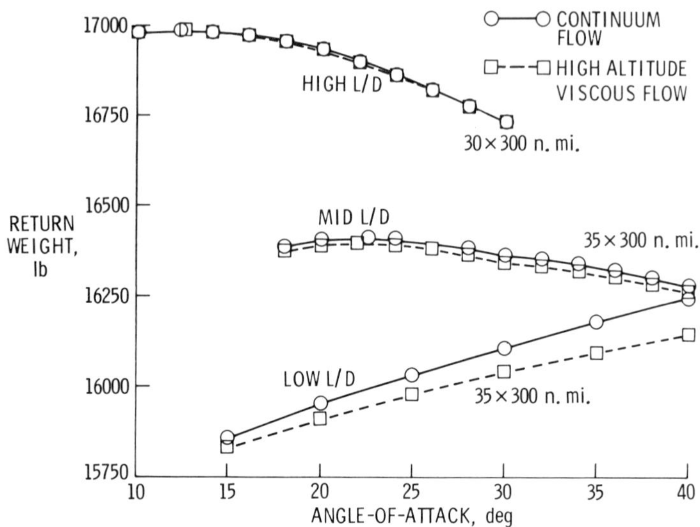

Fig. 8 AOTV return weights to Shuttle.

return weights, heating rates, total heat loads, dynamic pressures, accelerations, plane changes, and bank angle histories. Comparisons of vehicle performance can also be made for the assumption of continuum flow ($\overline{V}'_\infty$ constant) where vehicle aerodynamics remain constant versus high-altitude viscous interaction flow ($\overline{V}'_\infty$ variable) where vehicle aerodynamics vary, for each vehicle for a range of angles of attack. Comparison of the performance of vehicles with respect to one another can also be made.

The second part of the study includes the effects of off-nominal atmospheres on vehicle performance. The most optimum vehicle trajectories in terms of maximum return weights were selected from the previous analysis to represent baseline cases for each of the vehicle configurations. These three baseline trajectories provided atmospheric entry conditions and bank angle histories to be used in attempting to fly five selected nonstandard atmospheres. Three of these atmospheres were defined by data collected from Shuttle entries on flights STS-2, STS-4, and STS-6. Constant multipliers of 1.25 (25% high) and 0.75 (25% low) applied to the 1962 U.S. standard atmosphere resulted in the remaining two off-nominal atmospheres considered.

With the same atmospheric entry conditions and bank angle histories found earlier for the three nominal cases as input conditions, three-dimensional POST calculated new

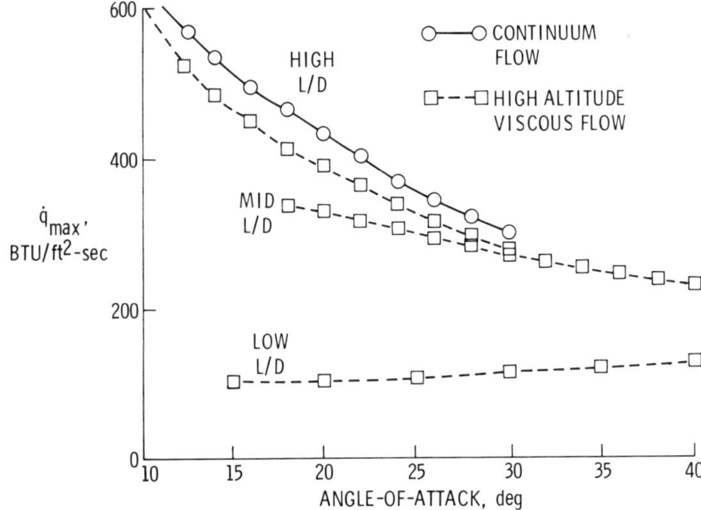

Fig. 9 Reference maximum heating rates on AOTV.

vehicle trajectories through the off-nominal atmospheres. The objective was to see whether the vehicles deorbited during the atmospheric pass or attained the target phasing orbit. In the last part of the analysis, the same entry conditions were used, but POST was allowed to seek out new bank angle histories to target the vehicles to the final phasing orbit conditions. Performance variations from the nominal cases are compared.

Results and Discussion

Viscous Interaction Effects

Maximum return weights to the Shuttle are presented in Fig. 8 as a function of vehicle angle of attack and under assumptions of both hypersonic continuum flow (constant aerodynamics) and high-altitude viscous interaction flow (variable aerodynamics). For each class of AOTV, the orbits attained upon atmospheric exit were constant as noted on the figure.

For the low L/D vehicle, Fig. 8 shows increasing return weights for increasing angle of attack for the range considered (15-40 deg). As Fig. 5 demonstrates, this is consistent with the increasing L/D capabilities of this vehicle as the angle of attack is increased. High-altitude viscous interaction effects, resulting in variable aerodynamics, depicted in Figs. 5-7, decrease the return

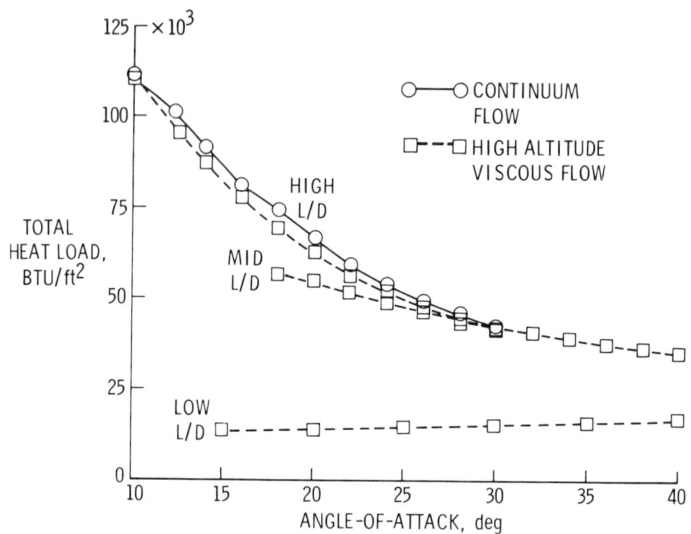

Fig. 10 Reference total heat loads on AOTVs.

weights by between 25 and 95 lb depending on the angle of attack. The ordinate on this plot has been expanded to show these differences with clarity. For the moderate L/D vehicle, the return weights also decrease when high-altitude viscouseffects are considered, but these reductions are less (between 11 and 21 lb). Here, it is interesting to note that the peak return weights occur near the angle of attack for $(L/D)_{max}$ (Fig. 6) under continuum flow assumptions irregardless of whether high-altitude viscous effects are accounted for or not. For the high L/D vehicle, these same trends are evident, but the weight differences between the two curves are reduced further to between 0.1 and 8 lb. Maximum reference heating rates, $\dot{q}_{max}$, are shown in Fig. 9. Viscous interaction flow calculations are shown for all three vehicles, but for clarity, continuum results are presented only for the high L/D vehicle. For the continuum flow case ($\bar{V}_\infty' = 0.007$), the peak heating rates are only up to 3.8 and 0.7% more for the low and moderate L/D vehicles, respectively. Notable are the low peak heating rates for the low L/D vehicle, the high heating rates for the high L/D vehicle at low angles of attack, and the approach of the high L/D vehicle heating rates to those of the moderate L/D vehicle at higher angles of attack.

Total reference heat loads are depicted in Fig. 10. The higher heat loads for the higher L/D vehicles are

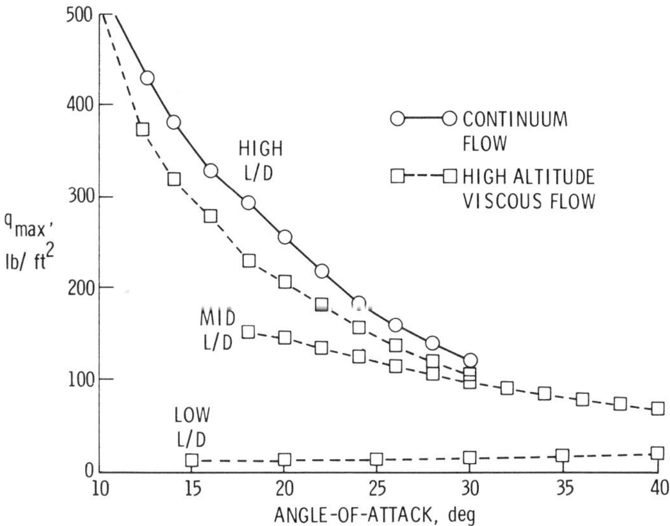

Fig. 11 Maximum dynamic pressures on AOTVs.

caused not only by higher heating rates, but are also due to the longer times actually spent flying the atmospheric pass. Not shown are the total heat loads for the continuum flow cases ($\bar{V}'_\infty$= 0.007) for the low and moderate L/D vehicles which range up to 6.8 and 1.2% higher, respectively.

Maximum dynamic pressures are shown in Fig. 11 and follow the same trends established by the peak heating rates and total heat loads. Notable, however, are the much lower values for the low L/D vehicle. As will be shown later, this can be traced to the overall higher altitudes flown by the low L/D vehicles. The peak dynamic pressures under continuum flow assumptions for the low and moderate L/D vehicles range up to 4.9 and 1.5% higher, respectively.

Figure 12 presents the maximum accelerations on the vehicles. Not shown are the accelerations for the low and moderate L/D vehicles for $\bar{V}'_\infty$ constant which are within 1-3% of the results obtained when $\bar{V}'_\infty$ is variable. Notable are the significantly larger differences for the high L/D vehicle.

Finally, Fig. 13 demonstrates the plane-change capabilities of these three classes of vehicles. At its maximum L/D, near 12-deg angle of attack, the high L/D vehicle aerodynamically obtains approximately 26 deg of the required 28.5 deg of plane change. However, this

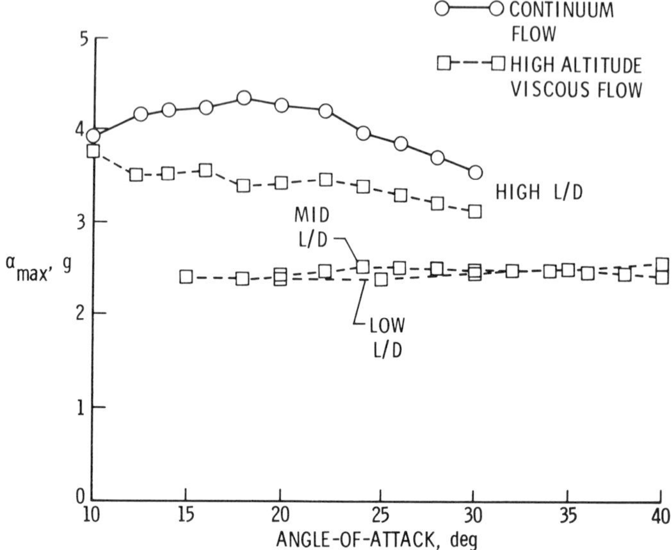

Fig. 12 Maximum acceleration on AOTVs.

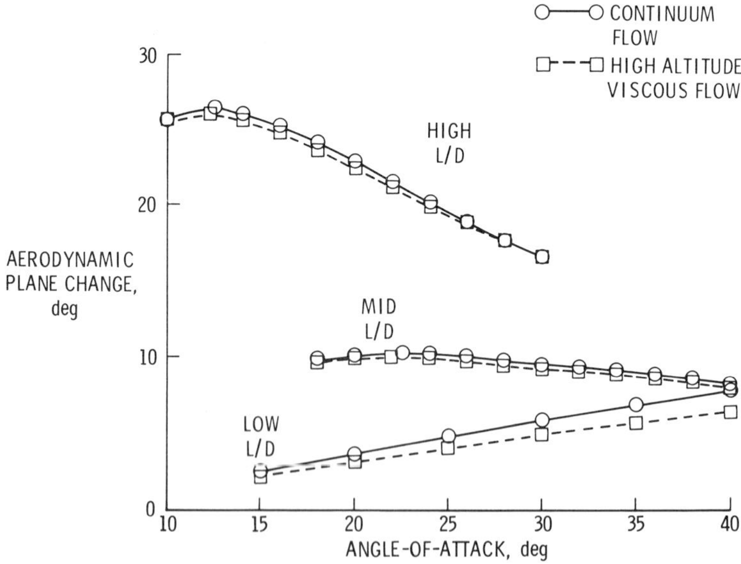

Fig. 13 Aerodynamic plane changes for AOTVs.

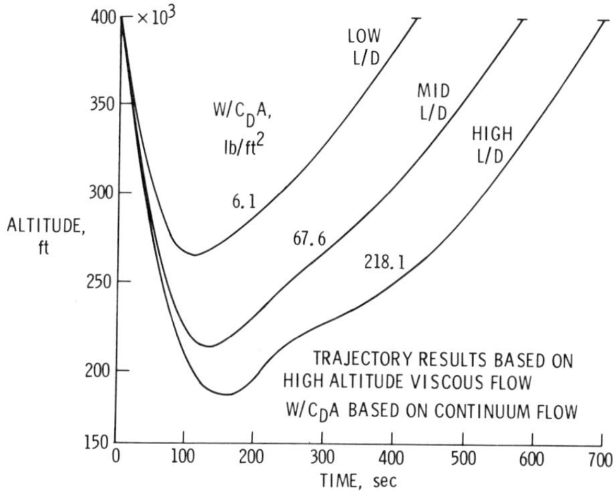

Fig. 14 Altitude histories for maximum return weight AOTVs.

capability is reduced considerably at higher angles of attack. The moderate L/D vehicle demonstrates between 8-deg and 10-deg plane-change capability, with the maximum near 22-deg angle of attack [very close to where $(L/D)_{max}$ is occurring (Fig. 6)]. The low L/D vehicle has only about 2-deg to 2.5-deg plane-change capability at low angles of attack, but approaches the capabilities of the moderate L/D vehicle at high angles of attack. Of the vehicles considered, the low L/D vehicle demonstrates the largest losses in plane-change capability because of high-altitude viscous effects.

Although Figs. 5-7 depict very significant reductions in L/D capability when high-altitude viscous effects are included, these do not directly translate into significantly reduced return weight capabilities for the AOTVs (Fig. 8). To understand this, it should be realized that all the vehicle configurations considered were capable of attaining the required low Earth orbit in terms of exit perigee and apogee by drag modulation effected by bank angle control of the lift. However, beyond this, any lift not required for this purpose is used to effect as much of the 28.5-deg plane-change aerodynamically as possible, with the higher L/D vehicles capable of much greater synergetic-maneuvering. Earlier studies by other investigators have shown these maneuvers are best carried out at $(L/D)_{max}$[3,9,10]. The POST simulations conducted here indeed show a maximum amount of plane change occurring near $(L/D)_{max}$.

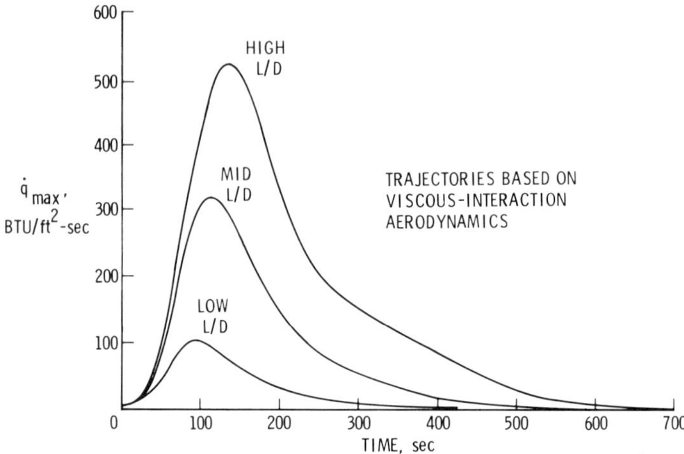

Fig. 15 Reference heating rate histories for maximum return weight AOTVs.

One then must consider how much of the plane-change capability of each vehicle class is lost by the introduction of high-altitude viscous effects and how these reductions affect the return weight capability. To examine these questions and provide input for the second part of these analyses dealing with off-nominal atmospheres, a baseline case is established for each vehicle class.

For the low L/D vehicle, an angle of attack of 15 deg was selected as an upper bound to limit possible flow impingement effects on the payload. POST optimized the angles of attack for the moderate and high L/D vehicles at 21.8 deg and 12.3 deg, respectively. The orbits upon exiting the atmosphere for the low, moderate, and high L/D vehicles optimized at 35x300 n.mi, 35x300 n.mi., and 30x300 n.mi., respectively. Trajectories could be found that had higher exit perigees for the low and moderate L/D vehicles, and these would reduce the propellant requirements for the phasing and rendezvous orbit propulsive burns that followed. However, these higher exit perigees came at the expense of reduced plane-change capability, which, in turn, increased propellant requirements for the GEO transfer burn. Thus, optimum exit perigees existed for these lifting vehicles that minimized overall propellant expenditures and maximized return weight.

Figures 14-20 present POST simulation results for the three baseline cases with high-altitude viscous interaction effects included. Figure 14 depicts the altitude histories. The reference values of ballistic coefficient, W/C_DA, are based on hypersonic continuum values of C_D

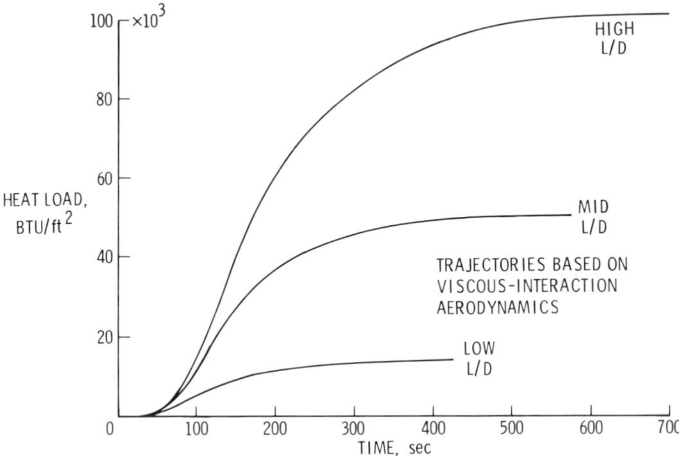

Fig. 16 Reference heat load histories for maximum return weight AOTVs.

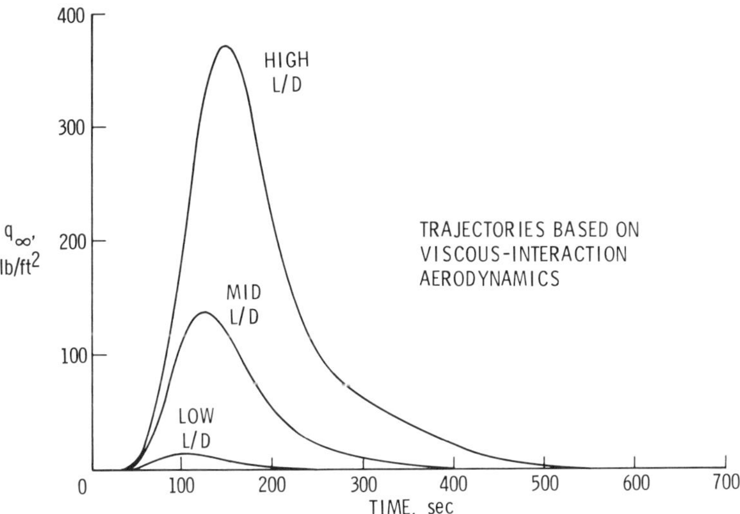

Fig. 17 Dynamic pressure histories for maximum return weight AOTVs.

and will vary as high-altitude viscous effects are encountered.

As the vehicle L/D capabilities increase, the AOTVs are shown to fly lower in the atmosphere for much longer periods (Fig. 14). Thus, it is not surprising that heating rate, total heat load, dynamic pressure, and total acceleration histories become more severe with increases in L/D as shown in Figs. 15-18. Plane-change histories for these maximum return weight cases are shown in Fig. 19.

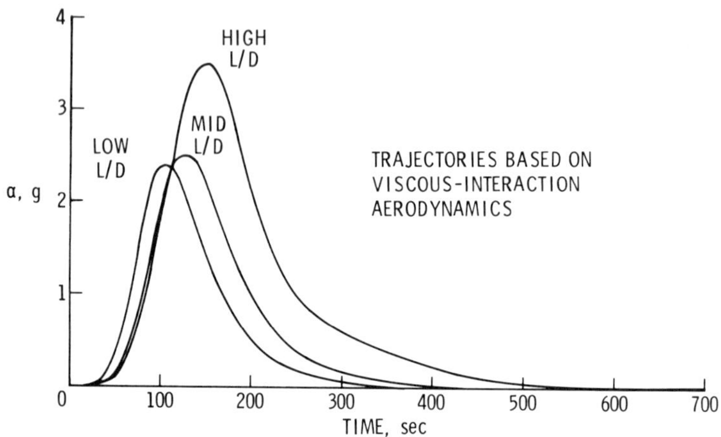

Fig. 18 Acceleration histories for maximum return weight AOTVs.

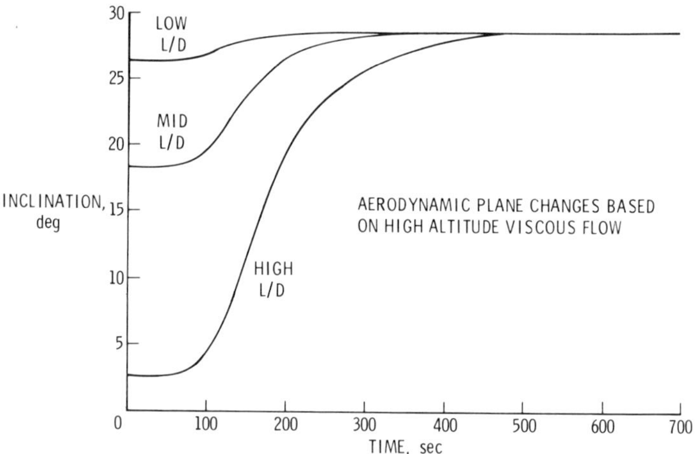

Fig. 19 Plane-change (inclination) histories for maximum return weight AOTVs.

Propulsive plane changes of 26.3 deg, 18.3 deg, and 2.6 deg are accomplished at GEO for the low, moderate, and high L/D vehicles, respectively. In these instances, the aerodynamic plane changes are 2.2 deg, 10.2 deg, and 25.9 deg for the low, moderate, and high L/D vehicles, respectively, during the atmospheric pass. Figure 20 presents the bank angle histories used by these vehicles during the atmospheric pass. The bank angles are all confined to one side (0 deg to -180 deg, where 0 deg is up), which indicates the use of lift to effect the plane

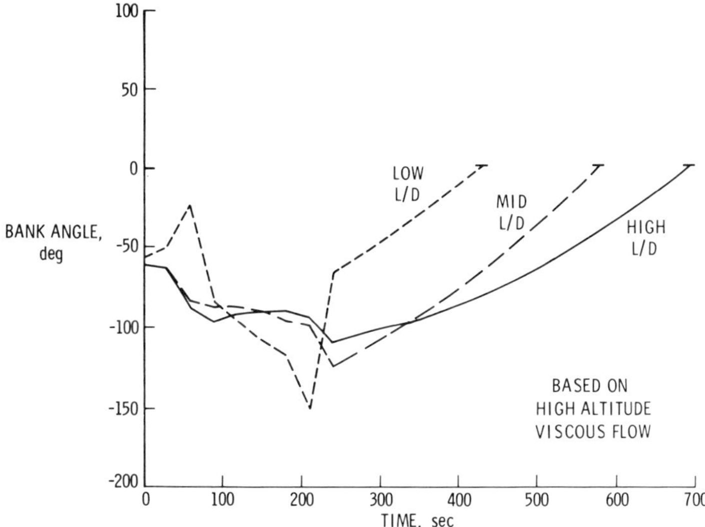

Fig. 20 Bank angle histories for maximum return weight AOTVs.

change. The larger excursions in bank angles occur for low L/D vehicle. The question of whether the vehicle can actually be controlled through bank angle histories of this type is addressed in Ref. 11.

Figure 21 shows the L/D histories for each vehicle. Significant reductions in L/D do indeed occur both early and later in the atmospheric passes, at the higher altitudes, as a result of variable $\overline{V}'_\infty$ effects. However, most of the deceleration, heating, and plane changes are occurring in the period between 50 s and 200-400 s, depending on the vehicle L/D class. During these times, the aerodynamics are not varying as greatly. Indeed, the higher L/D vehicles, flying lower in the atmosphere, have nearly continuum flow aerodynamics over these periods. Thus, the integrated effect of reduced L/D capabilities caused by high-altitude viscous effects does not appreciably penalize the maximum return weight capabilities (Fig. 8).

The observed reductions in return weights caused by high-altitude viscous effects are related to the losses in plane-change capability presented in Fig. 13. Figure 22 shows propellant requirements per degree of plane change, using ideal ΔV calculations, for propulsive plane changes up to 28.5 deg at GEO altitudes (19,323 n.mi.). The dashed curve represents propellant usage for a plane change maneuver alone. The decrease at larger total

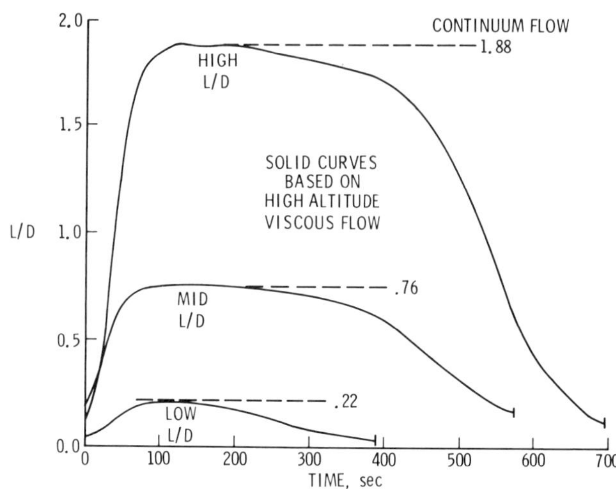

Fig. 21 Lift/drag histories of maximum return weight AOTVs.

plane changes occurs because of the total weight reduction as propellant is expended. The solid curve is based on a combined propulsive maneuver at GEO that places the AOTV on a GEO-LEO return trajectory while performing the plane change. The curve represents the propellant requirements per degree of plane change, <u>in excess</u> of the GEO-LEO transfer burn, to accomplish the plane change. Under these conditions, propellant expenditures for small total plane changes are low, but increase to much higher values when a larger total plane change is required. Conversely, a high L/D vehicle, accomplishing most of a 28.5-deg plane change aerodynamically, is not penalized as much for a loss in plane-change ability (extra pounds of propellant needed per degree loss of plane change) as a low L/D vehicle that accomplishes only a few degrees of plane change by aerodynamic means (i.e., most of it propulsively). To demonstrate these results, the weight losses observed in Fig. 8 caused by high-altitude viscous effects were divided by the observed losses in plane-change capability from Fig. 13 for the range of angles of attack studied. Figure 23 shows the results. The results confirm the return weight penalties dependence on both loss of plane-change capability and the L/D class vehicle used.

Before turning to the effects of off-nominal atmospheres on the baseline AOTV POST simulations, an overall comparison of performance between the three AOTV types and an all-propulsive OTV is summarized.

Table 1 Performance comparisons of an all-propulsive OTV with AOTVs of differing L/D capabilities for GEO round trip missions

OTV type	All-propulsive[a]	Low L/D = 0.215[b]	Moderate L/D = 0.755[b]	High L/D = 1.878[b]
Initial weight, lb	66,000	66,000	66,000	66,000
Weight returned to Shuttle, lb	9,666	15,833	16,396	16,987
$\dot{q}_{max}$,[d] BTU/ft^2-s	•••	102	317	520
Total heat load,[d] BTU/ft^2	•••	13,794	49,855	100,870
q_{max}, lb/ft^2	•••	14.1	137	372
a_{max}, g	1.55[c]	2.40	2.51	3.50

[a]Chemical propulsion specific impulse of 456 s; engine thrust of 15,000 lbs; finite burn analysis.
[b]Continuum flow conditions; 15-deg angle of attack for low L/D; operating at $(L/D)_{max}$ for moderate and high L/D vehicles.
[c]At end of last propulsive burn.
[d]Reference heating.

Table 2 Atmospheric entry flight conditions at 400,000 ft for three AOTV baseline trajectories

AOTV capability	Low L/D = 0.215[a]	Moderate L/D = 0.755[a]	High L/D = 1.878[a]
Entry weight, lb	16,889	17,485	18,122
W/C_DA,[a] lb/ft^2	6.1	67.6	218.1
Velocity,[b] ft/s	32,435.4	32,350.2	32,270.2
Flight-path angle,[b] deg	-4.44	-5.15	-5.39
Azimuth,[b] deg	62.8	71.2	87.3
Entry orbit apogee, n.mi.	19,323	19,323	19,323
Entry orbit perigee, n.mi.	43.0	35.2	32.5
Orbit inclination, deg	26.33	18.32	2.63

[a]Based on continuum flow conditions ($\overline{V}'_\infty = 0.007$).
[b]Atmosphere relative in local horizontal reference frame.

A three-dimensional POST simulation of a round trip to GEO and return to the Shuttle was made using all-chemical propulsion and finite burns for the orbital maneuvers. Table 1 summarizes the final OTV weight returned to the Shuttle as compared with the three baseline cases for the various L/D AOTVs. In addition, for the AOTV's, comparisons of peak heating rate, heat load, dynamic pressure, and total acceleration are presented.

A 64% increase in return weight over an all-propulsive system results by using the low L/D capability AOTV. This increase is 70% for the moderate L/D vehicle and 76% for the high L/D vehicle. Thus, most of the increases observed in return weight using AOTVs can be obtained with a small amount of lift capability. Although somewhat higher return weights occur for the higher L/D vehicles, it is at the expense of markedly higher heating rates, heat loads, dynamic pressures, and accelerations. The relative levels of payloads, associated with the various aeroassist systems, require additional analysis and are not addressed in this study. However, it is questionable whether any payload increases are possible for these vehicles considering the increased thermal protection and load-bearing system weights needed to meet the challenges of the more severe environments. Figures 9-12 demonstrate the less severe environments are possible for the higher L/D vehicles by flying them at higher angles of attack, not at $(L/D)_{max}$. This results in some return weight penalities. For example, by operating the high L/D vehicle at a constant 30-deg angle of attack, the return weight is reduced by about 260 lb, but the peak heating rate is reduced 47%, the total heat load by 58%, the peak dynamic pressure by 71%, and the peak acceleration by 11%. This also results in a more than 9-deg reduction in plane-change capability.

Effects of Off-Nominal Atmospheres

The 1962 U.S. standard atmosphere was assumed for the previous aerobraking studies. It is well known, however, that the actual atmosphere varies temporarily and spatially from this norm[3,8,12]. Figure 24 shows the density structure derived from Space Shuttle flights STS-2, STS-4, and STS-6 for the altitudes of interest in these AOTV studies.[8] The observed densities were ratioed to the 1962 U.S. standard and smoothed to provide input for subsequent

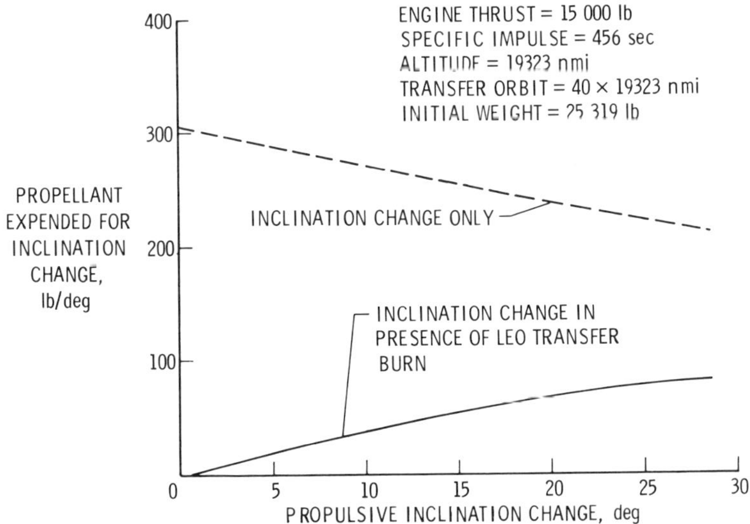

Fig. 22 Propellant expenditures for plane changes at GEO.

POST analysis. As this is along-track data, it is likely that the density structure reflects combined variations in both a vertical and horizontal extent. Two additional atmospheres considered were a 25% increase and a 25% decrease in the 1962 U.S. standard density over all altitudes. These two maximum/minimum density atmospheres essentially bracket the observed maximum changes observed in the STS atmospheres.

To examine whether or not the three AOTV configurations can successfully fly these five off-nominal atmospheres, the same entry conditions and bank angle histories from the baseline cases of Figs. 14-20 were used. Table 2 presents the atmospheric interface entry conditions. Figure 20 shows the corresponding bank angle histories, which are assumed fixed for the present studies. Table 3 presents the results of POST simulations in terms of the exit orbit parameters. The results of using the 1962 U.S. standard are presented as the nominal cases.

None of the vehicles, which sucessfully negotiated passage through the 1962 standard atmosphere, would survive a pass through a 25% higher density situation. Too much energy is dissipated during the pass initiated under nominal entry conditions and the vehicle deorbited. Conversely, all the vehicles exited the 25% lower density atmosphere, but ended up in high apogee orbits with the higher L/D vehicles missing the required plane changes by wide margins. Significantly more propulsive maneuvers than the nominal case are then required to return these vehicles to the Shuttle.

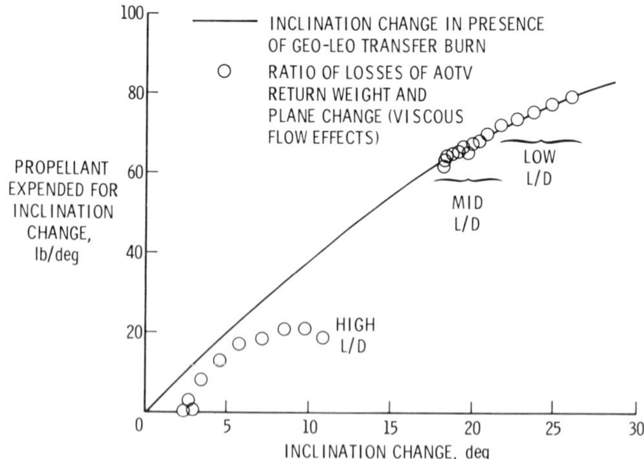

Fig. 23 Propellant expenditures for GEO plane changes compared with AOTV losses.

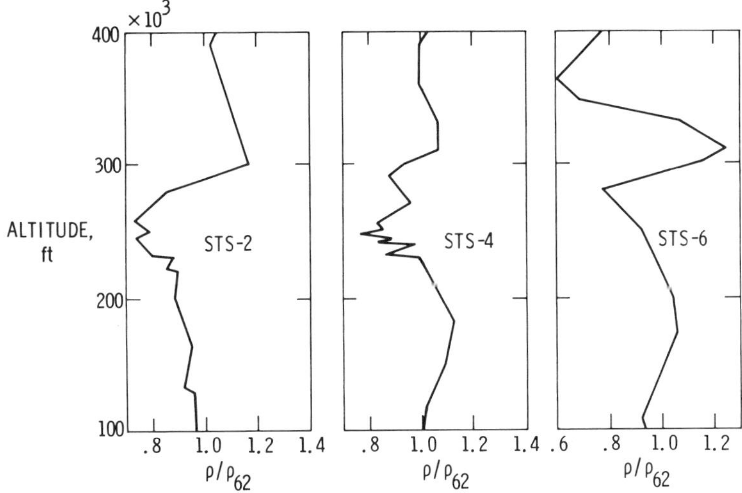

Fig. 24 Shuttle-derived densities compared to 1962 U.S. standard atmosphere.

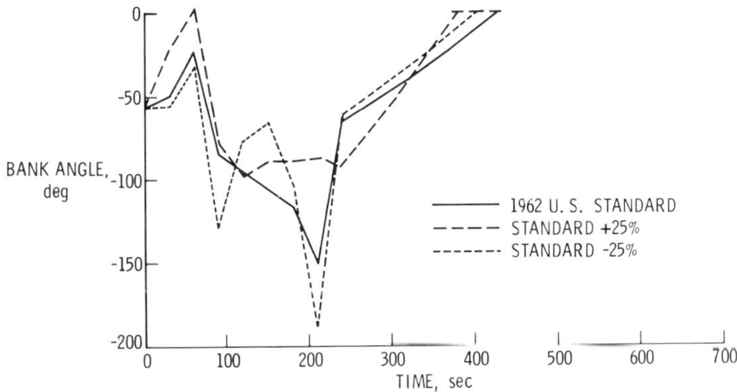

Fig. 25 Bank angle histories in off-nominal atmospheres (low L/D).

Turning to the Shuttle-derived atmospheres, it is seen from Fig. 24 that the density structure is variable, being greater than the nominal at some altitudes while less than the nominal at other altitudes. The low and moderate L/D vehicles passed through all three atmospheres, although the resulting orbits vary considerably. The STS-4 atmosphere appears to give results closest to the U.S. standard. However, the high L/D vehicle deorbited while flying the STS-4 or STS-6 atmospheres. This can be traced to the higher-than-nominal densities at the lower altitudes where peak deceleration is occurring. Too much energy is lost to aerobraking under these conditions.

Table 3 Effects of off-nominal atmospheres on baseline trajectories of three AOTV configurations

Atmosphere type	Exit orbit parameter[a]	Low L/D	Moderate L/D	High L/D
1962 U.S. standard	Perigee, n.mi.	35	35	30
	Apogee, n.mi.	300	300	300
	Inclination, deg	28.50	28.50	28.50
Standard +25%	Perigee, n.mi.	Vehicle deorbited	Vehicle deorbited	Vehicle deorbited
	Apogee, n.mi.			
	Inclination, deg			
Standard -25%	Perigee, n.mi.	43.3	35.9	32.4
	Apogee, n.mi.	1806	2357	3753
	Inclination, deg	27.89	24.90	15.58
STS-2	Perigee, n.mi.	43.2	36.1	33.1
	Apogee, n.mi.	1460	1535	2302
	Inclination, deg	27.97	26.09	19.64
STS-4	Perigee, n.mi.	41.1	36.1	Vehicle deorbited
	Apogee, n.mi.	660	495	
	Inclination, deg	28.33	28.07	
STS-6	Perigee, n.mi.	43.1	36.3	Vehicle deorbited
	Apogee, n.mi.	1266	530	
	Inclination, deg	28.05	27.99	

[a]Atmospheric exit at 400,000 ft.

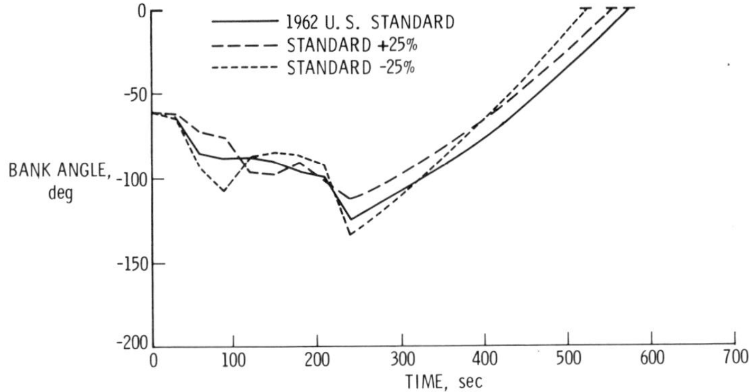

Fig. 26 Bank angle histories in off-nominal atmospheres (moderate L/D).

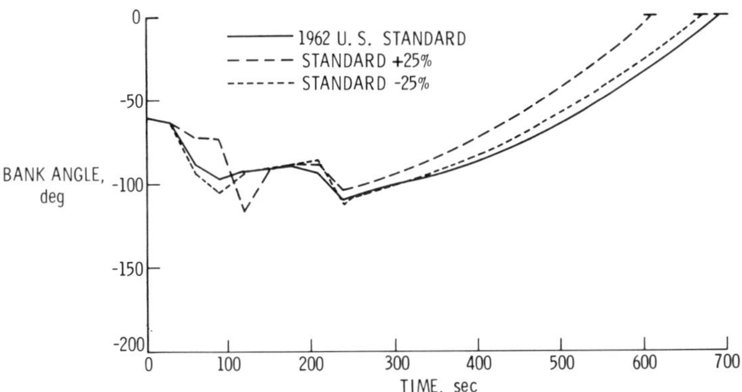

Fig. 27 Bank angle histories in off-nominal atmospheres (high L/D).

Thus, AOTV entry trajectories and bank angle histories, preset for flight in a nominal atmosphere, do not guarantee a successful pass through and a desirable exit for off-nominal atmospheres. If the vehicle does accomplish the atmospheric pass, the resulting orbit may require significant propulsive maneuvers to return it to the Shuttle.

The next consideration was to use the same nominal entry conditions but allow POST to seek out new bank angle histories that would target the vehicles to the final required phasing orbit. The goal was to examine any changes in performance from the nominal cases. Tables 4-6 compare these performance results with the nominal pass through a 1962 U.S. standard atmosphere.

The results show that all the AOTVs could be targeted to the final orbit desired (300-n.mi. apogee at 28.5-deg inclination and same ascending node longitude as the Shuttle) by redefinition of the bank angle histories. Changes in the exit perigee, however, occurred. In both the +25% and -25% dispersed atmospheres, the exit perigees were lower than the nominal, leading to somewhat larger propulsive maneuvers to attain the proper phasing and rendezvous orbits. For the STS-derived atmospheres, the exit perigees were dispersed around the nominals. Some atmospheres improved the maximum return weight to the Shuttle (low L/D STS-4, STS-6; high L/D STS-4, STS-6, for example). Tables 4-6 also show that the peak heating rates, total heat loads, peak dynamic pressures, and accelerations disperse about the nominal values but do not differ greatly. In general, the STS-derived atmosphere results are closer to the nominal results than for the fixed +25% or -25% atmospheres.

Figures 25-27 show how the new bank angle histories, derived by the POST simulations for the +25% and -25% dispersed atmospheres, compare with the nominal histories. In general, the bank angle histories show significant changes, with the low L/D changes being most pronounced. For clarity, the bank angles for the STS atmospheres have not been plotted. However, they show similar trends of change with respect to the nominal with smaller changes for the higher L/D vehicles.

These results demonstrate that the AOTVs can be made to fly off-nominal atmospheres with little loss and sometimes a gain in performance. The implication is, however, that the vehicle must still be given the updated control history prior to atmospheric entry to successfully negotiate the off-nominal atmosphere.

Summary

GEO return missions for three classes of AOTVs were studied using the three-dimensional POST trajectory analysis program. These AOTVs differed in terms of their L/D capabilities. Although high-altitude viscous interaction effects appreciably reduced the L/D capabilities of the vehicles, the return weight performance was not reduced in like measure compared with continuum flow results. This was because most of the aerobraking occurred at or near continuum flow altitude conditions.

The analyses also showed that heating rates, heat loads, dynamic pressures, and accelerations increase dramatically with the vehicle L/D capability. Large

Table 4 Low L/D AOTV performance in flight through off-nominal atmospheres using new bank angle histories

	Atmosphere type					
	1962 Standard	Standard +25%	Standard -25%	STS-2	STS-4	STS-6
Exit perigee, n.mi.	35	30.48	29.30	34.67	36.87	37.99
ΔW,[a] lb	...	-8.9	-11.2	-0.6	+7.6	+5.9
$\dot{q}_{max}$,[b] BTU/ft^2-s	101.8	104.5	101.2	96.9	101.2	100.6
Total heat load,[b] BTU/ft^2	13,794	13,261	13,314	13,975	13,870	14,113
q_{max}, lb/ft^2	14.12	14.85	14.80	13.31	14.01	13.94
a_{max},g	2.40	2.52	2.51	2.26	2.38	2.37

[a]Weight change from nominal.
[b]Reference heating.

Table 5 Moderate L/D AOTV performance in flight through off-nominal atmospheres using new bank angle histories

	Atmosphere type					
	1962 standard	Standard +25%	Standard -25%	STS-2	STS-4	STS-6
Exit perigee, n.mi.	35	33.22	28.60	32.29	34.96	34.88
ΔW,[a] lb	...	-3.6	-13.0	-5.5	0	-0.2
$\dot{q}_{max}$,[b] BTU/ft^2-s	316.8	310.8	322.0	321.3	322.7	320.4
Total heat load,[b] BTU/ft^2	49,855	50,182	47,511	48,744	49,569	49,549
q_{max}, lb/ft^2	136.9	126.0	147.7	144.2	141.7	140.7
a_{max}, g	2.51	2.31	2.70	2.64	2.59	2.58

[a]Weight change from nominal.
[b]Reference heating.

Table 6 High L/D AOTV performance in flight through off-nominal atmospheres using new bank angle histories

	Atmosphere type					
	1962 Standard	Standard +25%	Standard -25%	STS-2	STS-4	STS-6
Exit perigee, n.mi.	30	23.19	27.86	29.21	30.71	30.70
ΔW,[a] lb	...	-14.4	-4.5	-1.7	+1.5	+1.5
$\dot{q}_{max}$,[b] BTU/ft^2-s	520.4	485.2	526.9	522.1	513.9	520.6
Total heat load,[b] BTU/ft^2	10,870	98,297	97,818	99,597	100,717	100,799
q_{max}, lb/ft^2	372.2	357.7	394.5	380.5	360.8	370.0
a_{max}, g	3.50	3.37	3.71	3.58	3.39	3.48

[a]Weight change from nominal.
[b]Reference heating.

plane changes were possible with the high L/D vehicle. Maximum return weights and plane changes were associated with flight at $(L/D)_{max}$. Significantly smaller heating and aerodynamic loads were possible for the high L/D vehicle by flying at higher angles of attack, hence lower L/D, at the expense of smaller plane changes and a lower return weight capability.

When compared with an all-propulsive OTV, most of the return weight increase of an AOTV was obtained using a low L/D vehicle. The higher L/D AOTVs considered increased this return weight by smaller percentages, but suffered large increases in heating and aerodynamic loads.

Effects of off-nominal atmospheres on AOTVs were analyzed. ATOVs targeted under the assumption of standard atmosphere entry conditions and bank angle histories, when encountering an off-nominal atmosphere, either deorbited during the atmospheric pass or exited widely dispersed off of a target phasing orbit. However, under the same entry conditions, new bank angle histories could be found to allow successful negotiation of the off-nominal atmosphere with little loss or even a gain in return weight performance.

References

[1]Heald, D. A., "Economic Benefits of the OTV Program," Acta Astronautica, Vol. 8, No. 11-12, 1981, pp. 1237-1249.

[2]London, H. S., "Change of Satellite Orbit Plane by Aerodynamic Maneuvering," Journal of the Aerospace Sciences, Vol. 29, No. 3, March 1962, pp. 323-332.

[3]Walberg, G. D., "A Review of Aeroassisted Orbit Transfer," AIAA Paper 82-1378, 9th Atmospheric Flight Mechanics Conference, San Diego, Calif., Aug. 1982.

[4]Cruz, M. I., "Trajectory Optimization and Closed Loop Guidance of Aeroassisted Orbital Transfer," AAS Paper 84-413, AAS/AIAA Astrodynamics Specialist Conference, Lake Placid, N. Y., Aug. 1983.

[5]Rehder, J. J., "Multiple-Pass Trajectories for an Aeroassisted Orbital Transfer Vehicle," AIAA Paper 84-0407, 22nd Aerospace Sciences Meeting, Reno, Nev., Jan. 1984.

[6]Wilhite, A. W., Arrington, J. P., and McCandless, R. S., "Performance Aerodynamics of Aeroassisted Orbital Transfer Vehicles," AIAA Paper 84-0406, 22nd Aerospace Sciences Meeting, Reno, Nev., Jan. 1984.

[7]Brauer, G. L., Cornick, D. E., and Stevenson, R., "Capabilities and Applications of the Program to Optimize Simulated Trajectories (POST)," NASA CR-2770, Feb. 1977.

[8]Findlay, J. T. and McConnell, J. G., "Atmospheres for Aero-Assisted Orbital Transfer Vehicles Based on Shuttle Flight Experience," AMA Report 83-18, Analytical Mechanics Associates, Inc., Hampton, Va, Sept. 1983.

[9]Lau, J., "Implications of Maneuvering-Range Constraints on Lifting-Vehicle Design," Journal of Spacecraft, Vol. 4 May 1967, pp. 639-643.

[10]Maslen, S. H., "Synergetic Turns with Variable Aerodynamics," Journal of Spacecraft, Vol. 4, Nov. 1967, pp. 1475-1482

[11]Powell, R. W., Stone, H. W., and Naftel, J. C., "Performance Evaluation of the Atmospheric Phase of Maneuvering Orbital Transfer Vehicles," AIAA Paper 84-0405, 22nd Aerospace Sciences Meeting, Reno, Nev. Jan. 1984.

[12]U. S. Standard Atmosphere Supplements, 1966. Published by the ESSA, NASA, and USAF.

Analytical Characterization of AOTV Perigee Aerothermodynamic Regime

Dick Desautel*
San Jose State University, San Jose, California

Abstract

In preliminary design studies of aeroassisted orbital transfer vehicle (AOTV) configurations, it is desirable to have a simple analytical method for characterizing the aerothermodynamic regime at skip trajectory perigee as a function of vehicle L/D and ballistic coefficient m/C_DA. The present study derives an approximate perigee solution from the generalized equations of motion. For a prescribed entry velocity vector, the solution determines a Zeta function (proportional to density divided by m/C_DA) at perigee as a parametric function of the exit velocity vector and constant L/D. The perigee Zeta function then determines perigee density or altitude as a parametric function of m/C_DA. The solution allows the following classic aerothermodynamic parameters to be determined at perigee as parametric functions of m/C_DA, L/D, and the exit velocity vector: Reynolds number (viscous effects), Mach number (compressibility effects), Knudsen number (rarefaction effects), sphere and disk bow shock standoff distance, Damköhler number (relaxation effects), viscous correlation parameter (viscid-inviscid interactions), and Stanton number (convective heat transfer). Results of the analysis are given for low L/D and moderate L/D AOTV configurations on return from geosyn-

---

Presented as Paper 84-1713 at AIAA 19th Thermophysics Conference, Snowmass, Colorado, June 25-28, 1984.

*Associate Professor, Department of Mechanical Engineering.

chronous and L5 orbits. It is concluded that the method successfully provides preliminary estimates of the aerothermodynamic parameters through the use of simple algebraic equations and plots.

Nomenclature

a = atmospheric speed of sound
AOTV = aeroassisted orbital transfer vehicle
C = Chapman-Rubesin viscosity parameter
Cp = specific heat at constant pressure
Da = Damköhler number
h_o = freestream total enthalpy
h_w = enthalpy at the wall
H = atmospheric scale height
k = angle weighting factor
k_s = density ratio, ρ/ρ_s
Kn = Knudsen number
ℓ_r = relaxation reference length
ℓ_m = molecular mean free path
L = flow reference length
L/D = lift-to-drag ratio
LEO = low Earth orbit
log = natural logarithm
m/C_DA = ballistic coefficient
M = Mach number
ODE = ordinary differential equations
p = pressure
PDE = partial differential equations
Pr = Prandtl number
$\dot{q}$ = convective heat transfer rate
r = trajectory radial position, $R_o + y$
R = nose radius
Re = Reynolds number
R_o = planet radius
s = trajectory path length
St = Stanton number
$t_{0.1}$ = time to 90% radiation relaxation
T = temperature
T_o = stagnation temperature
T_v = viscous-interaction reference temperature
T_w = wall temperature
TPS = thermal protection system

u	= dimensionless velocity squared, $V^2\cos^2\gamma/V_c^2$
$\bar{u}$	= dimensionless velocity, $V\cos\gamma/V_c$
V	= velocity
V_c	= circular orbit velocity, $\sqrt{gr}$
$\bar{V}$	= viscous correlation parameter
x	= distance from reference point
y	= altitude
Z	= Vinh Zeta function, $(H/Z)(C_D A/m)(\sqrt{r/H})\rho$
$\bar{Z}$	= Chapman Zeta function, $\bar{u}Z$
γ	= flight path angle or ratio of specific heats
λ	= pitch plane lift component
μ	= dynamic viscosity coefficient
μ_v	= viscous-interaction reference viscosity
ν	= kinematic viscosity coefficient
σ	= lift vector bank angle
ρ	= density
Δ/R	= normalized sphere standoff distance
Δ/R_d	= normalized disk standoff distance

Subscripts and Superscripts

i	= initial (entry) condition
f	= final (exit) condition
p	= perigee
s	= average or effective value on shock-layer stagnation streamline
( )'	= "unit" parameter, i.e., based on L = 1 m (Note: unsubscripted flow parameters are freestream)

Introduction

Recent studies have demonstrated the substantial payload advantage to be gained by use of the aeroassist technique for orbital change maneuvers on future space transport missions operating from a low Earth orbit (LEO) base.[1-3] In the aeroassist concept, high-altitude aerodynamic forces are used during sortie return for producing transition to base orbit, in order to avoid the propellant load required for transition by retrothrust.

Aeroassist orbital transfer vehicle (AOTV) design concepts tend to emerge into three categories: 1) low L/D, low ballistic coefficient aerobraking vehicles intended for essentially planar orbit operation; 2) moderate L/D, medium ballistic coefficient vehicles with capability for moderate orbital inclination changes; and 3) high L/D, high ballistic coefficient vehicles with the capability for producing significant changes in orbital inclination.[4,5]

The space basing generic to the AOTV concept requires that the thermal protection system (TPS) be reusable. As with guidance and control, structural as well as aerodynamic subsystems, the preliminary TPS design process and determination of enabling technologies are complicated by the wide variety of design categories, of possible missions (especially target orbit inclinations), along with their influence on the transatmospheric portion of the return trajectory. In particular, design of several AOTV subsystems is driven by the unique AOTV aerothermodynamic environment of high-altitude, high-enthalpy flow. Studies have shown that specific return skip trajectories for each of the three vehicle categories occur primarily in complex viscous-interaction and transitional flow regimes where the air ratio of specific heats and shock-layer gas chemistry are variable and dependent on speed and altitude.[3,4] These factors determine the most significant AOTV flowfield features such as bow shock position and shape, base region recirculation, and wake recompression. It has also been shown that the AOTV aerothermodynamic regime, especially for the low L/D configuration, features significant nonequilibrium flow and gas radiation in the shock layer.[6-10] These factors have a dominant influence on the aerodynamic forces and heat transfer imposed on the vehicle.

As described, the aerothermodynamic environment is bounded by the conditions at skip trajectory perigee. The environment can be characterized in terms of several dimensionless parameters such as Reynolds, Knudsen, Stanton, and Damköhler numbers, by the normalized bow shock standoff distance, and the viscous correlation parameter. The perigee values of these parameters depend on perigee altitude and speed, which in turn nominally depend on the initial entry conditions and vehicle L/D and $m/C_D A$.

Clearly, it would be very useful in preliminary design evaluations to have a simple and rapid analytical method for parametrically characterizing the perigee aero-

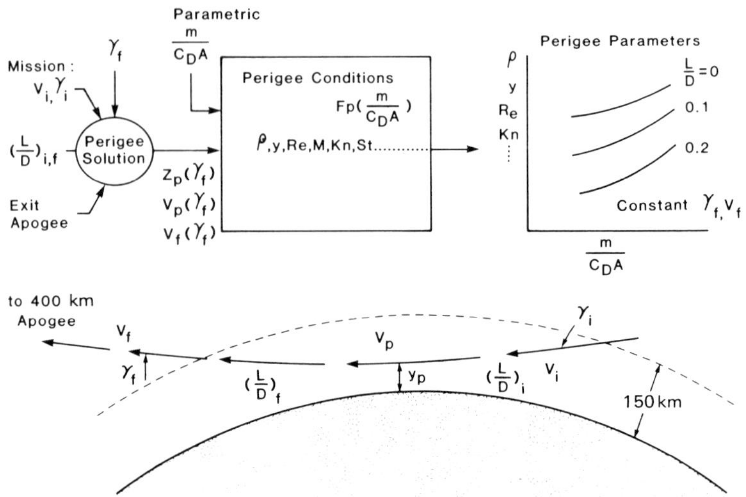

Fig. 1 Aerothermodynamic characterization by analytical perigee solution.

thermodynamic regime as a function of initial and final orbits as well as vehicle properties. Such a method hinges on a parametric solution of perigee altitude and speed; as described below, such a solution is not available in classical analytical trajectory solutions. The purpose of this paper is to derive an approximate analytical perigee solution method and to incorporate it into a parametric characterization of AOTV aerothermodynamic regimes at perigee. Fig. 1 shows the conceptual procedure for aerothermodynamic characterization of perigee conditions by the analytical method derived in this paper.

Trajectory Perigee Solution

Various attempts have been made since the late 1950s to reduce the coupled, nonlinear partial differential equations (PDE) governing lifting entry trajectory motion to a simpler set of ordinary differential equations (ODE) for which either analytical solutions could be found or parametric results could easily be numerically generated.[11-13] Most of these attempts were based on various approximations together with regroupings and normalizations of trajectory variables, sometimes resulting in newly defined special functions such as the Chapman Zeta function.[11] Descriptions of several of the schemes and their incorporation into "unified" theories is given by Loh[12] and by Vinh et al.[13]; the former in terms of an analytical solution, the latter in terms of differential equations. The solutions of Chapman and Loh will be briefly presented and noted to be incorrect for AOTV skip trajectories. The governing differential equations given by Vinh will be used for derivation of a new analytical solution for skip perigee altitude and velocity. It should be noted that past attempts focused on solution of the entire trajectory, while the present solution determines the conditions only at the perigee location.

Classical Solutions

Chapman[11] derived the following unified ODE from the planar planetary entry PDE written in horizontal, vertical axes:

$$\bar{u}\bar{Z}'' - \left(\bar{Z}' - \frac{\bar{Z}}{\bar{u}}\right) - \frac{1 - \bar{u}^2}{\bar{u}\bar{Z}} \cos^4 \gamma + \sqrt{\frac{r}{H}} \frac{L}{D} \cos^3 \gamma = 0 \tag{1}$$

where

$$\bar{Z} = \frac{H}{2}\left(\frac{C_D A}{m}\right)\sqrt{\frac{r}{H}}\, \rho \bar{u} \quad \text{Chapman Zeta function}$$

$$\bar{u} = \frac{V \cos \gamma}{V_c}$$

$$\bar{Z}', \bar{Z}'' = \frac{d\bar{Z}}{d\bar{u}}, \quad \frac{d^2\bar{Z}}{d\bar{u}^2}$$

The principal assumptions made in the work are valid for AOTV skip trajectories with low or moderate L/D. The following closed-form solution was also derived for the specific case of skip entry with constant L/D (assuming $\cos \gamma = 1$, $\sin \gamma = \gamma$):

$$\bar{Z} = \frac{\bar{u}}{\bar{u}_i} \bar{Z}_i + \bar{u} \sqrt{\frac{r}{H}} \gamma_i \log\left(\frac{\bar{u}}{\bar{u}_i}\right) - \frac{\bar{u}}{2} \sqrt{\frac{r}{H}} \left(\frac{L}{D}\right) \log^2\left(\frac{\bar{u}}{\bar{u}_i}\right) \tag{2}$$

Given L/D, initial entry speed $\bar{u}_i$, and angle γ_i, the general result is tabulations of $\bar{Z},\bar{u}$ as trajectory histories for any entry vehicle. The tables have been extended to less restrictive assumptions and nonplanar (constant bank angle) cases by Vinh et al.,[13] who have shown that Chapman's choice of $\bar{u}$ as an independent variable renders his solution totally inaccurate for high-altitude skip trajectories. In fact, both Eq. (2) and the numerical integration of Eq. (1) fail to predict the occurrence of perigee for a typical set of AOTV entry and vehicle conditions.

Loh[12] has derived a "unified" trajectory solution which consists of a coupled pair of first-order nonlinear ODE (trajectory axes) that lead for the case of constant L/D and m/C_DA to closed-form coupled $\gamma - \bar{u}$ solutions. When evaluated at perigee, these reduce to:

$$L/D = 0: \quad \frac{H}{r}\left(\frac{1}{\bar{u}_p^2} - 1\right) = \cos\gamma_i - 1 \tag{3}$$

and

$$L/D \neq 0: \quad 0 = \gamma_i + \frac{1}{2}\frac{L}{D}\log\left(\frac{\bar{u}_i^2}{\bar{u}_p^2}\right) \Big/ \left[\frac{H/r(1/\bar{u}_p^2 - 1)}{\cos\gamma_i - 1} - 1\right] \tag{4}$$

assuming $\rho_p \gg \rho_i$. For both cases, at any point in the trajectory,

$$\log\left(\frac{\bar{u}^2}{\bar{u}_i^2}\right) = \frac{H(C_DA/m)(\gamma - \gamma_i)}{H/2(L/D)(C_DA/m) - H/r(\cos\gamma/\rho)(1/\bar{u}^2 - 1)} \tag{5}$$

The method's assumptions which correspond with Chapman's theory are: 1) planar motion, 2) exponential atmosphere, and 3) small altitude compared to planetary radius. It is easy to show that Eq. (3) fails to predict even the occurrence of perigee solutions ($\gamma = 0$) for initial entry angles steeper than about -2 deg. Further, Eq. (4) is uselessly inaccurate for typical AOTV cases. For example, a GEO return (V_i = 9.8 km/s, γ_i = -6.10 deg) with L/D = 0.25 yields V_p = 6.2 km/s from Eq. (4) and Y_p = 69 km from Eq. (5) (for m/C_DA = 11 kg/m^2).

These compare very poorly with values determined from numerically integrated trajectories: $V_p = 8.6$ km/s, $Y_p = 82$ km. However, note that the Loh solution indicates (as does Chapman's) that the entry vector and L/D determine V_p, and then m/C_DA determines Y_p.

Perigee Solution

The present study has successfully derived an approximate analytical solution for perigee conditions as follows. Vinh et al.[12] derived exact three-dimensional entry equations of motion in an advantageous form independent of vehicle parameters. Assuming locally exponential atmosphere, the principal equations are

$$\frac{dZ}{ds} = -\frac{r}{H} Z \tan \gamma$$

$$\frac{du}{ds} = -\frac{2\sqrt{r/H}(Zu)}{\cos \gamma}\left[1 + \lambda \tan \gamma + \frac{1}{2Z}\sqrt{\frac{H}{r}} \sin \gamma\right]$$

$$\frac{d\gamma}{ds} = -\frac{\sqrt{r/H}(Z)}{\cos \gamma}\left[\lambda + \sqrt{\frac{H}{r}}\frac{\cos \gamma}{Z}\left(1 - \frac{\cos^2 \gamma}{u}\right)\right]$$

where

$$Z = \frac{H}{2}\left(\frac{C_DA}{m}\right)\sqrt{\frac{r}{H}}\,\rho \quad \text{Vinh Zeta function}$$

$$u = \frac{V^2 \cos^2 \gamma}{V_c^2}$$

$$\lambda = \frac{L}{D} \cos \sigma$$

σ = bank angle, and s = nondimensional arc length. Note that Z is proportional only to the exponential of altitude, i.e., a velocity factor does not appear as it does with the Chapman Zeta function. Further, u is proportional to V^2 unlike the Chapman parameter $\bar{u}$.

Specifying $\sqrt{r/H} = 30$ (Earth), $\sigma = 0$ (planar trajectory), and the small flight path angle case ($\cos\gamma = 1$, $\sin\gamma = \tan\gamma = \gamma$), we have

$$\frac{dZ}{ds} = -900\,\gamma Z \tag{6}$$

$$\frac{du}{ds} = -60uZ - 60uZ\gamma\left(\frac{L}{D}\right) - u\gamma \tag{7}$$

$$\frac{d\gamma}{ds} = 30Z\left(\frac{L}{D}\right) + 1 - \frac{1}{u} \tag{8}$$

Dividing Eq. (6) by Eq. (7), rearranging and integrating in two sections (from initial entry i to perigee p and from perigee p to exit f), we obtain

$$\int_{Z_i}^{Z_p}\left(\frac{1}{15\gamma} + \frac{1}{15}\frac{L}{D}\right)dZ + \frac{1}{900}\log\left(\frac{Z_p}{Z_i}\right) = \log\left(\frac{u_p}{u_i}\right) \tag{9}$$

and

$$\int_{Z_p}^{Z_f}\left(\frac{1}{15\gamma} + \frac{1}{15}\frac{L}{D}\right)dZ + \frac{1}{900}\log\left(\frac{Z_f}{Z_p}\right) = \log\left(\frac{u_f}{u_p}\right) \tag{10}$$

For constant m/C_DA and equal entry/exit altitudes, the addition of Eqs. (9) and (10) gives

$$\int_{Z_i}^{Z_p}\left(\frac{1}{15\gamma} + \frac{1}{15}\frac{L}{D}\right)dZ + \int_{Z_p}^{Z_f}\left(\frac{1}{15\gamma} + \frac{1}{15}\frac{L}{D}\right)dZ = \log\left(\frac{V_f}{V_i}\right)^2 \tag{11}$$

The two integrals in Eq. (11) do not cancel, even for constant L/D, because the variation of γ will be different in each case. The easiest way to evaluate the integrals is to assume a constant value of flight path angle within each integral represented as a weighting factor k times the initial or final (exit) angle. Typical k values are found from solution comparisons with numerically integrated trajectories to be 0.30 to 0.35. With this approach and noting that for constant m/C_DA (and equal entry/exit altitudes) $Z_i = Z_f$, there results

$$Z_p: \left[\frac{1 + k\gamma_i (L/D)_i}{15k\gamma_i} - \frac{1 + k\gamma_f (L/D)_f}{15k\gamma_f}\right] Z_p = 2 \log\left(\frac{V_f}{V_i}\right) \quad (12)$$

Eq. (12) represents the Z_p solution for prescribed L/D, entry, and exit conditions. The perigee Z value then determines the relationship between vehicle m/C_DA and perigee density, i.e., altitude; naturally, higher ballistic coefficients correspond to lower perigee altitudes (increased densities) for a given Z_p value. Perigee velocity can be found from either Eq. (9) or (10) for known Z_p. For example, from Eq. (9) (noting $Z_p >> Z_i$)

$$V_p: \log\left(\frac{V_p}{V_i}\right) = \frac{1}{2}\left[\frac{1 + k\gamma_i (L/D)_i}{15k\gamma_i}\right] Z_p \quad (13)$$

In summary, for an angle weighting factor k calibrated by comparison with generated trajectories, Eq. (12) yields solution of Z_p as a function of vehicle L/D and entry/exit V,γ values. With Z_p determined, perigee density (or altitude) can be determined as a function of vehicle m/C_DA. Finally, perigee velocity is found from Z_p using Eq. (13).

Note the form of the present solution as compared with previous solutions. Both Chapman[11] and Vinh et al.[13] found the important variables to be velocity ratioed to circular orbit velocity (or the ratio squared), and a Zeta function containing the ratio of air density to ballistic coefficient. In the present solution, like those of Chapman[11] and Vinh,[13] the specifying of vehicle L/D and entry vector determine Z as a function of u (velocity). However, the present solution shows that the exit vector may be specified with the result that the perigee Z_p value is uniquely determined. For example, Fig. 2 shows the exit V_f,γ_f relationship parametrically with Z_p for given k, L/D, and entry V_i,γ_i values. The dashed line superimposed on Fig. 2 represents the V_f,γ_f curve for producing a 400-km apogee after exit. Clearly, an exit V_f,γ_f pair may be chosen in Eq. (12) to give the desired post-exit apogee for AOTV circularization in LEO orbit. The exit conditions chosen, along with the entry V_i,γ_i pair and L/D determine Z_p. Table 1 shows perigee

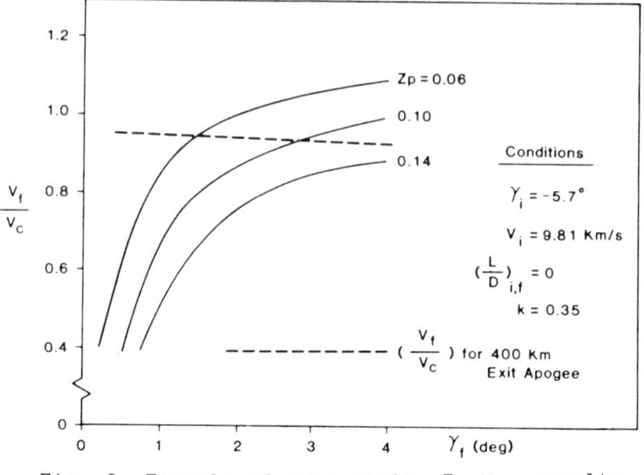

Fig. 2 Example of parametric Z_p, γ_f results.

Table 1 Comparison of analytical and simulation perigee solution

Return Mission	$\frac{L}{D}$	$\frac{m}{C_D A}$ (kg/m^2)	Entry Vector γ_i(deg)	Entry Vector V_i(Km/s)	Exit Apogee (Km)	Exit Vector γ_f(deg)	Exit Vector V_f(Km/s)	Perigee Solution-Analytical (Simulation) Density (gm/cm^3)	Altitude (Km)	Velocity (Km/s)
GEO	0.25	11.0	-6.10	9.81	402	2.63	7.32	1.02×10^{-8} (1.33×10^{-8})	84.9 (81.9)	8.99 (8.61)
	0.0	36.2	-5.79	9.81	399	2.24	7.35	2.86×10^{-8} (2.64×10^{-8})	77.4 (77.6)	9.05 (8.42)
	0.0	11.7	-5.51	9.81	452	2.31	7.37	9.23×10^{-9} (8.91×10^{-9})	85.5 (84.3)	9.02 (8.41)
	-0.1	11.0	-5.30	9.81	448	1.36	7.41	5.67×10^{-9} (5×10^{-9})	89.0 (87.6)	9.26 (8.82)
5×GEO	0.25	11.0	-6.71	10.41	400	3.00	7.29	1.38×10^{-8} (1.85×10^{-8})	82.6 (79.9)	9.34 (8.85)
L5	0.25	11.0	-5.79	10.95	404	2.89	7.76	1.51×10^{-8} (1.72×10^{-8})	81.9 (80.3)	9.68 (9.23)

parameters for several numerically simulated 3-DOF trajectories compared with corresponding parameters from the present analytical solution. The single k value (0.35) chosen for the present analytical solution results shown is the one which gives the best overall perigee density match. It is seen that the density/altitude matches are excellent; velocity from the present solution is often biased by about 0.5 Km/s. In any case, the simple analytical solution presented is impressive in view of the fact that previous classical solutions fail to even predict occurrence of a perigee. Also, the present solution is uniquely configured to allow choice of exit vector to

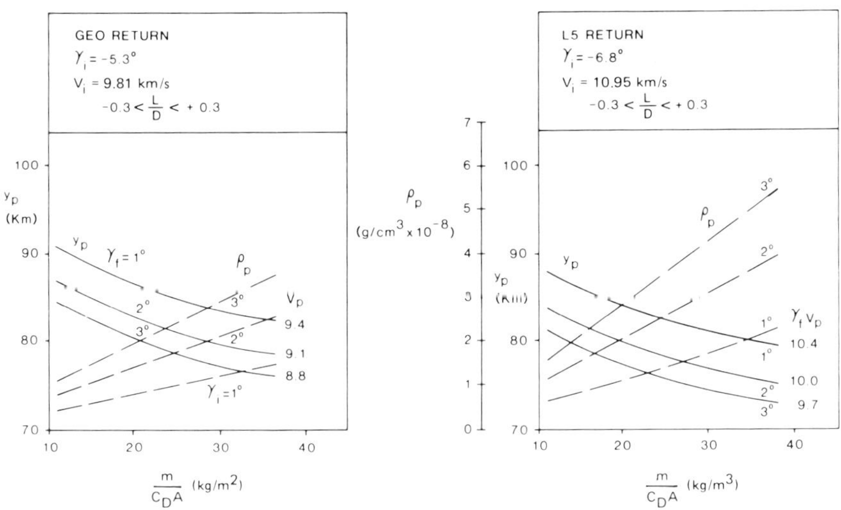

Fig. 3 Examples of perigee altitude and velocity results for GEO and L5 return.

provide a desired post-exit apogee. Fig. 3 shows GEO and L5 mission results from the present analytical method. Shown are perigee altitude, density, and velocity as parametric functions of exit angle and m/C_DA for constant 400-km post-exit apogee. Within the indicated L/D range, little sensitivity to L/D was noted.

Aerothermodynamic Parameters

With the present perigee solution, perigee freestream density and velocity can be studied parametrically as functions of vehicle L/D and m/C_DA for chosen entry V_i, γ_i and exit V_f, γ_f conditions. We capitalize on this solution by extending it to parametric evaluation of the perigee values of classic aerothermodynamic parameters. This proves valuable for preliminary vehicle aerothermodynamic design because the perigee conditions represent bounds on parameter values, i.e., the parameters generally vary monotonically from entry to perigee. Presented here is a brief discussion of each parameter included presently in the analysis; others may easily be added in future work.

A standard atmosphere table (1966 Fall/Spring, Mid-latitude Model) is used in this work for lookup values of

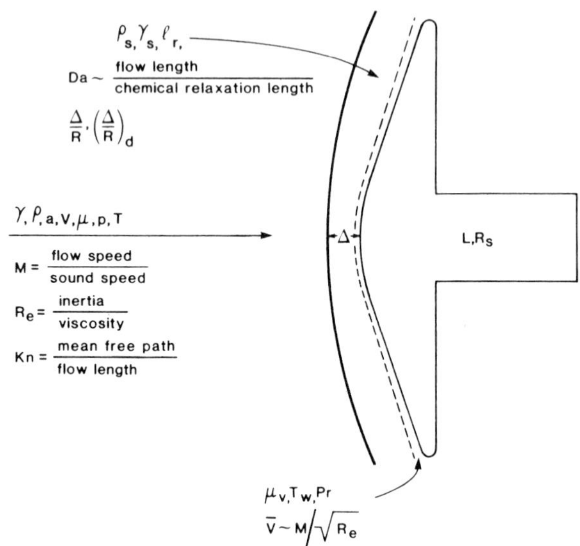

Fig. 4 Aerothermodynamic similarity parameters.

ambient atmospheric pressure, kinematic viscosity, and speed of sound as functions of altitude.

Fig. 4 indicates the dimensionless aerothermodynamic parameters considered and discussed in the present analysis.

Compressibility and Viscosity Effects

Overall flowfield compressibility and viscosity effects are essentially characterized by the freestream Mach number M and Reynolds number Re, respectively:

$$M = V/a$$
$$Re = \rho VL/\mu = VL/\nu \tag{14}$$

M represents the ratio of flow (bulk) speed to disturbance propagation speed (speed of sound), or equivalently, the ratio of "inertial" (momentum) forces to elastic forces. It can also be shown to equal the ratio of kinetic energy of the flow (bulk) speed to the kinetic energy of the mean molecular translational speed. All AOTV perigee regimes involve uniformly large Mach numbers

(20 to 40). Thus, the flow speed is sufficiently large, so that when converted by the bow shock into thermal (temperature) energy, substantial thermochemical changes occur in the gas.

The freestream Re represents the ratio of "inertial" to viscous forces and is useful for distinguishing laminar from turbulent flow regimes, as well as indicating the validity of the boundary-layer concept. For convenience, a unit Re (per meter flow characteristic length) is used:

$$Re' = Re/L = \rho V/\mu \qquad (m^{-1}) \tag{15}$$

Reynolds number regimes Re > 10,000, based on the flat plate analogy, provide assurance of separation of the shock layer into inviscid and viscous boundary-layer regions. Lower values, Re < 10,000, imply an entirely viscous flowfield.[14]

Low-Density and Rarefaction Effects

The Knudsen number Kn is often used to characterize low-density and rarefied flow regimes.[15] It represents the ratio of molecular mean free path length to flowfield characteristic length: $Kn = \ell_m/L$. The one extreme of free molecular flow involves isolated interaction between each individual gas molecule and the aerodynamic body. The opposite extreme of continuum flow involves complete molecular equilibration due to intermolecular collisions over the time/distance scales of wall impact. In between are the difficult regions typically delineated as "transition" and slip flow. Order-of-magnitude Kn estimates for the various regimes are:

continuum	$Kn < 0.01$
slip	$0.01 < Kn < 0.1$
transition	$0.1 < Kn < 1$
free molecular	$1 < Kn$

Kn can be shown to be a function of Mach and Reynolds numbers.[15] When based on the unit Re, the unit Kn is

$$Kn' = Kn \times L = 1.26\sqrt{\gamma}M/Re' \qquad (m) \tag{16}$$

Shock-Layer Viscous Interaction

High-altitude continuum and transitional flow aerodynamic data have been correlated for hypersonic shock-layer viscous interactions by various scalings of the power ratio form M^n/Re^m, of which the Kn mentioned previously is a classic example.[16] $\bar{V}$, the viscous-interaction parameter, is another such correlation parameter; it has been particularly useful for correlations of orbiter (shuttle) aerodynamic data in the viscous-interaction, high-altitude continuum regime.[4] It is directly related, in fact, to the Kn in the transitional flow regime and serves as a useful alternative parameter in the viscous-interaction regime. $\bar{V}$ effectively adjusts freestream temperature and viscosity to proper shock-layer values, and further accounts for adiabatic wall temperature in the thermal boundary-layer conditions. It is modeled here in the following form[4] where a Prandtl number Pr of 1 has been assumed and the use of unit Re leads to a unit $\bar{V}'$:

$$\bar{V}' = \bar{V}x\sqrt{L} = M\sqrt{C_v/Re'} \qquad (m^{1/2}) \tag{17}$$

where

$$C_v = \left(\frac{\mu_v}{\mu}\right)\left(\frac{T}{T_v}\right) = \sqrt{\frac{T_v}{T}}\left[\frac{T + 122.1(10^{-5}/T)}{T_v + 122.1(10^{-5}/T_v)}\right] \tag{18}$$

and

$$\frac{T_v}{T} = 0.468 + 0.532\,\frac{T_w}{T} + 0.195\,\frac{\gamma - 1}{2}\,M^2 \tag{19}$$

As indicated, the viscous-interaction temperature T_v is a function of freestream T and vehicle adiabatic wall temperature T_w. In the present calculations, T (the ambient atmospheric temperature at perigee) is inferred from the standard atmosphere speed of sound, and T_w is chosen as an input to the calculations. $\bar{V}$ ranges in value from about 0.005 to 0.1 for high-altitude continuum flow, and from about 0.1 to 4 for transitional flow.

Convective Heat Transfer

The Stanton number St proves useful for characterizing convective wall heat transfer regimes. Assuming unit thermal accommodation coefficient, it has the basic definition

$$St = \dot{q}/[\rho v(h_o - h_w)]$$

Physically, it represents the actual convective heat transfer rate to the wall ratioed to the available heat transfer rate based on the difference between freestream total enthalpy and the adiabatic wall enthalpy. It may be calculated for continuum and free-molecular flow regimes in terms of the unit St' as continuum[17] (Re' > 6000):

$$St' = St \times \sqrt{L} = 0.332 \frac{Pr^{-2/3}}{\sqrt{Re'}} \qquad (m^{1/2}) \qquad (20)$$

and free-molecular[18] (Re' < 6000):

$$St' = \frac{\gamma + 1}{4\gamma} \sqrt{\frac{2}{\pi\gamma}} \frac{1}{M} \qquad (-) \qquad (21)$$

Degree of Shock-Layer Relaxation

The important parameter characterizing the extent of relaxation processes in the shock layer is the first Damköhler number Da.[16] It represents the ratio of two characteristic time scales, flow time to relaxation time, or equivalently, the ratio of characteristic lengths $Da = L/\ell_r$. For Da >> 1, equilibrium for the process corresponding to the chosen relaxation time or distance is obtained very quickly, i.e., in a fraction of the flowfield characteristic length. This is the case of equilibrium flow. For Da << 1, the flow is "frozen" in that very little relaxation toward equilibrium for the process of interest occurs within flowfield distances. An essential data set for calculation of Da for nonequilibrium radiation is the nonequilibrium air radiation relaxation time under the freestream speeds and pressures relevant to

AOTV perigee. These data are available[19] in terms of the product ($pt_{0.1}$), i.e., freestream pressure times the time to 90% radiation relaxation in air. The time $t_{0.1}$ is experimentally based on the freestream velocity (shock velocity in the determining experiments). Thus,

$$Da = pL/(pt_{0.1})V$$

As with previous parameters, the calculation of Da may be performed for L = 1 m to give the unit Da'. However, it is more relevant to calculate unit Da' based on unit bow shock standoff distance (see below). When rewritten in terms of a unit vehicle equivalent spherical nose radius (based on actual bow shock standoff distance), this gives

$$Da' = Da \times \left(\frac{\Delta}{R}\right)\left(\frac{1}{L}\right) = \frac{\Delta}{R}\,\frac{p}{(pt_{0.1})v} \qquad (m^{-1}) \qquad (22)$$

Bow Shock Standoff Distance

Bow shock standoff distance may be estimated based on normalized spherical and flat nose disk standoff distances. In particular, the AOTV equivalent spherical nose radius based on bow shock standoff distance is of great interest. Published data have confirmed the basic mass flux conclusion that normalized standoff distance (Δ/R) is solely a function (ignoring viscous effects) of the ratio of freestream density to the average stagnation streamline density in the shock layer, $k_s = \rho/\rho_s$. For hypersonic flow, $k_s = (\gamma - 1)/(\gamma + 1)$; it is useful to use an "effective" shock layer γ_s as an input to the present calculations. The chosen γ_s value determines k_s which is then used in the published correlations for sphere and disk bow shock standoff distances[20]

$$\text{sphere:} \quad (\Delta/R) = k_s(1 - 2k_s + 5.5k_s^2) \qquad (-) \qquad (23)$$

$$\text{disk:} \quad (\Delta/R)_d = \sqrt{k_s}(1 + k_s) \qquad (-) \qquad (24)$$

where

$$k_s = \rho/\rho_s = (\gamma_s - 1)/(\gamma_s + 1) \qquad (25)$$

The bow shock standoff distance has global effects on the shock-layer flow conditions and volume (important to AOTV radiation heat transfer), and thus is a parameter of great interest. Typical values are 9% of sphere radius and 30% of disk radius for AOTV perigee conditions.

Examples of Aerothermodynamic Parameter Results

Results shown here for the present method are calculated as follows. Inputs to the calculations include: entry V_i, γ_i; a set of γ_f values; a desired post-exit apogee; vehicle L/D and a set of m/C_DA values; a choice for angle factor k; and Pr and adiabatic wall temperature T_w.

The method first solves for V_f values that correspond to γ_f values and the desired post-exit apogee, and then solves for Z_p parametrically with γ_f using Eq. (12). Equation (13) is then used to find V_p for each exit angle case. To complete the trajectory solution, perigee altitudes corresponding to the chosen m/C_DA values for each exit angle are found from the Vinh Zeta function definition. The various aerothermodynamic parameters discussed are calculated next, using the remaining inputs as well as Eqs. (14-25).

Figures 5-8 illustrate typical results derived by the present method. All four results correspond to choices of a 400-km post-exit apogee, Pr = 0.7 for Δ/R calculation, and T_w = 1000 K for $\bar{V}$ calculation. The results for the low L/D vehicle are shown for GEO and L5 return in Figs. 5 and 6, respectively; Figs. 7 and 8 show corresponding moderate L/D vehicle results. The utility of the present method for rapid characterization of the AOTV aerothermodynamic regime in preliminary design is illustrated by the conclusions that can be drawn from these example figures. Within a vehicle category:

1) Perigee parameter values and variations are not significantly different for the two missions.

2) Perigee altitude has only a moderate sensitivity to m/C_DA (<10 km), whereas almost as large a sensitivity occurs with exit angle (7 km). Low L/D perigees run 76 to 91 km and moderate L/D perigees run 59 to 75 km.

3) Unit Re roughly triples with the tripling of m/C_DA and doubles with the increase of exit angle from 1 to 3 deg. Re based on diameter is about 100,000 to

1,000,000 for a 20-m low L/D vehicle, about 150,000 to 1,500,000 for a 5-m moderate L/D vehicle. It is seen that laminar flow (Re < 300,000) dominates for low L/D vehicles on GEO return and exiting with γ_f < 2 deg. (Note that perigee Re is the maximum for the trajectory.) In contrast, the moderate L/D vehicle GEO return will involve significant turbulent flow even at γ_f = 1 deg.

4) For a 1-m shock standoff distance, the low L/D vehicle Da indicates that significant nonequilibrium flow will occur in the shock layer (Da < 1). The moderate L/D vehicle results indicate that equilibrium flow will dominate.

5) For both vehicle types, convective heat transfer as indicated by the low St will only be a small fraction of freestream enthalpy (<1%).

6) For L = 1 m, results for the viscous-interaction parameter and Kn indicate low L/D returns involve primarily viscous-interaction ($\bar{V}$ < 0.2), quasicontinuum low-density effects (Kn < 0.01) flow regimes. The moder-

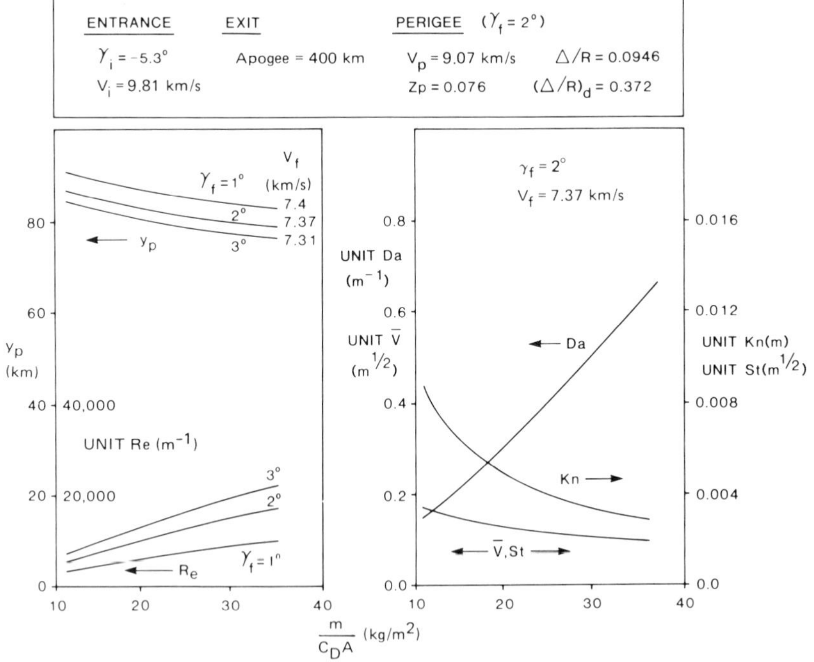

Fig. 5 Example of GEO-return results (low L/D).

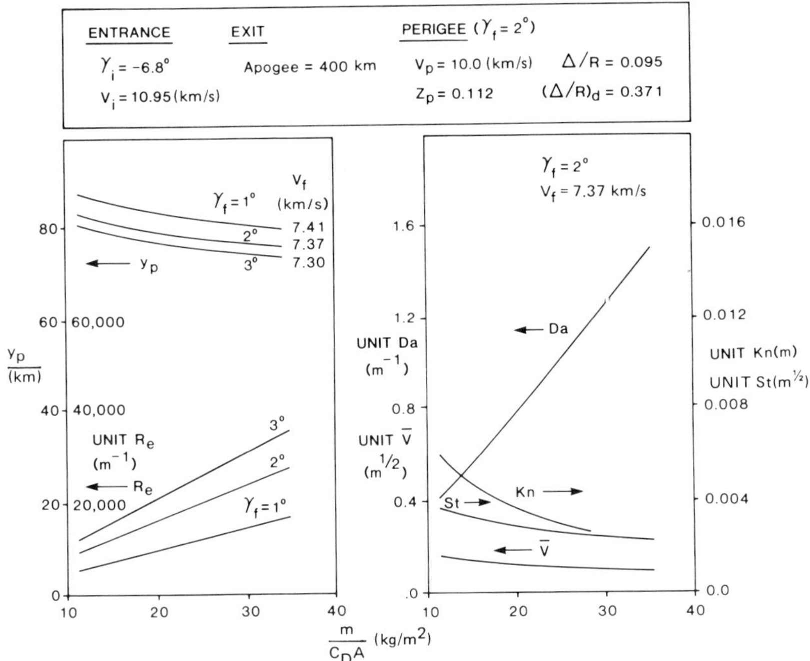

Fig. 6 Example of L5-return results (low L/D).

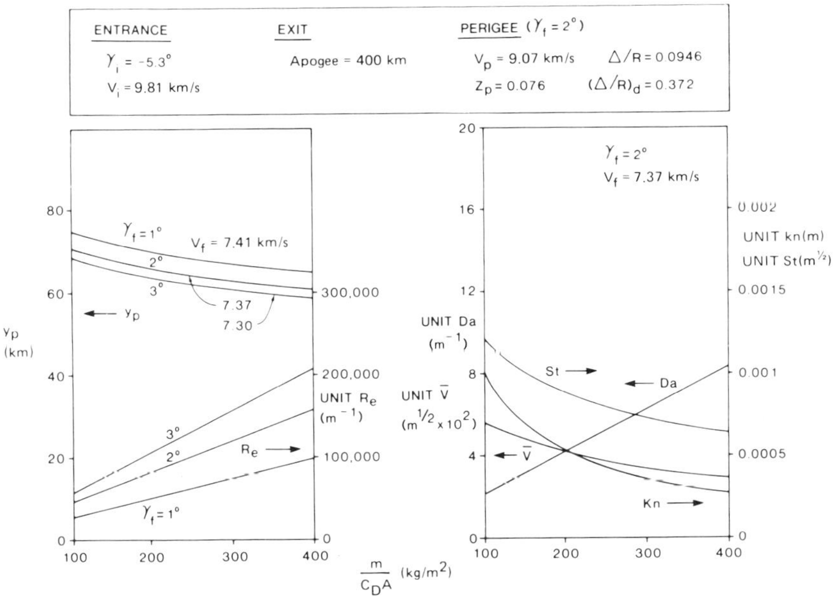

Fig. 7 Example of GEO-return results (moderate L/D).

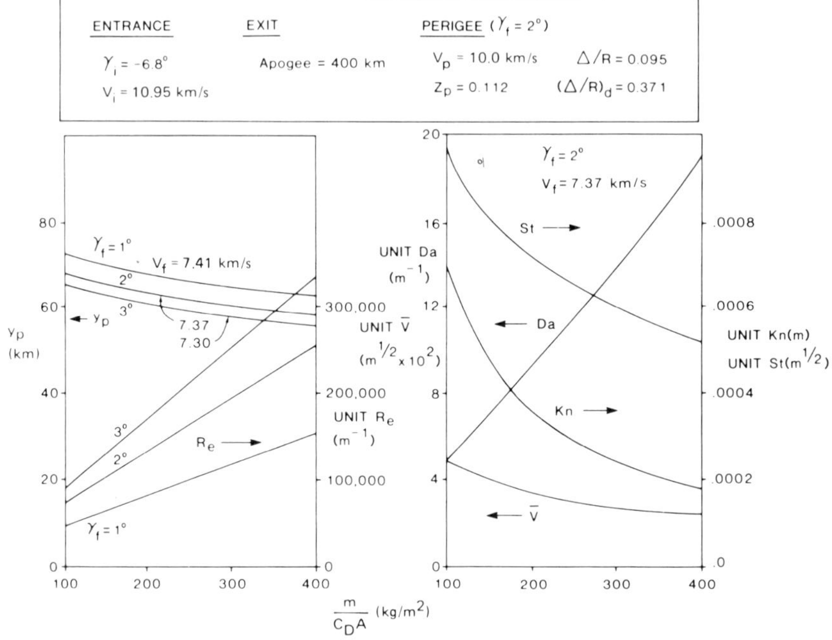

Fig. 8 Example of L5-return results (moderate L/D).

ate L/D results indicate dominance of viscous-interaction ($\bar{V} < 0.08$) and continuum ($Kn < 0.001$) flow regimes at perigee.

Conclusions

It is concluded that the present analytical method, through simple equations and plots, provides useful, accurate estimates of the perigee aerothermodynamic parameters for typical AOTV skip trajectories. These estimates show parametric dependencies on vehicle properties and entry/exit velocity vectors. The solution derived herein provides for the first time an employable, accurate analytical perigee velocity and altitude solution for nonlifting and small L/D skip trajectories. Previous classical solutions are either very inaccurate or fail to even predict occurrence of perigee.

Based on example results shown, the following contrasts in perigee conditions can be emphasized for the moderate L/D and low L/D AOTV categories:

1) Laminar flow will dominate for the low L/D vehicle; turbulent flow may dominate for the moderate L/D vehicle.

2) Nonequilibrium shock-layer flow will dominate for the low L/D vehicle; equilibrium flow for the moderate L/D vehicle.

3) Viscous-interaction and low-density quasicontinuum flow will dominate for the low L/D vehicle; low density effects will not be significant for the moderate L/D case.

Acknowledgments

With great pleasure, I thank Dr. Chul Park for suggesting the problem and for his encouragement and many kind suggestions. The work was sponsored in part by a NASA/ASEE Summer Faculty Fellowship (summer 1983, NASA Ames) and in part by NASA Ames University Consortium Joint Interchange NCA2-1R675-401.

References

[1]Walberg, G. D., "A Review of Aeroassisted Orbit Transfer," AIAA Paper 82-137, Aug. 1982.

[2]Howe, J. T., "Introductory Aerothermodynamics of Advanced Space Transportation Systems," AIAA Paper 83-0406, Jan. 1983.

[3]Talay, T. A., White, N. H., and Naftel, J. C., "Impact of Atmospheric Uncertainties and Viscous Interaction Effects on the Performance of Aero-Assisted Orbital Transfer Vehicles," AIAA Paper 84-0408, AIAA 22nd Aerospace Sciences Meeting, Reno, Nev., Jan. 1984; published elsewhere in this volume.

[4]Wilhite, A. W., Carrington, J. P., and McCandless, R. S., "Performance Aerodynamics of Aero-Assisted Orbital Transfer Vehicles," AIAA Paper 84-0406, AIAA 22nd Aerospace Sciences Meeting, Reno, Nev., Jan. 1984; published elsewhere in this volume.

[5]Powell, R. W., Stone, H. W., and Naftel, J. C., "Performance Evaluation of the Atmospheric Phase of Aeromaneuvering Orbital Transfer Vehicles," AIAA Paper 84-0405, AIAA 22nd Aerospace Sciences Meeting, Reno, Nev., Jan. 1984; published elsewhere in this volume.

[6]Menees, G. P., "Trajectory Analysis of Radiative Heating for Planetary Missions with Aerobraking of Spacecraft," AIAA Paper 83-0407, Jan. 1983.

[7]Park, C., "Radiation Enhancement by Nonequilibrium in Earth's Atmosphere," AIAA Paper 83-0410, Jan. 1983.

[8]Menees, G. P., "Thermal-Protection Requirements for Near-Earth Aeroassisted Orbital-Transfer Vehicle Missions," AIAA Paper 83-1513, AIAA 18th Thermophysics Conference, Montreal, Canada, June 1983; published elsewhere in this volume.

[9]Menees, G. P., Park, C., and Wilson, J. F., "Design and Performance Analysis of a Conical Orbital Transfer Vehicle Concept," AIAA Paper 84-0410, AIAA 22nd Aerospace Sciences Meeting, Reno, Nev., Jan. 1984; published elsewhere in this volume.

[10]Park, C., "Calculation of Nonequilibrium Radiation in AOTV Flight Regimes," AIAA Paper 84-0306, AIAA 22nd Aerospace Sciences Meeting, Reno, Nev., Jan. 1984; published elsewhere in this volume.

[11]Chapman, D. R., "An Approximate Analytical Method for Studying Entry into Planetary Atmospheres," NACA TN 4276, May 1958.

[12]Loh, W. H. T., Ed., Re-entry and Planetary Entry Physics and Technology, Vol. 1, Springer-Verlag, 1968, pp. 26-29.

[13]Vinh, N. X., Busemann, A., and Culp, R. D., Hypersonic and Planetary Entry Flight Mechanics, University of Michigan Press, 1980, pp. 226-245.

[14]White, F. M., Fluid Mechanics, McGraw-Hill, 1979, Chap. 7.

[15]Schaaf, S. A. and Chambre, P. L., Flow of Rarefied Gases, Princeton Aeronautical Paperbacks, Princeton University Press, 1961, pp. 3-7.

[16]Zierep, J., Similarity Laws and Modeling, Vol. 2 of Gasdynamics, edited by P. Wegener, Marcel Dekker, 1971, pp. 121-127.

[17]Kays, W. M., Convective Heat and Mass Transfer, McGraw-Hill, 1966, p. 138.

[18]Pan, Y. S. and Probstein, R. F., "Rarefied-Flow Transition at a Leading Edge," in Fundamental Phenomena in Hypersonic Flow, edited by J. Gordon Hill, 1966, p. 293.

[19]Allen, R. A., Rose, P. H., and Camm, J. C., "Nonequilibrium and Equilibrium Radiation at Super-Satellite Re-entry Velocities," AVCO Research Rept. 156, Sept. 1962.

[20]Park, C., "Calculation of Radiation from Argon Shock Layers," J. Quant. Spectrosc. Radiat. Transfer, Vol. 29, No. 1, Jan. 1982.

Chapter III. Thermal Protection

Thermal Protection Requirements for Near-Earth Aeroassisted Orbital Transfer Vehicle Missions

Gene P. Menees*
NASA Ames Research Center, Moffett Field, California

Abstract

The thermal protection required for decelerating and maneuvering spacecraft by aerodynamic forces is determined for return missions from geosynchronous to low Earth orbits. The effect of vehicle configuration on surface heating rates and selection of heat shield materials is analyzed. The effects of the current widespread estimates in the structure of atmospheric density are also evaluated. It is shown that nonequilibrium radiation can be a major source of surface heating during atmospheric entry and a significant factor for heat shielding requirements. It is also demonstrated that drag-brake concepts have application to a broad range of orbital transfer missions because of the favorable tradeoffs with aeromaneuvering vehicles in volumetric efficiency, retrothrust plane-change capability, and heat protection requirements. In addition, the results of this study indicate that the aeroassist technique produces acceptable penalties in vehicle payload capacity for drag-brake concepts, because of the system's heat protection requirements, and is highly attractive relative to all-propulsive orbital change maneuvers.

Nomenclature

AOTV = aeroassisted orbital transfer vehicle
d = diameter
GEO = geosynchronous orbit
H = altitude

Presented as Paper 83-1513 at AIAA 18th Thermophysics Conference, Montreal, Canada, June 1-3, 1983.

*Research Scientist.

I_p = specific impulse of liquid rocket engine
L/D = lift-drag ratio
LEO = low Earth orbit
l = length
Q = total heat load over flight trajectory
q = heat flux
r_b = base radius
r_n = nose radius
s = surface distance
T = temperature
t = flight time
V_∞ = flight velocity
V_1 = apogee velocity of GEO-LEO descent orbit
V_2 = orbital velocity of aeroassist orbit at LEO
V_C = orbital circular velocity
ΔV_1 = decircularization velocity pulse at GEO (retrothrust) = $|\vec{V}_1 - \vec{V}_C|$
ΔV_2 = circularization velocity pulse at LEO (thrust) = $|\vec{V}_2 - \vec{V}_C|$
α = angle of attack
α_λ = spectral absorptivity
β = ballistic coefficient
ε_λ = spectral emissivity (= α_λ by Kirchoff's law)
λ = wavelength
ϕ = bank angle of AOTV lift-plane vector
θ = orbital inclination from equatorial plane
$\Delta\theta$ = change in orbital inclination

Subscripts

c = convective
e = emitted by equilibrium black-body radiation
i = incident to surface
R = reflected
r = radiative
s = stagnation point
w = wake

Introduction

One of the most difficult operational problems required for advanced space transportation systems to achieve high-frequency, low-cost access to near-Earth space is to efficiently descend from distant high orbits (e.g., geosynchronous) to low Earth orbits (e.g., Space Shuttle). Such missions are comparable in difficulty to lunar return and require a large propellant weight to accomplish deceleration to low orbital velocities with retrothrust. Previous studies[1] have shown that the use of aerodynamic forces to achieve the velocity decrement required for orbital transfer by a single pass or successive grazing passes through the upper atmosphere is not only feasible but highly attractive. Substantial gains in payload capacity are obtained when compared with all-propulsive maneuvers on a total-system weight basis because of the much smaller expenditure of propulsion energy.

A variety of configurations have been proposed, in conceptual design studies spanning the last two decades, for transport of material and personnel between Earth and low Earth orbits and thence between the low Earth and higher orbits. For vehicles with aeroassisted orbital transfer applications (AOTVs), the aerothermal environments have generally been analyzed to determine thermal protection requirements for convective and equilibrium radiative heating rates.[2] Recent studies[3,4] using contemporary state-of-the-art spectral physics and radiative-transport prediction techniques, however, have shown that nonequilibrium radiative heating may be a significant factor in the thermal protection system design. The nonequilibrium radiative-heating rates were found to be enhanced over those for equilibrium for atmospheric entry trajectories encompassing a wide range of mission scenarios and vehicle shapes. The enhancement effect occurs because AOTVs are typically large-scale vehicles designed to perform aerobraking maneuvers in hypervelocity, low-density flight regimes of the upper atmosphere where thermochemical relaxation times are significant.

The purpose of this paper is to extend the previous exploratory studies[4] to include AOTV concepts that may have early operational capability for return missions from geosynchronous to low Earth orbits. Parametric comparison studies of previously proposed vehicle shapes are made to identify the most attractive AOTV design features on the basis of volumetric efficiency and aerodynamic characteristics, as well as aerothermodynamic heating. As a result of the initial tradeoff studies, one configuration is selected for detailed analysis of the heat protection requirements. Design constraints for materials selection are reusability and current state-of-the-art knowledge for fabrication and application to the predicted heating environments and conditions of hard space (i.e., no new technology). The required vehicle system's heat shield weight is estimated, and potential aerothermodynamic problem areas are discussed. The results of this study provide guidelines for the design of new AOTV concepts to meet specific mission requirements.

Scope of Analysis

Mission Requirements

The study is conducted for entry conditions corresponding to aeroassisted returns from geosynchronous to low Earth orbits at 400 km altitude (GEO-LEO). Two cases are considered, which are accomplished in single atmospheric passes to minimize turnaround time: 1) a simple coplanar return using aerobraking, and 2) a more complex synergistic plane-change return using aeromaneuvering to achieve simultaneous altitude and inclination orbital transfers. The GEO-to-LEO mission is selected because the bulk of near-future space activity will undoubtedly occur within this altitude range. The location of a space station at GEO is a future certainty because of its importance for scientific, commercial, and strategic applications. Consequently, frequent commuting between LEO and GEO for supply and maintenance will be essential. The corresponding thermal protection requirements will also serve to satisfy those of many in-between orbital change missions.

In the course of this work, it was discovered that orbital plane-inclination changes encompassing about 30° could be obtained by retrothrust maneuvers at GEO with a penalty in vehicle payload capacity of only about 10%, and inclinations as great as 60° with a penalty of about 30%. These results indicate that drag-brake concepts are competitive with lifting vehicles in achieving the 28.5° plane change required for Shuttle rendezvous and, perhaps, for greater inclinations, depending on the tradeoffs in volumetric efficiency and heat protection requirements. It is well known, for example, that lift is generally obtained at a significant sacrifice in volumetric efficiency.[1] In addition, more severe thermal protection requirements occur because of complex shock-interaction effects, which produce high, localized heat fluxes at the leading edge of lifting surfaces. Consequently, a conservative estimate is that AOTV configurations with significant plane-change capability are most advantageously used for orbital change missions with inclinations greater than that of the Shuttle orbit (e.g., satellites having inclinations as great as 90° for polar orbiters). This concept is illustrated in Fig. 1, which shows schematics for both aerobraking and aeromaneuvering orbital transfers relative to the Shuttle orbit. The penalties in vehicle payload capacity caused by propulsive plane-inclination changes at GEO are shown in Fig. 2. The retrothrust velocity increments for inclinations as great as 90° are shown in Fig. 2a, and the corresponding payload mass sacrificed because of increased propellant requirements are given in Fig. 2b for two liquid rocket engines (e.g., current and future storable propellant capability).

Atmospheric Uncertainties

An attempt is also made in this study to account for the random, unpredictable variations in atmospheric properties on the aerothermodynamic heating rates. Considerable uncertainty still exists regarding the extent of changes in temperature, pressure, and density distributions in the upper atmosphere as a function of time of day, season, and solar activity. This issue was discussed previously[1] and recommendations were given to compensate for atmospheric variations in typical AOTV flight regimes. These results were incorporated into the present study by determining the effect of ±50% dispersions in the 1962 U.S. standard atmosphere density distribution on the predicted aerothermodynamic heating rates.

Aerothermodynamic Heating

Radiation. Complete details of the computational procedure for obtaining the present predictions of the nonequilibrium and equilibrium radiative emission from a reacting gas mixture were given previously.[3,4] In general, the method involves the numerical calculation of the chemical reactions for inviscid flow behind oblique shock waves at conditions simulating specified locations on entry vehicle configurations. Only stagnation point heating rates are determined in the present work. This case corresponds to the flow behind a normal shock wave and, consequently, simulates the flow downstream of the shock front in a shock tube. The computational procedure is approximately correct for axisymmetric entry-body configurations. It is the only method currently available for this work, since fully coupled CFD codes modeling all the relevant physics of the shock-layer flowfield are not yet available.

The radiative transfer due to the nonequilibrium chemistry is determined by accounting for the deviation from equilibrium of the electronic state populations, using state-of-the-art spectral physics. The present results represent the incident radiative flux at the boundary-layer edge, since the calculations concern only the inviscid region of the shock layer. The effect of boundary-layer thickness in truncating the radiative emission below the optically thick limit is not accounted for in this study.

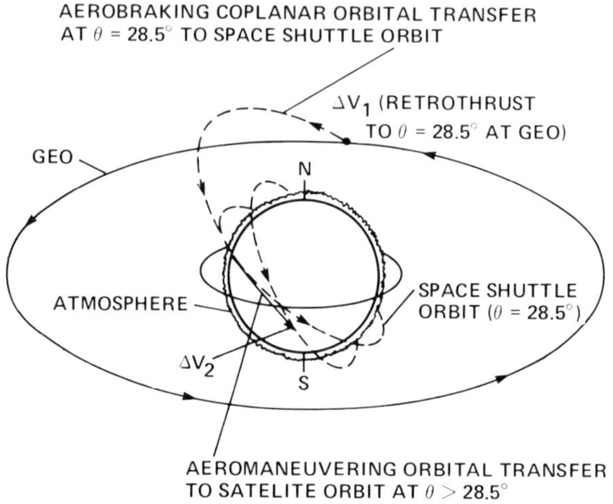

Fig. 1 Schematic of aeroassisted return missions from geosynchronous orbit to low Earth orbits.

However, the effect of the high absorptivity of the cooler boundary-layer gases in blocking the incident radiative flux below 0.2 μm is estimated by specifying this value as a lower limit in the calculations. Experience with entry-body flowfields indicates that most of the radiative flux below this spectral frequency range is absorbed in the boundary layer and that above transmitted through the boundary layer. In addition, the present method does not account for low-density phenomena of upper atmosphere flight regimes. However, the method is applicable over the important heating range of the entry cases considered herein, since an examination of Knudsen numbers shows the flow to be in the continuum or pseudocontinuum regime.

The predicted results have been verified by comparisons with pre-Apollo-era shock-tube and ballistic-range data for blunt bodies at reentry velocities spanning the mission range of Earth orbital, lunar return, and return from the near planets. The method is generally considered capable of determining the radiative heating to within a factor of 2 for hypervelocity flight in air. In addition, the calculations are based on the nonequilibrium shock-layer shape. Consequently, the equilibrium radiative results are overpredicted by about a factor of 2.

Convection. The stagnation point heat-transfer rates were determined from the Fay-Riddell expression[5] for dissociated air because of the hypervelocity entry conditions.

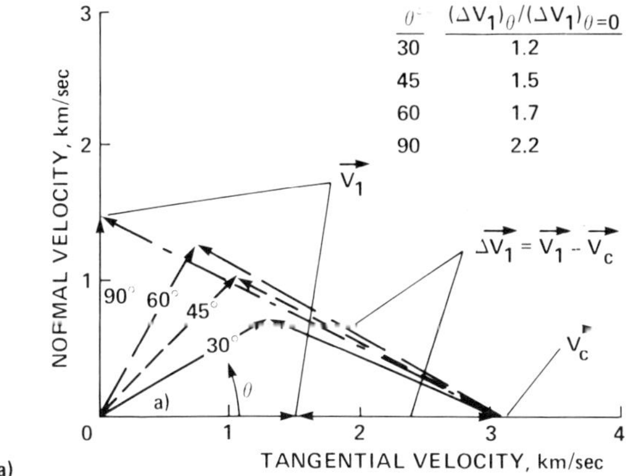

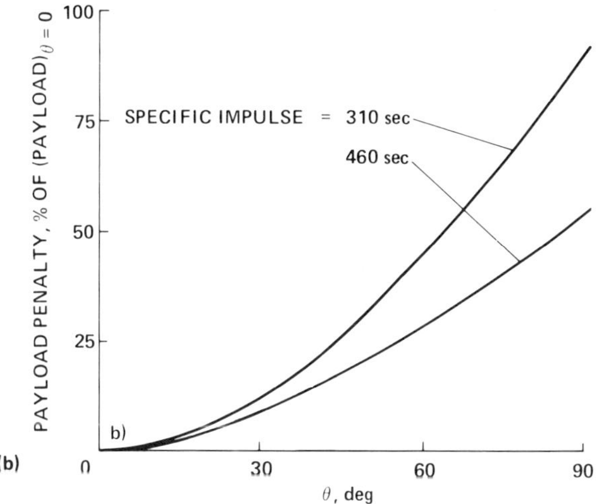

Fig. 2 Retrothrust requirements for GEO in-orbit plane-inclination changes for LEO descent orbits: a) velocity vector diagram; b) propulsion fuel mass penalties.

For downstream locations on selected configurations, heating rates were estimated from the results of calculations for similar configurations and flight regime conditions.

A major factor in the convective heating calculations is transport properties as expressed through the Prandtl and Schmidt numbers. These parameters are strongly affected by the nonequilibrium relaxation phenomena in AOTV flight regimes, and the wide variation in magnitude has never been adequately resolved. An extensive study is in progress to upgrade our knowledge of these important quantities. The

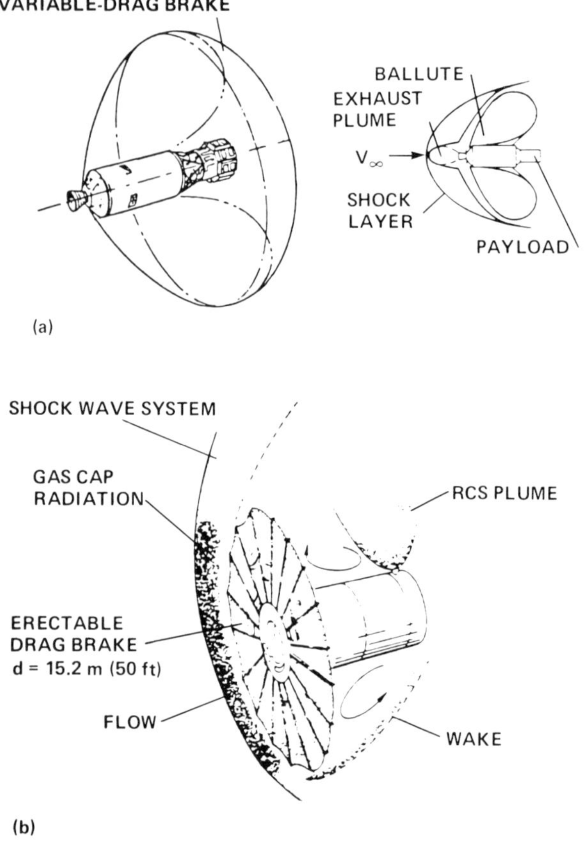

Fig. 3 Proposed AOTV configurations: a) ballute (Ref. 6); b) lifting brake (Ref. 7).

values used herein are typical of those that have been used traditionally for air.

Results and Discussion

Vehicle Parametric Comparisons

The selection of AOTV configurations considered in this study was restricted to the open literature, because this process ensures some detailed design analysis. In addition, the study was limited to shapes constrained to fit the Space Shuttle cargo bay to enhance early operational feasibility. The four configurations subsequently chosen for aerothermodynamic analysis are illustrated in

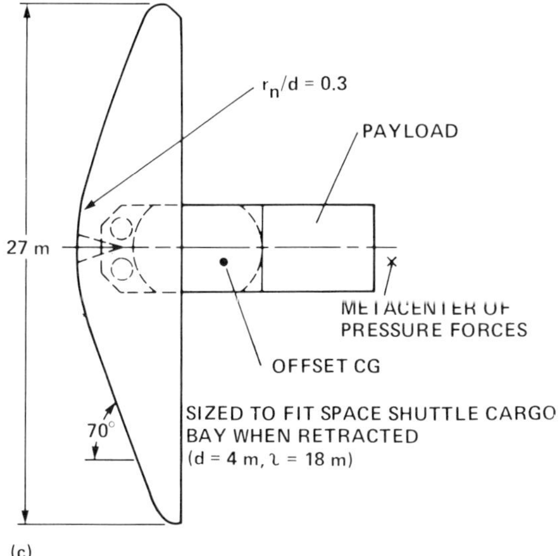

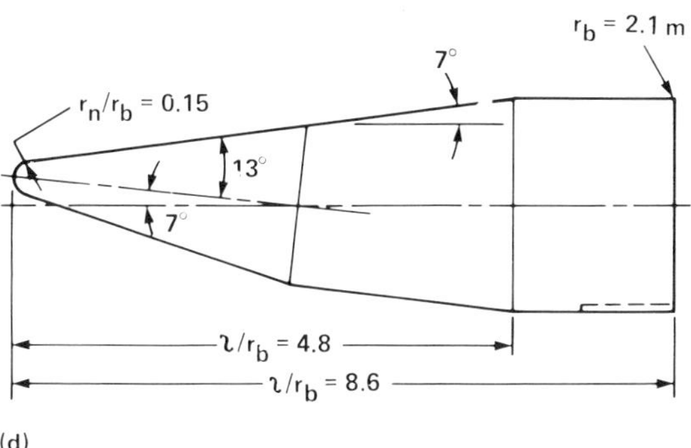

Fig. 3 (continued) Proposed AOTV configurations: c) 3° asymmetric conical lifting brake (Ref. 8); d) bent biconic (Ref. 9).

Fig. 3. Three of the vehicles use large-diameter deployable drag devices and are intended only for altitude-change aeroassisted orbital transfers. Such designs provide low lift-drag ratios and ballistic coefficients to produce flight trajectories with perigee locations in the upper rarefied regions of the atmosphere to minimize aerothermal heating effects. They also have maximum volumetric efficiency, because the vehicle shape is cylindrical and sized to the Shuttle cargo bay. These "aerobraking" configurations are the ballute,[6] lifting brake,[7] and conical lifting

Table 1 AOTV aerodynamic parameters[a]

Configuration	L/D	β, kg/m²	α, deg	Significant dimension, m
Ballute	0	39	0	d = 18.3 drag brake
Lifting brake	±0.2	30	±25	d = 15.2 drag brake
Conical lifting brake	±0.3	11	±20	d = 27 drag brake
Bent biconic	±2	800	±25	r_n = 0.32

[a]All vehicle entry weights = 10,886 kg (12 tons); all vehicles sized to fit Space Shuttle cargo bay with drag brakes retracted.

brake[8] and are shown in Figs. 3a, 3b, and 3c, respectively. The other AOTV concept included in the study is the bent biconic,[9,10] shown in Fig. 3d. This configuration has moderately high lifting capability to provide some synergistic plane-inclination change with altitude orbital transfer, which is obtained at the expense of volumetric efficiency. Such "aeromaneuvering" vehicles must also have significant drag to expend the large excess velocity required for descent from high altitudes. However, the ballistic coefficients are much larger than those of the drag brakes because of the comparatively smaller drag surfaces. The perigee locations of the flight trajectories occur much deeper in the higher-density regions of the lower atmosphere. Consequently, aeromaneuvering designs are penalized by much higher aerothermodynamic heating rates than the drag brakes, which places severe requirements on the thermal protection systems. For example, the only other aeromaneuvering concept extensively analyzed in the literature [i.e., the aeromaneuvering orbit-to-orbit shuttle (AMOOS)] was found to require ablating heat shield materials.[11] The significant aerodynamic parameters of the various AOTV configurations required in the course of this study are given in Table 1.

Aerodynamics. The ballute is a nonlifting, inflatable, balloonlike structure that surrounds the AOTV during atmospheric entry and serves as a variable-drag device that must respond on demand to accommodate atmospheric variations. Longitudinal stability problems can result if the c.p. is not carefully controlled relative to the c.g. during the drag-modulation procedure. A more significant shortcoming, however, is that the variable-drag technique provides only one-dimensional aerodynamic guidance capability. Consequently, severe directional stability problems may occur due to the turbulence and density dispersions resulting from atmospheric instabilities. The additional degree of freedom provided by lifting vehicles for guidance corrections is unavailable in the ballute design. A reactive control system must be used to compensate for this shortcoming but this seems difficult to implement, since the ballute essentially envelopes the spacecraft during the atmospheric pass.

The lifting brake is a constant drag concept, which has a small variable-lift capability to provide the maneuverability to compensate for atmospheric dispersions. It is inherently unstable because of the position of the erectable drag brake. This produces a negative static margin resulting from the front pressure surface and rear cargo location. Nevertheless, the simplicity of this design makes it one of the most viable AOTV candidates for achieving early operational capability. The stability problem can be corrected by inclining the brake backward at an angle which ensures that the metacenter of pressure forces is positive (i.e., aft of the center-of-gravity location). This provides longitudinal stability in the same manner that a ship is roll-stabilized. In addition, the original design concept required the drag brake to be foldable, which is difficult to achieve since it deforms under heat loads and aerodynamic forces. This requirement seems unnecessary, because the vehicle will remain in orbit outside the atmosphere between missions. Consequently, the brake may remain deployed until it is no longer useful and then be discarded in space. A new brake may possibly be installed in space or the vehicle returned to Earth for refurbishment.

The above features are incorporated into the conical lifting brake, which, in addition, has other design improvements. The frustum is contoured, for example, to alleviate the high edge-heating effects. Furthermore, the geometry incorporates asymmetry, which overcomes the roll-instability characteristic of symmetric shapes. Moreover, the diameter of the drag brake is selected to prevent impingement of the wake flow on the vehicle surface; this avoids increased TPS requirements caused by high local heat fluxes.

The bent biconic represents a family of vehicle geometries, which were developed as generic planetary aerocapture vehicles and have received widespread theoretical and experimental analysis.[4,8-10,12-14] The design achieves a variable range of lift-drag ratios, between about ±2, by roll-control modulation of the flight-plane lift vector. The stability problems are severe because of the rear cargo location and forward center-of-pressure position, which results from the high angle-of-attack necessary to achieve maximum lift-drag ratio. In addition, the vehicle is constrained to a narrow stability range and essentially a

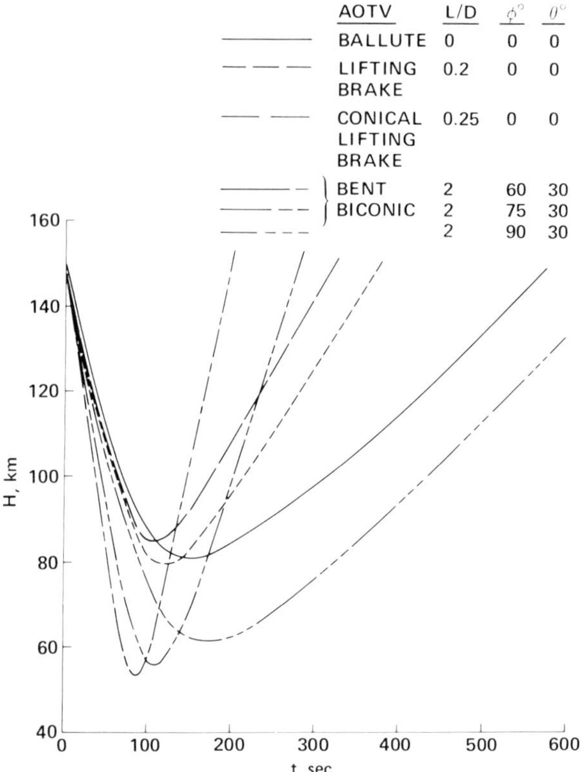

Fig. 4 Flight trajectories for a 50% overshoot of the 1962 U.S. standard atmosphere density distribution.

single trim angle of attack, because typical AOTV mission profiles require a variable cargo capacity. This causes large shifts in the c.g. position that must be accommodated by the aft-end control flap. The larger rear cone does, however, act as a vertical stabilizer to improve lateral stability.

Aerothermodynamic Heating. The surface heating characteristics of the various AOTV configurations are obtained from the flight histories shown in Fig. 4. The calculations were obtained for the same entry mass for all cases [10,886 kg (12 tons)] to provide a common basis for comparison. This value is somewhat a future expectation, but is used herein to provide conservative estimates and a design cushion for the more severe future requirements of returns from distances greater than GEO. In addition, results are shown for the 50% overshoot of the 1962 U.S.

standard atmosphere density distribution, since this case produces the worst heating conditions in the study. The trajectories of the three drag-brake vehicles have much higher perigee altitudes than the bent biconic because of the large differences in ballistic coefficient. The three cases shown for the bent biconic were selected to illustrate the differences in flight characteristics for various bank angles, which produce a plane-inclination change of 30°. This value was selected as a practical goal to justify the sacrifice in volumetric efficiency characteristic of aeromaneuvering vehicles. The important effect obtained is that increasing the bank angle raises the perigee altitude. This has a significant impact on surface heating rates, as will be shown subsequently.

Predicted stagnation point aerothermodynamic heating rates, which are the maximum surface values, are given in Figs. 5-8. The present analysis showed that the effect of dispersions in the atmospheric density distribution varies inversely with altitude. The differences between the results for the 50% overshoot and undershoot of the standard atmosphere are small at high altitudes (about 10% above 80 km) and relatively large at low altitudes (as much as 50% below 50 km). Consequently, the effect of atmospheric density variations on heat protection requirements is generally within the accuracy of prediction techniques for the altitude range where AOTVs are constrained to operate.

The results for the three drag-brake vehicles are compared in Fig. 5 and are seen to be of the same relative magnitude for both radiative and convective heating. Noncatalytic convective fluxes are given here, since it is expected that typical drag-brake surface materials will have low catalysis. The nonequilibrium radiative flux distributions greatly exceed those for equilibrium radiation and convection over the range of the heating pulse. This effect is expected because the hypervelocity, rarefied, upper-altitude flight regimes of these large-scale bodies provide the conditions that make nonequilibrium effects dominant. For the ballute, the combined magnitude of the nonequilibrium radiative and convective heat fluxes near peak heating (~35 W/cm^2) is well in excess of the capability of the material proposed for the thermal protection of this structure.[6] This occurs because the bulk of the heating intensity is radiative, which is not significantly affected by the derived cooling of the reverse propulsion motor flow. In fact, there is no existing flexible-reusable material that will accommodate the predicted surface heat fluxes. For the lifting brake, the present results for the nonequilibrium radiative flux exceed the

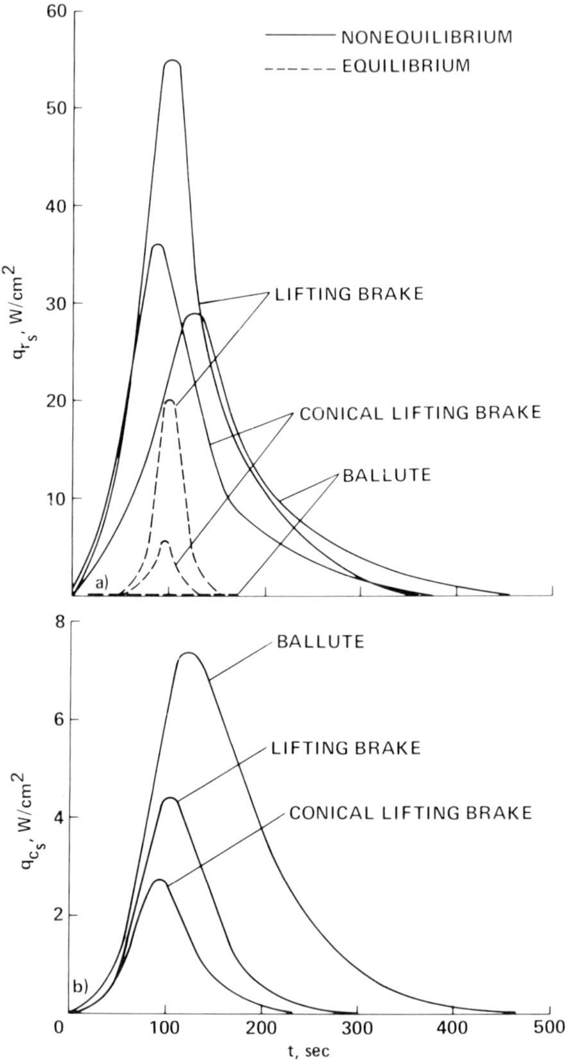

Fig. 5 Stagnation point aerothermodynamic heating rate distributions for drag-brake AOTV configurations: a) radiative; b) noncatalytic laminar convective.

previous predictions for the equilibrium case by more than an order of magnitude. In addition, the combined value of the peak nonequilibrium radiative and convective fluxes is more than a factor of 4 greater than the previous results.[7] For the conical lifting brake, no previous results are available for comparison, since the present aerothermal analysis is the first for this shape.

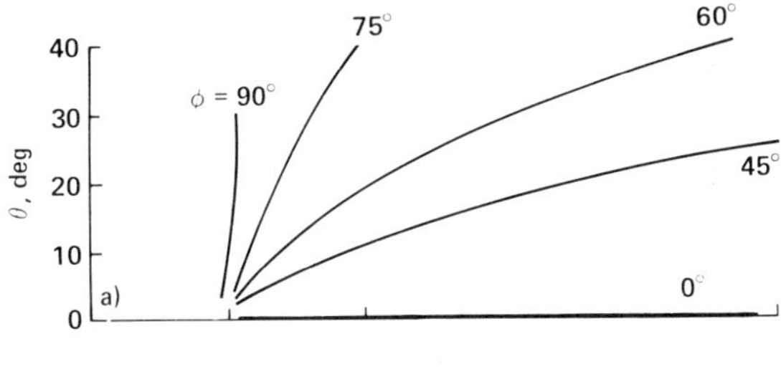

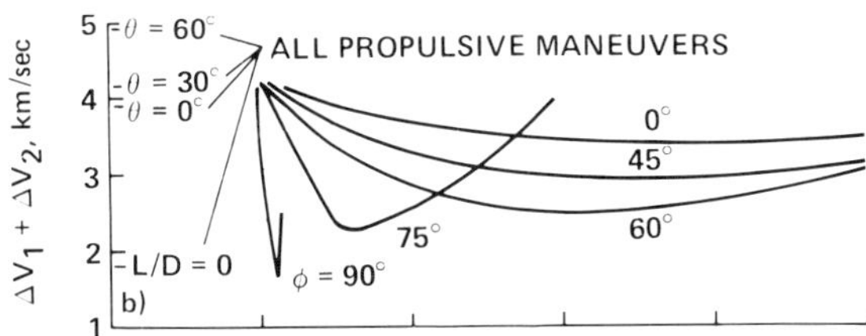

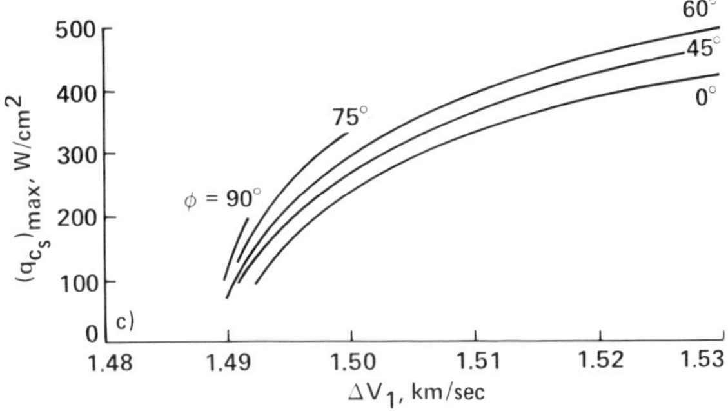

Fig. 6 Plane-inclination change characteristics of bent biconic for GEO-LEO return at L/D = 2: a) plane-inclination change; b) propulsive thrust requirements; c) stagnation point noncatalytic convective heating.

Exploratory studies of the plane-inclination change characteristics of aeromaneuvering vehicles showed that the aerothermodynamic heating effects are strong functions of the retrothrust velocity pulse at GEO (ΔV_1) and the vehicle bank angle (ϕ). These two factors determine the flight trajectories, of which there are an infinite variety. Consequently, it is essential to define the conditions that provide the minimum surface heating effects with optimum plane-change capability. Results for the bent biconic at its maximum lift-drag ratio of 2 are given in Fig. 6. The sensitivity of plane-inclination change (θ) to variations in ΔV_1 and ϕ is illustrated in Fig. 6a. The total pro-

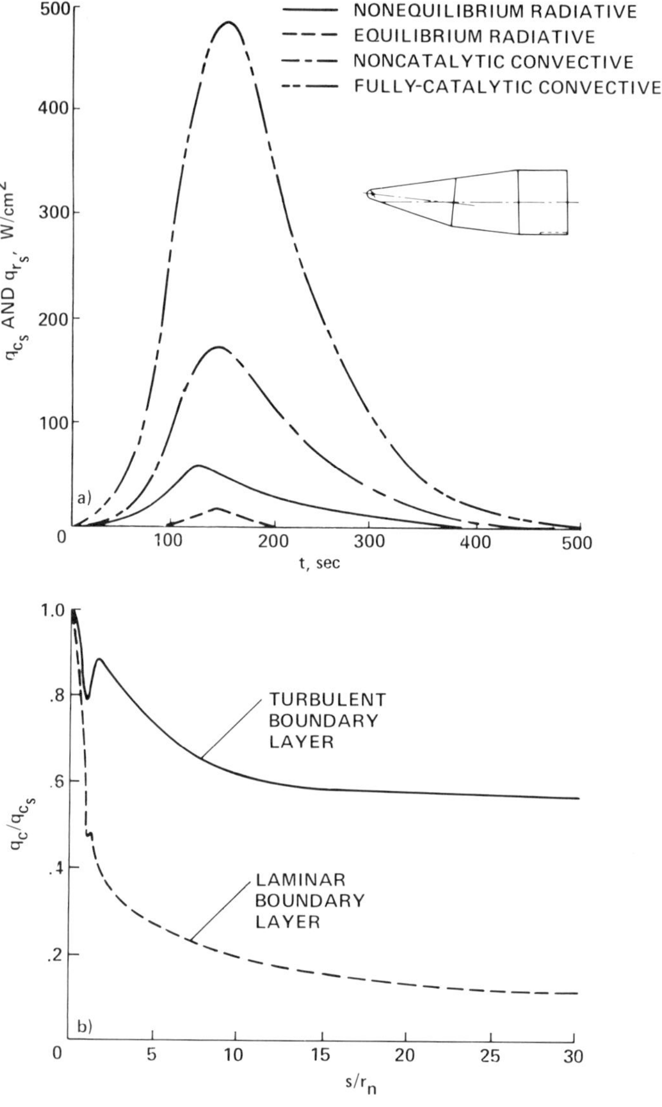

Fig. 7 Bent biconic aerothermodynamic heating rate distributions for minimum heating trajectory for $\theta = 30°$ ($\Delta V_1 = 1491.26$ m/s, $\phi = 90°$): a) stagnation point; b) surface convective-heating distribution near peak heating.

pulsive thrust requirements ($\Delta V_1 + \Delta V_2$) for the orbital change trajectories defined by ΔV_1 and ϕ are shown in Fig. 6b. The corresponding requirements for all-propulsive maneuvers are also indicated in the figure, including the aerobraking case (L/D = 0), which provides the baseline for

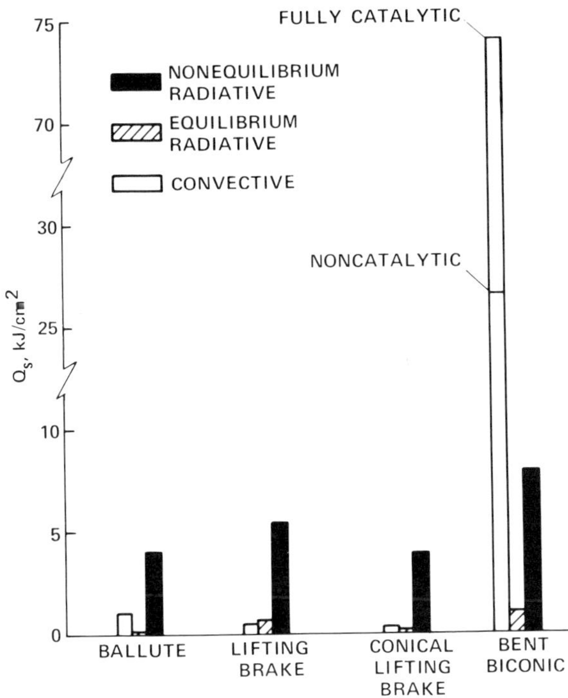

Fig. 8 Total heat loads over the flight trajectories of the various AOTV configurations.

comparison. The velocity increments are related to the propulsion fuel-mass requirements by an inverse exponential function. Consequently, the conservation of propulsion fuel mass and vehicle payload capacity increase substantially with the differences between thrust requirements for the aeroassist and all-propulsive maneuvers. An optimum orbital change trajectory occurs for each vehicle bank angle, which corresponds to the minimum value of $\Delta V_1 + \Delta V_2$. There is a tradeoff in the selection of this condition, however, because of the plane-change capability shown in Fig. 6a and the surface heating characteristics shown in Fig. 6c. The maximum values of the stagnation point convective heating rates are shown here, because this is the dominant surface heating mechanism for the bent biconic. In addition, noncatalytic results are given since they correspond to the minimum possible heat fluxes and, therefore, are most favorable to the design. The actual surface conditions are, however, closer to the fully catalytic case, as will be shown later. This occurs because of the large magnitude of even the noncatalytic heating rates and properties of contemporary TPS materials.

The best overall performance is obtained for $\phi = 90$, which combines maximum plane-change capability with minimum heating effects. The $\theta = 30$ case, for example, corresponds to the minimum thrust requirements ($\Delta V_1 + \Delta V_2$), which provides an increase in payload capacity of about 2 relative to the all-propulsive maneuver. This is also about the same gain in performance achieved by the drag-brake concepts (L/D = 0). A more significant comparison, however, is $\theta = 60$, because it was shown previously that 30 of plane change can be obtained efficiently by in-orbit propulsive thrust at GEO. In this instance, the payload capability is increased by nearly a factor of 3. The volumetric efficiency of the bent biconic, however, is about one-third less than that of the cylindrical shapes of the drag brakes. Consequently, the tradeoff in payload capability is about even, because the all-propulsive maneuver for $\theta = 60°$ decreases the payload capability by about one-third (see Fig. 2b). Another factor of importance is the extreme sensitivity of the $\phi = 90°$ case to ΔV_1, which may impose severe requirements on the vehicle guidance and control system.

The aerothermodynamic heating rate distributions corresponding to the minimum heating rate trajectory for $\phi = 90°$ and $\theta = 30°$ are given in Fig. 7. The fully catalytic case is also shown here and is seen to increase the peak-heating rat by about a factor of 3 (to the order of 0.5 kW/cm^2). The convective rates exceed the radiative over the entire range of the heating pulse, because of the small nose radius and deep atmospheric penetration caused by the large ballistic coefficient. However, the nonequilibrium radiative results are substantially greater than those for the equilibrium case. The magnitude of the convective rates is much larger than that of the drag brakes, which is also a result of the nose radius and ballistic coefficient. The radiative rates, however, are of the same order, which is a result of the much smaller shock volume in the nose region.

Because of the large magnitude of the stagnation point convective heating rates, the surface distributions were considered essential and are shown in Fig. 7b for peak heating conditions. These results were estimated from unpublished calculations for the Titan aerocapture mission but are considered applicable herein because of the similar composition of Titan's atmosphere and air [i.e., both are nitrogen rich (≈80%)].[15] Results for both laminar and turbulent boundary-layer flows were obtained using a fully coupled CFD code for an equivalent body simulating the bent biconic at trim angle of attack. The turbulent heating rates are greater, by more than a factor of 4, than those of the laminar case and are nearly 60% of the stagnation point value on the vehicle flank. The turbulent case is considered most likely because of several factors. First, the deep atmospheric penetration and high trim angle of attack of the biconic produce strong crossflow effects that cause the vehicle to shed streaks or sheets of vortex filaments. This thins the boundary layer and causes high shear-layer stresses, which promotes transition to turbulence. In addition, fully catalytic surface conditions are considered most likely, as pointed out previously. The high heat fluxes for this case require ablating TPS materials. Much experimental data exist that show even a moderate amount of ablation will cause immediate transition to turbulent flow.[16,17] The transition Reynolds number decreases substantially with blowing, with the effect of the freestream Reynolds number being negligible from about 0.5 to 2.0×10^6. Finally, even if ablating materials are not used, a pyrolyzing material such as carbon will also cause transition to turbulence. This happens because oxidation causes the surface to become so rough and pitted after the first atmospheric pass that transition will undoubtedly occur on the second pass (e.g., witness the Shuttle nose region). Moreover, roughness effects further enhance surface heating by destroying the laminar sublayer. This causes more oxygen to reach the surface, which reacts in a combustion process with carbon to form carbon dioxide.

The total surface heat loads over the range of the flight trajectories for the various AOTV configurations are shown in Fig. 8. These results further emphasize the dominance of the nonequilibrium radiative heating for the drag-brake vehicles and the convective heating for the bent biconic. The magnitude of the heating rates and loads for the latter case causes major thermal protection problems.

Heat Protection Analysis

As an outgrowth of the foregoing parametric tradeoff studies, two configurations are selected for analysis of the heat shielding requirements: the conical lifting brake because of its favorable aerothermodynamic characteristics, and the bent biconic, because of its aeromaneuvering capability and current widespread interest. The principal guidelines for the further studies follow: 1) The heat protection system must be reusable; 2) it must be constructed of materials within the current state-of-the-art knowledge for fabrication and applicability; and 3) it must have proven heat protection capability for the magnitude

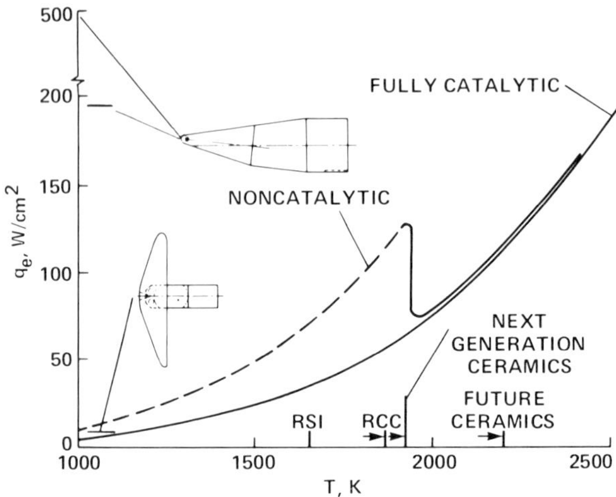

Fig. 9 Radiative heat rejection characteristics of ceramic thermal protection materials.

and duration of the predicted heat fluxes and loads in the high-vacuum, low-temperature environment of space.

The heat-rejection characteristics, by equilibrium radiative emission, of modern reusable ceramic materials are summarized in Fig. 9, and the heat-rejection requirements for both the bent biconic and conical lifting brake are indicated on the figure. The high stagnation point convective heating rates for both noncatalytic and fully catalytic surface chemistry place the bent biconic well beyond the capability of contemporary, reusable materials. This indicates that materials with nearly full-catalysis properties, such as carbon or ablators, must be used for thermal protection in the nose region. Furthermore, this condition persists over most of the vehicle surface, since it was argued previously that, for this configuration, turbulent flow conditions are most probable. Even if the laminar case occurs initially, however, the downstream surface heating rates are so high that reusable materials are not possible. Thus carbon or ablators are also required for the laminar case as well, which ensures transition to turbulence as discussed earlier. Since the reusable requirements of AOTVs have not been clearly specified, calculations of the weight penalty for the heat protection system of the biconic were made assuming ten flights as a minimum capability. A weight penalty of the order of 50% of the entry mass was obtained using carbon as the heat shield material with reusable space insulation as the supporting insulator. TPS penalties of this magnitude are unacceptable for an AOTV, since no payload advan-

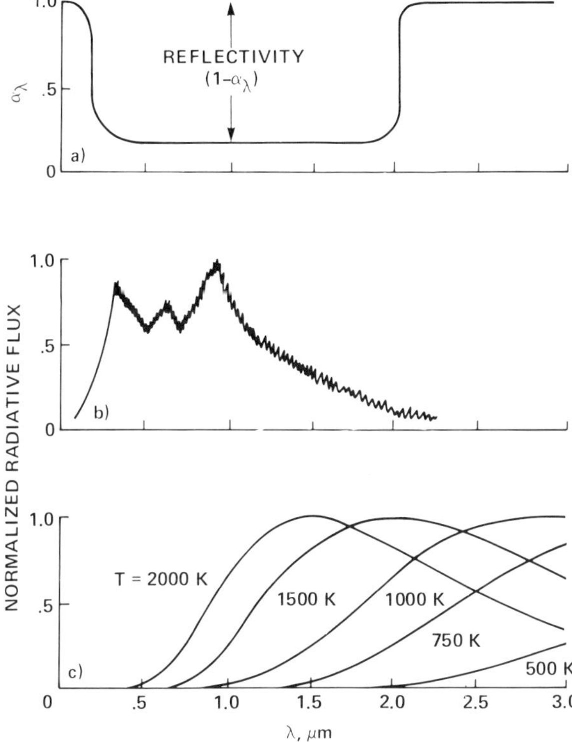

Fig. 10 Typical spectral optical properties and surface heating and rejection requirements of impure silica: a) spectral absorptivity; b) incident spectral heat flux from air; c) emitted spectral heat flux from heat shield.

tage is achieved. Although the issue of turbulence could still be considered somewhat of an unknown factor, reusability of the bent biconic is highly questionable until further research provides satisfactory answers. The stagnation point heating rates could be alleviated by increasing the nose bluntness; however, this increases the radiative heating and decreases the lift-drag ratio and volumetric efficiency. Consequently, there are many trade-offs to consider in making design changes. In addition, transpiration or derived cooling techniques must be applied over large areas of the surface, which results in mechanical complexities with large volume and weight penalties for the necessary apparatus.

The relatively low heating rates of the conical lifting brake are well within the capability of the reusable materials and justify detailed development of a heat protection system. The dominant component of the incident

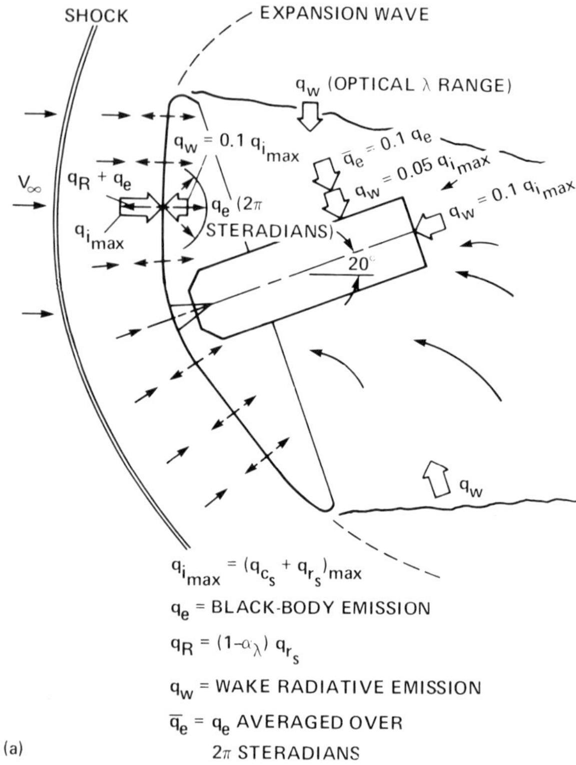

Fig. 11 Heat protection analysis of conical lifting brake for worst heating conditions: a) heating environment.

surface heat flux for this configuration results from nonequilibrium radiative emission. The principal means of heat rejection for this condition is reflection. It has been shown that ceramic thermal protection materials can be tailored to reflect incident radiation while efficiently reradiating convective heat flux. In addition, the noncatalytic nature of ceramics lowers convective heating by about 50%. Furthermore, state-of-the-art ceramic materials are available that are flexible, load supporting, and have low sensitivity to thermal stresses. It is proposed, therefore, to construct the deployable drag brake from thin sheets of silica cloth, about 0.38 mm (15 mil) thick. This serves the dual purpose of providing aerodynamic deceleration and heat protection for the spacecraft body. The back surface of the cloth is coated with a white thermal-control paint to prevent transparency and enhance reflection. Such paints typically have similar optical properties to the silica cloth. Surface optical properties

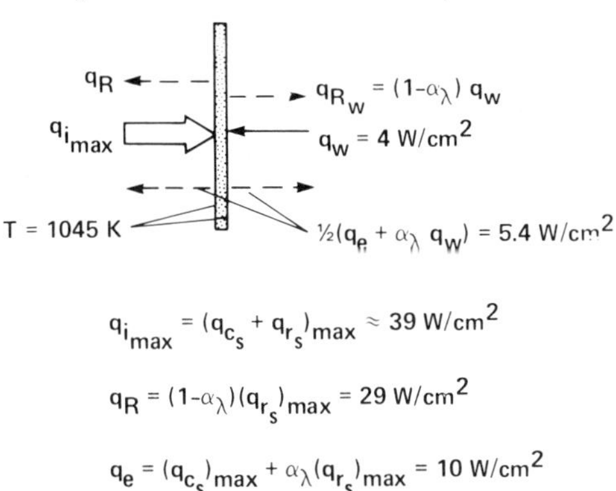

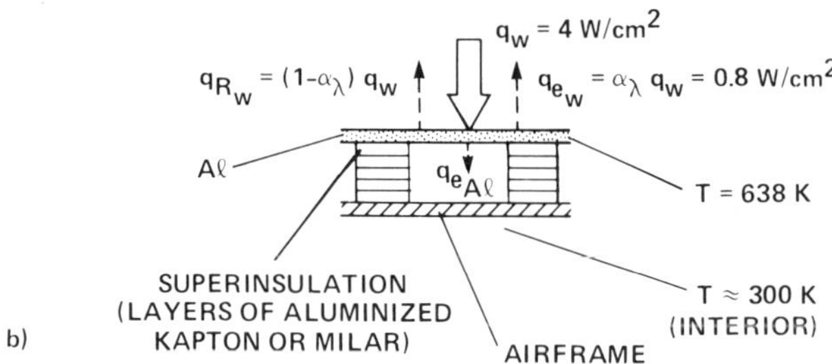

Fig. 11 (continued) Heat protection analysis of conical lifting brake for worst heating conditions: b) surface energy balances and materials selection.

and heat-rejection characteristics which are typical of impure silica are illustrated in Fig. 10. The spectral distributions of absorptivity,[18] incident radiative flux from high-temperature air,[19] and surface radiative emission[20] are given in Figs. 10a, 10b, and 10c, respectively. The absorptivity is low (high reflectivity) at the short wavelengths, where the intensity of the incident flux is greatest, and high (also high emissivity by Kirchhoff's law) at the long wavelengths, where surface emission is greatest. This occurs because the source of the incident flux is always at much higher temperatures (>10,000 K) than that of the heat shield surface (≈1000 K). Consequently, the surface optical properties are nearly optimized for the thermal environment.

The details of the heat protection scheme are illustrated in Fig. 11. The worst heating case occurs for the drag brake oriented normal to the flight path at the peak heating conditions in the trajectory. This produces the strongest shock front and greatest radiating shock-layer volume because of the large effective radius of the flat surface. The heating environment is shown schematically in Fig. 11a, and the principal contributions to the surface heat fluxes are identified. Most of the incident flux is reflected because its major component is radiative, which emits in the spectral range of the maximum reflectivity of the surface material. The maximum incident heat flux is determined from Fig. 5 to be 39 W/cm^2 (i.e., 36 radiative and 3 convective). After reflection, the surface heat flux is about 10 W/cm^2, because only about 20% of the radiative flux is absorbed. The attenuation effect on the convective heating resulting from the porosity of the drag-brake material has also been included in the analysis. This consists of two contributions caused by the seepage of air through the thin silica cloth. First, the boundary-layer temperature gradient is increased due to the suction across the surface; second, chemical energy is released within the material walls during flow through. The modulation effects on the convective heating are estimated to be only about 10% for the present flight conditions. The convective heat flux is about 50% higher at the drag-brake edge than the stagnation point. This increase is offset, however, by the lower radiative heating at this location caused by the greater inclination of the bow shock. The radiative flux incident to the back surface of the drag brake and the vehicle body, from the optical emission of the wake-flow gases, is also accounted for in the analysis. This amounts to about 10% of the maximum incident shock-layer flux for the backface of the drag brake and aft end of the vehicle and about 5% for the cylindrical surface, because of its curvature and inclination.[21] In addition, the convective heating on the vehicle surface is considered negligible, since it is only about 1% of the incident convective heating on the forebody.

The heat fluxes absorbed by the front and back surfaces of the drag brake are rejected by black-body emission. Both surfaces radiate equally (about 5.4 W/cm^2) because of the high heat transfer through the material, which results from its thinness and relatively high thermal conductivity. The backface emission radiates through a volume of 2π steradians and is incident on the cylindrical surface of the vehicle but is blocked from the aft end. The magnitude is attenuated by the inverse square of distance and, because of the solid angles and view factors involved in the radiative transport, provides only a small contribution to the surface heat flux. An overconservative estimate of this is about 10% of the backface emission. Consequently, the worst heating condition on the vehicle surface occurs at the aft end which must accommodate a maximum heat flux of about 4 W/cm^2. This can be easily accomplished by an airframe design incorporating a highly reflecting surface metal with contemporary insulation materials between the outer surface and inner structure.

The details of the surface energy balances for the applied heat fluxes to the selected heat protection materials are given in Fig. 11b. The maximum surface temperature of the silica cloth comprising the drag brake is about 1000 K for the worst case of peak power emission. This condition is short-lived in the duration of the heating pulse (see Fig. 5) but is still well within the range of the maximum surface emissivity for the infrared spectral range of the emitted radiation (see Fig. 10). Thermal protection for the vehicle surface is provided by a stand-off radiation heat shield consisting of natural-state aluminum-alloy foil about 0.05 mm (2 mil) thick. The shield is supported by a structural material of low-thermal conductivity arranged in a pattern separated by large regions of the vacuum of space. Superinsulation materials such as aluminized Kapton or Milar (i.e., plastics of very low conductivity) that have proven to be very effective in minimizing heat transfer in space applications are suitable for this purpose. The outer surface of the shield is coated with a white thermal-control paint which has an average emissivity and reflectivity of about 80% in the optical spectral range of the incident heat flux. Consequently, the surface is required to reject about 0.8 W/cm^2, which is only about a factor of 5 greater than the solar constant (0.145 W/cm^2). Thermal controlled vehicles for space operations have already been designed to withstand heating rates an order of magnitude greater than the solar constant (e.g., Pioneer Mercury and Venus).

The heat-rejection requirement by equilibrium radiative emission causes a uniform temperature of about 638 K in the thin aluminum shield. The outer surface radiates at the emissivity of the thermal control paint ($\approx$0.8); however, the inner surface radiates at the much lower emissivity of aluminum ($\approx$0.05). All of the heat from the underside is assumed to be absorbed by the spacecraft body. The mass-to-projected-area ratio of the present vehicle design is about 6 g/cm^2, and the average specific heat of the structure and insulation system is conservatively estimated at about 1 J/g-K. Considering a maximum heating pulse of 100-sec duration with the heat load being dumped into space after the atmospheric pass, all the factors involved in the calculations result in an average vehicle temperature rise of only about 1 K. In addition, estimates of the mass penalty for the thermal protection system are negligibly small compared to the vehicle entry mass. Furthermore, mass penalty estimates for the drag-brake support apparatus based on minimum mass beam optimization techniques used for aircraft structures is also relatively small (<15% of the entry mass).

It was discussed earlier that the present analysis is considered overly conservative. Nevertheless, a mass penalty of 15% sacrificed for the drag-brake system's design still allows a substantial gain in payload capability compared to all-propulsive orbital change maneuvers. It appears, therefore, that aeroassist is not only feasible but highly attractive when considered on a total system's weight basis. This is also generally true throughout cislunar space, since the previous exploratory studies[4] indicate that reusable thermal protection materials are adequate for return missions from the stable libration centers. These unique points in space are located at the lunar distance (more than an order of magnitude farther than GEO) and cause the most severe entry conditions in near-Earth space (comparable to escape velocity). Thus the drag-brake concept may extend NASA/DOD mission requirements (5×GEO) by a factor of 2.

Concluding Remarks

The results of this study have uncovered the following important factors that provide guidelines for the design of new AOTV concepts:

1) Plane-inclination changes encompassing 30° (e.g., rendezvous with the Shuttle) can be obtained efficiently by in-orbit retrothrust maneuvers at GEO; this may be extended to about 60° depending on the tradeoffs in volumetric efficiency, plane-change capability, and heat protection requirements.

2) Aeromaneuvering plane-change capability is best applied as an adjunct to at least the 30° propulsive change; this implies that moderate-lift vehicles (i.e., $L/D \approx 2$) are not useful and high-lift concepts (i.e., $L/D \approx 4$) are needed.

3) The effects of the current widespread estimates in the atmospheric density dispersions are within the accuracy of heating rate predictions.

4) Nonequilibrium radiative heating is the major factor driving heat shield design requirements for aerobraking vehicles and may be significant for aeromaneuvering concepts.

5) Bent biconic configurations may be impractical for reusable, space-based applications because the reduced volumetric efficiency is not competitive with propulsive plane-change capability, and severe local heat fluxes cause unacceptable thermal protection weight penalties.

6) Drag-brake concepts have promise for early operational service. They may satisfy a broad range of orbital transfer mission scenarios because of the favorable combination of volumetric efficiency, propulsive plane-change capability, and heat shielding requirements which may be accommodated without advanced materials technology for the most severe entry conditions in near-Earth space.

7) The aeroassist technique produces acceptable penalties in payload capacity for drag-brake concepts because of the system's heat protection requirements.

The overall conclusion is that aeroassist is not only feasible but highly attractive when compared with all-propulsive orbital change maneuvers. Obviously, additional studies of other missions and AOTV concepts involving more technical depth are required. In particular, the development of high-lift concepts and further refinement and experimental verification of the radiative transport prediction techniques are needed. This includes determining nonadiabatic shock-layer effects, the effects of boundary-layer cooling in truncating shock-layer radiative emission, wake-flow radiation effects, updating the spectral properties of air species, and developing a CFD flow-field code incorporating all these factors. In addition, significant inroads in materials technology seems unnecessary for drag brakes but may be required for high-lift concepts.

Acknowledgments

The author is grateful to John Wilson for providing the aeromaneuvering orbital plane-inclination change calculations and to Dr. Chul Park for helpful discussions.

References

[1]Walberg, G. D., "A Review of Aeroassisted Orbit Transfer," AIAA Paper 82-1378, 9th Atmospheric Flight Mechanics Conference, San Diego, Calif., Aug. 1982 (also to be published in the Journal of Spacecraft and Rockets; in addition, a synopsis appears in Astronautics and Aeronautics, Nov. 1983, pp. 36-43).

[2]Howe, J. T., "Introductory Aerothermodynamics of Advanced Space Transportation Systems," AIAA Paper 83-0406, 21st Aerospace Sciences Meeting, Reno, Nev., Jan. 1983 (also to be published in the Journal of Spacecraft and Rockets).

[3]Park, C., "Radiation Enhancement by Nonequilibrium in Earth's Atmosphere," AIAA Paper 83-0410, 21st Aerospace Sciences Meeting, Reno, Nev., Jan. 1983 (also to be published in the Journal of Spacecraft and Rockets).

[4]Menees, G. P., "Trajectory Analysis of Radiative Heating for Planetary Missions with Aerobraking of Spacecraft," AIAA Paper 83-0407, 21st Aerospace Sciences Meeting, Reno, Nev., Jan. 1983 (also to be published in the Journal of Spacecraft and Rockets).

[5]Fay, J. A. and Riddell, F. R., "Theory of Stagnation-Point Heat Transfer in Dissociated Air," Journal of the Aeronautical Sciences, Vol. 25, Feb. 1978, pp. 73-85.

[6]"Orbital Transfer Vehicle Concept Definition Study," The Boeing Company, Seattle, Wash., Report D180-26090-3, Vols. 1-6, 1980.

[7]"Orbital Transfer Vehicle (OTV) Concept Definition Study," General Dynamics-Convair Division, San Diego, Calif., Report GDC-ASP-80-012, Feb. 1981.

[8]Davies, C. B. and Park, C., "Aerodynamic Characteristics of Generalized Bent-Biconic Bodies for Aeroassisted Orbital-Transfer Vehicles," AIAA Paper 83-1512, 18th Thermophysics Conference, Montreal, Canada, June 1983 (also to be published in the Journal of Spacecraft and Rockets).

[9]"Generic Aerocapture Atmospheric Entry Study—Final Report, Vol. 1," General Electric Company, Reentry Systems Division, Philadelphia, Pa., Document 80SDR2226, July 1980.

[10]Florence, D. E., "Aerothermodynamic Design Feasibility of a Generic Planetary Aerocapture/Aeromaneuvering Vehicle," AIAA Paper 81-1127, 16th Thermophysics Conference, Palo Alto, Calif., June 1981.

[11]"Feasibility and Tradeoff Study of an Aeromaneuvering Orbit-to-Orbit Shuttle (AMOOS) Final Report," NASA CR-129431, June 1973.

[12]Miller, C. G., Blackstock, T. A., Helms, V. T., and Midden, R. E., "An Experimental Investigation of Control Surface Effectiveness and Real-Gas Simulation for Biconics," AIAA Paper 83-0213, 21st Aerospace Sciences Meeting, Reno, Nev., Jan. 1983.

[13]Stephenson, B. L. and Hassan, H. A., "Heating Analysis of Bent-Nose Biconics at High Angles of Attack Using the Parabolized Navier-Stokes Equations," AIAA Paper 83-1507, 18th Thermophysics Conference, Montreal, Canada, June 1983.

[14]Nicol, J. R. and Miller, C. G., "Heat Transfer Distributions on Biconics at Incidence in Hypersonic-Hypervelocity He, N, Air, and

CO Flows," AIAA Paper 1508, 18th Thermophysics Conference, Montreal, Canada, June 1983.

[15]Moss, J. N. (private communication), June 1982.

[16]Park, C., "Injection-Induced Turbulence in Stagnation-Point Boundary Layers," AIAA Journal, Vol. 22, Feb. 1984, pp. 219-225.

[17]Park, C., "Ablation of Galileo Probe Heat-Shield Models in Ballistic Range," to be published in AIAA Journal.

[18]Park, C., "Preliminary Design Study of Solar Probe Heat Shields," Progress in Aeronautics and Astronautics: Spacecraft Radiative Transfer and Temperature Control, Vol. 83, edited by T. E. Horton, AIAA, New York, 1982, pp. 439-471.

[19]Allen, R. A., Rose, R. H., and Camm, J. C., "Nonequilibrium and Equilibrium Radiation at Super-Satellite Re-Entry Velocities," AVCO Corporation-AVCO Everett Research Laboratory, Everett, Mass., Research Report 156, Sept. 1962.

[20]American Institute of Physics Handbook, 3rd edition, D. E. Gray (coordinating editor), McGraw-Hill Book Company, New York, 1972.

[21]Park, C., "Problems of Radiative Base Heating," AIAA Paper 79-0919, AIAA/NASA Conference on Advanced Technology for Future Space Systems, Langley Research Center, Hampton, Va., May 1979.

Design and Performance Analysis of a Conical Aerobrake Orbital Transfer Vehicle Concept

Gene P. Menees* and Chul Park*
NASA Ames Research Center, Moffett Field, California
and
John F. Wilson†
Informatics General Corporation, Palo Alto, California

Abstract

A Shuttle-compatible systems design based on the core concept of attachable modules for the major vehicle components is proposed. The principal features include a disposable cargo/extra-propellant tank module; a porous, radiative, back-scattering drag-brake surface material of thin silica cloth; and a lightweight carbon-composite support structure. The mission payload capability for delivery, retrieval, and combined operations is determined for a broad range of missions including NASA/DOD requirements and extending through cislunar space. The effects of finite-rate surface catalysis, negative lift, and multiple atmospheric passes in reducing the aerothermodynamic heating rates are also investigated. In addition, the structural and thermal protection problems of the drag-brake support apparatus are analyzed, and recommendations are proposed for future design refinements.

Nomenclature

H = altitude
I_{sp} = specific impulse of liquid rocket engine
i = orbital inclination from equatorial plane
L/D = lift-drag ratio
M_{∞} = flight Mach number
p = pressure
q = incident heat flux to drag-brake surface
t = transatmospheric flight time from 150-km altitude
V_{∞} = flight velocity

Presented as Paper 84-0410 at AIAA 22nd Aerospace Sciences Meeting, Reno, Nev., Jan. 9-12, 1984.

*Research Scientist.
†Consultant, Professional Services Operations West.

α = angle of attack
β = ballistic coefficient
γ = ratio of specific heats
Δi = change in orbital inclination
ΔV = propulsive impulse velocity increment

Subscripts

c = convective
r = radiative
s = stagnation point

Introduction

Studies of the enabling technology for future Earth-centered space transport missions have shown the substantial payload advantage obtained by spacecraft that use the aeroassist technique for orbital change maneuvers.[1-3] This plan calls for using the aerodynamic forces produced by passage through the upper atmosphere to achieve the transition to local orbit with a minimum use of propellant; earlier methods relied exclusively on retropropulsion.

The variety of proposed aeroassist design concepts can generally be categorized as 1) nonlifting, low ballistic coefficient, "aerobraking" vehicles and 2) lifting, high ballistic coefficient, "aeromaneuvering" vehicles that have aerodynamic plane-inclination change capability. The former employ large-diameter drag devices that provide flight trajectories with perigee locations in the high, rarefied regions of the upper atmosphere to minimize aerothermal heating effects. For the latter design, however, the high ballistic coefficients cause the vehicle to plunge much deeper into the lower atmosphere to obtain the deceleration required for the orbital transition, which results in severe thermal protection requirements.

An essential operational quideline of aeroassist-orbital transfer vehicles (AOTVs) is space basing, which requires that the heat protection system be reusable. Consequently, the surface heating rates and temperature characteristics must be within the capabilities of contemporary, normal-growth-technology, thermal protection system (TPS) materials. Derived cooling techniques may be practical for some applications, but the complexity of the necessary apparatus results in additional sacrifices in weight and volume.

Recent exploratory studies[3] indicate that a conical lifting brake AOTV design may be feasible for early operational service, because the thermal protection requirements are accommodated by current reusable materials for the predicted heating environments and conditions of hard space. It was also shown that the aeroassist technique produces acceptable penalties in vehicle payload capacity and is highly attractive compared with the fuel mass requirements for all-propulsive orbital transitions. In addition, it was demonstrated that drag-brake designs may have generic application to a broad range of mission scenarios, a result of the favorable trades with aeromaneuvering vehicles in volumetric efficiency, retrothrust plane-change capability, and heat protection requirements. Furthermore, the drag-brake-type design performs all of its aerodynamic maneuvering in the far-outer extent of the atmosphere. Moreover, it is currently the leading candidate for proposed flight demonstration tests.

The purpose of this study is to extend the previous work to obtain greater depth in the feasibility analysis of the conical lifting brake concept. The principal issues addressed are the refinement of the aerothermodynamic surface heating predictions, a more detailed thermal and structural analysis of the support apparatus and drag-brake surface material, and a performance analysis of the capability to satisfy the diverse mission requirements of NASA and the DOD. Included in the updated radiative heating calculations is the determination of shock-layer thickness with varying angle of attack from computational fluid dynamic (CFD) flowfield calculations, which give improved accuracy over the previous empirical estimates. This is essential to define the volume of radiating shock-layer gases, because nonequilibrium radiation was shown previously to be the dominant surface heating mechanism. In addition, the effects of boundary-layer thickness in cooling the shock-layer gases and truncating the radiative emission are estimated from Apollo and Fire flight test data. The convective heating predictions are also updated by the CFD shock-layer calculations and, in addition, by estimating the attenuation effects owing to the catalysis and porosity of the drag-brake fabric structure. Implicit in this study is the ability of the autonomous guidance, navigation, and control (GNC) system to establish a target orbit. Details of the stability, control response, damping, and dynamic characteristics of the vehicle will be presented in a future paper.

Vehicle Systems Design

Results of initial studies of the geometry and aerodynamic characteristics of the conical lifting brake concept were given previously.[4] A schematic illustrating the con-

Table 1 Vehicle specifications for mission performance analysis, kg

Gross (zero payload) weight	23,822
Nominal fuel weight (97% of total, i.e., 3% reserves)	18,480
Payload weight	5,662
Dry weight of optional disposable external tank	1,000
Maximum fuel weight in optional disposable external tank	4,662
Total allowed liftoff weight (gross weight plus payload or extra fuel tank)	29,484

figuration and systems design technique is shown in Fig. 1. It has the advantage of generic application for the entire range of possible mission combinations (i.e., manned or unmanned, with or without cargo). The vehicle design is based on the core concept, which provides for the separate packaging of attachable modules incorporating the propulsion, command/control, and cargo systems. The vehicle weight specifications for integration with the Shuttle cargo bay are given in Table 1. The propulsion module has a fuel capacity of about 19,000 kg of LH_2/LO_2 and is equipped with normal-growth-technology rocket engines delivering a specific impulse of 480 sec. The command/control module is attached to the propulsion moduel; it contains the avionics and reaction control system (RCS) motors for attitude control and orbital maneuvering. Attached next in tandem is the cargo/extra-tank module, which has the option of accommodating additional fuel or payload, depending on the mission requirements. For Shuttle interface, the maximum payload capability is about 5,600 kg, or alternatively the extra fuel capacity is about 4,600 kg, allowing 1,000 kg for the dry tank weight. The extra tank is disposable and may be jettisoned to enhance operational performance. Space-based operations will allow the eventual alleviation of the Shuttle-compatible weight restrictions to approximately that of the propulsion module, which greatly improves mission capability.

The drag brake is designed as a 70° spherical cone because of the reduced surface heating and favorable aerodynamic stability of that geometry. The current diameter is very large and was conservatively selected to prevent impingement of the wake flow on the vehicle surface, since this phenomena causes high local heat fluxes.[5] Design optimization, however, must await future research on base-flow characteristics, which it is anticipated will allow some reduction in size. For the present feasibility study, axial symmetry of the brake is assumed, which is considered sufficient for the structural and thermal protection analysis. The advantages of asymmetry in providing improved roll stability, however, were pointed out in the initial configura-

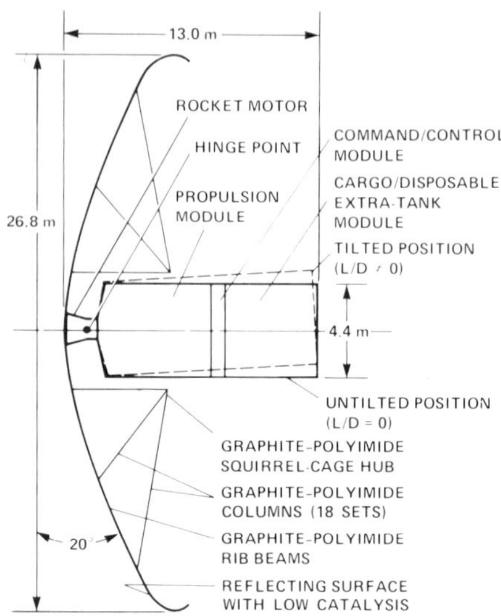

Fig. 1 Schematic of conical lifting brake systems design.

tion studies and will be incorporated in the final design. Attitude control is achieved by revolving the drag brake around the hinge point illustrated in Fig. 1. The displaced position of the vehicle causes a shift in the center-of-gravity location, which produces a finite yawing motion and resulting angle of attack. This provides a small lifting capability to compensate for the unpredictable dispersions in the atmospheric structure.

The drag brake comprises five major parts: 1) the surface fabric, 2) the rib beams supporting the fabric, 3) the insulator between the surface fabric and support beams, 4) the columns supporting the rib beams, and 5) the mechanism supporting the entire apparatus, which is referred to as the "squirrel-cage hub" because of its similar structure. The surface fabric is made of reflective materials such as silica or Nextal, which are woven into a cloth having a thickness of the order of 0.25 mm (10 mils). The support structure, consisting of the ribs, columns, and hub, is made of a lightweight carbon composite such as graphite polyimide. All exposed surfaces of these components are coated with white thermal control paints, which reject incident radiation in the visible and near-infrared spectral ranges and efficiently emit absorbed energy in the far-infrared frequency range. The vehicle surface is also coated with similar paints or wrapped in highly reflective mirror-finished materials such as aluminum foil.

In addition to the foregoing, the stagnation region of the brake surface must have provision for a variable opening to allow operation of the rocket motors. It is anticipated that the components of the vehicle will be transported into space by the Shuttle and assembled at a space station located in a low Earth orbit. The vehicle will remain in its fully assembled operational state until its return to Earth for refurbishment. During inactive periods between missions, the vehicle will be housed in a hangar located at the Space Station to protect the optical properties of the surface materials from degradation by solar radiation.

Mission Performance Analysis

The drag-brake concept is optimized for high Earth orbit (HEO) sortie missions, since it is a high-drag, low-lift device that achieves its aeroassist capability primarily by aerobraking. Consequently, the design effectively obtains the rapid decelerations and large velocity decrements required for orbital-change maneuvers from high altitudes. It is incapable, however, of accomplishing the large plane-inclination changes characteristic of sortie missions from low Earth orbits (LEO) exclusively by aeroassist, because the design lacks significant aeromaneuvering or lifting capability. The drag brake can satisfy the LEO mission requirements by implementing propulsive thrust as an adjunct to its aeroassist capability as will be discussed subsequently and, therefore, achieve generic application for all operational requirements in Earth-Moon space. The ability to effectively utilize cislunar space may have important future implications for various reasons involving strategic and scientific interests.

The previous exploratory studies[3] determined the propellant requirements for geosynchronous orbit (GEO) return missions with large plane-inclination changes. The present work includes not only the performance capability for the entire range of NASA/DOD mission requirements,[6] but further extends the analysis throughout cislunar space by including the stable libration centers (e.g., L_5),[2] which occur at the lunar distance. The analysis includes LEO sorties from the Shuttle orbit involving large multiple-plane changes and HEO sorties to GEO, the 5×GEO-polar orbit (i.e., a factor of 5 greater than GEO distance and having an inclination of 90°), and L_5. The former has the greatest implications for military operations of any AOTV missions and the most technology crossover with advanced military spacecraft (AMSC) and transatmospheric-vehicle (TAV) launch systems. The present Shuttle orbit-to-Shuttle orbit sortie analysis is considered an average of the proposed NASA/DOD LEO sortie operations, that involve both multiple-plane and moderate-altitude orbital changes. These include round trips with payload delivery to the Molniya orbit (i.e., the principal Russian satellite orbit) and a satellite orbit located at 75° inclination and 185 km altitude, and an eastern test range (ETR) launch to a 6-h polar orbit at 10,378 km with return to the Shuttle orbit. The altitude differences from the Shuttle orbit are easily accommodated by small changes in the aerobraking maneuvers, with negligible propellant penalties. The GEO sortie mission has current widespread interest for a number of applications, has been extensively analyzed, and probably has the highest initial priority because of its importance for a space operations center (SOC) location. The more distant HEO missions will play important roles in the future efforts to colonize cislunar space.

The results of the performance analysis are shown in Figs. 2 and 3. The transfer orbits and propulsive-burn maneuvers for the multiple-plane Shuttle orbit and GEO sortie missions are illustrated schematically in Figs. 2a and 2b, respectively. The latter is also representative of the higher HEO missions. The strategy of the LEO orbital-change maneuvers is to minimize the large propellant penalties required for plane changes at low orbits because of the high circular velocities. This makes such missions analogous to the HEO sorties and is accomplished by propulsive thrust maneuvers involving three burns: the first burn decircularizes at LEO and achieves elliptical transfer orbits of high eccentricity and apoapsis; the second burn produces the required plane-inclination change at the higher apogee with reduced propellant requirements because of the lower circular velocity; and the third burn recircularizes the vehicle into the target LEO after the aeroassist maneuver. The selection of acceptable transfer orbits is governed, however, by cost-time tradeoffs in the mission requirements.

The performance capability for delivery, retrieval, and combined delivery/retrieval operations for the various mission scenarios is illustrated in Fig. 3. These results represent the maximum payload capability for the Shuttle-compatible vehicle weight constraints specified in Table 1 (i.e., 32.5 tons). Furthermore, the analysis is based on normal-growth-technology liquid rocket engines having a specific impulse of 480 sec and includes generous allowances for propellant losses resulting from GNC corrections, docking maneuvers, and boil-off over the duration of the mission. Moreover, various techniques were investigated for distributing the propellant in the propulsion and extra-tank modules. The optimum method consisted of fully loading the

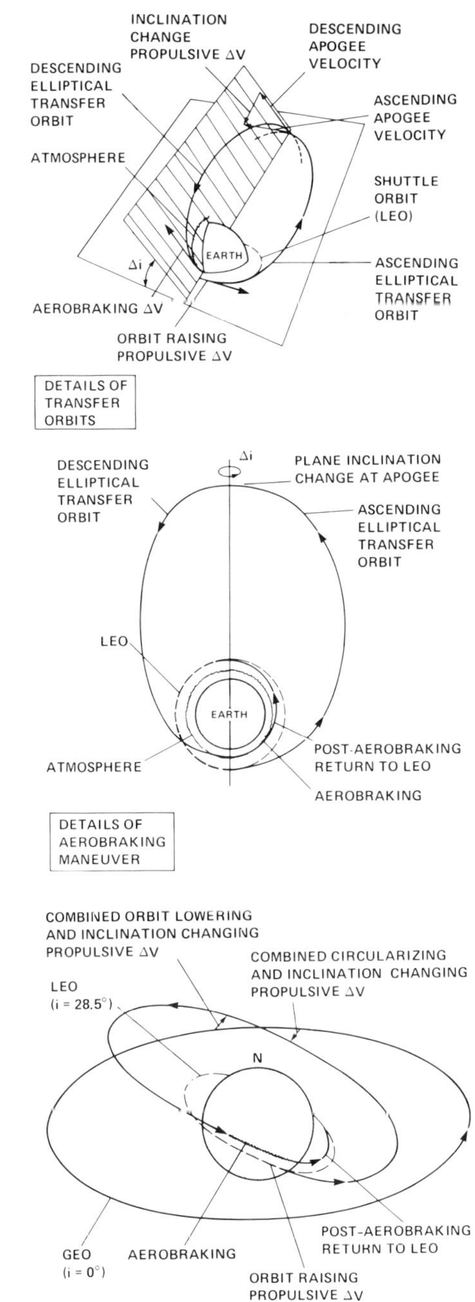

Fig. 2 Schematics of transfer orbits and all-propulsive plane-inclination change maneuvers for AOTV mission scenarios: a) Shuttle orbit sorties; b) Shuttle orbit/GEO sortie mission.

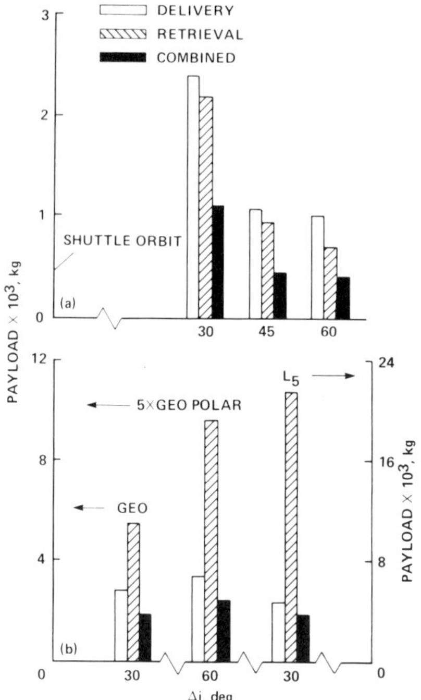

Fig. 3 Maximum payload capability for delivery, retrieval, and combined delivery/retrieval round-trip sortie missions for liquid rocket engine with I_{sp} = 480 sec: a) Shuttle orbit sorties with 41-h round-trip orbits for Δi = 30° and 45° and 104-h orbit for Δi = 60°; b) return missions to Shuttle orbit from GEO, 5×GEO polar, and L_5.

propulsion module tanks, but only loading the extra tank with sufficient fuel to complete the proposed mission. The extra-tank propellant was consumed first; then the tank was jettisoned to reduce the vehicle mass.

The payload for the Shuttle orbit sorties (Fig. 3a) is substantial for plane changes of the order of 30° and significant for changes as great as 45°. These results correspond to round-trip transfer orbits having a duration of about 41 h (i.e., less than a day to achieve the target orbit). For the case of a polar orbit (Δi = 60°), no significant payload capability is possible with the 41-h orbit, unless the Shuttle compatibility constraint is alleviated and the capacity of the extra-fuel tank increased. Alternatively, the required conservation of propellant mass, which provides payload capability, is achieved by resorting to transfer orbits of higher apogee. As a consequence, however, compromises must be made with the greater time duration of the mission. The results shown for this case correspond to 104-h , round-trip transfer orbits (i.e., about 2 days each way), but obtain about the same payload capacity as the 41-h orbit for the $\Delta i = 45°$ case. This is sufficient to perform most service/maintenance operations or to accomplish strategic objectives. The apogee altitudes of the 41-h and 104-h Shuttle orbit sorties are between GEO and 5×GEO; consequently, the severity of the entry conditions are between these two limits. The foregoing results involve relatively moderate time penalties and are especially significant, since missions of this type are envisioned for high-lift aeromaneuvering vehicles that must sacrifice payload capacity for faster response time.

The payload capability for the HEO sorties (Fig. 3b) is substantial throughout cislunar space. The magnitude increases with orbit altitude because the total propulsion energy requirements are dominated by circular velocity, which decreases as the inverse square root of distance. This more than compensates for the larger propellant penalties required at LEO to initiate the HEO missions. Although the velocity pulse required for recircularization at LEO is comparatively small, it has cumulative effects in the round-trip scenario and must be minimized by optimal aeroassist-orbital transfers.

A significant feature of Fig. 3b is that the ratio of retrieval to delivery payload increases with HEO altitude. For example, this factor is about 3 for the 5×GEO-polar orbit and is even greater for an equatorial orbit. This fact has important strategic implications. A proposed use of this location is that of a storage region or "warehouse" in space for hardware which may be required as "replacements" in lower orbits during emergencies when return to Earth is impractical. The enhanced retrieval capability provides important operational time advantages for this requirement. The least-expensive orbit to achieve propulsively and the greatest payload advantages occur for the L_5 mission. The retrieval/delivery ratio here exceeds a factor of 4. This location offers many advantages for scientific experiments because of the minimal occultation and null relative force environment produced by the overlapping gravitational fields of the Earth-Moon system. Theoretical analyses predict that a body is suspended in orbit with zero-energy penalty at the stable libration centers; consequently, minimal propulsion energy is required to initiate orbital change maneuvers.

Trajectory Analysis

Studies of transatmospheric flight trajectories during the return phase of the specified missions were conducted under the following assumptions: 1) the vehicle ballistic coefficient is constant (β = 11 kg/m^2); 2) only aerodynamic forces are employed during atmospheric passes; and 3) the apoapsis altitude of the final aeroassist transfer orbit is tangent to the target LEO where circularization is accomplished propulsively. The calculations were obtained for the same entry mass for all cases [10,886 kg (12 tons)], which corresponds to a practical upper limit for the reusable TPS requirements and Shuttle-compatible design constraints. Results are given for the 1962 U.S. standard atmosphere model, since previous work[3] indicated that the effects of the recommended unpredictable variations in the atmospheric structure were within the accuracy of the aerothermodynamic heating rate prediction methods.

Exploratory studies showed that the optimum aerodynamic strategy for minimizing surface heat fluxes and pressures consists of flying the vehicle at the maximum negative lift-drag ratio for all atmospheric passes. This is impractical because of GNC requirements, however, since a sufficient margin must be maintained to provide the aeromaneuvering capability necessary to accommodate atmospheric dispersions. This is a major design driver and one that cannot be determined without improved knowledge of the upper atmosphere. Efforts should be focused on developing adaptive-guidance algorithms that are tailored to the aerodynamic characteristics of proposed AOTV concepts. Moreover, consensus must be obtained on the issue of accommodating the average or worst-case spikes in the atmospheric pothole density fluctuations. The maximum lifting capability of the conical lifting brake is approximately $L/D = \pm 0.3$. A design margin of about $L/D = \pm 0.2$ is considered adequate to account for atmospheric effects because of the low ballistic coefficient. In addition, this margin provides the means for minimizing the exit flightpath angle from the atmosphere to achieve optimal orbital transfers.[7] Consequently, a negative lift capability of the order of $L/D = -0.1$ is available for trajectory shaping to minimize operational conditions. The corresponding calculations were obtained for as many as three atmospheric passes for the aforementioned range of missions extending through cislunar space. Extending the analysis beyond three passes becomes comparatively unrewarding, since only incremental benefits are obtained in alleviating the TPS and structural requirements.

Results that are illustrative of all missions analyzed are shown in Fig. 4 for the GEO-Shuttle orbit return trip. The baseline for relative design comparisons is considered the $L/D = 0$ trajectory for a single atmospheric pass. This case causes the most severe structural and TPS requirements for practical operational constraints. In all probability, the vehicle should be able to fly the L/D = 0 trajectory, but operated mainly at negative lift to provide additional safety margins. Negative lift alleviates the design conditions by forcing the vehicle to fly higher-altitude trajectories, which seems contradictory since positive lift is measured up from the Earth's surface. The puzzle results from the constraint imposed on the aeroassist maneuver that requires the target LEO to be achieved tangentially to expend minimal fuel for the circularization burn. This forces the vehicle into a shallower entry angle and flightpath to counteract the downward effect of negative lift. The resulting lower-density flight regimes caused by this strategy decrease surface heat fluxes and aerodynamic forces. Total heat loads are raised, however, because the atmospheric pass must be of longer duration to achieve the deceleration for the orbital change maneuver. Nevertheless, the overall relative tradeoffs are beneficial in reducing TPS requirements. Additional benefits are obtained from multiple atmospheric passes, since the perigee is raised further on the first pass and the flight velocity is decreased for the succeeding passes. The effects in shaping the flight trajectories are shown in Fig. 4 for up to three atmospheric passes. The time penalties resulting from the longer duration of multiple passes for the missions under analysis are given in Table 2 and must be evaluated relative to overall requirements.

A potential hazard that must be considered in multipass missions is the additional time the vehicle is exposed to the radiation of the Van Allen belt. This consists primarily of X-rays produced by the solar wind particles trapped in the Earth's magnetic field. Unless properly shielded, the radiation is hazardous to manned vehicles and certain types of payloads. The significance to the missions analyzed herein is determined from Table 2, which also gives the apogee altitudes for the multiple atmospheric passes. The extent of the Van Allen belt is about 2.5 to 7 Earth radii. The results of Table 2 show that the effect is relatively minor for the GEO return, since the only encounter occurs for the first pass of the three-pass case, which grazes the lower edge of the belt. For the more distant HEO missions, however, the effect becomes more important because the first pass of the two-pass cases travels through the lower region, and the first pass of the three-pass cases extends deep into the interior. Nevertheless, it is expected that significant unmanned payload mission requirements will develop for which the Van Allen belt radiation is unimportant.

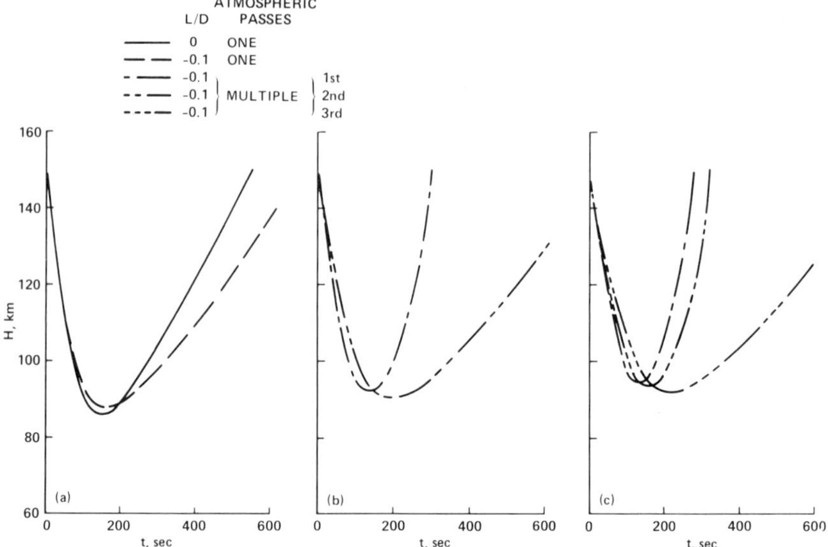

Fig. 4 Flight trajectories illustrating multiple atmospheric passes for aeroassisted return missions from GEO to Shuttle orbit (β = 11 kg/m^2): a) one pass; b) two passes; c) three passes.

Table 2 Duration and apogee altitude for multipass aeroassist return missions

Number of atmospheric passes		GEO to Shuttle orbit		5×GEO polar to Shuttle orbit		L_5 to polar LEO	
		Time, hr	Alt, km	Time, hr	Alt, km	Time, hr	Alt, km
1		6.1	400	42.1	400	122.3	400
2	#1	10.0	11,661	47.5	18,516	128.2	20,299
	#2		400		400		400
3	#1	14.1	16,773	54.7	30,974	136.5	35,335
	#2		7,670		10,874		11,673
	#3		400		400		400

Structural Analysis

The calculations sizing the trusses for the support apparatus were obtained using minimum-mass, beam-optimization techniques employed in aircraft design[8] and structural properties of typical graphite-polyimide composites.[9] Implicit in this analysis is the assumption that the beam temperatures could be maintained below 650 K to

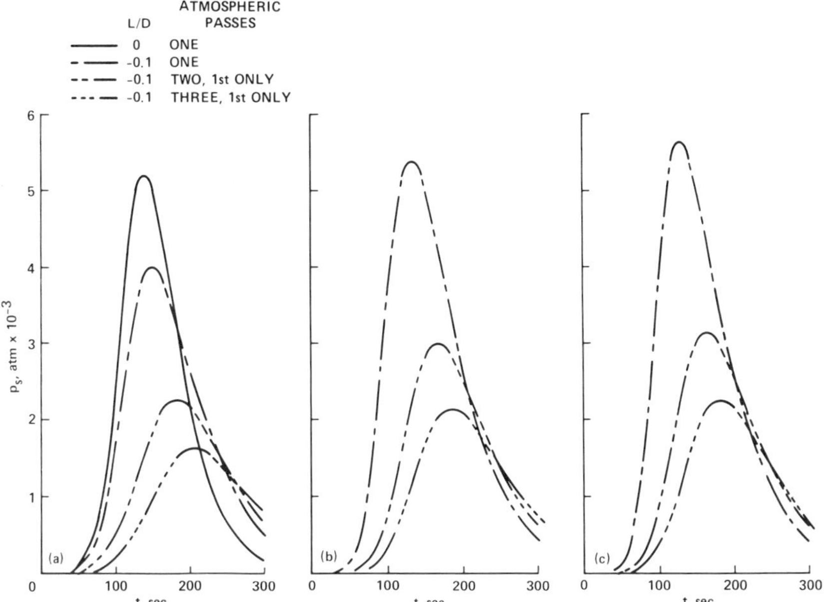

Fig. 5 Stagnation point pressure distributions along flight trajectories for aeroassisted return missions to LEO: a) GEO to Shuttle orbit; b) 5×GEO polar to Shuttle orbit; c) L_5 to polar LEO.

prevent thermal deterioration of the structural properties. In addition, the transient effects of aerodynamic flutter on the structural requirements are considered negligible because (to a first order) there is no nonlinear element in the aerodynamic forces. The baseline design aerodynamic loads were determined from the maximum stagnation point pressure for the L/D = 0 case of the GEO return, which is only about 0.005 atm. This provides a design safety margin of about 2 for the material structural properties. The variation of stagnation point pressure corresponding to the missions under analysis is shown in Fig. 5. Results are not shown for the LEO multiplane Shuttle orbit sorties, since it was noted earlier that the entry conditions are spanned by the GEO-5×GEO range.

The L_5 case given here differs from the performance analysis in that the return is to a polar LEO at the Shuttle orbit altitude. This scenario was selected because it produces the highest entry velocity in cislunar space and provides the upper limit for structural, aerothermodynamic heating, and TPS requirements. The benefits of multiple passes, however, reduce the pressure forces to the level of the design baseline and make the support apparatus generic for the entire range of missions in cislunar space. The

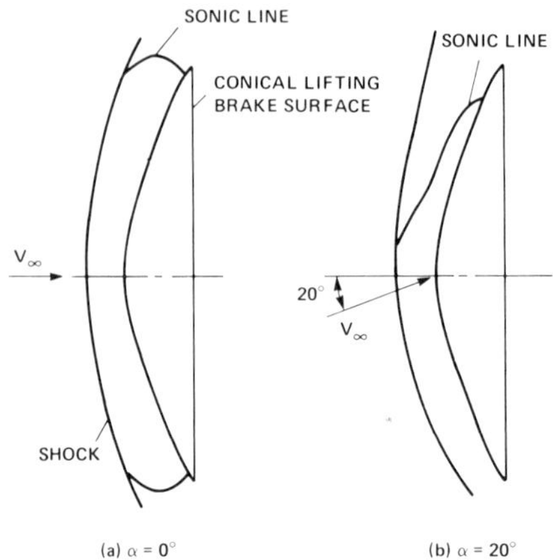

Fig. 6 Shock-shape predictions for ideal inviscid flow: drawn to scale, $M_\infty = 30$, $\gamma = 1.4$.

estimated weight penalties resulting from the support structure and drag-brake surface material, which are considered somewhat conservative, are less than 15% of the entry mass.

Aerothermodynamic Heating

Real-gas CFD codes that couple all the relevant physics of the shock-layer flow environment (i.e., thermodynamics, chemistry, radiative transport, and viscous effects) are not currently available but are forthcoming. During the interim, approximate methods are used to predict the surface heat fluxes for conceptual design studies. Reliable estimates of the shock-layer shape and equivalent body nose radius are required in the computational procedures.

Shock-Layer Calculations

Inviscid, ideal-gas results have been obtained for the forebody flowfield using a fully implicit, time-marching CFD code based on the Beam-Warming algorithm. The variation of shock shape with angle of attack for entry conditions approximating typical requirements (i.e., $M_\infty = 30$ and $\gamma = 1.4$) were determined. For the heating rate calculations, however, the shock-shape results were further extrapolated to $\gamma = 1.2$, which is more representative of the actual state of the shock-layer flow environment. Initial results of this work are shown in Fig. 6 for $\alpha = 0°$ and 20°. Analysis is currently underway to extend the calculations to include viscous real-gas effects. New

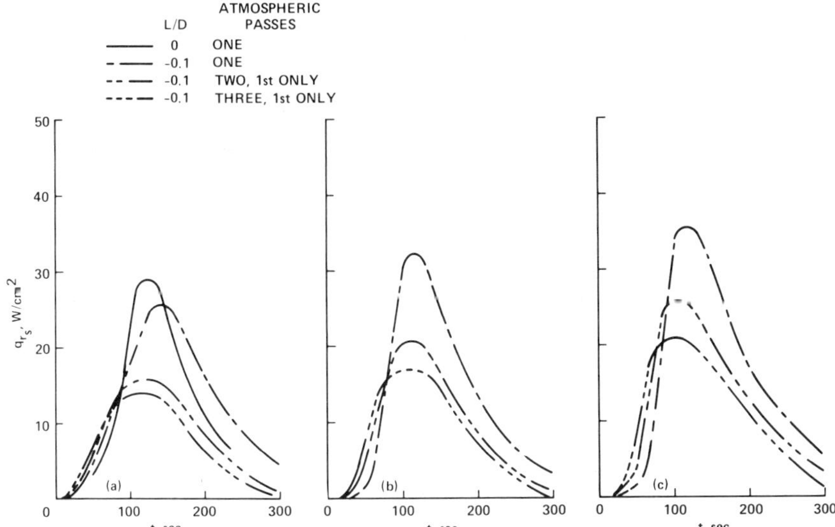

Fig. 7 Stagnation point nonequilibrium radiative heating rate distributions along flight trajectories for aeroassisted return missions to LEO: a) GEO to Shuttle orbit; b) 5×GEO polar to Shuttle orbit; c) L_5 to polar LEO.

results and details of the computational procedure are presented in a separate paper.[10] Previous empirical estimates of shock-layer thickness and equivalent nose radius were found to agree well (<20%) with the current CFD predictions.

Radiation

The predicted stagnation point radiative heating distributions for the range of missions analyzed are given in Fig. 7. Only the nonequilibrium condition is shown, since this is the dominant surface heating mechanism (typically an order of magnitude greater than the equilibrium case). This occurs because of the large-scale vehicle and highly energetic rarefied flow regimes, which provide the long relaxation times making nonequilibrium effects important. In addition, results are given only for the first pass of the multipass cases, since this produces the worst heating conditions. Complete details of the radiative transport computational procedure have been reported previously.[2,3,11]

The present results are considered overconservative estimates because of the necessity of basing the calculations on adiabatic shock-layer conditions. This is unrealistic for the low-density hypervelocity AOTV flight regimes, because there are significant shock-layer energy losses to space. In addition, lower limits have been used for the reduction of radiative flux by boundary-layer absorption. The proper treatment of these issues is very difficult but should substantially reduce the radiative fluxes. The present calculations also incorporate new estimates of the radiative emission truncation effects caused by the cooling of the shock-layer gases owing to the extent of the boundary layer. For AOTV flight conditions, the boundary-layer thickness was estimated from Apollo and Fire data by adjusting the shock standoff distance until agreement was obtained between the predicted and experimental results. Further refinement of this issue, and possibly estimates of nonadiabatic effects, will be provided by future CFD calculations that couple all of the relevant physics of the prototype flight regime. In addition, the radiative transport calculations are being upgraded by including nonequilibrium excitation phenomena,[12] which is expected to further reduce the level of radiative emission.

The present results for the GEO return mission (Fig. 7a) are less severe than the initial predictions.[3] Most of this reduction is provided by the different flight trajectories and the technique for estimating the boundary-layer truncation effects in reducing radiative flux that is incorporated into the present analysis. The remainder results from the use of the standard atmosphere model, instead of the 50% overshoot condition used previously. The effects of negative lift and multiple passes are highly beneficial in providing substantial reductions in the radiative flux. This amounts to as much as 50% for the comparison of the $L/D = 0$ and three-pass $L/D = -0.1$ cases. Higher heating rates are obtained for the more distant HEO missions (Figs. 7b and 7c). Multiple passes, however, reduce these below the level of the GEO return case for $L/D = 0$, which is the baseline for comparison. Consequently, a generic TPS design may be possible for all missions extending through cislunar space.

Convection

A modified version of the Fay-Riddell equation[13] incorporating finite-rate surface catalysis was used for the convective heating calculations. In addition, the heating effects caused by the suction or flow through the porous drag-brake material were determined. These include the enhancement in surface heating owing to the attenuation of the boundary-layer temperature profile and the interior wall heating resulting from release of chemical energy during flow through. The analysis is identical but inverse to that

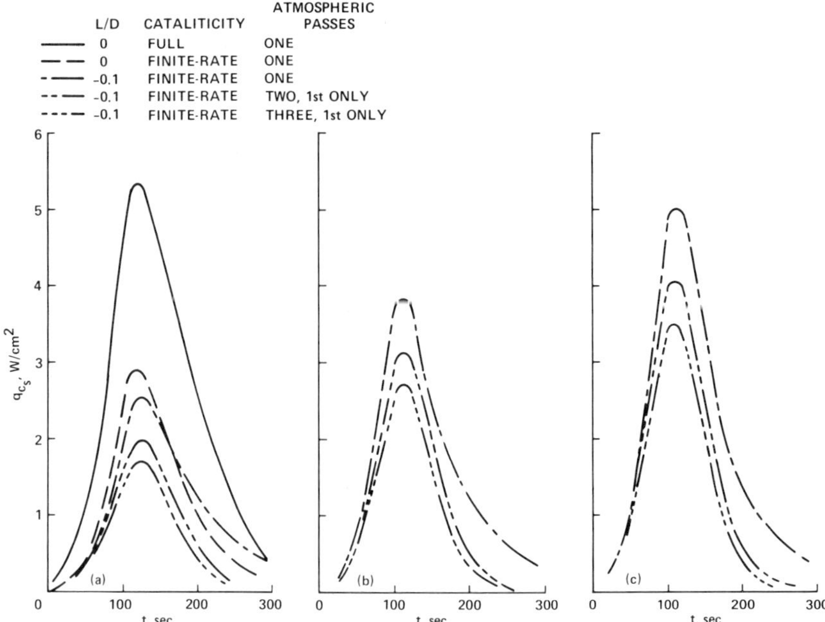

Fig. 8 Stagnation point convective heating rate distributions along flight trajectories for aeroassisted return missions to LEO: a) GEO to Shuttle orbit; b) 5×GEO polar to Shuttle orbit; c) L_5 to polar LEO.

for boundary-layer blowing or mass injection.[14] The current results are considered very conservative, since all the heat due to the chemical energy is released within the porous medium. It is most probable, however, that the flow will exit the rear drag-brake surface without significant release of energy because of the thin material. Nevertheless, the overall effect is generally less than 10% for the present flight conditions and relatively unimportant. A major factor in the convective heating calculations is the transport properties, which are expressed through the Prandtl and Schmidt numbers. These parameters are strongly affected by the nonequilibrium relaxation phenomena in AOTV flight regimes, and the wide variation in magnitude has never been adequately determined. An extensive study is in progrsss, however, to upgrade our knowledge of these important parameters. The values used herein are typical of those used traditionally by others for air.

The stagnation point heating rate distributions are given in Fig. 8. The values near the frustum can be greater at higher angles of attack, but are counteracted by lower radiative fluxes at this location because of the weaker shock front. Consequently, the overall effect is compensating, and the stagnation point values are considered appropriate for design purposes. The finite-rate surface catalytic results were obtained for a wall rate parameter of 50 cm/sec (Ref. 15). This value was selected to be conservative for the low catalysis of the silica cloth proposed for the drag-brake material. The substantial effect of the finite-rate chemistry in reducing the convective heating is demonstrated for the GEO return in Fig. 8a, which also shows the fully catalytic case for $L/D = 0$. The convective fluxes are decreased by almost 50%, and the additional benefits of negative lift and multiple passes provide further reductions amounting to a total of almost 70% for the three-pass case. Moreover, recent studies[16,17] have shown that even further reductions are obtained by accounting for nonequilibrium effects in the surface chemistry. It is also observed that the benefits of multiple passes again cause small design penalties for the more distant 5×GEO (Fig. 8b) and L_5 (Fig. 8c) missions. Further refinement of the convective heating characteristics, which include viscous and angle-of-attack effects, will be provided by the future real-gas CFD calculations.

Total Heat Loads

The present results for the baseline GEO mission are lower than those of the earlier studies because of the decreased radiative heating. They are also typical of the more distant HEO cases because of the benefits of multiple passes in reducing the heating requirements. The heat loads resulting from succeeding passes of multiple-pass missions are of the same order as the first pass. This is not expected to cause additional TPS penalties, however, since at least 4 h are available to reject the heat to space between passes of the various missions (see Table 1). The circumstance occurring here is that for the ideal case of heat transfer between a black-body radiator and an infinite heat sink of negligible temperature. The foregoing discussion pertains to the incident heat fluxes, which are uncorrected for the reflective characteristics of the silica cloth proposed for the drag-brake surface material. This is of the order of 80% and reduces the total heat loads by more than 60%. Consequently, the effect of successive heat loads is undoubtedly negligible when only the absorbed heat fluxes are considered.

Thermal Protection Analysis

The combined magnitude of the incident radiative and convective heat fluxes and loads are less than the initial predictions[3] for the design baseline GEO return mission. The benefits of multiple atmospheric passes reduce TPS requirements for the more distant HEO sorties to the level of the design baseline. Consequently, the initial heat protection analysis remains valid and is applicable to the entire range of missions in cislunar space. Additional design refinements in progress include updating the spectral optical properties of appropriate materials for the predicted aerothermal environment.

In addition, a detailed surface-to-surface analysis of radiative transport effects on the drag-brake support structure caused by wake-flow emission and radiative backscattering from the drag-brake surface is underway. This work includes new details in the construction of the drag brake, such as coating the back side with a thermal-control paint or metal (e.g., nickel or chromium through a plasma-jet spraying process) that enables the cloth to efficiently reflect the incident shock-layer radiation by both volume and back-face reflection. The reflectance of such a system is very high (typically 80%) for the wavelength range of the incident radiation, which is expected to be about 0.3 to 1.5 μm. The short wavelengths below 0.3 μm are likely to be absorbed in the boundary layer. A metallic surface typically has a very low emissivity in the infrared wavelength range of the reradiated heat flux, which was absorbed on the front face. Consequently, the metal-coated back face of the aerobrake will cast only a weak radiation field to the wake-flow region where the support structure and vehicle body are heated.

The supporting beams are thermally insulated from the metallic back-face surface by appropriate materials such as silica. However, the beams will be heated somewhat by both the back-face radiation and conduction through the insulation. Another mechanism (and possibly the primary one) heating the afterbody is the radiative emission of the wake-flow gases, since previous studies show the convective contribution to be relatively unimportant. The beams are cooled by reradiating the absorbed heat. This process is enhanced by coating the beam surfaces with a thermal control paint that reduces absorption of radiation and increases emission. A cursory order-of-magnitude examination of the heat fluxes involved indicates that an equilibrium state is reached between the heat intake from the aerobrake and rejection by radiation at temperatures tolerable for graphite polyimide. Complete details will be provided by the comprehensive surface-to-surface radiative transport analysis currently in progress. Definitive results will not be obtained, however, without suitable experiments, since adequate analytical models of the wake-flow region are not possible.

Concluding Remarks

A preliminary design and performance analysis of a conical lifting brake AOTV concept has been conducted for a broad range of missions, including NASA and DOD requirements, that extend through cislunar space. The performance capability for payload delivery, retrieval, and combined operations is substantial given the proper implementation of aerobraking and three-impulse propulsive thrusts for aeroassist maneuvers involving altitude and plane-inclination changes. The ratio of retrieval-to-delivery payload was found to increase significantly for the more distant missions beyond GEO, which may have important strategic applications. This capability will be further enhanced by the alleviation of Shuttle-compatible weight constraints in accordance with normal-growth technology.

The beneficial effects of finite-rate catalysis, negative lift, and multiple atmospheric passes in reducing aerothermal heating rates and thermal protection requirements were demonstrated. Future refinements, which are expected to give further reductions, include CFD flowfield calculations of the fully coupled shock-layer physics; nonadiabatic effects; nonequilibrium excitation phenomena; and improved estimates of the chemical reaction rates, transport, and spectral properties of air species.

A greatly improved understanding of the wake-flow heating environment and radiative transport characteristics is also urgently needed. This is the crucial factor in the thermal analysis and design of the support apparatus, because the aerodynamic forces and loads present no structural problems. The overall results of the feasibility analysis are promising for the continued development of the drag-brake concept without advanced technology TPS materials development.

References

[1]Walberg, G. D., "A Review of Aeroassisted Orbital Transfer." AIAA Paper 82-137, 9th Atmospheric Flight Mechanics Conference, San Diego, California, Aug. 1982 (also to be published in the Journal of Spacecraft and Rockets; in addition a synopsis appears in Astronautics and Aeronautics, Nov. 1983, pp. 36-43).

[2]Menees, G. P., "Trajectory Analysis of Radiative Heating for Planetary Missions with Aerobraking of Spacecraft," AIAA

Paper 83-0407, 21st Aerospace Sciences Meeting, Reno, Nevada, Jan. 1983 (also to be published in the Journal of Spacecraft and Rockets).

[3]Menees, G. P., "Thermal-Protection Requirements for Near-Earth Aeroassisted Orbital-Transfer Vehicle Missions," AIAA Paper 83-1513, 18th Thermophysics Conference, Montreal, Canada, June 1983 (also published in the 1984 Thermophysics Volume of the AIAA Progress Series).

[4]Davies, C. B. and Park, C., "Aerodynamic Characteristics of Generalized Bent Biconic Bodies for Aero-Assisted Orbital-Transfer Vehicles," AIAA Paper 83-1512, 18th Thermophysics Conference, Montreal, Canada, June 1983 (also to be published in the Journal of Spacecraft and Rockets).

[5]Hair, L. M. and Engel, C. D., "Low L/D Aerobrake Test at Mach 10," AIAA Paper 83-1509, 18th Thermophysics Conference, Montreal, Canada, June 1983.

[6]"Aeroassist Technology Development Plan," Staff, NASA Aeroassist Technology Working Group, Oct. 1982.

[7]Cruz, M. I., Kechichian, J. A., Rinderle, E. A., and Vihn, N. X., "Optimization and Closed Loop Guidance of Drag-Modulated Aero-Assisted Orbital-Transfer," AIAA Paper 83-2093, Atmospheric Flight Mechanics Conference, Gatlinburg, Tennessee, Aug. 1983.

[8]Niles, A. S. and Newell, J. S., Airplane Structures, 4th ed., John Wiley and Sons, Inc., New York, 1954.

[9]"Composites for Extreme Environments," ASTM Special Technical Publication, PCN04-768000-33, edited by N. R. Adsit, Nov. 1980.

[10]Balakrishna, A., "Flowfield Computation of an Aerobrake Orbital Transfer Vehicle," AIAA Paper 84-1697, 19th Thermophysics Conference, Snowmass, Colorado, June 1984.

[11]Park, C., "Radiation Enhancement by Nonequilibrium in Earth's Atmosphere," AIAA Paper 83-0410, 21st Aerospace Sciences Meeting, Reno, Nevada, Jan. 1983 (also to be published in the Journal of Spacecraft and Rockets).

[12]Park, C., "Calculation of Nonequilibrium Radiation in AOTV Flight Regimes," AIAA Paper 84-0306, 22nd Aerospace Sciences Meeting, Reno, Nevada, Jan. 1984 (also published in the 1984 Thermophysics Volume of the AIAA Progress Series).

[13]Fay, J. A. and Riddell, F. R., "Theory of Stagnation-Point Heat Transfer in Dissociated Air," Journal of the Aeronautical Sciences, Vol. 25, No. 2, Feb. 1958, pp. 73-85.

[14]Dorrance, W. H., Viscous Hypersonic Flow, McGraw-Hill Book Company, Inc., New York, 1962.

[15]Rakich, J. V. and Stewart, D. A., "Results of a Flight Experiment on the Catalytic Efficiency of the Space Shuttle Heat Shield,"

AIAA Paper 82-0944, AIAA/ASME 3rd Joint Thermophysics, Fluids, Plasma and Heat Transfer Conference, St. Louis, Missouri, June 1982.

[16]Shinn, J. L., "Chemical Nonequilibrium Effects to Flow Fields for Aeroassist Orbital Transfer Vehicles," AIAA Paper 83-0214, 21st Aerospace Sciences Meeting, Reno, Nevada, Jan. 1983.

[17]Scott, C. D., "Effects of Nonequilibrium and Catalysis on Shuttle Heat Transfer: A Review," AIAA Paper 83-1485, 18th Thermophysics Conference, Montreal, Canada, June 1983.

An AOTV Aeroheating and Thermal Protection Study

Carl D. Scott,* Robert C. Ried,† Robert J. Maraia,*
Chien-P. Li,* and Stephen M. Derry‡
NASA Lyndon B. Johnson Space Center, Houston, Texas

Abstract

The aerothermodynamics and thermal protection of aerobraking orbital transfer vehicles are studied with a view toward design methodology and technology requirements. A particular class of geometries is investigated: the ellipsoidally blunted raked-off elliptic cone. Discussed are the influence of various geometrical parameters on the trajectory and on the heat flux as well as heating prediction methods ranging from an engineering correlation formula to computational solutions of the Navier-Stokes equations. Finite-rate catalytic thermal protection materials and gas radiation are considered. Estimates of the specific mass of possible thermal protection systems indicate that support structure is a large contributor to the aerobrake mass as compared with the thermal protection insulation itself. Sizing curves are included which enable the designer to estimate the size aerobrake required from heating considerations for different mass vehicles having aerobrakes of several specific masses.

Nomenclature

A = reference area
ADI = alternate direction implicit

Presented as Paper 84-1710 at AIAA 19th Thermophysics Conference, Snowmass, Colorado, June 25-28, 1984. The work presented in this paper is a major revision of but, to some extent, repeats work presented at the NASA Symposium on Recent Advances in TPS and Structures for Future Space Transportation Systems, NASA Langley Research Center, December 13-15, 1983, and published in NASA CP-2315, July 1984.

*Research Engineer.
†Special Assistant to Director of Engineering.
‡Senior Engineer.

AOTV = aerobraking orbital transfer vehicle
C_D = drag coefficient
C_H = Stanton number
CFD = computational fluid dynamic
CFL = Courant-Friedrich-Lewy convergence criterion
D = diameter
D = drag
EC = engineering correlation formula
f = heating factor to account for deviations from fully catalytic heating and for shape
F-R = Fay and Riddell
GEO = geosynchronous Earth orbit
h = altitude
K = Knudsen number
L = lift
LEO = low Earth orbit
M = Mach number
m = mass
m_B = aerobrake mass
m_o = system mass without aerobrake
N-S = Navier-Stokes
p = pressure
$\dot{q}$ = heat flux
R_o,R_N = nose radius of curvature at stagnation point
Re = Reynolds number
R_{eff} = effective radius of equivalent sphere
T = temperature
TPS = thermal protection system
$\bar{t}$ = equivalent thickness
V_∞ = velocity of free stream
y = distance from cone axis in y-direction; vertical coordinate
y_c = geometrical quantity defined in Fig. 5
β = m/C_DA, ballistic coefficient
γ = energy transfer catalytic recombination coefficient
γ = ratio of specific heats (Fig. 11)
γ_o = initial entry angle
δ = rake angle
ε = emittance
ε_b = ellipticity of bluntness
θ_{xy} = cone angle in x-y plane
ρ_∞ = free-stream density
ρ_B = specific mass

Subscripts

c = convection
FC = fully catalytic
max = maximum
NC = noncatalytic, $\gamma = 0$
o = stagnation point
r = radiation
ref = reference
T = total
W = wall
∞ = free stream

Introduction

Numerous concepts for orbit transfer vehicles have been proposed over the last decade. The aerobraking concept has received much attention since it does not require propulsion fuel for its reduction in velocity. The subject of aerobraking and aerocapture vehicles has been reviewed by several authors including Walberg.[1,2] The aerothermodynamics and thermal protection of aerobraking vehicles has been studied by, among others, Andrews et al.,[3] Florence,[4] Menees,[5,6] and Howe.[7] Concepts range from vehicles having lift-to-drag ratio $L/D = 0$, such as the inflatable ballute concept, to high-L/D vehicles, such as slender-winged craft that can maneuver extensively in the atmosphere. Low-L/D concepts that would apply to a space-based, reusable, geosynchronous Earth orbit (GEO) to low Earth orbit (LEO), aerobraked mission are addressed. The technology discussed here should be considered as one step beyond the Space Shuttle and Apollo Programs and thus assumes some improvement in the accomplishments of those programs.

To achieve a viable candidate aerobraking orbital transfer vehicle (AOTV) requires a vehicle system study, which in turn requires a study of the interrelation among the heating, the entry configuration, and the thermal protection system (TPS). This latter study serves as input along with propulsion, structures, and payloads requirements to an overall system study to provide a preliminary design.

The aerothermodynamics and TPS are addressed from the standpoint of parametric variations of the vehicle aerodynamic characteristics, mass, and size, and of the thermal protection mass and thermal response. The influence of such technology issues as flight flow regimes on the heating calculational techniques and TPS characteristics on overall system size and mass are discussed.

Although various mission scenarios have been proposed for aeroassisted orbital transfer vehicles, the space-based version is addressed in this paper. This emphasis is consistent with the recent U.S. national commitment to build a space station in LEO. The space station would be configured for AOTV servicing capability. It is envisioned that some of the AOTV missions would be manned so that satellites in GEO could be repaired effectively. The space-based mission requires extremely reliable and simple subsystems. The less refurbishment needed, the better. The influence of the aerodynamic parameters such as lift-to-drag ratio, initial entry angle, and ballistic coefficient on the trajectory and, therefore, on the maximum heat flux is addressed. Factors that influence the heating include the effect of geometry on heat flux distributions by simple Lees theory and Navier-Stokes (N-S) calculations, the nonequilibrium flow chemistry and catalytic recombination on the TPS surfaces, the delineation of flow regimes and their influence on the heating methods required, and nonequilibrium radiative heating. A parametric study of TPS mass as a function of structural thermal mass and incident heat load will show relative insensitivity to insulator thickness and a sensitivity to support structure. Finally, a vehicle brake sizing philosophy is described, and conclusions are drawn on the basis of those curves as well as the other results presented.

Analysis and Discussion

In this analysis, use is made of some of the following assumptions about the mission of the AOTV considered in this paper. The vehicle would serve as a cost-effective transportation system from LEO to GEO and return. Any significant plane change is assumed to be obtained by a propulsive maneuver at apogee. Therefore, a low-L/D vehicle is sufficient for control during atmospheric braking. Because the vehicle will be space based and possibly manned, there will be a requirement for high reliability and little or no refurbishment permitted. Thus, ablative TPS and ballutes are undesirable, whereas a fixed structure and a reusable TPS are desirable. A space-based system initially assembled in space is permitted, although it would obviously be preferable to carry the entire vehicle into LEO as a single package in the Space Shuttle. In this study, little reliance is placed on technological breakthroughs, especially in thermal protection concepts; therefore, a modest extension of Apollo and Space Shuttle technology is assumed.

Parametric Trajectory Influence on Heating

To determine the influence of L/D, ballistic coefficient m/C_DA, and entry angle γ_o on the peak heat flux and heat load of an aerobraked OTV, a number of trajectory cases were run using the techniques of Gamble and Cerimele[8] and Gamble et al.[9] employing the adaptive guidance logic method of Hill.[10] The primary constraints on the trajectory calculations were entry and exit velocities of 10.27 km/s and 7.86 km/s, respectively, and maintenance of sufficient control margin to allow for uncertainties of 20% in the atmospheric density. The minimum entry angle, based on the assumption of $\rho_\infty = 0.8\rho_{\infty_{ref}}$, was selected for each case, where $\rho_{\infty_{ref}}$ is that of the 1962 U.S. Standard Atmosphere.

An examination of the maximum reference sphere heat flux dependence on the trajectory input variables L/D and m/C_DA indicated that the peak heat flux is most sensitive to m/C_DA and that dependence is well represented by the following relation:

$$\dot{q}_{max}\sqrt{R_N} = 7.3(m/C_DA)^{0.467}(L/D)^{-0.242} \ (\mathrm{W/cm^2})(\mathrm{m})^{1/2} \qquad (1)$$

where m/C_DA is expressed in kilograms per square meter. The heat flux is calculated by the engineering correlation (EC) formula for a sphere.

$$\dot{q}\sqrt{R_N} = 18\,300\rho_\infty^{1/2}(V_\infty/10^4)^{3.05} \ (\mathrm{W/cm^2})(\mathrm{m})^{1/2} \qquad (2)$$

where free-stream density ρ_∞ is expressed in kilograms per

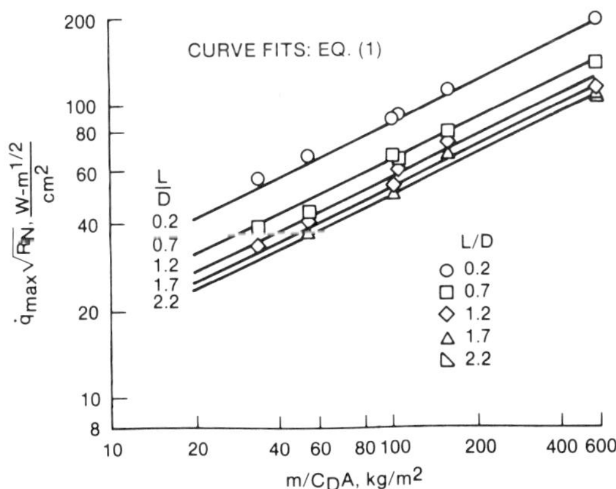

Fig. 1 Correlation of peak reference heat flux with ballistic coefficient and lift-to-drag ratio for AOTV trajectories.

cubic meter and free-stream velocity V_∞ is expressed in meters per second.

The least squares curve fit and the data from which it is derived are shown in Fig. 1. The peak heat flux does depend on entry angle γ_0. However, because the AOTV enters the atmosphere with the shallowest entry angle consistent with maintaining control assuming $\rho_\infty = 0.8\rho_{\infty_{ref}}$, its dependence is not shown in Fig. 1.

Geometry

A proposed aerodynamic shape used in this study is an ellipsoidally blunt raked-off elliptic cone, represented in Fig. 2, which has the following characteristics. The conical afterbody is raked off at angle δ to provide lift at zero angle of attack. The base in the rake plane is circular for packaging efficiency and minimum ratio of perimeter to area. An alternate approach would be to employ a circular cone and thus provide an elliptic base. A circular base requires that the cone be elliptic. The cone is blunted to minimize the stagnation-point heat flux; for mathematical convenience, an ellipsoid with variable ellipticity was chosen. To help reduce computational problems at the ellipsoid-cone junction, the surfaces are made mutually tangent in a plane normal to the cone "axis." The intersection is then an ellipse having the same ellipticity as the cone. The cone's ellipticity is defined as the ratio of vertical to horizontal axis lengths of an elliptic cross section. Blunting the cone with an ellipsoid permits the Newtonian stagnation point to be on the surface having a larger radius of curvature than a sphere and, thus, helps reduce the maximum heat flux. The trailing edge of the cone is shown sharp in Fig. 2. A real vehicle would probably be rounded somewhat to reduce the trailing-edge heat flux and to provide better dynamic stability. With a suitably placed center of gravity, this configuration will have lift and will trim at zero angle of attack. The Newtonian aerodynamic characteristics of a similar class of shapes is given in Ref. 11, where it was found that for large rake and cone angles, the relation $L/D = \cot \delta$ holds for all bluntnesses for large-angle circular cones and for all ellipticities for sharp elliptic cones. It has been assumed for preliminary purposes that this relation also holds for blunted elliptic cones. One advantage of large rake and cone angles is the greater latitude in center of gravity placement[11]; i.e., the center of gravity may be farther aft.

A measure of the bluntness of the configuration is the radius of curvature R_0 of the ellipsoid at the Newtonian

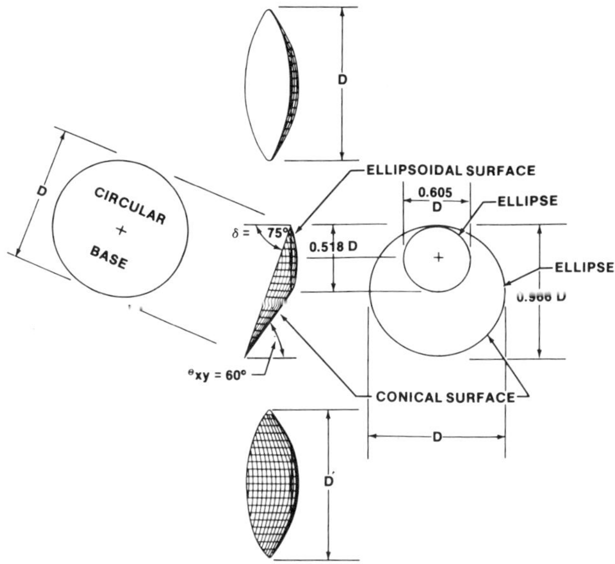

Fig. 2 Geometry of a raked-off elliptic cone with ellipsoidal bluntness; $\delta = 75°, \theta_{xy} = 60°$.

stagnation point in the x-y (pitch) plane. The effect of rake angle and cone angle on the radius of curvature is illustrated in Fig. 3. That the radius of curvature is a minimum means the ellipsoid is the sharpest for the given δ and cone angle θ_{xy}. It is also a minimum in the sense that the transverse radius of curvature (in the x-z plane) is larger. Larger R_o values are possible for given δ and θ_{xy} by increasing the bluntness ellipticity ε_b. As ε_b increases, the blunt surface becomes flatter. It is seen in Fig. 3 that R_o/D increases with rake angle and that for a given rake angle, there is a corresponding optimum cone angle for maximum R_o/D. An optimum ε_b is possible from the standpoint of a uniform heat flux distribution on the blunt surface, as is discussed further in the section on heat flux distributions.

Engineering Methods of Heat Flux Distributions

The influence of configuration on the heat flux is unquestionably one of the most important considerations. As mentioned earlier, the simple EC formula was used to estimate the reference heat flux for the trajectories of this study. However, the limitations of that formula should be noted. It does not take into account the effects of actual geometry of the vehicle, low-density environment, nonequilibrium chemical reactions, wall catalytic recombination, and thermal conduction within the shock layer. These effects may be considered by more sophisticated models based on the Navier-Stokes equations and experimental studies.

To assess the influence of some geometrical parameters on the heat flux distribution, the method of Lees[12] combined with Newtonian pressure distributions was used. Results for several bluntnesses were obtained for 40° and 60° *circular* cones as shown in Fig. 4. The heat flux distributions are normalized by the stagnation-point heat flux of a sphere with radius R_o expressed in terms of R_o/D, where D is the diameter of the circular base of the associated raked-off *elliptic* cone. This normalization then allows one to compare the effect of geometric parameters on the relative heat flux. The side views of these shapes are shown in Figs. 5(a) and 5(b). It is seen that a considerable reduction in stagnation-point heat flux can be obtained by making the ellipsoid more blunt. However, there is a point at which the heat flux near the ellipsoid-cone junction exceeds the stagnation-point value; thus, it would seem that an optimum ε_b might exist where the heat flux distribution is relatively flat. Decreasing the cone angle θ_{xy} lowers the heat flux on the conical surface but increases the heating on the ellipsoid. Moreover, the total surface

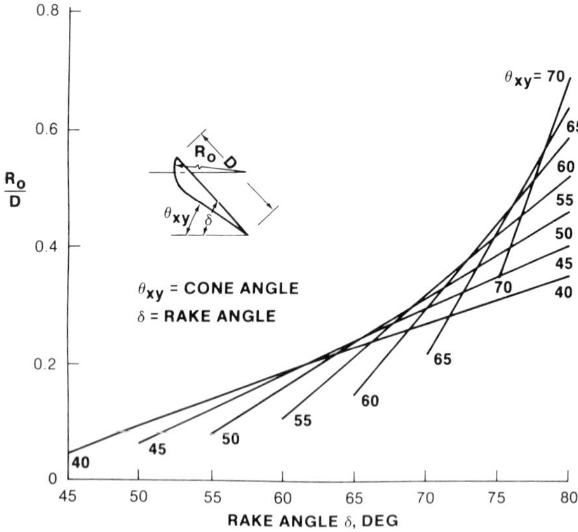

Fig. 3 Minimum radius of curvature in the x-y plane for an ellipsoidally blunt, raked-off elliptic cone with circular base.

area of a sharper cone is larger than that of a blunter cone so that the weight may be greater.

Because the Lees method is developed from the boundary-layer theory that cannot take the downstream influence of the vehicle shape into account, nor can it account for boundary-layer/shock-layer interaction, a comparison was made with a Navier-Stokes solution. The computation was for an ideal gas and a very blunt axisymmetric body as was used in the Lees calculation. The results are compared in Fig. 6, where it can be seen that the N-S solution predicts higher heat flux near the ellipsoid-cone junction. Likewise, the heat flux at the trailing edge of the cone was very large in the N-S prediction. This result should serve as a warning that the simple theories are not adequate and higher order solutions should be obtained. A similar indication of the influence of the extent of the body on the pressure distribution is given in Fig. 7, where axisymmetric ideal-gas N-S solutions are presented for an elliptically blunted circular cone. Two pairs of cases are shown in Fig. 7; in each case, the body was extended to assess that effect on the upstream pressure. It is seen that by truncating the body, one obtains a lower pressure distribution on the remaining body.

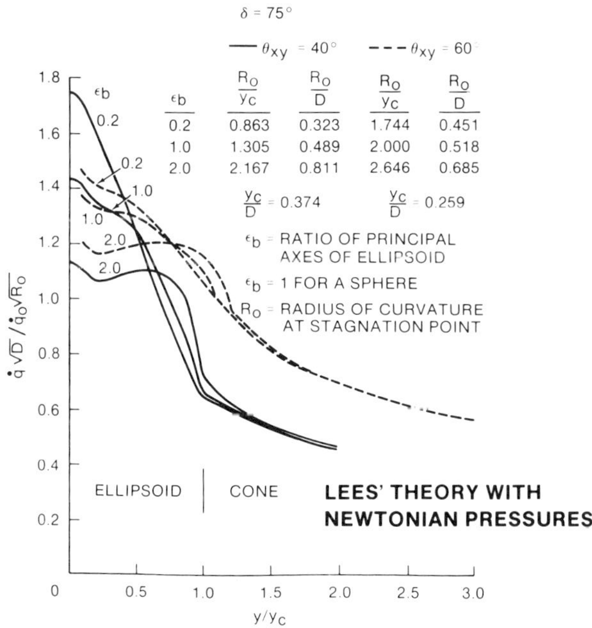

Fig. 4 Relative heat flux distributions on an elliptically blunted circular cone.

Navier-Stokes Solutions for Wide-Angle Bodies. The prediction of heat flux and L/D ratios is a very challenging problem for the type of wide-angle body flying at altitudes considered in this study since the flowfield is strongly influenced by both the shoulder and inviscid-viscous interactions. The conventional approach of decoupling the inviscid shock layer and the boundary layer applicable to the Space Shuttle Orbiter and at lower altitudes becomes inadequate for an AOTV. A more accurate approach appears to be needed even in the preliminary design stage for sizing the vehicle configuration and for development of correlation equations used later in the design cycles. Since the peak heating and maximum velocity changes occur in the continuum flow regime (Fig. 8), the solution from the N-S equations can be used for our intended objectives. A description of the method of solution is briefly given, followed by a discussion of the N-S results compared with the FIRE II data.

The body and the bow shock are schematically shown in Fig. 9. Note that only the flowfield next to the frontal face of the body is of present interest. It is seen that the body corner b-h is a circle in the raked-off plane across the cone. The body wall is denoted by b-a-h but is extended downstream on an imaginary cone (dashed line) in order to enclose the computational region. Hence, outflow conditions are imposed on b-c-d and f-g-h. The windward boundary conditions are specified by Rankine-Hugoniot relations, which are adequate for the considered large body

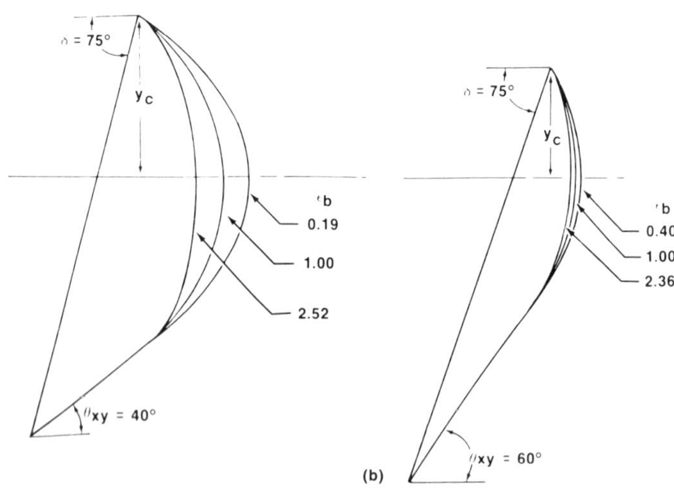

Fig. 5 Ellipsoidally blunted raked-off elliptic cone with several bluntness parameters.

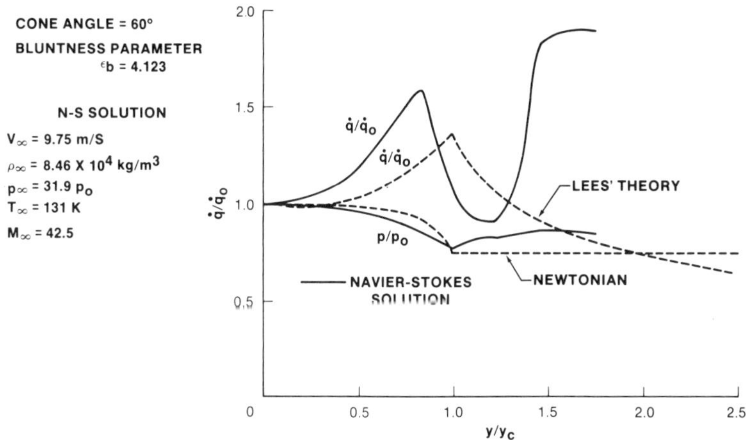

Fig. 6 Heating and pressure distributions on an ellipsoid cone.

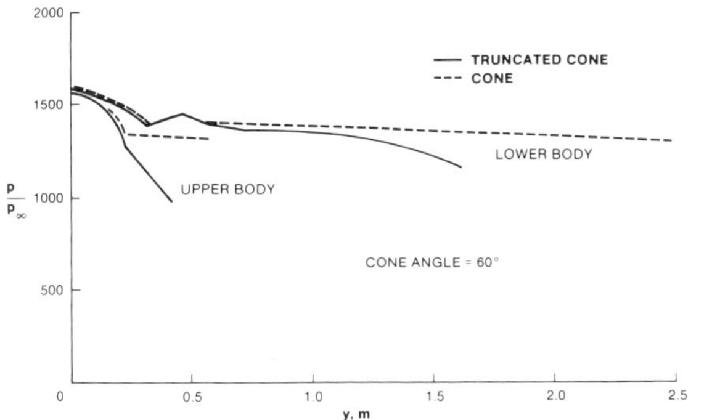

Fig. 7 Pressure distributions on blunted cones from Navier-Stokes solutions.

size and altitudes. The computational grid is based on the spherical-polar system centered on the axis a-e, so selected as to allow equal space (angular) resolution on either side of the axis. An enhanced version of the three-dimensional N-S code to be described in detail in Ref. 13 was applied to this new problem. Minor changes of the code are required because of the assumed fictitious wall conditions, the body geometry defined on an arbitrary axial orientation, and the difficulty of defining reasonable initial conditions. The earlier version of the code using an alternate direction implicit (ADI) factorization technique to advance the time-dependent solution until steady results are obtained has demonstrated high efficiency and accuracy for ideal and equilibrium airflow around the Orbiter nose at wind-tunnel and flight conditions for angles of attack of 30° and 40.2°, respectively. The enhanced version has an even faster convergence rate and lower computation time and has been validated for low- and high-Reynolds-number conditions.

To demonstrate the feasibility of the present method of handling wide-angle cones with a sonic corner, axisymmetric solutions were obtained from the N-S code at flight conditions of the FIRE project,[14] for which the radiant heat flux and nonequilibrium chemistry are insignificant; and comparisons are made of the stagnation-point heat fluxes between the results and the measurements. Shown in Fig. 10 are two sets of N-S results made for the spherical nose radius at R_N = 0.725 m and for the truncated spherical nose at R_N = 0.936 m. The truncated nose represents the FIRE vehicle actually flown, whereas the effective nose radius was used in Fay and Riddell[15] (F-R) and EC equations. It is seen from the figure that the ideal-gas N-S result agrees well with the convective heating measurements at a high altitude but has a lower value at lower altitude. Furthermore, the effective nose radius calculation shows a higher value and lower value than the measurements at the two altitudes. More work should be done on the validation of the N-S code for low-density environments before reaching firm conclusions, yet it is indicated that the N-S results can be used to obtain the effective radius of any given truncated body. The correlation formulas give the correct trend but show consistently higher heating. Since the convective heating is deduced from the total heat flux by subtracting the radiant heat flux and since there is a strong possibility that the measured radiant magnitude is less than its true value, the difference between the prediction and the measurement may be very small.

At the high speed and high altitude of an AOTV reentry, the nonequilibrium dissociation and viscous interactions are prevalent. To estimate their order of magnitude, we examine the temperature profiles across the shock layer for an ideal and an equilibrium air model obtained from N-S solutions. Figure 11 depicts these profiles along the stagnation streamline for h = 71 km. The large ratio of temperatures and shock-layer thicknesses obtained from ideal and equilibrium air calculation suggests that a great amount of translational energy is used to dissociate the air molecules. In reality, the major portion of the stagnation region is out of equilibrium. Both calculated profiles show that the wall influence due to viscous dissipation comprises approximately a quarter of the shock layer. Similar calculations made for h = 85 km predict that 65% of the shock layer is viscous. Hence, the

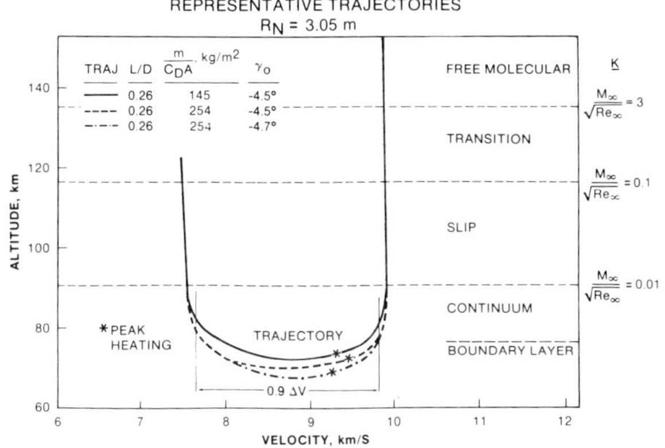

Fig. 8 AOTV trajectories and flow regimes.

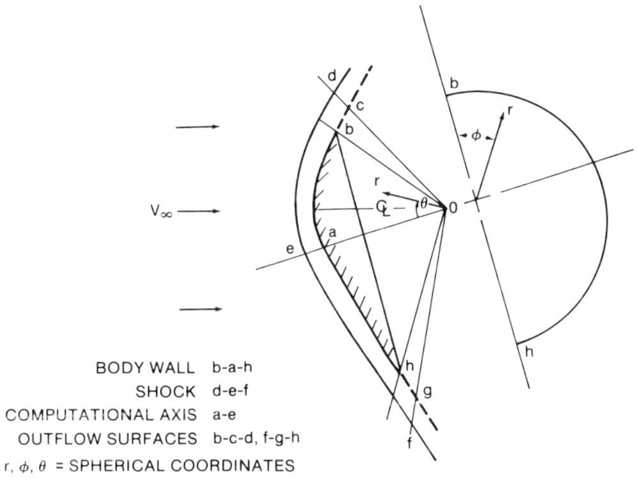

Fig. 9 Computational region on the spherical coordinate system for the ellipsoid/60° raked-off cone.

validity of the N S approach is justified. Note that the results in Fig. 11 were obtained for a spherical nose with R_N = 0.725 m. The Stanton number for the equilibrium air calculation is 0.0156 compared with 0.037 for the ideal gas and with 0.053 for the F-R formula. The large disparity between the FIRE data and the N-S results has called for a series of detailed comparisons for free-stream conditions ranging from M_∞ = 5.92 to 35.4 and Re_{R_N} = 431 to 10^6. The study has shown that the N-S results agree well with the boundary-layer, other N-S, and the viscous shock-layer results, except for the conditions in the neighborhood of $3000 < Re_{R_N} < 6000$. The discrepancy is being investigated further.

Convective Heating and Pressure Distribution. The geometric effect of the body corner was examined by applying the N-S code to the upper portion of the ellipsoid-cone. Considering the upper body as a body of revolution, the convective heat flux and pressure were obtained at zero angle of attack. Figure 12 displays the body shapes as well as the shock locations for the short and the long body. The flow expansion at the sonic corner has caused the bow shock to be more concave toward the body near its end and increased the velocity gradient over the frontal face. Therefore, higher heat fluxes resulted on the raked-off body than on the infinitely long body. The first peak on the heating distribution is due to the body slope changes on the juncture point between the ellipsoid and the cone. Similar reasoning applies to the second peak on the curve as the flow is nearing the body corner. As expected, only one heating peak appears on the curve for the long body. Since the velocity gradient all over the raked-off body is higher than on the long body, even the stagnation heating is higher. The pressure at the stagnation point is relatively insensitive to the body shape, yet strong influence is seen on the distribution near the corner. The juncture point also exerts influence on both pressure distributions; its effect appears to be local as opposed to upstream propagation seen for the heating dis-

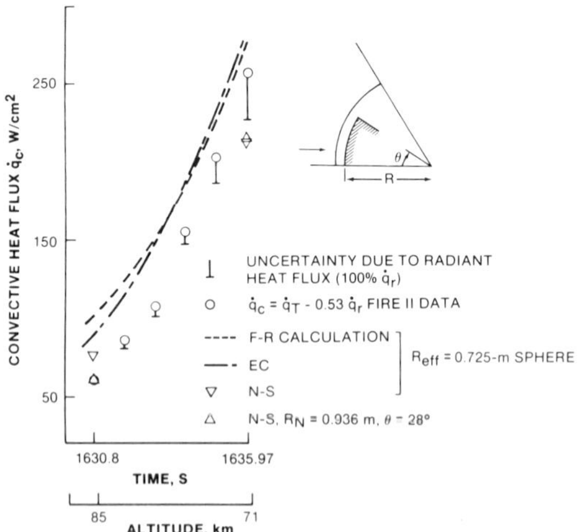

Fig. 10 Stagnation-point convective heat flux during early data period of FIRE II flight experiment re-entry.

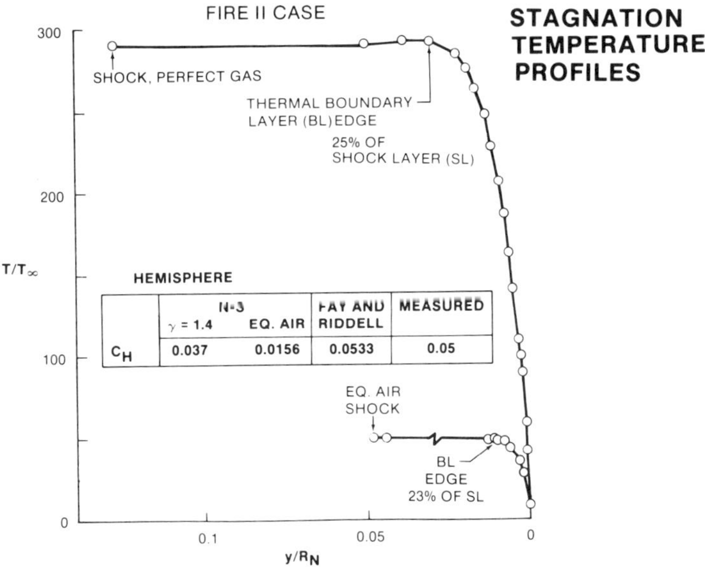

Fig. 11 N-S calculated shock-layer temperature distribution at stagnation point for FIRE II at an altitude of 71 km and a velocity of 11.4 km/s.

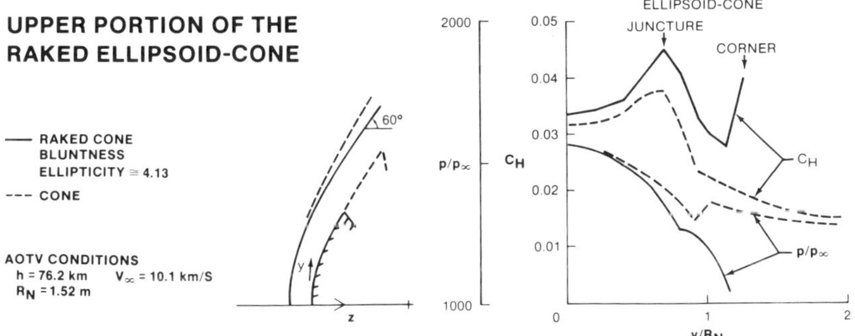

Fig. 12 Boundary condition effects on heating distributions from Navier-Stokes solutions.

tributions. The results shown in Fig. 12 were obtained with a grid of 28 by 16 and after 600 iterations. The high temperature gradient at the wall was resolved by a clustered grid having $\Delta y_W = 0.04 R_N/(Re_\infty)^{1/2}$; the CFL number was as high as 100 without any stability problems. The computation time was 20 min on a Univac 1182.

Catalysis Effects. In the present catalysis model, one conceives of a certain fraction of atoms that strike a

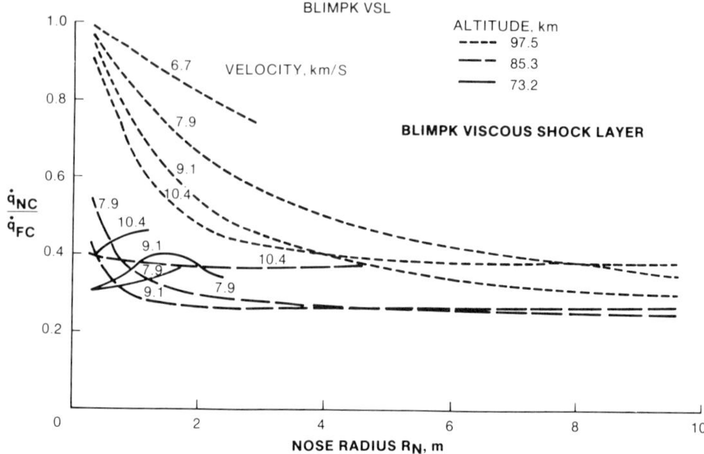

Fig. 13 Noncatalytic to fully catalytic heat flux ratio at stagnation point of sphere.

surface and recombine into the molecular ground (or equilibrium) state and leave all the energy of dissociation on the wall and thus contribute to the heat flux. If the wall is noncatalytic, then no atoms recombine on the surface and hence there is no chemical contribution to the heat flux. This model shall be used in the present study, although models involving excited species production on the surface are being considered as future refinements.

An estimate of the effect of a noncatalytic wall on the heat flux to an AOTV is given in Fig. 13. The noncatalytic to fully catalytic heat flux ratio is given as a function of effective nose radius for a set of altitudes and velocities. These results were obtained using the viscous shock-layer option of the reacting boundary-layer code BLIMPK.[16,17] At high altitude and small nose radius, the shock layer is out of dissociation equilibrium (lower dissociation than for equilibrium); therefore, there is not much reduction in heating on a noncatalytic surface. As the altitude is decreased, the dissociation goes to completion much sooner because of the higher collision frequency and, therefore, $\dot{q}/\dot{q}_{FC}$ reaches a minimum. As the altitude decreases further, there are sufficient collisions for recombination in the gas phase to increase and, hence, $\dot{q}/\dot{q}_{FC}$ starts to increase for larger bodies. It is then that the solution method breaks down and the solution no longer converges. There is a potential for a significant reduction in heat flux if the surface is noncatalytic. In the AOTV peak heating regime, $\dot{q}/\dot{q}_{FC}$ is about 0.3 to 0.4.

For a real TPS surface ($\gamma \neq 0$), Shinn and Jones[18] calculated $\dot{q}/\dot{q}_{FC}$ for a body having a nose radius of 2.5 m at a velocity of 8.65 km/s over a range of altitudes and different recombination coefficients γ. In the peak heating regime near an altitude of 80 km, the $\dot{q}/\dot{q}_{FC} \approx 0.55$ to 0.9 for $\gamma \approx 0.01$ to 0.05. Thus, we see that even for existing materials,[19] there is a substantial influence due to reduced catalytic behavior. If a TPS material has a low catalycity, the temperature requirement can be reduced; likewise, if a material has a higher temperature capability, then a high catalycity is permitted, all other things being equal. This trend is indicated in Fig. 14.

Nonequilibrium Chemistry Effects. That chemical nonequilibrium exists in reentry flowfields is an experimentally demonstrated fact of the Orbiter experiment described in Refs. 20 and 21. If the nonequilibrium also extends into the realm of electronic and vibrational nonequilibrium, the models now used in finite-rate chemistry flowfield calculations may be inadequate to correctly partition the energy. These calculations all assume that the thermodynamic properties are described by equilibrium thermodynamics for the internal degrees of freedom. This assumption applies to both gas-phase and surface-related phenomena. The fact that both nitrogen and oxygen molecules have long-lived (metastable) electronic states may be useful in assessing whether ground-test (arc jet) and flight flowfields have nonequilibrium excited states. An AOTV flight experiment with spectrometric

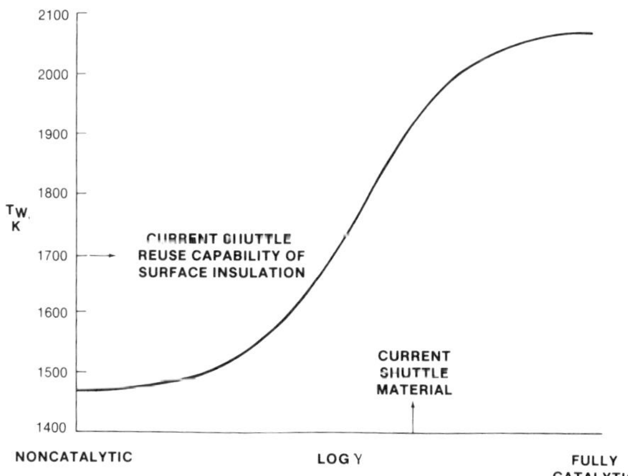

Fig. 14 Representative AOTV peak temperature dependence on catalycity.

instrumentation would help resolve this question. Calculations and experiments of Park[22,23] have shown that nitrogen atom excited states are not in thermal equilibrium with the electron temperature. Detailed calculations that include electronic state nonequilibrium will be required to predict the extent of the nonequilibrium and its influence on convective and radiative heat flux.

The production of excited species by catalytic recombination on the TPS surface is a possibility and is suggested by the seemingly conflicting recombination coefficient results of Breen et al.[24] and Scott.[19] In the former, γ was determined by direct atom titration measurements, whereas in the latter, γ was determined from heat-transfer measurements. These two methods can be reconciled if atoms recombine on the surface and form excited molecules that do not give up all their dissociation energy.

Flow Regimes. Upon examining the AOTV trajectories calculated[8,9] using the adaptive guidance logic of Hill,[10] it was apparent that the vehicle would be spending considerable time in the noncontinuum flow regimes. The different computational fluid dynamic (CFD) codes — boundary layers, Navier-Stokes, viscous shock layer, Monte Carlo — are limited in their application to different flow regimes. It is important to determine the flow regimes in which the aerobrake configuration of interest is flying and the location in that trajectory at which the events of interest (heat flux, heat load, aerodynamic forces, etc.) occur. Therefore, the trajectories were considered in terms of the flow regimes (Fig. 8). The parameter used to define the flow regimes is the mean free path compared with a characteristic length (the Knudsen number, K), discussed in several references (Refs. 25 and 26). These boundaries are not strict and should only be used as guidelines. The flow regimes were calculated on the basis of a nose radius of 3.05 m. It should be noted that the trajectories were calculated on the basis of continuum drag and could be different if the noncontinuum drag forces were substantially different.

As is pointed out later, the aerobrake design considered in this study is not very sensitive to heat load but is sensitive to peak heating rate. The points of peak heating for a 3.05-m-radius sphere for the various trajectories are indicated in Fig. 8; they all occur within the boundary-layer portion of the continuum flow regime, where boundary-layer and Navier-Stokes codes are applicable. An attempt was made to evaluate the significant region of aerodynamic forces — defined as the region in which 90% of the total change in velocity ΔV occurred —

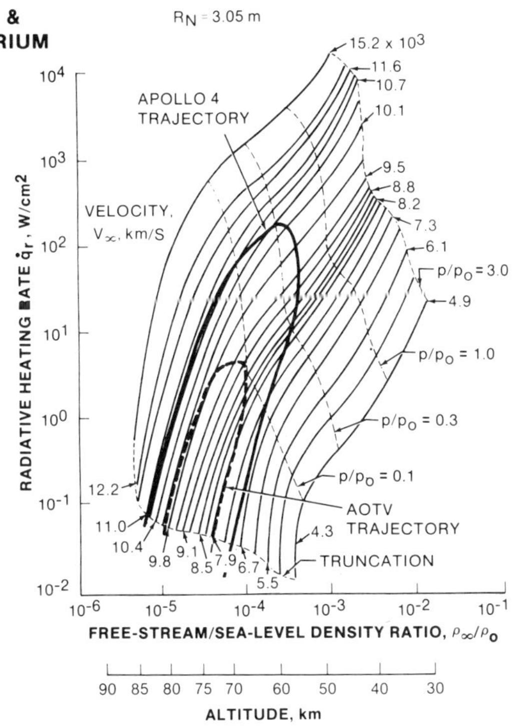

Fig. 15 Stagnation-point total radiative heat flux.

and the result is marked on the figure. This ΔV occurs within the continuum regime, where the Navier-Stokes codes can be reliably used, but above the boundary-layer portion. The heating correlation used in the engineering code is probably sensitive to flow regimes and should be evaluated using measurements and calculational (CFD) results before applying the correlation outside the boundary-layer regime.

Radiative Heating. Consideration of the higher velocities of the AOTV trajectories led to the concern of radiative heating. Radiative heating was extensively addressed during the Apollo Program. In a report by Ried et al.,[27] the total radiation heating — three-dimensional, nonequilibrium, nonadiabatic, for the entire spectrum — is evaluated over a large range of trajectory conditions. Figure 15 has been taken from this reference, and on it is plotted an AOTV trajectory. Fortunately, because the radiative heating is a complex function of size, these curves were developed for a stagnation point of 3.05-m radius, the same as that assumed here for the AOTV aerobrake under consideration. Note that the AOTV trajectory is enveloped by the Apollo trajectory.

There are conflicting views in the literature as to the importance of nonequilibrium for AOTV trajectories. References 5 and 6 show considerably higher radiative heat fluxes than seen here. This issue must be resolved by obtaining better experimental data, especially reaction rates and improved flowfield calculations.

A comparison of the convective and radiative heating is shown in Fig. 16. The convective heating was calculated for the stagnation point of a 3.05-m-radius sphere by the Fay and Riddell[15] correlation. A note on the curve of convective heating: the table of equilibrium air properties of the engineering code does not extend above an enthalpy of 35 MJ/kg or below a pressure of 20 Pa. Therefore, the code uses ideal-gas equations beyond this range, resulting in heating rates about 25% higher than those for equilibrium air. The radiation was obtained from Fig. 15 assuming a surface absorption coefficient of 1. At the lower densities and the lower velocities, the radiative heating is truncated because the shock layer is thinner than the nonequilibrium zone. The peak radiative and convective heating rates occur at about the same trajectory condition. Thus, based on Apollo experience, for the AOTV configuration under study, it appears that the peak radiative heating rate and heat load are less than 10% of the comparable convective heating. There is much uncertainty in evaluating the radiative heating, particularly in the ultraviolet region, and this subject should be carefully considered.

TPS Thermal Analysis for AOTV Brake Mass

Since the requirements for an AOTV design depend on the sensitivity of the TPS to its environment, an estimate

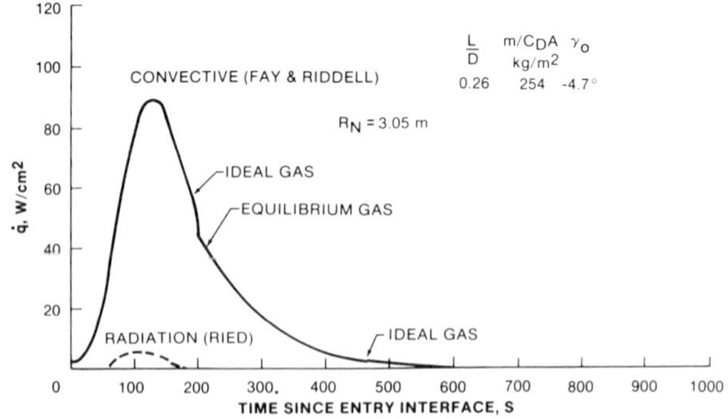

Fig. 16 Heat flux history for an AOTV trajectory.

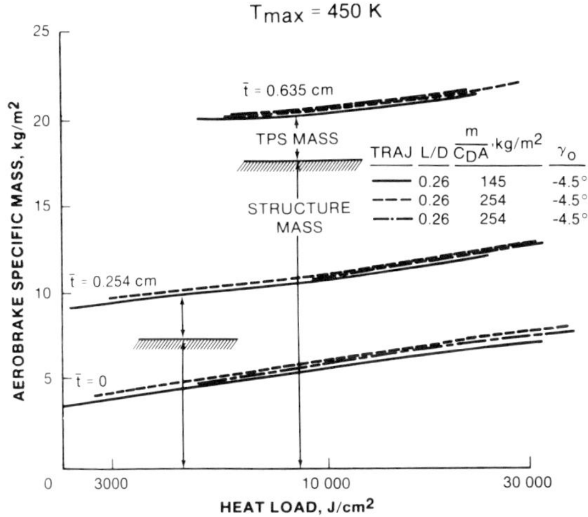

Fig. 17 Specific mass of LI-900 TPS on aluminum structure aerobrake.

is needed for determining the TPS and support-structural mass based on environmental factors and assumed materials. This value will then be used later to obtain an optimum vehicle size that minimizes total system mass. This evaluation is only partly addressed here in that the supporting structure is only crudely assessed.

Material Selection

In this initial analysis of a thermal protection system for the aerobrake, only present-day or near-future materials and design constraints were considered. For this study, low-density, porous, reinforced silica-fiber (LI-900) tiles were chosen. A large amount of thermal and strength data has been accumulated from Orbiter experience. A stronger, denser version, LI-2200, was analyzed, but it led to higher insulation mass for the same protection.

The structure of the brake currently is the most uncertain of the variables involved in estimating mass. Two materials were considered: aluminum (450 K temperature maximum) and graphite-polyimide honeycomb (617 K maximum). For each structure, only the mass, not the actual design of brake structure, is needed for a one-dimensional thermal sizing and mass analysis. It was assumed that a heavy aluminum structure would be 6.4-mm equivalent thickness $\bar{t}$ (that includes skin, any stiffeners, and contribution of nearby frame to local heat-sink capability), and a thin estimate would be 2.5 mm (representing possibly just a skin). These values are based on Orbiter thermal analysis. For the graphite-polyimide honeycomb, equivalent specific masses were represented by thicknesses of 3.8 mm and 1.9 mm. These values would include the total mass of the two facesheets and honeycomb core. The back of each structure was assumed to be adiabatic, which is slightly conservative, since some heat will be lost by conduction and/or radiation.

Figures 17 and 18 show the brake mass required for each of three representative missions so as not to exceed the appropriate maximum structure temperature. This mass per unit area includes the insulator, its coating, the required strain isolation pad, two layers of adhesive, and the assumed structure thickness. It can be seen for all cases that the structural mass is the majority of the brake weight.

These specific masses are plotted against local heat load, which was considered to be a constant fraction of the heating to a reference sphere. The reference sphere heating was calculated from the film coefficient and the recovery temperature. The factor is a function of vehicle geometry and size, surface catalycity, the method of heating calculation, and the heat flux distribution on the surface.

The brake mass shown in Figs. 17 and 18 can be seen to be predominantly structural mass; the insulator contributes only a small portion to the system mass. The short heat pulse of the aerobraking maneuver ends before the temperature pulse fully penetrates the tile, then the heat flows outward in both directions to a warm sink (structure) and to a radiating surface. Obviously, minimizing the total

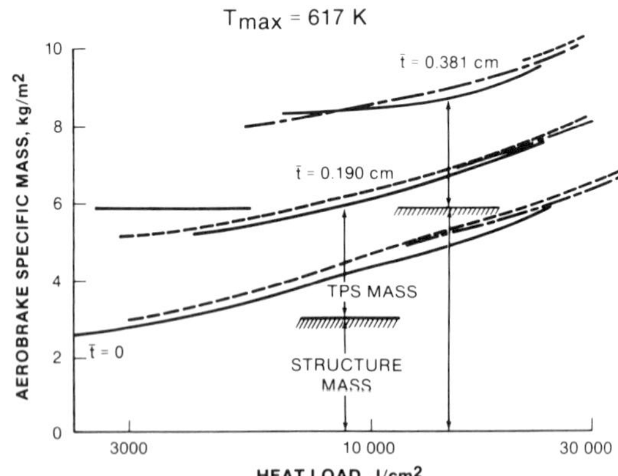

Fig. 18 Specific mass of LI-900 TPS on graphite-polyimide structure aerobrake.

structural mass of the brake is essential, even though a thicker insulator may be required.

The three trajectories considered are very similar in heating profile and duration of heating. This similarity makes the insulator mass very insensitive to the ballistic characteristics of the vehicle. In addition, the actual local heat load does not have a strong influence on local brake mass.

Sizing of Vehicle Based on Heat Flux

The factors that influence the heat flux to an AOTV have been discussed; therefore, attention is turned toward sizing a vehicle given the maximum temperature capability of a chosen TPS material. It was shown in the previous section that the mass of the TPS depends more on the supporting structure than on the thermal protection material itself, at least for an insulator-type TPS. A quantitative expression of this result is given in Figs. 19-21. Figure 19 corresponds to a hypothetical brake with a very lightweight structure and TPS having a specific mass of 1.0 kg/m^2. Figure 20 represents the heating to a system having a moderate specific mass of 15 kg/m^2, similar to that of the Space Shuttle, and Fig. 21 corresponds to a heavier TPS such as carbon-carbon or LI-2200 used on the Shuttle between the nose cap and the forward landing gear door. These peak heat flux curves are based on the ballistic

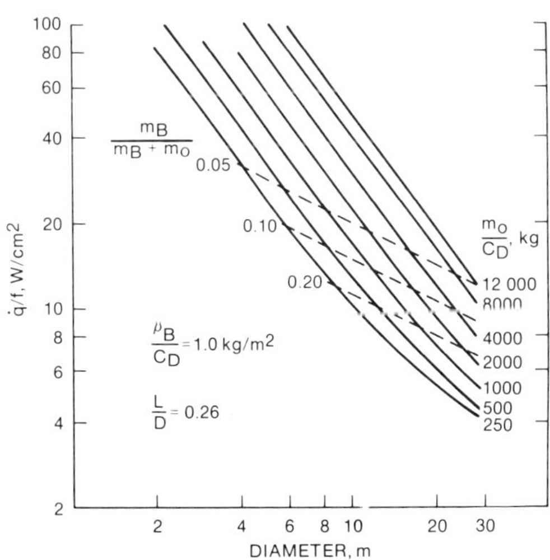

Fig. 19 Vehicle sizing curves based on convective heat flux: ultralight aerobrake.

coefficient influence on peak heat flux given in Eq. (1) and the correlation function given in Eq. (2) for a 1-m-radius reference sphere. To use Figs. 19-21, one must determine the deviation factor f that results from such effects as nonequilibrium flow/catalycity, deviations in heat flux from Eq. (2) such as that due to low density (high altitudes), and the vehicle shape. We consider a vehicle system of a given mass m_o (not including the TPS and its own support structures) and a drag coefficient C_D. The parameter ρ_B is the specific mass of the TPS and support structure (the brake). A given TPS material has an upper temperature limit which translates to a maximum heat flux limit $\dot{q}_{max}$ via radiation equilibrium. The factor f is the factor of deviation from the reference sphere EC heat flux. For a given brake specific mass ρ_B, system mass m_o, and drag coefficient C_D, one can then enter the appropriate curve in Fig. 19, 20, or 21 to obtain the diameter of the vehicle required that will result in temperatures no greater than the maximum permitted. For example, let us select a vehicle with mass m_o of 12 000 kg, with a blunted raked-off elliptic-cone aerobrake having C_D = 1.5, A TPS having a maximum temperature limit of 1700 K, and a brake mass per unit area ρ_B of 22.5 kg/m^2. The temperature limit corresponds to a radiation equilibrium heat flux $\dot{q}$ of 38 W/cm^2 (ε = 0.80). We assume that the net heating factor consists of the following: 1) a heating factor of 0.75 due to finite catalytic surface; 2) a geometry factor of

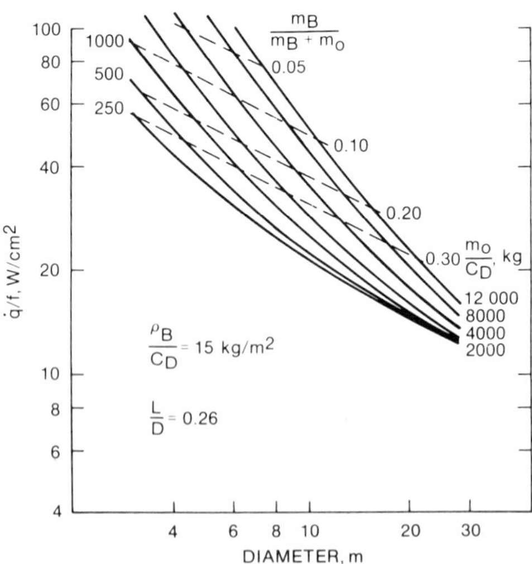

Fig. 20 Vehicle sizing curves based on convective heat flux: medium weight aerobrake.

1.2 corresponding to an elliptically blunted cone with δ = 75°, θ_{xy} = 60°, and ε_b = 2; and 3) radiation neglected. Then, the heating factor f = 0.75 × 1.2 = 0.9 and $\dot{q}/f$ = 42.2 W/cm^2; now, ρ_B/C_D = 15 kg/m^2 and m_o/C_D = 8000 kg. We then use Fig. 20 to obtain a brake diameter of 10.5 m. Also, note that the brake mass is about 0.14 of the total vehicle dry mass.

In this example, if the thermal protection material has a temperature capability of 1950 K, then the radiation equilibrium heat flux is 65 W/cm^2 and $\dot{q}/f$ = 72.2 W/cm^2. The required brake diameter would then be 6.8 m, and $m_B/(m_B + m_o)$ is about 7%. Thus, we can see that increasing the TPS material temperature capability can significantly improve the efficiency of the aerobrake. Brakes this small may not protect a very large payload. Therefore, the brake diameter may be sized on the basis of other considerations such as payload size or afterbody protection. This possibility points out the benefit that could be gained by designing a system to fit compactly behind the aerobrake shell. Shorter afterbodies would require smaller diameter aerobrakes and therefore would probably be lighter.

A more optimum aerobrake than that assumed in Figs. 19-21 might be one more tailored to the aeroheating environment. In such a system, one would employ a combination of materials and structures that would result in lighter

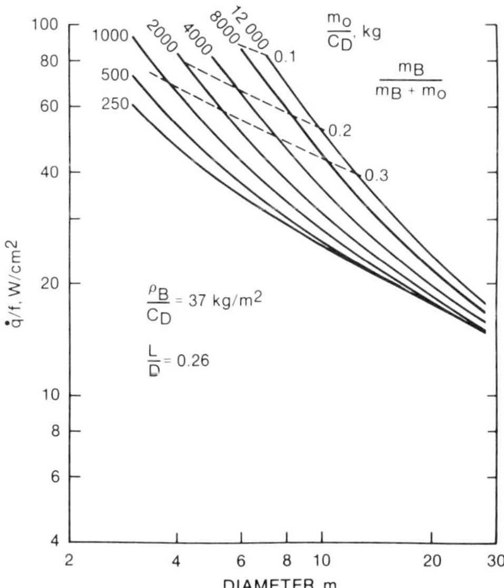

Fig. 21 Vehicle sizing curves based on convective heat flux: heavy aerobrake.

overall mass than having to use a heavy system with constant ρ_B assumed in Figs. 19-21. For example, one might use a heavier high-temperature system near the stagnation point of the aerobrake but use a lighter concept elsewhere. The type of curves in Figs. 19-21 would still be useful by incorporating the central heavier aerobrake mass to the system mass m_0 and the lighter aerobrake mass ρ_B elsewhere.

Of course, if nonequilibrium radiation contributed as much to the peak heat flux as did the convection, the factor f would be increased by a factor of 2 and $\dot{q}_{max}/f$ would then be halved; thus, the brake in the latter example must have a diameter of about 12 m. We therefore see that a reasonably accurate estimate of the peak heat flux is necessary to size the vehicle brake. Thus, such phenomena as catalysis, nonequilibrium chemistry and radiation, and effects of low density on the dynamics and heating should be investigated. Likewise, with improved TPS temperature performance, the size and mass can be reduced considerably.

The preceding calculations have been made based on the assumption of small TPS mass per unit area. Since structural mass is the dominant contribution to the TPS mass, attempts to lighten the supporting structure will directly affect the payload capacity. Attention should be paid to lightweight truss or integrated designs and to lighter and higher temperature structural materials. Also, packaging of the vehicle itself is important. Using structure to carry loads of the vehicle subsystems and simultaneously support the TPS would provide weight-saving dividends. Systematic and simultaneous consideration of the launch and LEO to GEO propulsive loads, as well as re-entry loads, might lead to better packaging and overall weight savings.

Conclusions

The following conclusions resulted from this aerothermodynamic and thermal protection study of a reusable, space-based, GEO to LEO aerobraking orbital transfer vehicle.

Within the constraints of the adaptive guidance logic and low L/D, the reference heat flux to an AOTV depends mainly on the ballistic coefficient. Lift is only needed to provide control of the trajectory.

Blunt raked-off elliptic cones have the following characteristics useful to an AOTV and an analysis of its heating and TPS characteristics: bluntness reduces the stagnation-point heat flux, and the heat flux distribution can be tailored by selecting the appropriate bluntness; analytical shapes make design and analysis simpler, especially for optimizing shape parameters such as rake angle, cone angle, and bluntness/ellipsoid ellipticity; and a large cone angle allows greater flexibility in the placement of the center of gravity.

A study of the trajectories showed that most of the heat load, the velocity decrease, and the peak heating occur within the continuum flow regime. Since the Navier-Stokes equations are applicable to this regime, numerical solutions can be made for these blunt shapes. It was shown that such solutions can be used to obtain heat flux and pressure distributions, and they define the sonic line at the downstream edge of the cone. This technique then yields the influence of the downstream edge on the upstream flow and heat flux. It will be necessary to include the nonequilibrium chemistry and excited species to obtain accurate results in the AOTV flight regime near peak heating. A noncatalytic surface in this flow environment would result in heating rates as low as 25% of that to a fully catalytic surface. Large (10%-45%) reductions in heat flux are also expected using Shuttle TPS materials.

Radiative heating is considered to be very small as based on Project FIRE and Apollo 4 experience applied to AOTV trajectories for a 6.10-m-diam vehicle.

It was shown that a blunt AOTV can be designed within the temperature limits of existing TPS materials, but some improvements in temperature capability would permit smaller and therefore lighter aerobrakes.

The mass of the TPS is small compared with that of the supporting structure. Minimizing structural mass by size reduction, use of low-density materials, and efficient design and layout of components will yield a greater payload-carrying capacity. Likewise, efficient design requires integration of the aerobrake and the rest of the system, i.e., one cannot simply attach a brake to an existing vehicle without paying a large penalty in mass and size.

Acknowledgment

The authors are indebted to C. J. Cerimele for running the trajectory cases.

References

[1]Walberg, G. D., "A Review of Aeroassisted Orbiter Transfer," AIAA Paper 82-1378, *9th Atmospheric Flight Mechanics Conference*, San Diego, Calif., Aug. 1982.

[2]Walberg, G. D., "Aeroassisted Orbit Transfer — Window Opens on Missions," *Astronautics and Aeronautics*, Vol. 21, No. 11, Nov. 1983, pp. 36-43.

[3]Andrews, D. G., Caluori, V. A., and Bloetscher, F., "Optimization of Aerobraked Orbital Transfer Vehicles," AIAA Paper 81-1126, *16th Thermophysics Conference*, Palo Alto, Calif., June 1981; published in *Thermodynamics of Atmospheric Entry, AIAA Progress in Astronautics and Aeronautics*, Vol. 82, 1982, pp. 455-476.

[4]Florence, D. E., "Aerothermodynamic Design Feasibility of a Generic Planetary Aerocapture/Aeromaneuver Vehicle," AIAA Paper 81-1127, *16th Thermophysics Conference*, Palo Alto, Calif., June 1981; published in *Thermodynamics of Atmospheric Entry, AIAA Progress in Astronautics and Aeronautics*, Vol. 82, 1982, pp. 477-519.

[5]Menees, G. P., "Trajectory Analysis of Radiative Heating for Planetary Missions with Aerobraking of Spacecraft," AIAA Paper 83-0407, *21st Aerospace Sciences Meeting*, Reno, Nev., Jan. 1983.

[6]Menees, G. P., "Thermal-Protection Requirements for Near-Earth Aeroassisted Orbital-Transfer Vehicle Missions," AIAA Paper 83-1513, *18th Thermophysics Conference*, Montreal, Canada, June 1983; published elsewhere in this volume.

[7]Howe, J. T., "Introductory Aerothermodynamics of Advanced Space Transportation System," AIAA Paper 83-0406, *21st Aerospace Sciences Meeting*, Reno, Nev., Jan. 1983.

[8]Gamble, J. D. and Cerimele, C. J., "Aerodynamic Performance Requirements for Aeroassist OTVs," NASA Symposium on Recent Advances in TPS and Structures for Future Space Transportation Systems, Langley Research Center, Dec. 1983; published in NASA CP-2315, July 1984.

[9]Gamble, J. D., Cerimele, C. J., and Spratlin, K., "Aerobraking of a Low L/D Manned Vehicle from GEO Return to Rendezvous with the Space Shuttle," AIAA Paper 83-2110, *10th Atmospheric Flight Mechanics Conference*, Gatlinburg, Tenn., Aug. 1983.

[10]Hill, O., "An Adaptive Guidance Logic for an Aeroassisted Orbiter Transfer Vehicle," AIAA Paper 83-357, *AAS/AIAA Astrodynamics Specialists Conference*, Lake Placid, N.Y., Aug. 1983.

[11]Mayo, E. E., Lamb, R. H., and Romere, P. O., "Newtonian Aerodynamics for Blunted Raked-Off Circular Cones and Raked-Off Elliptical Cones," NASA TN D-2624, May 1965.

[12]Lees, L., "Laminar Heat Transfer Over Blunt-Nosed Bodies at Hypersonic Flight Speeds," *Jet Propulsion*, Vol. 26, Apr. 1956, pp. 259-274.

[13]Li, C. P., "A Three-Dimensional Navier-Stokes/Euler Code for Blunt-Body Flow Computations," AIAA Paper 85-0361, *23rd Aerospace Sciences Meeting*, Reno, Nev., Jan. 1985.

[14]Cauchen, D. L., "Radiative Heating Results from the FIRE II Flight Experiment at a Reentry Velocity of 11.4 Kilometers per Second," NASA TM X-1402, July 1967.

[15]Fay, J. A. and Riddell, F. R., "Theory of Stagnation Point Heat Transfer in Dissociated Air," *Journal of Aeronautical Sciences*, Vol. 25, Feb. 1958, pp. 73-85.

[16]Bartlett, E. P. and Kendall, R. M., "An Analysis of the Coupled Chemically Reacting Boundary Layer and Charring Ablator, Pt. III, Nonsimilar Solution of the Multicomponent Laminar Boundary Layer by an Integral Matrix Method," NASA CR-1062, June 1968.

[17]Tong, N., Buckingham, A. C., and Morse, H. L., "Nonequilibrium Chemistry Boundary Layer Matrix Procedure," NASA CR-134039, July 1973.

[18]Shinn, J. L. and Jones, J. J., "Chemical Nonequilibrium Effects on Flowfields for Aeroassist Orbital Transfer Vehicles," AIAA Paper 83-0214, *21st Aerospace Sciences Meeting*, Reno, Nev., Jan. 1983.

[19]Scott, C. D., "Catalytic Recombination of Oxygen and Nitrogen on High Temperature Reusable Surface Insulation," *Aerothermodynamics and Planetary Entry, AIAA Progress in Astronautics and Aeronautics*, Vol. 77, edited by A. L. Crosbie, 1981, pp. 192-212.

[20]Rakich, J. V., Stewart, D. A., and Lanfranco, M. J., "Results of a Flight Experiment on the Catalytic Efficiency of the Space Shuttle Heat Shield," AIAA Paper 82-0944, *3rd AIAA/ASME Joint Thermophysics, Fluids, Plasma, and Heat Transfer Conference*, St. Louis, Mo., June 1982.

[21]Stewart, J. A., Rakich, J. V., and Lanfranco, M. J., "Catalytic Surface Effects Experiment on the Space Shuttle," *Thermophysics of Atmospheric Entry, AIAA Progress in Astronautics and Aeronautics*, Vol. 82, edited by T. G. Horton, 1982, pp. 248-272.

[22]Park, C., "Comparison of Electron and Electronic Temperature in Recombining Nozzle Flow of Ionized Nitrogen-Hydrogen Mixture, Pt. 1, Theory," *Journal of Plasma Physics*, Vol. 9, Pt. 2, 1973, pp. 187-215.

[23]Park, C., "Comparison of Electron and Electronic Temperatures in Recombining Nozzle Flow of Ionized Nitrogen-Hydrogen Mixture, Pt. 2, Experiment," *Journal of Plasma Physics*, Vol. 9, Pt. 2, 1973, pp. 217-234.

[24]Breen, J., Cibrian, R., Delgass, W. N., Krishnan, N. G., Nordine, P. C., and Rosner, D. E., "Catalysis Study for Space Shuttle Vehicle Thermal Protection System," Yale University, Final Rept. NAS 9-13058, NASA CR-134124, Oct. 1973.

[25]Keuthe, A. M. and Schetzer, J. D., *Foundations of Aerodynamics*, John Wiley and Sons, Inc., New York, N.Y., 1959, p. 379.

[26]Schaaf, S. A. and Chambre, P. L., "Flow of Rarefied Gases," *Fundamentals of Gas Dynamics*, edited by H. Emmons, Princeton University Press, Princeton, N.J., 1958, p. 688.

[27]Ried, R. C., Rochelle, W. C., and Milhoan, J. D., "Radiative Heating to the Apollo Command Module: Engineering Prediction and Flight Measurement," NASA TM X-58091, Apr. 1972.

Aerothermodynamic Heating Analysis of Aerobraking and Aeromaneuvering Orbital Transfer Vehicles

Gene P. Menees*
NASA Ames Research Center, Moffett Field, California
Carol B. Davies† and John F. Wilson†
Informatics General Corporation, Palo Alto, California
and
Kevin G. Brown‡
NASA Ames Research Center, Moffett Field, California

Abstract

The thermal-protection requirements of two aero-assisted orbital transfer vehicles (AOTVs) are analyzed for return missions between the geosynchronous and Shuttle orbits. One of the designs is a specialized version of a previously proposed generic aerobraking vehicle that is capable of only delivery-type operations. The other is a high-lift aeromaneuvering vehicle that is optimized for low Earth orbit sortie missions involving large, multiple plane-inclination changes. The aerothermal environment of the aerobraking vehicle is analyzed using state-of-the-art methods for nonequilibrium-radiative and convective heating that incorporate refinements unique to the configuration. The heating analysis of the aeromaneuvering vehicle required the development of a flowfield model for rarefied-hypersonic flow over a lifting surface at incidence. The predicted aerothermodynamic heating characteristics for both vehicles are correlated with thermal-control

---

Presented as Paper 84-1711 at AIAA 19th Thermophysics Conference, Snowmass, Colorado, June 25-28, 1984.

*Research Scientist.

†Consultant. Professional Services Operations West (PSOW).

‡Captain USAF.

requirements and flight performance capabilities for the specified mission guidelines. The results help identify technical issues related to the development of future operational systems.

Nomenclature

C_D = drag coefficient
C_L = lift coefficient
H = altitude
L/D = lift-drag ratio
q = heat flux
T = temperature
t = flight time
V_∞ = flight velocity
x,y,z = Cartesian coordinates
x_t = transition point from free-molecular to continuum flow
α = angle of attack
β = ballistic coefficient
ϕ = bank angle of lift-plane vector

Subscripts

c = convective
cl = centerline
e = equilibrium radiative emission
le = leading edge
r = nonequilibrium-radiative emission
s = stagnation point

Introduction

In conceptual studies extending over the past two decades,[1,4] the potential of aeroassisted technology for enhancing orbital operations and planetary missions has been widely recognized. This technique consists of using the aerodynamic forces generated during transatmospheric passes to achieve the transition to a local orbit, instead of relying exclusively on retropropulsion, as was done in earlier methods. The propellant mass saved by eliminating costly retropropulsion braking maneuvers not only makes possible missions that otherwise would be impractical, but also substantially increases the payload.

The development of aeroassisted orbital transfer vehicles (AOTVs) will extend the utility of the Space Shuttle by providing high-capacity space-transportation systems. This capability is the essential operational requirement that would make possible the effective use of cislunar space to achieve future national goals. For example, access to a variety of high and low Earth orbits for defense, scientific, and commercial purposes requires that material and personnel be transported to supply and service space stations and satellites or to provide unique payload delivery capability.

Recent exploratory studies[3,4] indicate that NASA/DOD program and mission requirements can be satisfied by two generic AOTV designs. One is useful primarily as a space freighter for transporting large payloads between two low Earth orbits or between a low Earth orbit and distant, high locations (e.g., geosynchronous orbit or the stable libration centers). This "aerobraking" vehicle performs its orbital change maneuvers by aerodynamic drag in the far-outer extent of the atmosphere to reduce aerothermodynamic heating and, consequently, thermal-protection requirements. It achieves plane-inclination changes propulsively at the transfer-orbit apogee where propellant requirements are minimal. The other generic design is a very high-lift "aeromaneuvering" vehicle. This is an essential operational capability if time-constrained aero-assisted maneuvers are to be made between low Earth orbits involving large multiple-plane changes. Such high-lift vehicles can achieve rapid response from one orbital plane to another, but have the inherent liability of a small payload fraction because of low volumetric efficiency. More advanced versions of this design would be able to descend through the atmosphere and land at designated locations on Earth. Fortunately, there is considerable technology crossover between space-based and orbit-on-demand designs (i.e., transatmospheric vehicles and advanced military spacecraft) because the early portions of the entry trajectories correspond.

The purpose of this study is to examine the aero-thermodynamic heating and thermal-protection requirements of two candidate AOTVs that can help achieve NASA/DOD goals in space. Both are derivatives of previously proposed "aerobraking" and high-lift "aeromaneuvering"[5] designs. The present aerobraking vehicle is a shortened version of the original design and is intended for more specialized applications (i.e., delivery-only missions to geosynchronous orbit). The initial vehicle design guidelines were more severe in requiring generic application throughout cislunar space (i.e., manned or unmanned missions with delivery, retrieval, or combined delivery and retrieval operations). The current aeromaneuvering design differs from the previous one by providing higher aerodynamic lift and more protected volume for cargo, propulsion, and command/control system components.

Mission Requirements

The study was conducted for aeroassisted return missions from geosynchronous Earth orbit (GEO) to a Shuttle orbit located at an altitude of 400 km [low Earth orbit (LEO)]. Two cases that correspond to the optimal orbital transfer capabilities of the aerobraking and aeromaneuvering vehicles under study are considered: 1) a simple coplanar return, using aerobraking for which the plane-inclination change (≈28.5 deg) required for Shuttle rendezvous is accomplished propulsively at GEO; and 2) a more complex, synergistic plane change return using aeromaneuvering to achieve simultaneous altitude and inclination orbital transfers. The latter case approximates more realistic entry conditions for high-lift AOTVs because such designs are optimized for LEO-sortie missions involving large multiple-plane changes. The higher entry velocities corresponding to returns from more distant orbits are beneficial in achieving enhanced lifting capability, and are obtained by applying propulsive thrust before re-entry as an adjunct to gravity.

Studies of the transatmospheric flight trajectories during the return phase of the GEO-LEO missions were conducted based on the following assumptions: 1) the vehicle ballistic coefficient is constant; 2) only aerodynamic forces are employed during atmospheric passes; and 3) the apoapsis altitude of the final aeroassist transfer orbit is tangent to the target LEO where circularization is accomplished propulsively. Calculations were obtained for the same entry mass for all cases (≈5400 kg), based on the 1962 U.S. standard atmosphere model; this was done because previous work indicated that the effects of the random-unpredictable dispersions in the atmospheric structure were within the accuracy of the aerothermodynamic heating prediction methods.

The GEO-LEO mission scenario was selected because the bulk of near-future space activity will undoubtedly take place within this altitude range. The location of a future space station at GEO is likely because of its scientific, commercial, and strategic importance. Consequently, frequent commuting between LEO and GEO for supply and maintenance will be essential. The corresponding thermal-protection requirements will also serve to satisfy those of many in-between orbital change missions.

Results and Discussion

Short Conical-Lifting-Brake Vehicle

Vehicle Design. The design is illustrated schematically in Fig. 1. Complete details of the vehicle systems layout (i.e., the drag brake, support structure, vehicle components, and mission performance capability in cislunar space) were given in the earlier work.[4] The short version differs from the generic configuration in that the drag brake is sized for delivery-only missions, but with the

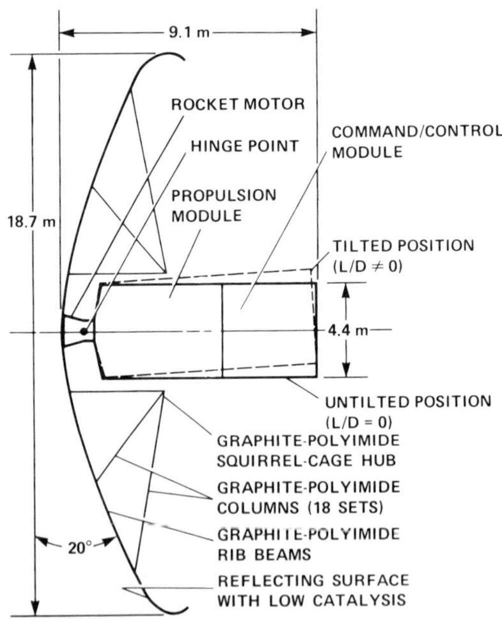

Fig. 1 Schematic of short conical-lifting-brake systems design.

constraint of retaining the same low ballistic coefficient (β = 11 kg/m^2). Consequently, no payload capability is required for re-entry, which allows the drag-brake diameter to be reduced by about 30%, the vehicle entry mass to be reduced by about 50%, and the total aerothermodynamic heating to be alleviated slightly. The diameter-to-length ratio is, however, the same as that of the generic configuration to prevent impingement of the wake flow on the vehicle surface, thus avoiding high localized heat fluxes.

Trajectory Analysis. Flight trajectories that are typical of optimal, aeroassisted GEO-return missions with minimum propellant requirements are shown in Fig. 2. The four trajectories show the effects of negative lift and multiple atmospheric passes in raising the perigee altitude relative to the design baseline trajectory (L/D = 0). Complete details of these strategies and the benefits obtained in alleviating surface heat fluxes and aerodynamic forces were discussed previously. Only the first pass of the multipass cases is shown, since it is the first pass that produces the most severe heating and pressure conditions. All of the trajectories are identical to those of the initial study because of the specified mission and the aerodynamic characteristics of the vehicle. Consequently, the design pressures that determine structural requirements remain unchanged, but the aerothermodynamic heating rates are different because of the reduced shock-layer volume of the smaller drag-brake geometry.

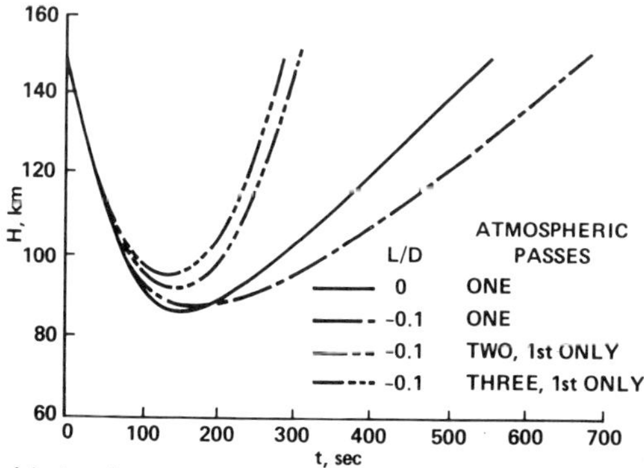

Fig. 2 Flight trajectories of short conical-lifting-brake AOTV for aeroassisted GEO-return missions (β = 11 kg/m^2).

Structural Analysis. An analysis of the drag-brake support mechanism was conducted, as in the generic study, using minimum-mass, beam-optimization techniques used in aircraft design. Implicit to the analysis are the assumptions that 1) the drag brake is assembled in space and remains deployed throughout its service lifetime (i.e., it is erected only once and remains erected), and 2) the temperature of the graphite-polyimide beams is kept below about 600 K to prevent thermal degradation of their structural properties. The baseline-design aerodynamic forces were determined from the maximum stagnation-point pressure in the $L/D = 0$ flight trajectory (≈0.005 atm). The results obtained for the mass penalty incurred because of the drag-brake apparatus (i.e., surface material and support structure), and the vehicle insulation system was about half that of the generic study, but about the same in terms of the entry mass (≈15%). This is a result of the smaller cross section and shorter support beams that are possible with the reduced diameter of the short configuration. The significance of this is that less of the vehicle launch mass (≈32.5 tons) is sacrificed for the aerobrake, which provides some improvement in the payload delivery capability.

The present analysis is considered to be conservative for two principal reasons: 1) the stagnation pressure is assumed to be uniformly distributed over the aerobrake surface, which neglects the reduction in pressure caused by surface inclination; and 2) the beneficial effects of the inertial forces, produced by atmospheric deceleration, in counteracting the aerodynamic loads is not accounted for. The crucial factors in the design of the support structure are, however, the wake-flow heating environment and radiative-transport characteristics. A greatly improved understanding of these factors is urgently needed; both experimental and analytical studies are planned or under way.[6,7]

Aerothermodynamic Heating. Time histories of the predicted stagnation-point aerothermodynamic heating-rate distributions are shown in Fig. 3, because the combined magnitude of the radiative and convective fluxes at this location approximate the highest on the drag-brake surface. Only the nonequilibrium component is given for the radiative fluxes because the previous work showed this to be the dominant surface heating mechanism and principal determinant of the heat-shielding requirements. These results are typically an order of magnitude higher than the equilibrium case, a consequence of the large-scale vehicle and highly energetic, rarefied-flow, hypersonic-flight regimes that provide the long relaxation times that make nonequilibrium effects important.

Real-gas computational fluid dynamics (CFD) codes that couple all the relavant physics of the prototype shock-layer environment (i.e., thermodynamics, chemistry, radiation transport, and nonadiabatic and viscous effects) are not now available but are forthcoming.[7,8] During the interim, approximate methods have been used to determine the surface-heating requirements. Complete details of the radiative-transport computational procedure have been reported previously.[2-4,9] The calculations include such refinements as CFD predictions of the shock-layer structure[8] and boundary-layer attenuation effects that reduce the radiative flux. The latter is caused by the absorption of radiation in the boundary layer (<0.2-μm spectral range) and the effect of its extent or thickness in prematurely cooling the shock-layer gases and truncating the

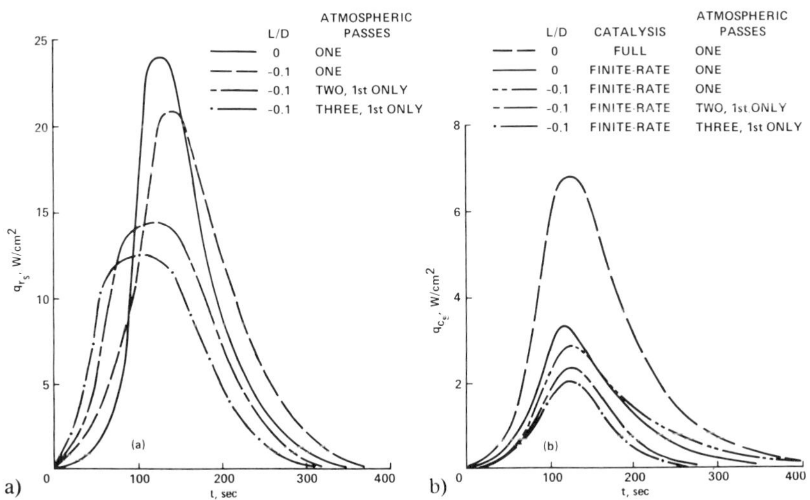

Fig. 3 Stagnation-point aerothermodynamic heating-rate distributions of short conical-lifting brake for aeroassisted GEO-return missions: a) nonequilibrium radiative; b) laminar convective.

radiative emission, which is not trivial for the rarefied AOTV flight regimes. Future refinements to the radiative heating predictions include nonequilibrium excitation effects,[10] updated rate chemistry and spectral properties of air species,[11] and the basic governing equations for three-temperature gases.[12] In addition to the foregoing, the convective-heating predictions incorporate refinements such as finite-rate surface catalysis and the attenuation effects of the suction, or flow through, the porous drag-brake surface material. Details of these calculations are also given in the earlier work. Future refinements important to both the convective and radiative heating include a more accurate evaluation of the transport properties of air species and the determination of nonadiabatic effects. The former are known to vary widely in low-density hypersonic flows, and the latter should provide substantial reductions in the aerothermodynamic heating since there are significant energy losses from the shock layer to the vacuum of space.

The results plotted in Fig. 3 show that the radiative fluxes are lower by about 20% and the convective fluxes higher by about 15% than those predicted for the larger generic vehicle. These differences are caused by the smaller shock-layer volume and effective nose radius, respectively, of the short configuration. In addition, the substantial reduction in both heating rates resulting from the benefits of negative lift, multiple atmospheric passes, and low catalysis of the drag-brake surface material (silica cloth) compared with the design baseline case

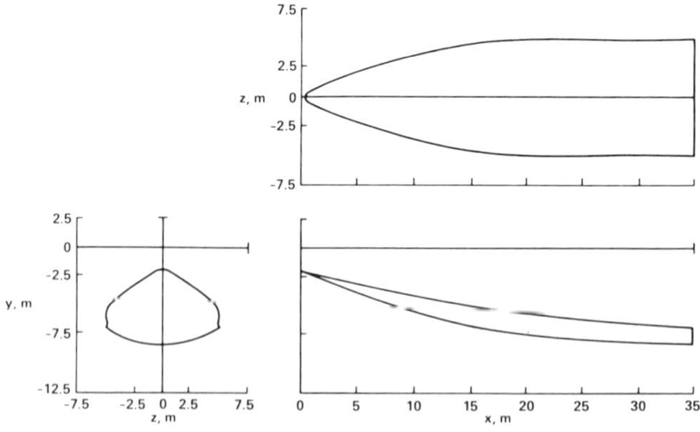

Fig. 4 Geometry of high-lift aeromaneuvering AOTV.

(L/D = 0) are evident. The net result of this analysis is that the magnitude of the total absorbed incident heat flux is reduced by about 1 W/cm^2 from that of the generic vehicle. This is a consequence of the high reflectivity of the surface material (≈80%) which allows absorption of only about 20% of the incident radiative flux. Consequently, the original thermal-protection analysis, which indicated that graphite polyimide would be able to withstand the temperatures produced in the wake-flow environment, remains valid since the drag brake heat-rejection requirements are reduced.

High-Lift Aeromaneuvering Vehicle

Vehicle Design. The basic geometry of the configuration is illustrated in Fig. 4. The design is completed by the addition of propulsion, command and control [including avionics and reaction control system (RCS) motors for attitude control], and cargo components to the upper surface. The vehicle is long (35 m), but the length is necessary to accommodate the same rocket engines that will be interchangeable with the drag-brake design. The shape is derived from hypersonic lifting-body theory, using truncated biconics with smoothing through the shock juncture point between the fore and aft cones to eliminate the discontinuity. The design avoids the complex shock interaction effects of winged configurations which cause high-localized heat fluxes and severe thermal-protection requirements. The aerodynamic characteristics are optimized in parametric studies by varying the angles of the fore and aft cones, the bend angle between the cones, and the upper surface curvature, which is truncated with second-order polynomials. The present shape is not completely optimized aerodynamically; however, this is unimportant for the major purpose of this study, which is to establish the magnitude of the heat-protection requirements. Details of the final design and analytical method will be given in a future paper.

Flow Phenomenology. Predictions of the aerodynamic and convective heat transfer characteristics of the vehicle were obtained using an approximate method based on rarefied-hypersonic flow over a sharp flat plate at zero angle of attack. The appropriate theories[13] that are applicable in the various low-density flow regimes[14] were adjusted to accommodate the present flow problem which simulates a flat plate at incidence, since this problem has not been analyzed previously. A physical model of the flow is illustrated in Fig. 5. It was assumed that the upper surface of the vehicle, with system components attached, is tangential to the oncoming free-molecular flow. This implies that kinetic conditions prevail in order to obtain finite values for skin friction and heat transfer. On the lower surface, however, free-molecular flow occurs at the leading edge, since the random fluctuations of kinetic effects are small relative to the scale of angle-of-attack effects. Downstream of the leading edge, the flow passes through a transition region and is assumed to merge into a strong, viscous interaction-effects layer having surface slip. Eventually, the flow develops into the classic continuum-hypersonic boundary layer far downstream. The current analysis terminates at the merged layer, however, since this phenomenon is considered to be the dominant physical mechanism for the flight regimes of interest in this study.

The transition point between the free-molecular and merged regions is determined by equating the heat transfer rates at the leading edge and a downstream postshock location in the merged layer. The leading-edge heating was

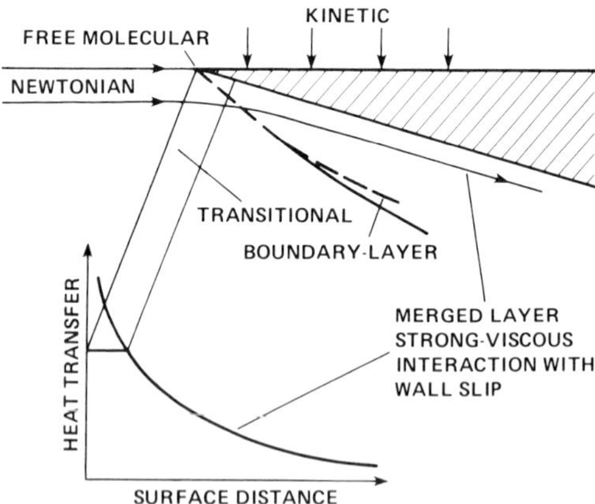

Fig. 5 Phenomenology of approximate method for skin-friction and heat transfer calculations of rarefied-hypersonic flow over a flat plate at incidence.

evaluated by assuming complete accommodation of the free-stream momentum and energy fluxes with appropriate corrections for sweep and incidence effects. The results were further adjusted by assuming uniform distribution over an area of 1 cm^2. This technique is considered to be a more realistic representation of the actual physics of the problem because the cooling effects produced by thermal conductivity and equilibrium-radiative emission are approximately accounted for. It was found that the surface area affected by the free molecular flow was relatively small and, therefore, that it had minimal effects on the overall friction-drag predictions. The heat transfer distribution through the transition region has yet to be evaluated because of the complexity of the problem, but is not expected to have a major effect on the total system heat-protection requirements. The streamwise heat transfer distributions on the lower surface were calculated using the formulation for the merged layer regime, but were estimated to be comparatively negligible in the kinetic-flow regime of the upper surface. In addition, the radiative heating for this configuration is negligible because of the weak shock waves produced by the shallow silhouette.

The aerodynamic parameters of the vehicle were also evaluated after the flowfield model was defined. Pressure forces were calculated from Newtonian theory and then corrected for low-density viscous effects using suitable theories for skin-friction coefficient in the various flow regimes. The results were resolved into the appropriate lift and drag components and then summed over the surface area to obtain configuration characteristics. It is expected that the aeromaneuvering capability will allow flight operations at sufficiently high altitudes that all-laminar flow conditions prevail on the windward surface. Consequently, turbulent transition criteria are unimportant for the present vehicle design. The foregoing flowfield model and calculation procedure provide conservative estimates for both the aerodynamic and surface-heating characteristics that approximate upper design requirements.

Aerodynamic Characteristics. The approximate operational altitude range, or entry flight envelope, of the

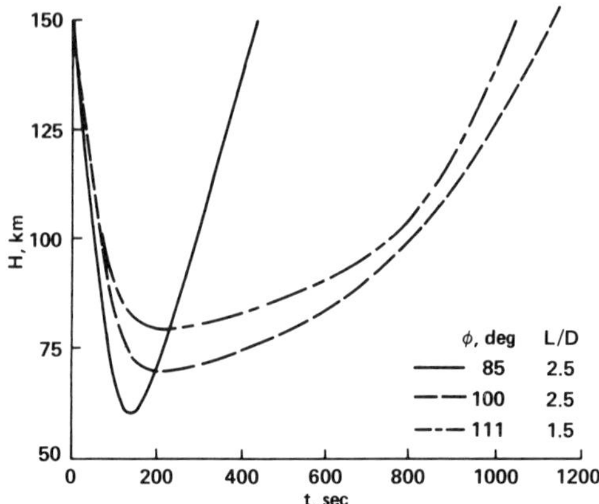

Fig. 6 Flight trajectories of high-lift AOTV for GEO-return entry conditions (β = 400 kg/m^2).

vehicle was estimated and is illustrated by the representative trajectories shown in Fig. 6. Three trajectories, which successively elevate the perigee location by about 10 km each, are shown. This is obtained by roll-control modulation of the flight-plane lift vector. The important result is that increasing the bank angle raises the perigee altitude, which has a significant effect on the heat transfer characteristics, as will be shown subsequently. There is some sacrifice in the lifting capability, however, because of the lower density range of the upper-altitude flight regimes.

This effect is illustrated by the example aerodynamic characteristics given in Fig. 7. These results correspond to the ϕ = 85-deg trajectory, which spans the approximate operational range of the vehicle. The variation of C_L, C_D, and L/D with α and H are shown over the peak-heating range of the trajectory. The lift-drag ratio, which is the dominant parameter in the plane change capability, approaches 3.0 in the lower altitude range of the trajectory and only degrades significantly above altitudes of about 85 km. In addition, the maximum L/D occurs at low α, which is beneficial to the stability since changes in the static margin are minimized. Sharp leading edge aerodynamic surfaces characteristically maintain high lift over a broader altitude range in low-density flows than do blunt leading-edge surfaces. This feature provides superior flexibility in selecting flight trajectories that

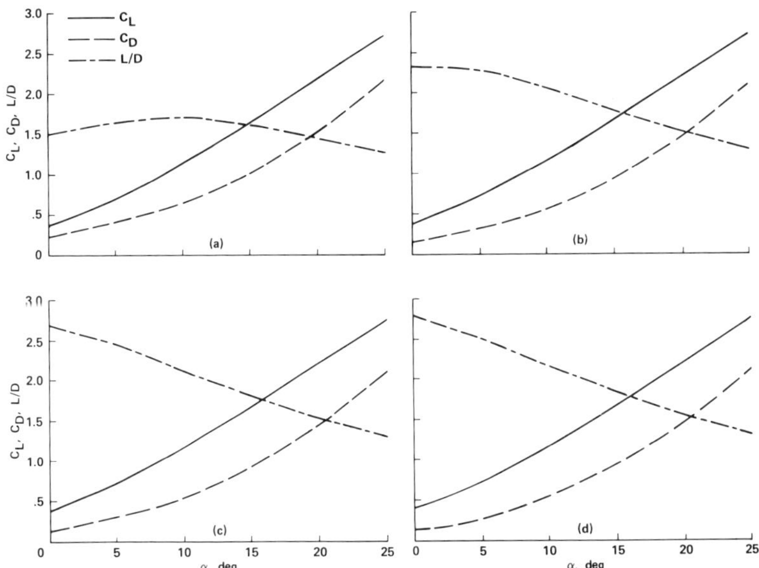

Fig. 7 Variation of viscous Newtonian aerodynamic characteristics with altitude of high-lift AOTV for ϕ = 85-deg trajectory: a) t = 60 s, H = 93.2 km; b) t = 80 s, H = 79.3 km; c) t = 100 s, H = 68.6 km; d) t = 120 s, H = 61.7 km.

achieve the same plane change capability for a wide range of altitudes. A simple analysis using only Newtonian mechanics shows that this occurs because the mass of air swept out during an atmospheric pass is independent of altitude. The boundary constraint on this somewhat surprising result is that the velocity decrement or deceleration caused by the orbital change maneuver remains constant. Consequently, the flight duration to obtain the same plane-inclination change increases with altitude. This also raises the total heat loads and may require some tradeoffs in the thermal-protection analysis. The aeromaneuvering capability further deteriorates as the vehicle decelerates during the exit phase of the atmospheric pass; however, the trajectory is optimally modulated to satisfy guidance, navigation, and control (GN+C) requirements.

Aerothermodynamic Heating. The heat transfer characteristics on the windward surface over the peak-heating range of the ϕ = 85-deg trajectory are illustrated in Fig. 8. This case provides the worst entry conditions and, therefore, the most severe heating requirements considered in this study. Distributions around the leading edge for the transition point between the two principal flow regimes and the free-molecular flow heat transfer rate are given in Figs. 8a and 8b, respectively. The present technique for determining the transition point becomes invalid beyond x-locations of about 25 m and produces fictitious results, since the flow becomes parallel to the leading edge and has no finite surface area to encounter. The heat transfer results are realistic, however, because the surface has a finite angle of attack and produces strong crossflow effects. Streamwise distributions along the vehicle centerline of the surface heat fluxes are given in Fig. 8c. All the results show the variation of the individual parameters with both entry time (or altitude) and angle of attack.

The transition point is observed to increase substantially at the higher altitudes and to decrease with angle of attack, whereas the converse behavior occurs for the heat transfer rates. The leading-edge heat fluxes are very high at the centerline, but decrease quickly around the leading edge because of the sweep angle. The decrease is even more rapid downstream, as seen in Fig. 8c, where the results fall below 40 W/cm^2 within 0.5 m of the leading edge and are less than 10 W/cm^2 over most of the vehicle surface. The transition distance is also very small (only of the order of a few millimeters) at the low altitudes where the leading-edge heat fluxes are large, and increases to only a few centimeters at high altitudes where the heat fluxes have dropped to moderately low values ($<50\ W/cm^2$).

The major features of the leading-edge heating characteristics were found to coalesce and show a consistent trend with altitude for all three of the flight trajectories. These results are summarized and illustrated more graphically in Fig. 9. The effect of sweep angle on the heat transfer and transition point are shown in Fig. 9a, and that of angle of attack is given in Fig. 9b for the most severe heating conditions [which occur at the vehicle tip (centerline)]. Again, it is evident that the transition distance is very small in the high-heating regime of the lower altitude flight range and only moderately large at higher altitudes, where the heating levels off and

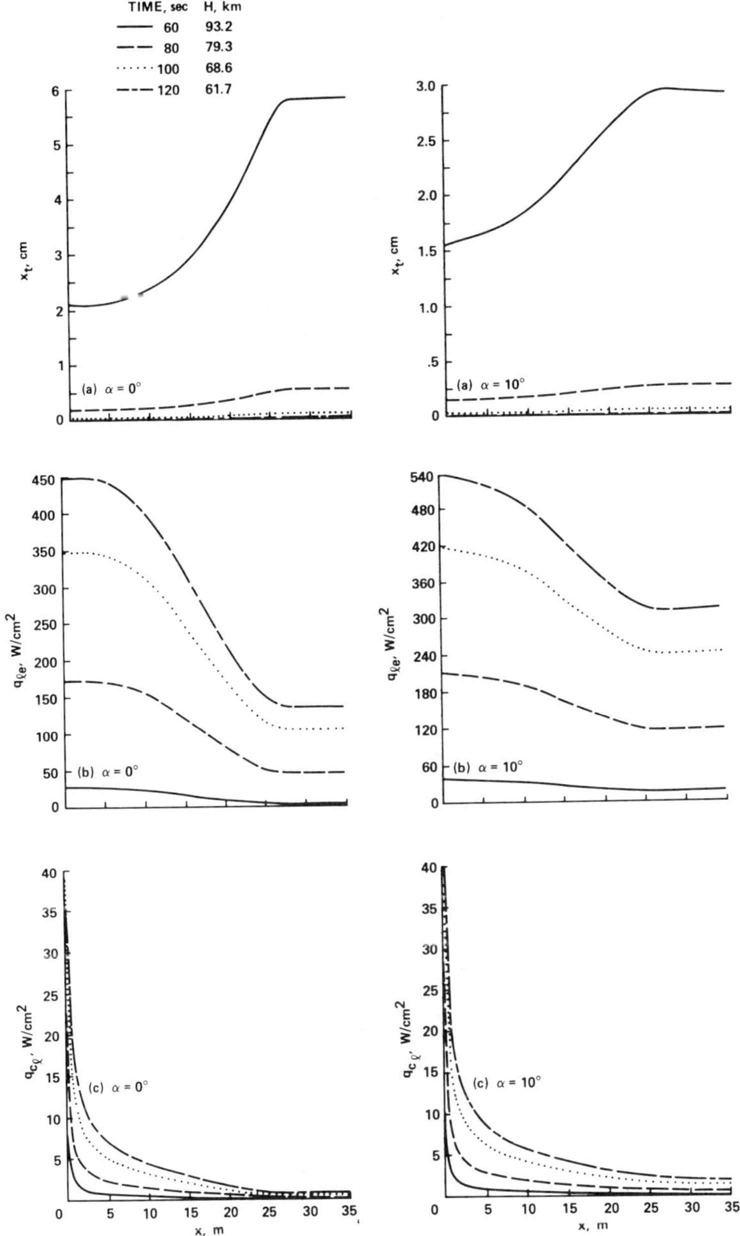

Fig. 8 Heating characteristics of high-lift AOTV showing altitude and angle-of-attack effects for ϕ = 85-deg trajectory: a) transition point; b) leading-edge heating; c) downstream heating along centerline.

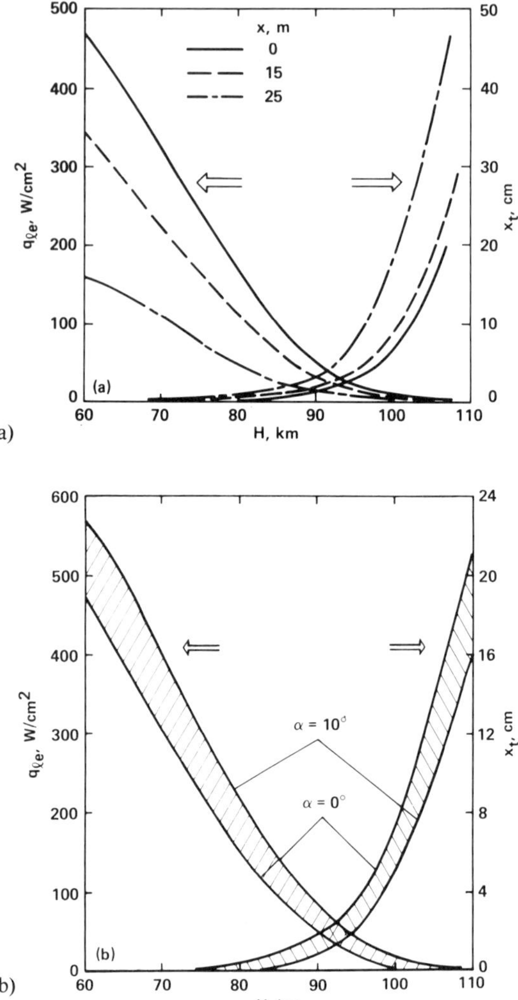

Fig. 9 Variation of leading-edge heat transfer and transition point with altitude for high-lift AOTV: a) sweep angle (curvature) effects at $\alpha = 0$ deg; b) angle-of-attack effects at centerline (tip).

becomes relatively insignificant. The total heat loads for the various trajectories were also evaluated and, except for the leading-edge region, were found to be less than those typical of the Shuttle flights (i.e., <5 kJ/cm^2).

Thermal Control. The heat-shielding requirements of the vehicle were analyzed in view of the constraints imposed by reusable space-based applications. The leading-edge heating is the dominant factor in restricting the operational ceiling, since the heat transfer and loads over most of the surface are modest (<10 W/cm^2) for a broad altitude range. Existing techniques in which lightweight aeroshell structural designs are used for both hot and insulated applications can accommodate the thermal-protection requirements for all of the vehicle surface except the small localized region around the leading edge. Thermal-control materials and methods for cooling the leading edge must provide for toleration of the highest possible temperatures, thus optimizing orbital change operations; therefore, their selection is critically important to high-lift AOTVs. There is some flexibility in selecting the flight envelope without incurring severe compromises in the plane change capability, however, because the aerodynamic characteristics degrade slowly with altitude from about 60 to 85 km.

The results of a preliminary screening analysis of candidate leading-edge materials[15-17] are given in Fig. 10. Although incomplete, this work may be helpful in guiding future efforts. The range of application of the various materials is indicated for upper temperature limits that are based on the mass loss for a 1-mm-thick shield and for 25 300-s flights. In addition to high melting temperatures, the key factors influencing the selection of materials are low oxidation and vaporization

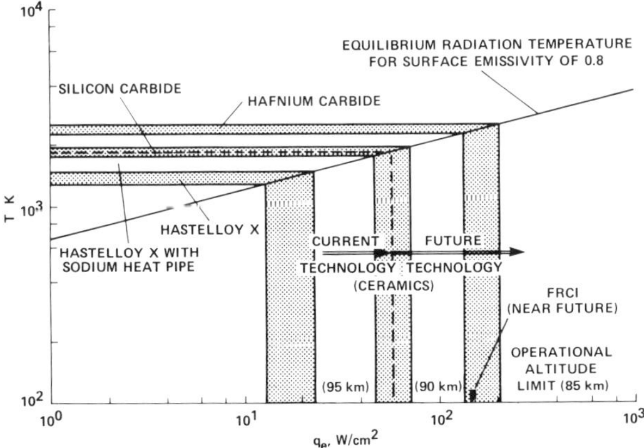

Fig. 10 Radiative heat-rejection characteristics of candidate leading-edge thermal-protection materials.

properties in an oxygen environment. The geometry of the lifting surface must also be considered; however, strength and thermal-shock effects are not considered critical issues because the aerodynamic loads are relatively moderate.

The lower temperature boundary is approximately represented by the nickel-based superalloy metals[15] such as Hastelloy X, which has an estimated temperature limit of about 1500 K (a heat transfer rate of ≈20 W/cm^2) in an oxidizing environment. This limit may possibly be extended somewhat by the future development of cobalt-based superalloys which provide the real upper limit for metals. For near-term application, the temperature capability of Hastelloy X can be extended dramatically by using a passive cooling technique incorporating thermal foil-gage leading edges and a liquid sodium-filled heat pipe.[16] This technique is very effective in controlling intense heating in localized areas, and raises the temperature limit to about 1800 K (a heat transfer rate of ≈45 W/cm^2). Additional increases are obtained with materials whose oxidation products are not volatile and, therefore, serve as protective coatings that resist further oxidation. Silicon carbide is a leading candidate in this category and raises the temperature ceiling to about 2000 K (a heat transfer rate of ≈65 W/cm^2). Such materials are currently under extensive development by NASA and the USAF in efforts to obtain protective coatings with significantly higher temperature capabilities [e.g., 2500 K (a heat transfer rate of ≈200 W/cm^2)]. Hafnium carbide, for example, offers superior performance because its melting temperature and vapor pressure are among the highest and lowest, respectively, of the refractory materials; however, its development remains an advanced technology issue. The practical near-future limit for ceramics (≈2 yr development time) will probably be provided by a fibrous-reusable-ceramic-insulation material (FRCI). This material, under development for space applications at NASA Ames Research Center, has a surface temperature and heat transfer capability of about 2500 K and 130 W/cm^2, respectively.

The best current candidate for leading-edge applications appears to be the system used for shielding the Shuttle nose-cap and leading-edge regions.[17] This consists of advanced carbon-carbon (ACC) panels that are coated with silicon carbide and then further overcoated with glass formers to seal cracks that develop at high temperatures. This system has mechanical properties that are superior to those of solid silicon carbide (e.g., lighter weight and higher specific strength, stiffness, and toughness). The maximum operating temperature, however, is limited to about 2000 K (a heat transfer rate of 65 W/cm^2) as mentioned previously. This controls the lower altitude limit on the flight envelope to about 90 km (see Fig. 9), which is close to the critical limit for marginally acceptable deterioration of the aerodynamic characteristics. The availability of FRCI, however, will lower this limit further to about 85 km.

Other strategies that may alleviate the constraints imposed by reusable materials include the use of ablating materials or forced and active cooling techniques that utilize the propulsion fuel (heat pipes were mentioned earlier). Since the severe heat transfer occurs only in small localized regions, it may be possible to replace the affected areas between missions, which significantly lowers the operational ceiling. If this proves impractical, the heat-rejection characteristics might be improved by using the heat capacity of the propulsion fuel in either forced or transpiration methods. Further development of these techniques may provide substantial improvements in the mission performance capabilities of high-lift AOTVs.

Concluding Remarks

A study has been conducted that establishes the approximate aerothermal heating and thermal-protection requirements of two AOTVs that are derived from previously proposed aerobraking and aeromaneuvering designs. The analysis was restricted to return missions from GEO to the Shuttle orbit. The weight penalty incurred for the thermal-control requirements of the short aerobraking vehicle was about half that for the original generic design, but was approximately equal when compared on an entry mass basis. For the aeromaneuvering design, the heat-protection requirements are modest over most of the vehicle surface for a broad operational envelope and can be accommodated by existing metals and ceramics. Severe heat transfer rates that exceed the temperature limits of current reusable materials occur only over relatively small, localized regions around the leading edge. Alternative solutions to this problem include ablating materials that are replaceable between missions, cooling by using the propulsion fuel or heat pipes, and limiting the flight regime to high altitudes where the heat transfer is reduced to levels commensurate with reusable materials.

The crucial factor in the drag-brake design is the wake-flow heating environment. A greatly improved understanding of this issue is urgently needed. In addition, CFD flowfield codes that couple all the relevant physics of the shock-layer environment are needed for definitive benchmark design predictions. These should include refinements to the aerothermodynamic heating calculations, such as nonadiabatic effects, nonequilibrium-excitation phenomena, and improved estimates of the chemical-reaction rates, along with the spectral and transport properties of important air species.

Future optimization of the aeromaneuvering design requires advancements in the aerothermal/aerodynamic prediction methods that account for rarefaction phenomena (i.e., Monte Carlo techniques) and real-gas effects. These features must be incorporated into a multidimensional CFD code (e.g., parabolized Navier-Stokes) because the vehicle's three-dimensional contour produces nontrivial effects that influence the aerothermal-heating and aerodynamic characteristics. The flowfield chemistry for this case is less complicated than that of the drag-brake design since radiation is unimportant. There are still important problems, however, such as nonequilibrium reaction rates and transport properties, which have a wide variation in the rarefied AOTV flight regimes and significantly affect surface catalysis and convective heating. The flow chemistry phenomena for the aeromaneuvering vehicle are, in fact, similar to those already analyzed for the Shuttle. In addition, leading-edge bluntness effects are required to obtain greater detail in the heat-protection analysis because some mass loss and contouring will occur over the operational lifetime of the vehicle.

References

[1]Walberg, G. D., "A Review of Aeroassisted Orbital Transfer," AIAA Paper 82-0137, Aug. 1982 (also to be published in the <u>Journal of</u>

Spacecraft and Rockets; a synopsis appears in Astronautics and Aeronautics, Nov. 1983, pp. 36-43).

[2]Menees, G. P., "Trajectory Analysis of Radiative Heating for Planetary Missions with Aerobraking of Spacecraft," AIAA Paper 83-0407, 21st Aerospace Sciences Meeting, Reno, Nev., Jan. 1983 (also to be published in the Journal of Spacecraft and Rockets).

[3]Menees, G. P., "Thermal-Protection Requirements for Near-Earth Aeroassisted Orbital-Transfer Vehicle Missions," AIAA Paper 83-1513, AIAA 18th Thermophysics Conference, Montreal, Canada, June 1983; published elsewhere in this volume.

[4]Menees, G. P., Park, C., and Wilson, J. F., "Design and Performance Analysis of a Conical-Aerobrake Orbital-Transfer Vehicle Concept," AIAA Paper 84-0410, AIAA 22nd Aerospace Sciences Meeting, Reno, Nev., Jan. 1984; published elsewhere in this volume.

[5]Davies, C. B. and Park, C., "Aerodynamic Characteristics of Generalized Bent Biconic Bodies for Aero-Assisted, Orbital-Transfer Vehicles," AIAA Paper 83-1512, AIAA 18th Thermophysics Conference, Montreal, Canada, June 1983 (also to be published in the Journal of Spacecraft and Rockets).

[6]Pitts, W. C. and Murbach, M. S., "Thermal Control for an AOTV with Conical Drag Brake," AIAA Paper 84-1712, AIAA 19th Thermophysics Conference, Snowmass, Colo., June 1984; published elsewhere in this volume.

[7]Lombard, C. K. and Venkatapathy, E., "Forebody and Baseflow of a Dragbrake OTV by an Extremely Fast Single Level Implicit Algorithm," AIAA Paper 84-1699, AIAA 19th Thermophysics Conference, Snowmass, Colo., June 1984; published elsewhere in this volume.

[8]Balakrishnan, A., "Flowfield Computation of an Aerobrake Orbital Transfer Vehicle," AIAA Paper 84-1697, AIAA 19th Thermophysics Conference, Snowmass, Colo., June 1984.

[9]Park, C., "Radiation Enhancement by Nonequilibrium in Earth's Atmosphere," AIAA Paper 83-0410, AIAA 21st Aerospace Sciences Meeting, Reno, Nev., Jan. 1983 (also to be published in the Journal of Spacecraft and Rockets).

[10]Park, C., "Calculation of Nonequilibrium Radiation in AOTV Flight Regimes," AIAA Paper 84-0306, AIAA 22nd Aerospace Sciences Meeting, Reno, Nev., Jan. 1984; published elsewhere in this volume.

[11]Park, C., "Problems of Rate Chemistry in AOTV Flight Regimes," AIAA Paper 84-1730, AIAA 19th Thermophysics Conference, Snowmass, Colo., June 1984; published elsewhere in this volume.

[12]Lee, J. H., "Basic Governing Equations for AOTV Flight Regimes," AIAA Paper 84-1729, AIAA 19th Thermophysics Conference, Snowmass, Colo., June 1984; published elsewhere in this volume.

[13]Wallace, J. E. and Burke, A. F., "An Experimental Study of Surface and Flow Field Effects in Hypersonic Low Density Flow over a Flat Plate," Rarefied Gas Dynamics, Supplement 3, Vol. 1, 1965, pp. 487-507.

[14]McCroskey, W. J., Bogdonoff, S. M., and McDougall, J. G., "An Experimental Model for the Sharp Flat Plate in Rarefied Hypersonic Flow," AIAA Journal, Vol. 4, No. 9, 1966, pp. 1580-1587.

[15]American Society of Metals Handbook: Properties and Selection of Metals. Vol. 1, 8th Ed., Metals Park, Ohio, 1961.

[16]Camarda, C. J. and Masek, R. V., "Design Analysis and Tests of Shuttle-Type Heat-Pipe-Cooled Leading Edge," Journal of Spacecraft and Rockets, Vol. 18, No. 1, Jan.-Feb. 1981, pp. 71-78.

[17]Curry, D. M., Latchem, J. W., and Whisenhunt, G. B., "Space Shuttle Leading Edge Structural Subsystem Development," AIAA Paper 83-0483, 21st Aerospace Sciences Meeting, Reno, Nev., Jan. 1983.

Thermal Response of an Aeroassisted Orbital Transfer Vehicle with a Conical Drag Brake

W. C. Pitts*
NASA Ames Research Center, Moffett Field, California
and
M. S. Murbach†
Informatics General Corporation, Palo Alto, California

Abstract

As an aeroassisted orbital transfer vehicle (AOTV) goes through an aerobraking maneuver a significant amount of heat is generated. In this paper, the thermal response of a specific AOTV to this aerobrake heating is examined. The vehicle has a 70-deg, conical drag-brake heat shield attached to a cylindrical body which contains the payload. The heat shield is made of ceramic fabric; its thickness is varied from that of a thin cloth to a 1.5-cm blanket. The fabric thickness, the radiation absorptivity of the vehicle surface materials, and radiation from the wake are all significant parameters in the thermal response to the heating produced by the braking maneuver. The maximum temperatures occur in the vicinity of the interface between the body and the conical heat shield.

Nomenclature

A = area, m^2
AFRSI = advanced flexible reusable surface insulation
AOTV = aeroassisted orbital transfer vehicle

Presented as Paper 84-1712 at AIAA 19th Thermophysics Conference, Snowmass, Colorado, June 25-28, 1984.

*Research Engineer, Thermal Protection Materials.

†Member of professional staff.

EAL = engineering analysis language
l = body length, m
NASTRAN = NASA structural analysis
$\dot{q}$ = heat flux, W
T = temperature, °C
x = axial distance from heat shield, m
α = absorptivity
ε = emissivity
σ = Boltzmann constant, $W/m^2/K^4$

Subscripts

c = convective
eq = equilibrium
g = radiation from hot gas
r = radiative
s = radiation from AOTV surface
w = radiation from wake

Introduction

The Space Shuttle is limited to delivery of payloads to low orbital altitudes. A new class of vehicles is needed to transfer payloads between the Space Shuttle or Space Station altitude and higher Earth altitudes. These vehicles are generally termed orbital transfer vehicles (OTVs). A bibliography of the work done on this subject and a discussion of the basic concepts are presented in Ref. 1. The several types of OTVs (Refs. 1-3) use two basic methods to remove the kinetic energy that must be dissipated in order to bring the vehicles down to lower orbits. In one class of vehicles, retropulsion is used for this maneuver; in the other class of vehicles, called aeroassisted orbital transfer vehicles (AOTVs), aerobraking is used during a transitory dip into the upper atmosphere. Recent studies[1-9] have shown that aerobraking can save a large amount of fuel, which translates into larger payloads or into making it possible to undertake otherwise impractical missions. This is true, however, only if the mass of the aerobraking system is not significant relative to the mass of the saved fuel. There are several AOTV designs[1,2] including low-density, zero-lift, inflatable ballute designs; high-density, lifting-body designs; and drag-brake designs. Heat fluxes during aerobraking maneuvers have been calculated for typical examples of each of these designs.[2,3,10]

The primary emphasis in AOTV studies has been on trajectory analysis and heat-flux calculations. Very little has been done to study the thermal response of AOTVs to these heat fluxes; this present paper is addressed to that thermal-response problem. The conical aerobrake design[11] was selected for this analysis. There are more heat-flux data available[3,10] in the literature for this design than for others, and the techniques used for calculating the heat fluxes are the most refined. Even so, there are still many unresolved questions,[12,13] such as whether the flow behind the shock is in equilibrium or nonequilibrium; whether the shock wave is adiabatic or nonadiabatic; and questions about wake radiation and heat-flux distribution over the heat-shield surface. In view of these uncertainties, the current analysis is limited to a study of the sensitivity of the vehicle temperature to the thermal and physical properties of the materials used to fabricate the conical aerobrake. Some thermally sensitive areas are indicated.

The ultimate goal of the ongoing program is to define requirements for modification or development of new thermal protection materials so that the AOTV thermal control system can be made light enough to take advantage of the fuel savings afforded by the aerobraking maneuver.

Thermal Analysis

The model used to calculate the heat fluxes during the braking maneuver is shown in Fig. 1. It is a realistic configuration for currently envisioned missions. The body is a cylinder with an estimated volume that is suitable for a typical payload with fuel and command and guidance hardware. The entry mass is 11,000 kg. The heat shield is approximately a sphere cone with an 8-m nose radius. The nominally 70-deg cone is designed to alleviate high edge-heating effects and to avoid hot-gas impingement on the body. Structure for supporting the heat shield is not shown and is not accounted for in the analysis.

This model was idealized to a simple cone cylinder (Fig. 2) for the thermal-response analysis. The thickness of the heat shield was varied from a thin cloth to a 1.5-cm-thick AFRSI blanket. Advanced flexible reusable surface insulation (AFRSI) is used on the Space Shuttle.

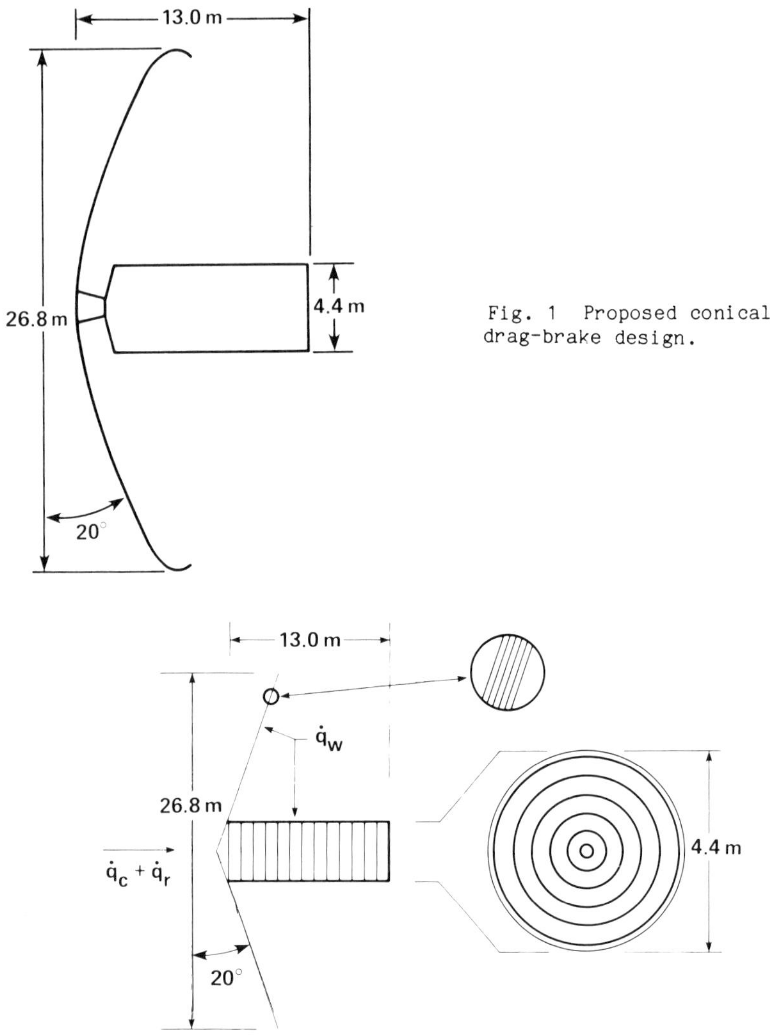

Fig. 1 Proposed conical drag-brake design.

Fig. 2 Conical drag-brake design modified for thermal analysis.

The body skin of the model is 3-mm-thick aluminum; the core is a homogeneous mass with the average density of the body and specific heat of 0.15 cal g^{-1} K^{-1}.

For thermal modeling, the body core was divided into seven concentric cylinders that were sliced into 13 disks. The blanket heat shield was divided into concentric cones. The number of conical elements depends on the thickness of the heat shield. Each conical element was assumed to be isothermal at any instant in time.

Three heat flux terms are shown in Fig. 2. The front surface of the heat shield is subjected to the sum of the convective ($\dot{q}_c$) and radiative ($\dot{q}_r$) heating. These heat fluxes were calculated for the stagnation point only. There are no values available for the flux distribution along the surface; therefore, stagnation-point values were used over the entire surface. (Thus, the assumption in the preceding paragraph that the heat-shield elements are isothermal. This has some basis because the changes in convective and radiative heating over the heat shield tend to be compensating.) The third heat flux term is the radiation heating from the wake $\dot{q}_w$. Again, there is no detailed distribution for wake radiation available. A conservative, average value of

$$\dot{q}_w = 0.06(\dot{q}_c + \dot{q}_r)$$

was used for the back surface of the heat shield and for the side of the body. This relation is based on a theoretical analysis[14] for radiation from the wake of the Galileo Probe as it enters the Jovian atmosphere.

The calculation of the thermal response of the body to these heat fluxes was done in two steps. First, the temperature on the back of the heat shield was calculated using a one-dimensional transient analysis. Then, the heat soak into the body was calculated using the axisymmetric model of Fig. 2, with the calculated heat-shield backface temperatures imposed as a time-dependent boundary condition. Both calculations were done with and without wake radiation imposed.

Several general-purpose computer codes were used to perform the thermal-response analysis. The engineering analysis language (EAL)[15] code was used for the one-dimensional analysis of the heat shield. The NASA structural analysis (NASTRAN)[16] thermal analysis code was used for the cone-cylinder model analysis. Both of these are finite-element codes. The EAL code provides for temperature-dependent thermal conductivity. This is important in calculations related to the heat shield because of the large temperature changes that occur within the material during the braking maneuver. A constant average value for thermal conductivity was used for the NASTRAN calculations. This was satisfactory because the temperature changes within the body are relatively small. Radiation view factors were calculated using a Monte Carlo code and then inserted into the NASTRAN model. This view-factor code can handle multiple reflections with arbitrary absorption coefficients, but for the present analysis, reflection was not considered.

Stagnation-point heat fluxes[3,12] to the conical drag brake are shown in Fig. 3. The maneuver brings the spacecraft down from geosynchronous orbit to a Shuttle orbit. The minimum altitude reached is about 85 km at about 150 s. Two curves are shown for the radiation heating (Fig. 3a). For one, the flow behind the shock was assumed to be in equilibrium, and for the other, it was assumed to be in nonequilibrium. In reality, the radiation-flux curve should lie between these extremes. Work is in progress to resolve this problem. As an alternative to the single pass braking maneuver, the braking can be done in stages on two or more successive orbits.[10] The short

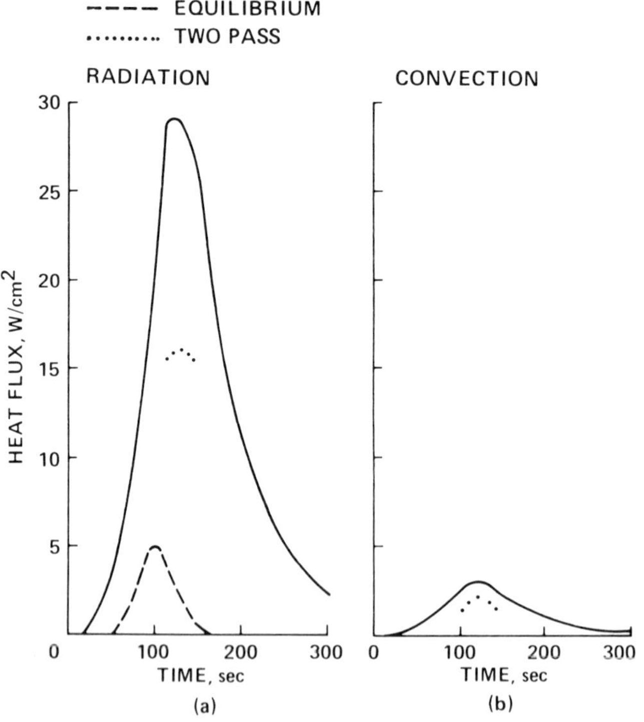

Fig. 3 Heat fluxes to conical drag brake during braking maneuver for transfer from geosynchronous to low Earth orbit.

dotted curves show the peak levels for a two-pass maneuver. This alternative significantly reduces the peak heating levels, but the time for the maneuver is increased by about 4 h per pass. For the present analysis, the nonequilibrium curve was used because it is the most conservative. For all cases, the convective heat flux (Fig. 3b) is a small fraction of the total flux to the surface. However, the front surface of the heat shield can be made highly reflective, whereas all of the convective heat must be absorbed by the surface. It is reasonable to expect 80% reflection of the radiation component.

Results

The results of the one-dimensional analysis of the heat-shield response to the braking maneuver thermal environment are shown in Fig. 4 for three cases. The structural temperature limits for polyimide graphite and aluminum are also shown because these materials are candidates for use in fabricating a heat-shield support structure. Of course, the local temperature will be altered somewhat by the presence of the support structure. The surface absorptivity and the wake radiation both have a significant effect on the backface temperatures. For $\alpha_g = 0.2$, the inclusion of wake radiation increases by 0.4 cm, the thickness of AFRSI required to limit the backface temperature to the polyimide-graphite limit. If the absorptivity is increased to 0.4, the maximum backface temperature is increased significantly and it cannot be reduced to the polyimide-graphite limit by increasing the heat-shield thickness. For very thick heat shields, the backface temperature limit is very nearly the thermal radiation equilibrium temperature for the wake flux:

$$T_{eq} = (\alpha_g \dot{q}_w / \varepsilon_s \sigma A)^{1/4}$$

The fact that the backface temperature exceeds the structural limit of the proposed materials does not preclude the use of thin heat shields. Local insulation or heat sinks can be utilized in the design of the attachment points for the heat-shield support structure. In fact, it has been proposed that a thin silica cloth be used.[3,10] The weight of the heat-shield fabric only is shown in

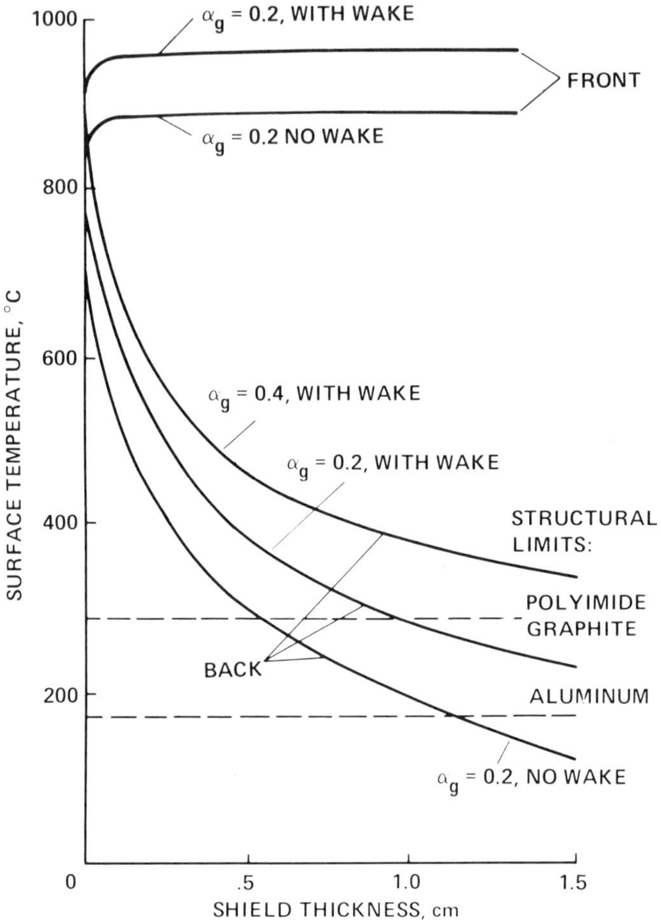

Fig. 4 Maximum surface temperatures of heat shield during braking maneuver.

Fig. 5. The AFRSI blanket used in this analysis is a silica glass felt with silica glass fabric fastened to each surface by 1-in.-square blanket stitching. The cloth mass is less than that of the AFRSI, but the support structure mass may be larger because of its greater flexibility and higher temperature. Selection of the optimum thickness awaits estimates of the mass of the entire thermal control system. For the present study, both a thin cloth and a 1-cm-thick AFRSI blanket are considered. The 1-cm thickness was chosen because it provides the limiting backface temperature for direct attachment to a polyimide-

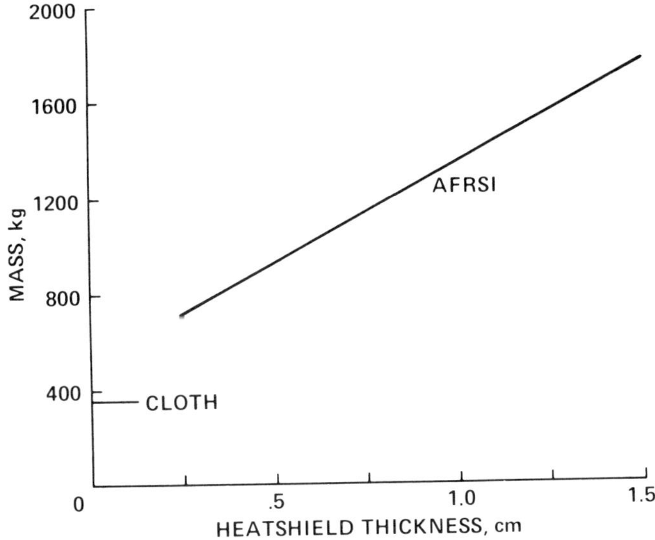

Fig. 5 Mass of heat-shield fabric only.

graphite structure. Although all-silica fabric materials are used in this analysis, other materials are also candidates. Ceramic textiles, such as aluminoborosilicate (Nextel) and silicon carbide (Nicalon) with or without coatings, may be more appropriate if the flight environment is too severe for silica.

As with the heat shield, the wake radiation can be a significant portion of the heat flux to the AOTV body. Figure 6 shows the peak flux to the body from the heat shield relative to the peak flux from the wake. The fluxes are averaged over the length of the body.

To calculate the AOTV body temperatures during the braking maneuver, the backface temperatures of the heat shield are imposed as a boundary condition. Typical backface temperatures are shown in Fig. 7.

Temperatures along the body for the 1-cm-thick AFRSI heat shield are shown in Fig. 8. Both the outer skin surface temperature and the temperature at 40 cm into the body core are shown. The temperature rise at the center is less than 2°C because of the large thermal mass and the short heat pulse. If the heat soak were uniformly distributed through the core, the average core temperature at 250 s would be 16°C for the no-wake case and 27°C for the wake cases. The average temperatures are quite moderate.

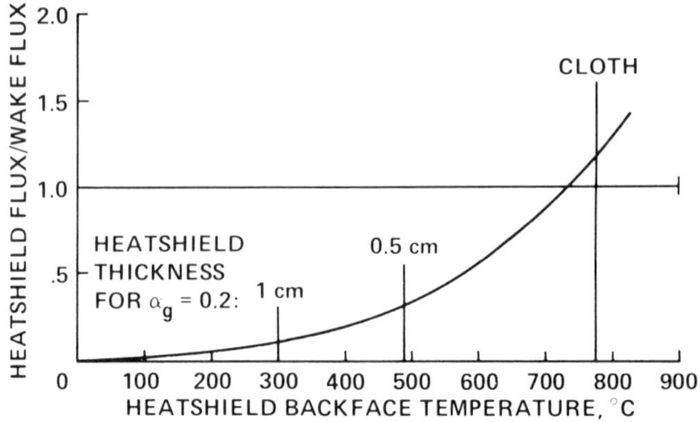

Fig. 6 Relative flux to body from heat shield and wake.

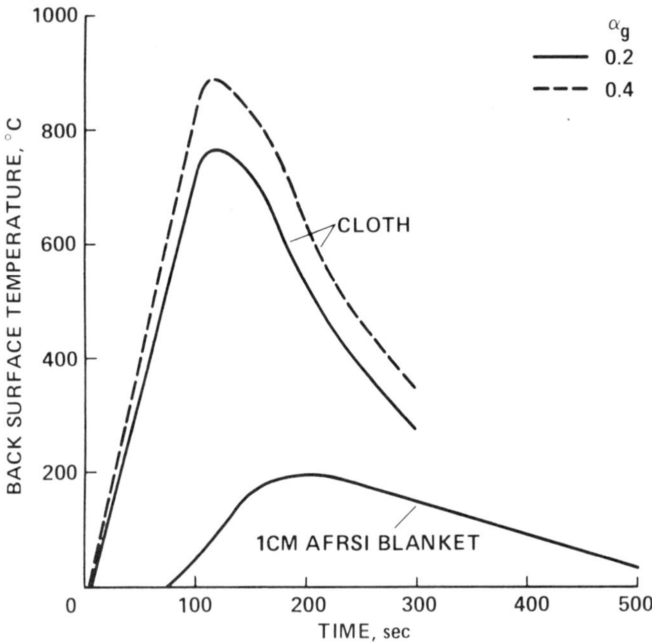

Fig. 7 Heat-shield backface temperatures during the braking maneuver.

The temperatures near the surface may cause some concerns, but there is adequate thermal heat sink available so that these can easily be alleviated by proper thermal management. The results shown in Fig. 8 are for the most probable value of surface absorptivity, $\alpha_g = 0.2$, but they apply to other values as well, as long as the backface

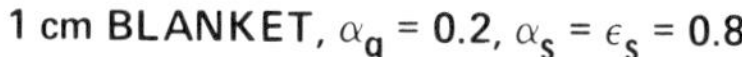

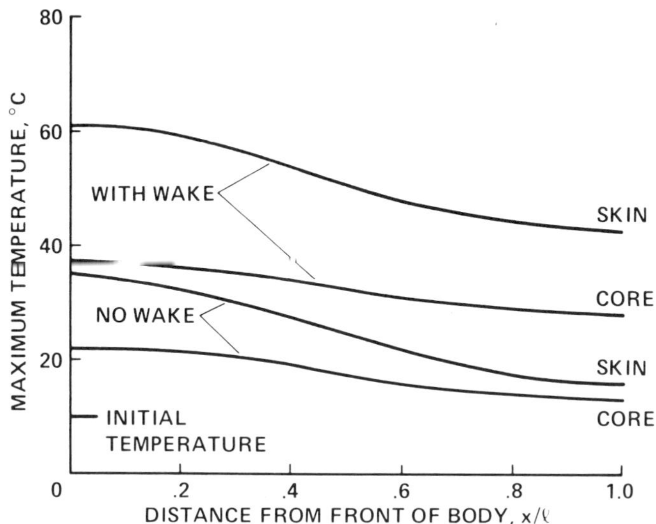

Fig. 8 Maximum temperatures for body during braking maneuver with 1-cm AFRSI blanket heat shield: $\alpha_g = 0.2$; $\alpha_s = 0.8$.

temperature criterion is maintained. For other values, the blanket thickness would be adjusted to give the same backface temperature.

Similar results are shown in Fig. 9 for the cloth heat shield. Because of the higher backface temperature of the heat shield, the body temperatures are higher by 10°C to 20°C. The average core temperature at 250 s is 34°C. For this case, $\alpha_s = 0.2$; it was 0.8 for the blanket heat-shield results shown in Fig. 8. An indication of the sensitivity of the body temperature to surface absorptivity is given in Fig. 10. Comparison of the bottom two curves shows that if α_s is increased from 0.2 to 0.8, the surface temperature is more than doubled. This is because, for the cloth heat shield, most of the heat to the body is infrared radiation from the back of the heat shield (see Fig. 6). It is, therefore, better to reflect (reflection = 1 - α) as much of the radiation in this wavelength range as possible than to cool the body by using a high emissivity ($\varepsilon_s = \alpha_s$ for a given wavelength). In contrast, for the thick-blanket case, a relatively small amount of heat comes from the heat shield (Fig. 6)

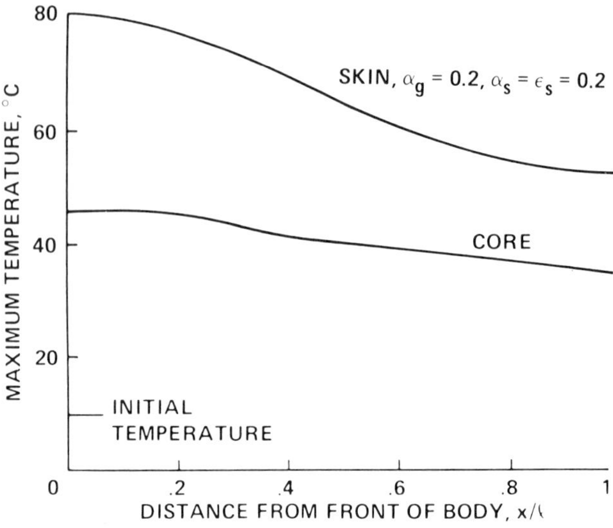

Fig. 9 Maximum temperatures for body during braking maneuvers with cloth heat shield: $\alpha_g = 0.2$; $\alpha_s = 0.2$.

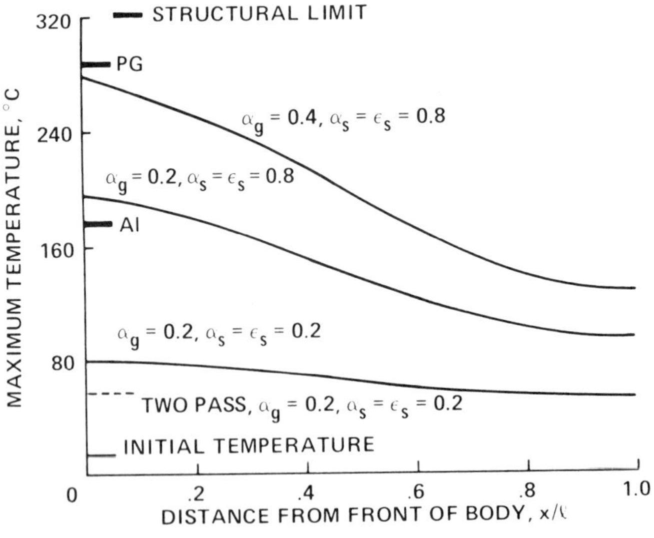

Fig. 10 Effect of surface absorptivity on body skin temperature with cloth heat shield.

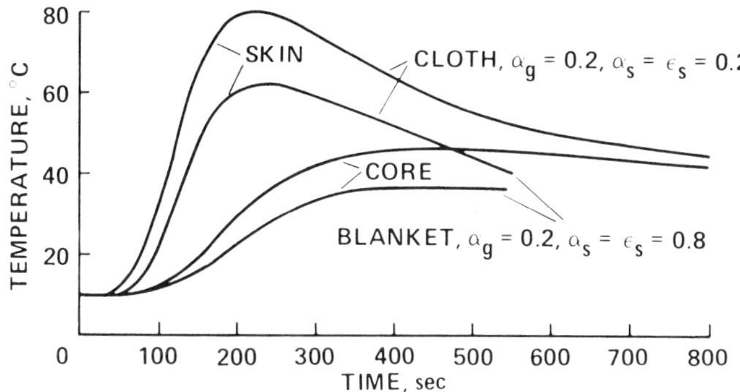

Fig. 11 Temperature history for conical drag brake during braking maneuver.

so it is better to absorb a large fraction of the radiation from the heat shield in exchange for a relatively high emission from the body surface. For each configuration there is an optimum $\alpha_s = \varepsilon_s$.

The top two curves in Fig. 10 indicate the sensitivity of the body temperature to α_g. For each of these curves the same α_s is applied to all surfaces. Also shown in Fig. 10 are the maximum skin temperature for the two-pass braking maneuver and the initial body temperature. All curves in this figure will shift up or down by nearly the amount of any shift in the initial body temperature.

The maximum temperature is not the only concern; its duration is also significant. Figure 11 shows the time variation for the temperatures corresponding to the results in Figs. 8 and 9 for $\alpha_g = 0.2$ with a wake. The core temperature is the average temperature at 40 cm below the skin. For this figure, $\alpha_s = 0.2$ on the body with the cloth heat shield. If a surface were selected with $\alpha_s = 0.8$ rather than $\alpha_s = 0.2$, the maximum skin temperature would be much higher, as shown in Fig. 10, but the cooling rate by radiation to space after the braking maneuver would be greater. Consequently, the vehicle with $\alpha_s = 0.8$ on the body would have a cooler skin after 1100 s and a cooler core to a depth of 40 cm after 1600 s.

Both Figs. 8 and 9 show that the body temperatures are higher near the heat shield. This is primarily because of the relative proximity of the body to the heat shield. Other factors, not included in the present analysis, may increase this temperature gradient. There are multiple reflections and emission exchanges between the heat shield and the body that will retain heat there. This problem will be enhanced by any spaces between the heat shield and the front of the body. Also the conical shield may focus some of the wake radiation on the body near the front.

Flight Experiment

There is very little experience with AOTVs and their missions, as well as with the thermal environment in which the aerodynamic braking will occur. As a result, NASA is seriously considering an experimental flight of a small-scale AOTV model that would be launched from the Space Shuttle. Instrumentation would be designed to give information about crucial aspects of mission operations and response to the thermal environment. In particular, the uncertainties discussed in the Introduction section of this paper would be examined.

The configuration for the proposed experimental flight model has not been selected and there are many alternatives in the mission that are being considered. One of the issues is the manner in which the vehicle is recovered. One option is to skip back into orbit after the braking maneuver and do an in-orbit recovery. The other option is for direct entry with either a low-altitude catch or an ocean recovery.

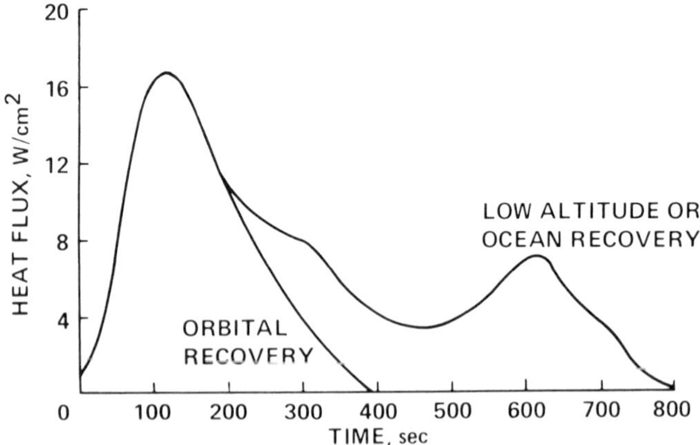

Fig. 12 Convective heat flux to a flight experimental model of an AOTV with conical drag brake for two recovery mode trajectories.

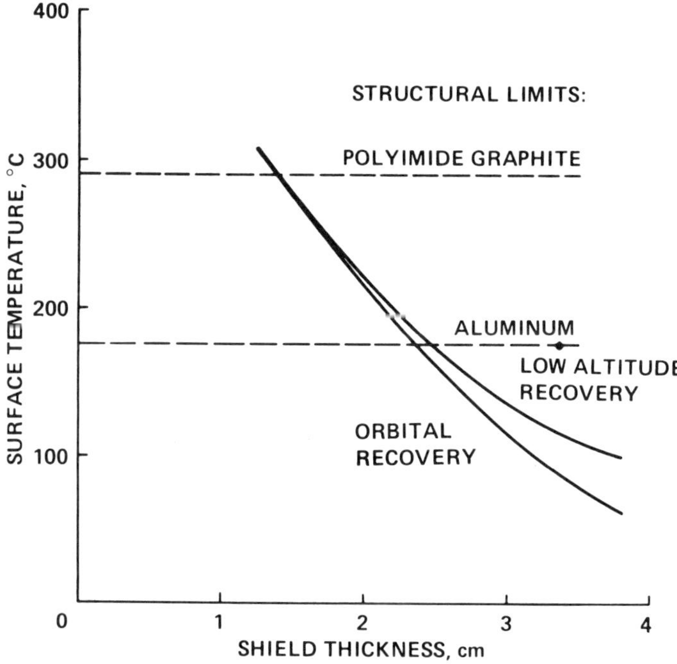

Fig. 13 Maximum backface temperatures for flight experimental model for two recovery modes.

Although the configuration has not been selected, an examination was made of the effect of these two recovery alternatives on the heat shield of a conical drag brake. The heat fluxes for this calculation were provided by R. W. Powell of Langley Research Center. The trajectory analysis was for a 5-m-diam model weighing 590 kg. The convective heating is shown in Fig. 12. Radiative heating was not calculated, but for this small model, convection is the primary heating mechanism. The trajectories are identical for the first 180 s.

The results of the one-dimensional analysis for the heat shield are shown in Fig. 13. They show that for a polyimide-graphite structure and for an AFRSI blanket design, the heat-shield requirement is about the same for both missions. Sufficient heat radiates away between the heat pulse peaks so that the temperature after the second pulse only slightly exceeds that after the first pulse. For a cloth heat shield the temperature at the second pulse would be much less than for the first pulse.

Concluding Remarks

A first analysis of the thermal response of a conical drag-brake AOTV to an aerobraking maneuver has been made. The results show that a ceramic fabric heat shield can be designed so that a support structure can be fastened directly to its back surface. With this design, the body temperatures are low and thermal management problems are minimal. Although the weight of this heat shield is significant, it is not prohibitive. The weight of the heat-shield fabric can be reduced by using a thin ceramic cloth, but the support structure might be heavier. Also, the cloth heat-shield temperatures are high so that the support structure attachment becomes more complex. More analysis is required to resolve these questions. The analysis to date has shown that temperatures are very sensitive to the surface material properties, that radiation from the wake may be a major contributor to body temperatures, and that body temperatures are highest in the body/heat-shield attachment area.

References

[1] Walberg, G. D., "A Review of Aeroassisted Orbital Transfer Vehicles," AIAA Paper 82-137, AIAA 9th Atmospheric Flight Mechanics Conference, San Diego, Calif., Aug. 1982.

[2] Menees, G. P., "Trajectory Analysis of Radiative Heating for Planetary Missions with Aerobraking of Spacecraft," AIAA Paper 83-0407, AIAA 21st Aerospace Sciences Meeting, Reno, Nev., Jan. 1983.

[3] Menees, G. P., "Thermal-Protection Requirements for Near-Earth Aeroassisted Orbital-Transfer Vehicle Missions," AIAA Paper 83-1513, AIAA 18th Thermophysics Conference, Montreal, Canada, June 1983; published elsewhere in this volume.

[4] Cruz, M. I., "Aerocapture Vehicle Mission Design Concepts for Inner and Outer Planets," AIAA Paper 79-0893, AIAA Conference on Advanced Technology for Future Space Systems, Hampton, Va., May 1979.

[5] French, J. R. and Cruz, M. I., "Aerobraking and Aerocapture for Planetary Missions," Astronautics and Aeronautics, Feb. 1980, pp. 48-71.

[6]French, J. R. and McRonald, A. D., "Thermophysical and Systems Integration Considerations in Aerobraking Design," AIAA Paper 80-1492, AIAA 15th Thermophysics Conference, Snowmass, Colo., July 1980.

[7]Florence, D. E., "Aerothermodynamic Design Feasibility of a Generic Planetary Aerocapture/Aeromaneuver Vehicle," AIAA Paper 81-1127, AIAA 16th Thermophysics Conference, Palo Alto, Calif., June 1981.

[8]French, J. R., "Trends in Unmanned Planetary Entry Systems," AIAA Paper 81-1125, AIAA 16th Thermophysics Conference, Palo Alto, Calif., June 1981.

[9]Andrews, D. G. and Caluori, V. A., "Optimization of Aerobraking Orbital Transfer Vehicles," AIAA Paper 81-1126, AIAA 16th Thermophysics Conference, Palo Alto, Calif., June 1981.

[10]Menees, G. P., Park, C., and Wilson, J. F., "Design and Performance Analysis of a Conical-Aerobrake Orbital-Transfer Vehicle Concept," AIAA Paper 84-0410, AIAA 22nd Aerospace Sciences Meeting, Reno, Nev., Jan. 1984; published elsewhere in this volume.

[11]Davies, C. B. and Park, C., "Aerodynamic characteristics of Generalized Bent Biconic Bodies for Aeroassisted Orbital Transfer Vehicles," AIAA Paper 83-1512, AIAA 18th Thermophysics Conference, Montreal, Canada, June 1983.

[12]Menees, G. P., Davies, C. B., Wilson, J. F., and Brown, K. G., "Aerothermodynamic Heating Analysis of Aerobraking and Aeromaneuvering Orbital-Transfer Vehicles," AIAA Paper 84-1711, AIAA 19th Thermophysics Conference, Snowmass, Colo., June 1984; published elsewhere in this volume.

[13]Balakrishnan, A., "Flow Field Computation of an Aerobrake Orbital Transfer Vehicle," AIAA Paper 84-1697, AIAA 19th Thermophysics Conference, Snowmass, Colo., June 1984.

[14]Park, C., "Problems of Radiative Base Heating," AIAA Paper 79-0919, AIAA Conference on Advanced Technology for Future Space Systems, Hampton, Va., May 1979.

[15]Marlowe, M. B., Moore, R. A., and Whetstone, W. D., "SPAR Thermal Analysis Processors Reference Manual System Level 16," Vol. I, NASA CR-159162, 1978.

[16]Lee, H. P., "NASTRAN Thermal Analyzer--Theory and Application Including a Guide to Modeling Engineering Problems," Vol. I, NASA TMX-3503, 1977.

Low Lift-to-Drag Aerobrake Heat-Transfer Test at Mach 10

Peter Kwang-Tien Shih* and Archie Gay†
General Dynamics Convair Division, San Diego, California

Abstract

Heat-transfer data for a low lift-to-drag ratio (0.25) aerobraking orbital transfer vehicle concept are presented. Tests were conducted in the NASA-Langley Continuous Flow Hypersonic Tunnel at Mach 10 to obtain heating data, and Schlieren runs were made in the Langley Hypersonic Flow Apparatus. Phase-change paint heat-transfer data were obtained, and the Schlieren photographs provided definition of shock patterns around the brake and afterbody. Five configurations were studied. High local heating was encountered at the nose and brake outer edges. For angles of attack α = 0 to 15 deg, the edge heating remains fairly constant but increases rapidly for α above 15 deg. Afterbody heating due to separated flow reattachment indicated high localized heating in this region. Heat flux was found to increase with increasing α, increasing Re, decreased edge radius, and decreased brake surface slope.

Nomenclature

AOTV = aerobraking orbital transfer vehicle
CFHT = Continuous Flow Hypersonic Tunnel (NASA-Langley)
D = distance along body axis from the edge of the brake
HFA = hypersonic flow apparatus
h = heat-transfer coefficient
h_{ref} = reference heat-transfer coefficient
k = mean density ratio across the shock
L = body length of option 2 measured from the edge of the brake
L/D = lift-to-drag ratio
M_∞ = freestream Mach number
P_o = total pressure
Q_o = stagnation heat flux

Presented as Paper 84-0309 at AIAA 22nd Aerospace Sciences Meeting, Reno,Nev., Jan. 9-12, 1984.

* Senior Engineering Specialist, Propulsion and Thermophysics.
† Chief, Propulsion and Thermophysics.

R	= body radius
Re	= Reynolds number
r	= nose radius
T_0	= total temperature
T_{RE}	= radiation equilibrium temperature
x	= distance along body axis
α	= angle of attack
Δ	= shock standoff distance

Introduction

Aerobraking orbital transfer vehicles (AOTVs) will be used to increase the Space Shuttle mission capability for delivering payloads to geostationary orbits and returning to low Earth orbit. Aerodynamic forces on the brake will replace retrothrust and eliminate the large amount of propellant required to accomplish deceleration to low Earth orbit conditions. Orbital transfer vehicle conceptual design studies completed in 1980 by General Dynamics Convair Division[1] indicated a large performance gain by using aerobraking concepts.

The primary configuration considered here is an umbrella-like aerobrake with a lift-to-drag (L/D) ratio of approximately 0.25, as shown in Fig. 1. The vehicle consists of a cylindrical body that can be fitted into the Space Shuttle cargo bay and a deployable aerobrake featuring ribs and Nextel cloth, a ceramic fabric manufactured by 3M. The brake is folded along the sides of the vehicle while in the Shuttle and is extended to provide a reentry shield on the return portion of the mission. After passage through the atmosphere, the brake is retracted and stowed alongside the stage for rendezvous with the Shuttle.

A joint General Dynamics Convair Division and Remtech wind tunnel test program under the sponsorship of NASA/MSFC was conducted to study the aerodynamic and heat-

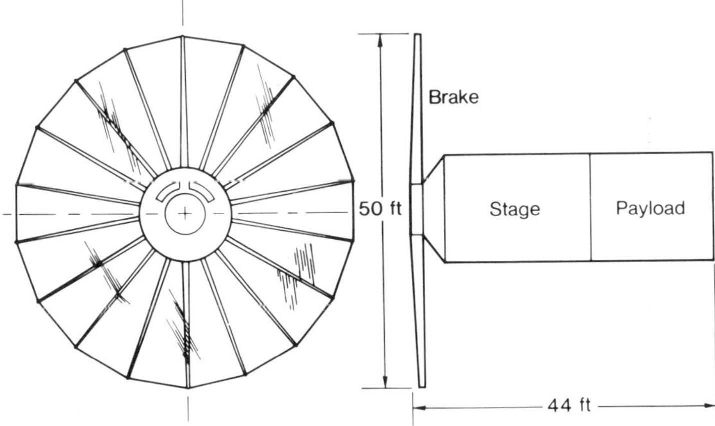

Fig. 1 Aerobraking OTV concept.

transfer characteristics of the aerobrake concept. Results of the phase I (Remtech) study were published in Ref. 2. The present paper presents heat-transfer results of the phase II study. Aerodynamic force and moment data obtained in a companion test were published previously in Ref. 3.

Experimental Methods

Facility Description

The phase II AOTV tests were conducted in the NASA-Langley Continuous Flow Hypersonic Tunnel (CFHT). This facility is currently operated in the blowdown mode at a nominal test Mach number of 10. The test medium is air-heated by an electrical resistance heater. Run times of up to 60 s are possible. This wind tunnel has a test section 31 in.^{2} and a test core of 12 to 14 in.^{2} at Mach 10. Reynolds numbers from $(0.3 \text{ to } 2.25) \times 10^6$/ft are available by varying the stagnation pressure.

This facility is designed for aerodynamic and aeroheating tests. Models are normally mounted on a sting that allows for variable angle of attack and yaw. The sting is incorporated into a pneumatic injection system that inserts and retracts the model from the test section. The sting is remotely controlled and can be programmed to traverse an angle of attack range in either a continuous or pitch-and-pause mode.

Since the CFHT does not have a Schlieren system, separate tests were conducted using the NASA-Langley hypersonic flow apparatus (HFA) to obtain Schlieren photographs. The HFA has a 10-in.-diam test section and tunnel parameters of $M_\infty = 10$, Re $= 1.4 \times 10^6$/ft, $T_o = 1600°R$, and $P_o = 1000$ psia.

Model Descriptions

Figure 2 shows the different configurations of the AOTV models used for the heat-transfer tests. Two stycast models were used to obtain forward face heating data. These

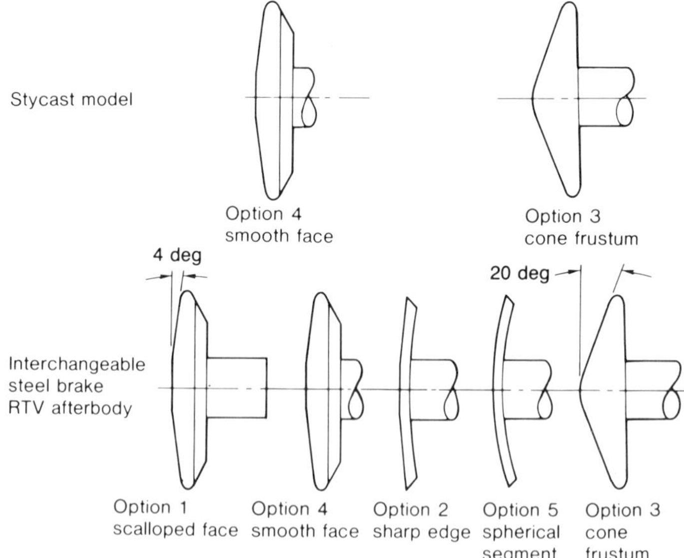

Fig. 2 AOTV test model matrix.

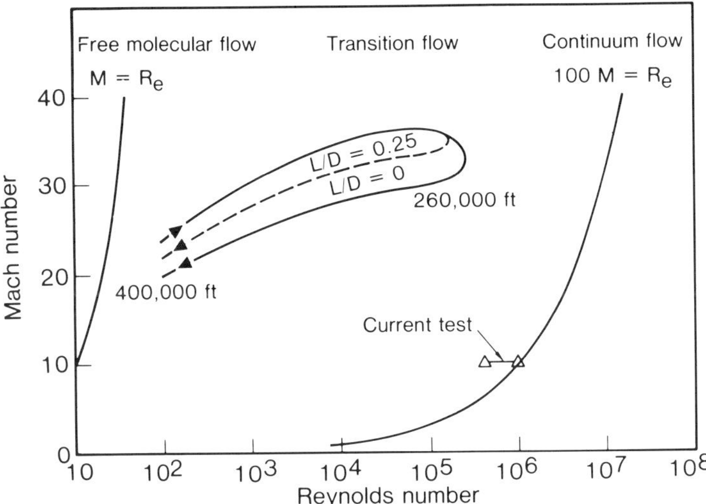

Fig. 3 Flight trajectories of AOTV.

models were equipped with an iron-constantan thermocouple buried in the face of the models to measure the initial temperature. To obtain heating data on the afterbody (stage and payload), each of five interchangeable steel brakes was mounted to a cylinder, which was constructed by casting RTV-60 silicone rubber over a skeleton of 17-4 PH steel. The cylinder is instrumented with four iron-constantan thermocouples to measure the initial temperature for phase-change paint testing. All of the heat-transfer models (both stycast and steel) had an overall face diameter of 4 in., corresponding to a 1/150th scale. Some details of each configuration are presented below.

Option 1. This model duplicates the model of MSFC/Remtech,[2] which consists of a very blunt cone-frustrum with a cone half-angle of 86 deg. The edges of the front face are round and the flat surface on the face is 1.0 in. in diameter. The front face of the frustrum has a series of flat surfaces to simulate a fabric stretched between support spokes.

Option 2. This configuration is identical to option 1 except that the front face is smoothed and carried out to a sharp edge.

Option 3. This configuration is based on a 70-deg half-angle cone-frustrum. The outer edges are rounded, and the front face has a small spherical segment instead of the flat portion of options 1 and 2. This configuration is very similar to the Viking aeroshell.

Option 4. This configuration is identical to option 1 with the flat surfaces removed. A stycast heat-transfer model of this configuration was also manufactured and tested.

Option 5. This configuration replaces the cone-frustrum with a spherical segment and keeps the sharp edge.

Limited by the test core size of the HFA tunnel, three models of 1.75-in. brake diameter (options 3, 4, and 5) were made for taking Schlieren pictures.

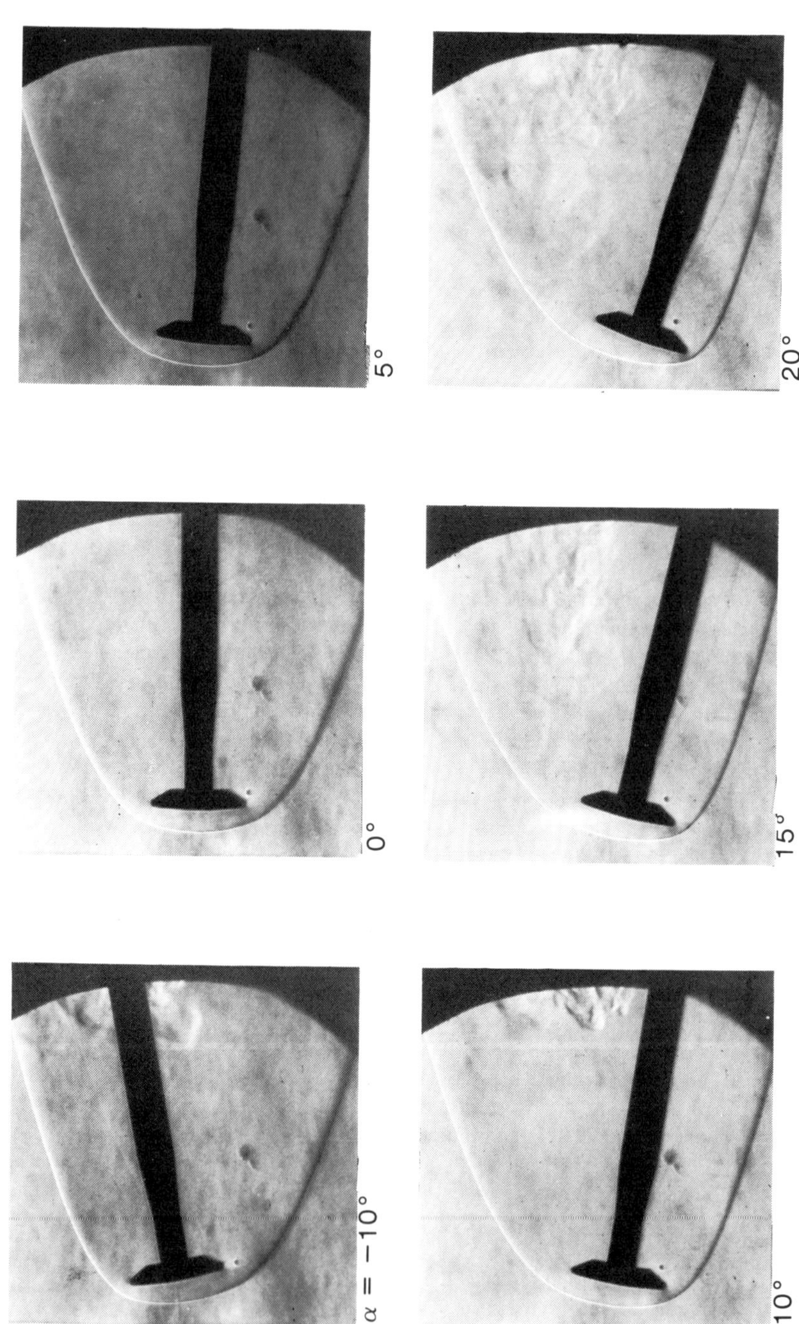

Fig. 4 Schlieren photographs of option 4.

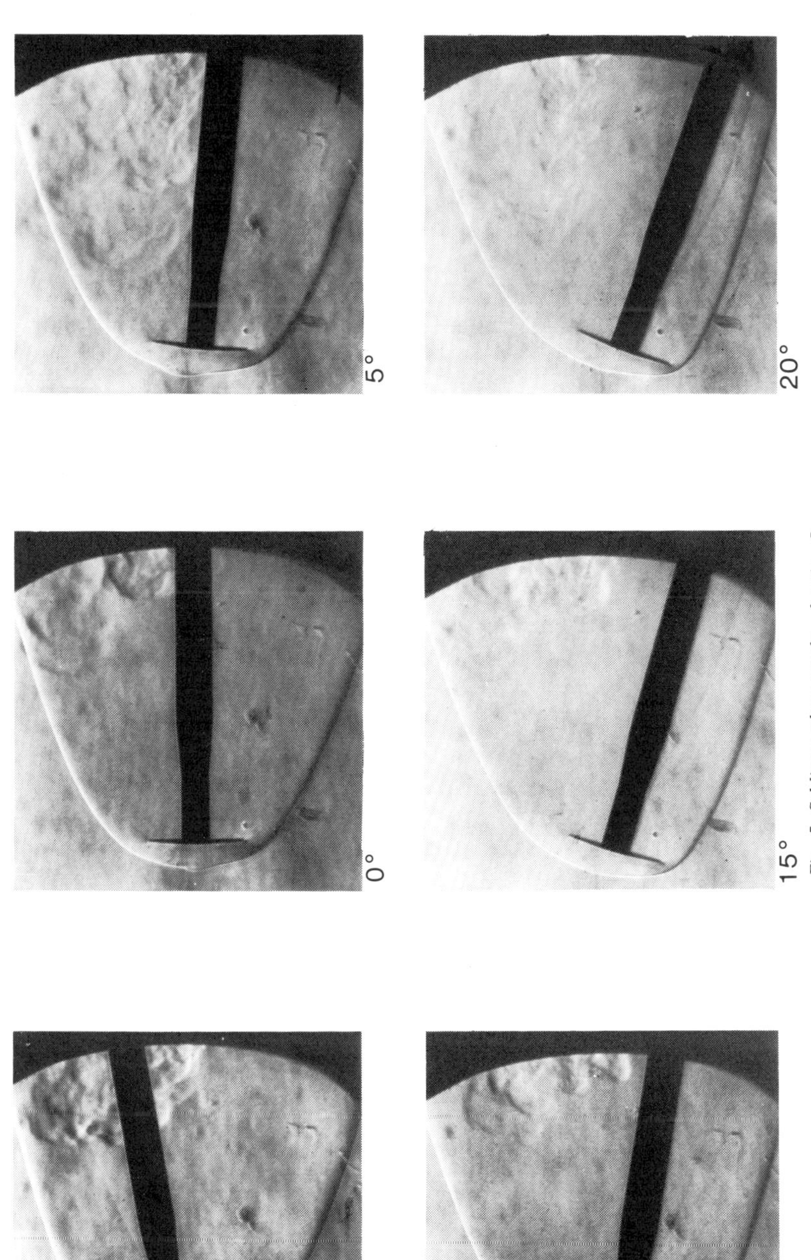

Fig. 5 Schlieren photographs of option 5.

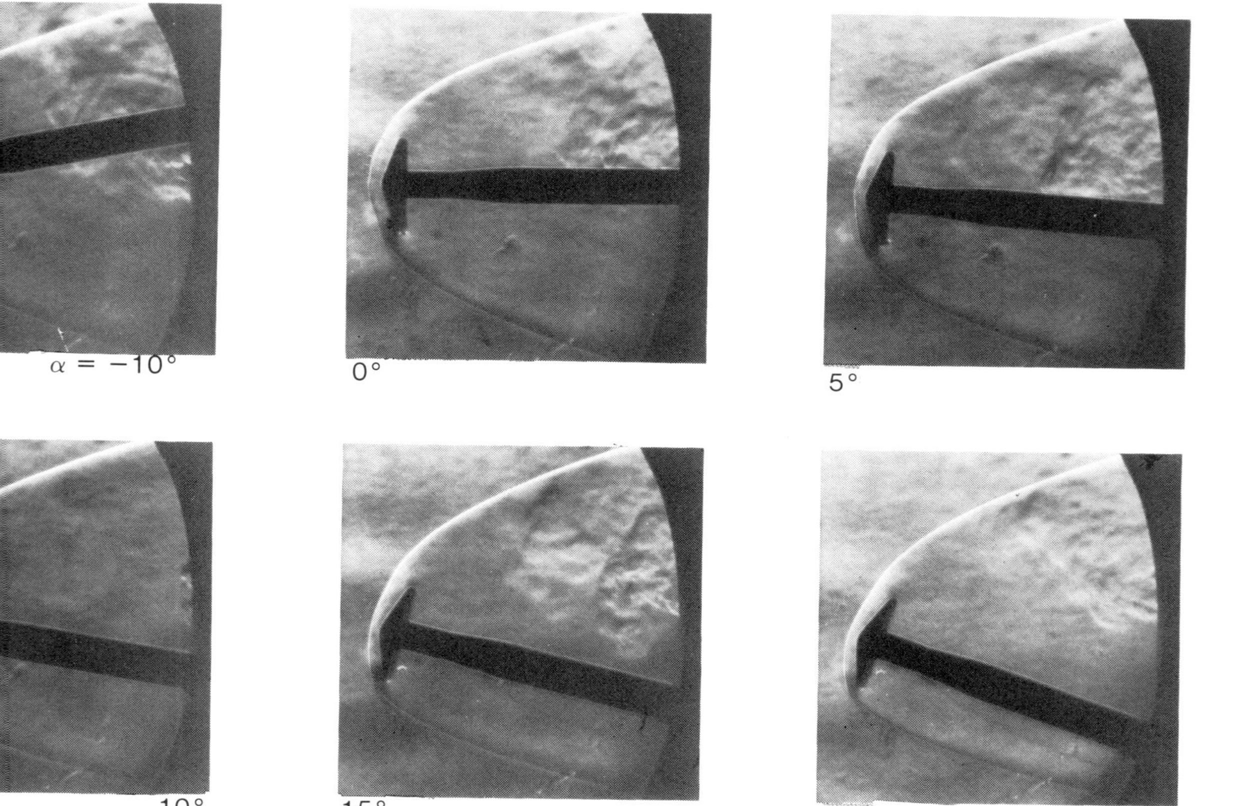

Fig. 6 Schlieren photographs of option 3.

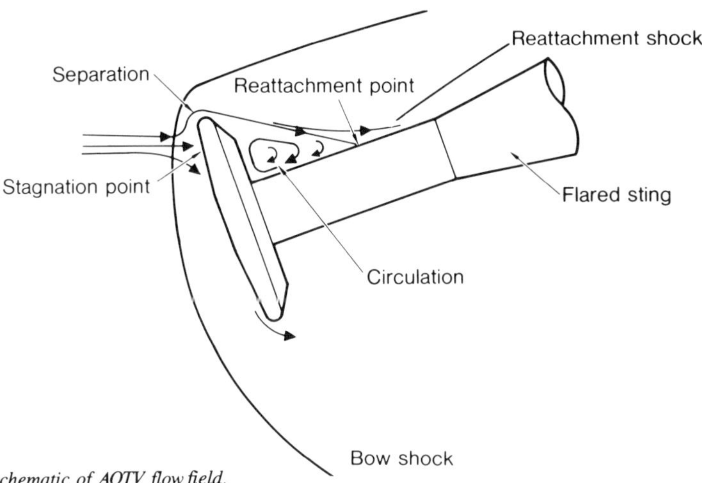

Fig. 7 Schematic of AOTV flowfield.

Test Procedures

Phase-change paint tests were conducted to study the forward face of configurations 3 and 4 and the afterbody for all five configurations. The primary test variables were configuration, Reynolds number, angle of attack, and paint temperature. One run consisted of obtaining motion picture photographic coverage of one configuration at one Reynolds number and angle of attack for one paint temperature. During each run, the thermocouple output was recorded, along with tunnel operating conditions.

Prior to each run, the model was cleaned and spray painted with the appropriate phase-change paint. The model was installed on the tunnel injection system and allowed to reach isothermal conditions. With the tunnel operating at the proper test conditions, the model was injected into the airstream at the desired angle of attack. Color motion picture coverage of the model was started prior to first motion of the injection system and stopped after the model had been extracted from the airstream. After a run was completed, residual paint was removed with solvent and the model was repainted.

The phase-change paint heat-transfer data were obtained for the different configurations at Mach 10, Reynolds numbers of 0.4 and 1.0 million/ft, and angles of attack from 0 to 30 deg. In addition, Schlieren photographs were taken to study the shock structures around the brake and afterbody at $M_\infty = 10$, and $Re = 1.4 \times 10^6$/ft. Figure 3 shows which portion of the AOTV flight regime the wind tunnel tests covered.

Data Reduction

Heat-transfer data were obtained in the form of phase-change paint patterns (isotherms), which were photorecorded at discrete time intervals during each test.

Motion picture data were reduced to heat-transfer rates by projecting individual frames of the film onto a drawing table, where the model outline and isotherm patterns were sketched to form a composite drawing illustrating the progression of the contours across the model surfaces. Experimental heat-transfer coefficients were then derived from the test data by using a one-dimensional transient heat-transfer analysis of a semi-infinite slab.

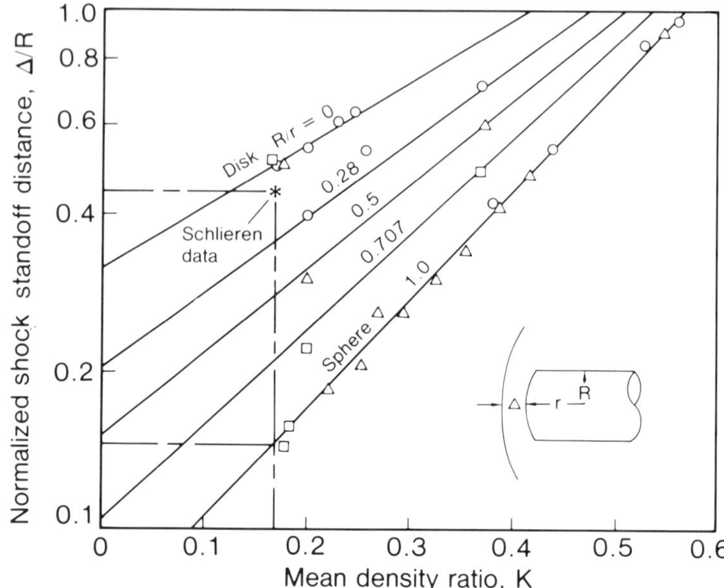

Fig. 8 Shock standoff distance as functions of mean density ratio.

Results

Schlieren Photographs

Figures 4 through 6 show the Schlieren photographs of options 4, 5, and 3, respectively, for angles of attack from −10 to 20 deg. The shape of the bow shock and the standoff distance provide valuable information for understanding the flowfield and heat-transfer characteristics of the aerobrake. Shocks on the afterbody can be observed at the angles of attack above 10 deg. However, it is questionable whether this secondary shock is generated by reattachment of separated flow from the brake or by the discontinuity between the afterbody and the flared sting. A new model with an alternative mounting system is currently being designed to investigate the secondary shock.

In Figs. 5 and 6, severe distortion of the bow shock can be seen for angles of attack of 0 and 5 deg. A similar phenomenon has been observed by Dunbar et al.[4] They believe it is caused by interaction of rebounding particles in the airstream with the shock layer. Figure 7 shows a schematic of the flowfield around the aerobrake illustrating the stagnation flow at the forward face. The flow accelerates and subsequently separates at the edge of the brake. This separated flow reattaches to the windward side of the afterbody, forming a reattachment shock. High localized heating occurs at the edge as the flow accelerates and when it reattaches to the afterbody.

Reference Heating Rate

The stagnation heat-transfer rate can be calculated by the theory of Fay and Riddell[5] for an equivalent sphere. Recently Park[6] has empirically curve-fit the shock standoff distance for spheres and disks as a function of the density ratio across the shock. He postulated that the stagnation heat-transfer rate is a function of the velocity gradient (hence, the standoff

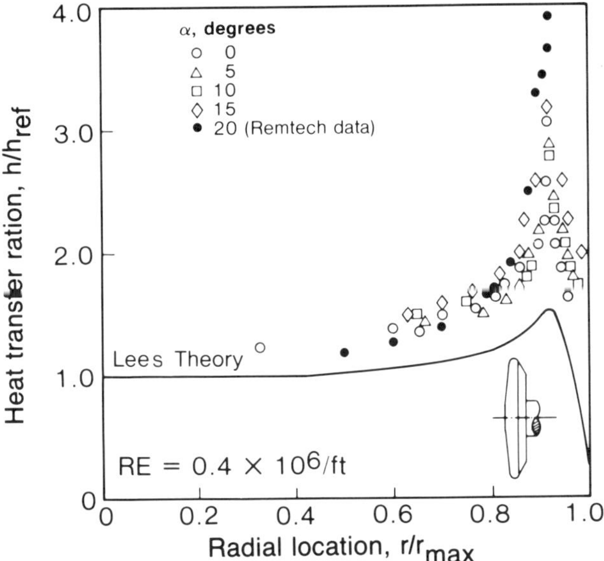

Fig. 9 Radial variation of heat transfer at low Reynolds number (option 4).

distance) at the stagnation point. An equivalent nose radius for the disk was then calculated for stagnation heat-transfer prediction. Figure 8 shows that the standoff distance data collected by Park for spheres and disks and some other geometric shapes[7] can be fitted by a simple empirical equation:

$$\ln (\Delta/R) = Ak - B$$

where

$$A = 2.74 + 2.05 (R/r)$$

$$B = 1.16 + 1.56 (R/r)$$

From Fig. 2, the stycast model of option 4 is nearly a disk, and the normalized standoff distance measured from Fig. 4 at a 0-deg angle of attack is $\Delta/R = 0.45$. At $M_\infty = 10$, $k = 0.175$. From Fig. 8, we have $\Delta/R \simeq 0.15$ for a sphere. Hence, the equivalent radius is $2 \times 0.45/0.15 = 6$ in. The stagnation point heat-transfer coefficients are given by the Fay-Riddell equation as

$$h = 0.0034 \text{ Btu/ft}^2\text{-R-s} \quad \text{for Re} = 0.4 \times 10^6/\text{ft}$$

$$= 0.0052 \qquad = 1.0 \times 10^6/\text{ft}$$

These values, which agree with the phase-change paint data to within 15%, are used as the reference heating rate (h_{ref}) for all dimensionless data.

Forward Face Heating

Figures 9 and 10 show the forward face heating data of option 4. This model is identical to the Remtech model except for the scalloped face. This difference apparently has no

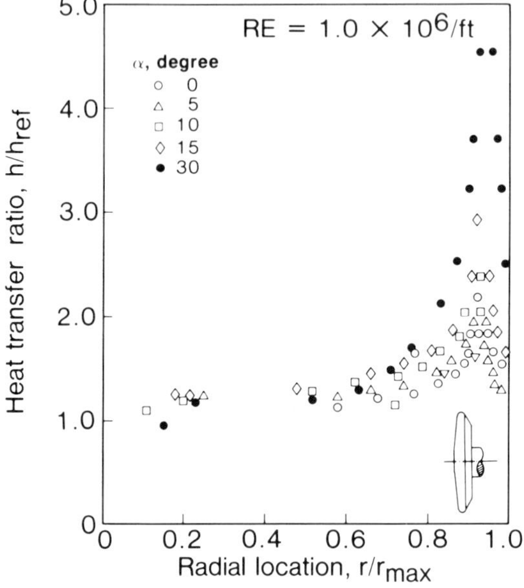

Fig. 10 Radial variation of heat transfer at high Reynolds number (option 4).

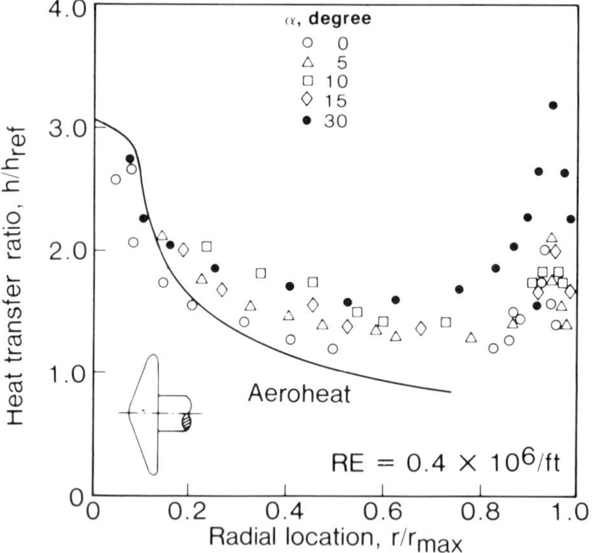

Fig. 11 Radial variation of heat transfer at low Reynolds number (option 3).

effect on the data; they agree well. Computations by Lees theory[8] using Remtech's pressure data at $\alpha = 0$ show a high heating rate at the edge but underpredict the magnitude. The measured stagnation point heating at $\alpha = 0$ and low Reynolds number is slightly higher than prediced by Fay-Riddell, which may be due to rarefied gas effects. In both high and low Reynolds number runs, the magnitude of edge heating does not vary too much from $\alpha = 0$ to $\alpha = 15$ deg. However, the value increases rapidly above $\alpha = 15$ deg.

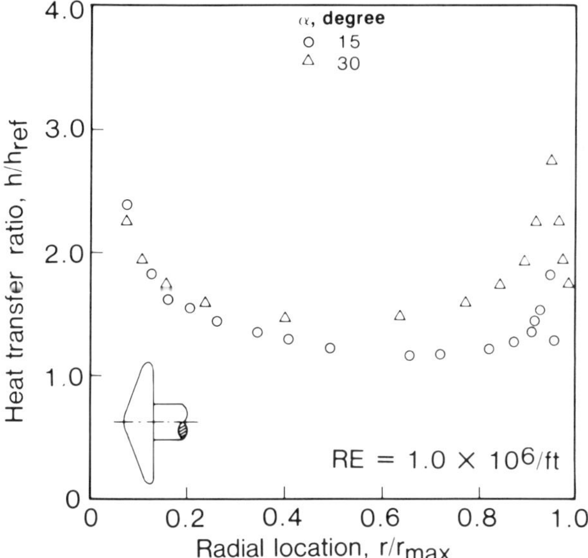

Fig. 12 Radial variation of heat transfer at high Reynolds number (option 3).

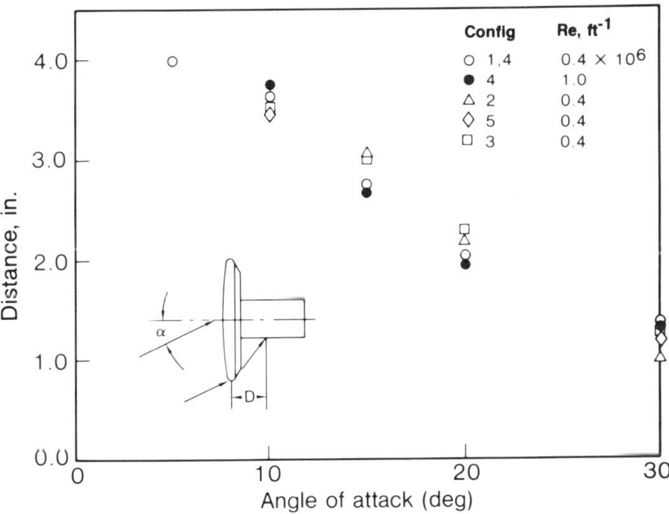

Fig. 13 Flow reattachment distance.

Figures 11 and 12 show the forward face heating data of option 3, which is characterized by a spherical nose, conical body, and round edge. The data show high nose and edge heating. At $\alpha = 0$, the values calculated by the Aeroheat program[9] agree with the experimental data for the forward half of the body. The general trend of the effects of Reynolds number and angles of attack are about the same as option 4.

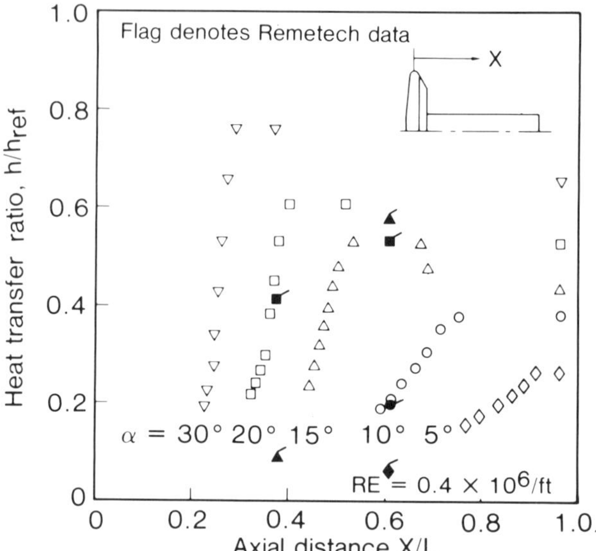

Fig. 14 Afterbody heat-transfer distribution (option 1).

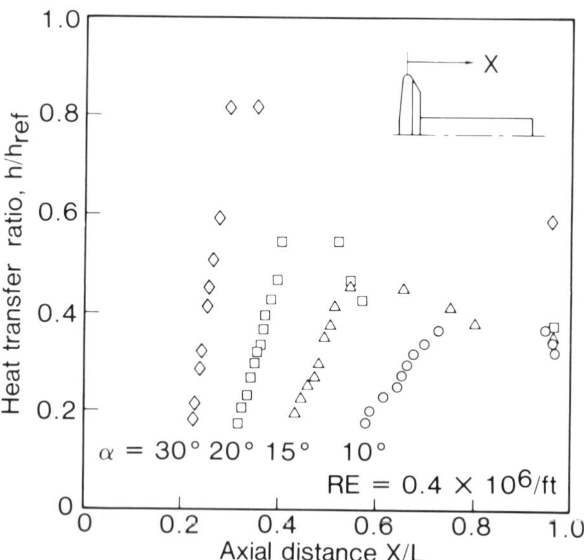

Fig. 15 Afterbody heat-transfer distribution (option 4).

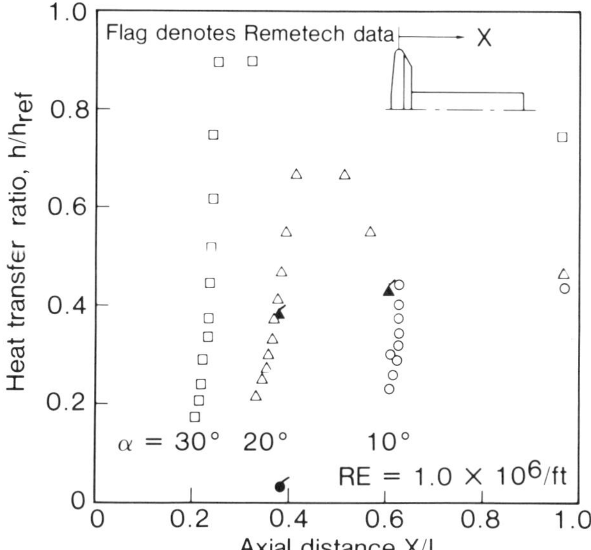

Fig. 16 Afterbody heat-transfer distribution (option 4).

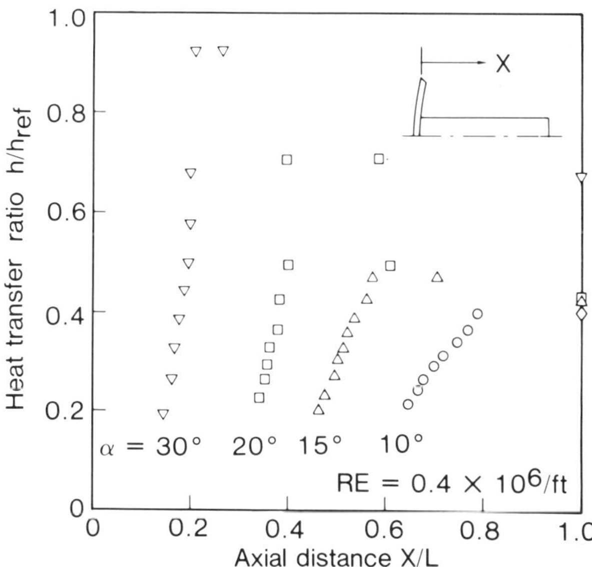

Fig. 17 Afterbody heat-transfer distribution (option 2).

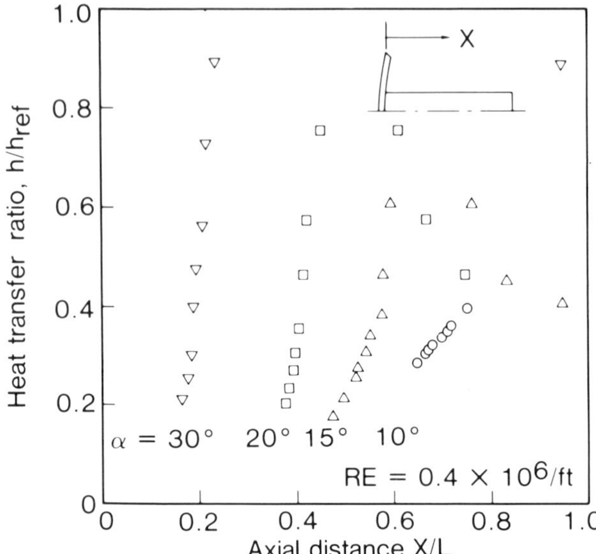

Fig. 18 Afterbody heat-transfer distribution (option 5).

Afterbody Heating

The characteristics of afterbody heating are generally controlled by the flowfield as described in Fig. 7. The complicated flowfield has prohibited any detailed analysis. On the other hand, the experimental data of reattachment location and heating magnitude are valuable, since they provide a guideline for the payload design. The question has been raised whether the flowfield was distorted by the sting in this test. In the real case, a vortex will be formed at the base of the afterbody, which may cause the reattachment point to occur farther downstream, hence, lessening the heating on the payload. A model to investigate the effects of such a vortex is currently under design.

Figure 13 shows the measured reattachment distance, the distance on the afterbody between the separation and reattachment point. Data indicate a decreased distance with increasing angles of attack. However, this distance seems insensitive to the aerobrake configuration and Reynolds number tested!

Figures 14 through 19 show the afterbody heating data for all five brake configurations. Heat-transfer data obtained from option 1 (Fig. 14) and option 4 (Fig. 16) are almost identical, as expected. Remtech thermocouple data are also included in Figs. 14 and 15 for comparison. The agreement is good. Overall, the experimental data indicate high localized heat flux at the reattachment point. This peak heat flux is increasing with 1) increased angle of attack (Figs. 14 through 19); 2) increased Reynolds number (Fig. 15); 3) decreased edge radius (Figs. 17 and 18); 4) decreased brake surface slope (Fig. 19).

Sample Analysis

We can use the information obtained from this test program to estimate the heating problem encountered by an AOTV (Fig. 1) that flies an L/D = 0.25 trajectory (Fig. 2) with α = 15 deg. Only convective heating is considered in this analysis. Stagnation

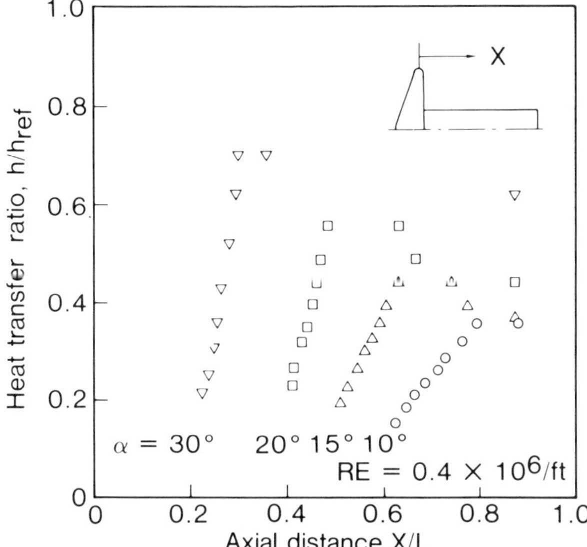

Fig. 19 Afterbody heat-transfer distribution (option 3).

heating is calculated by the Fay-Riddell equation using an equivalent diameter of 150 ft. Peak heating at the edge of the brake and on the payload are three times and one-half the stagnation value, respectively. The results indicate that maximum heating occurs near perigee. The maximum radiation equilibrium temperatures are 1850 F at the center of the brake, 2500 F at the edge of the brake, and 1430 F at the payload.

Conclusions

Heating data have been obtained for a low L/D aerobrake configuration. Data correlation indicated that the stagnation heating can be calculated by the Fay-Riddell equation using equivalent diameter as suggested by Park. For engineering analysis, estimates of peak heating at the edge of the brake and on the afterbody can be obtained from the experimentally determined heating ratios with respect to the stagnation levels.

Acknowledgments

The authors are grateful to Kenton Whitehead for helpful discussions.

References

[1]Heald, D.A. et al., "Orbit Transfer Vehicle (OTV) Concept Definition Study," General Dynamics Convair Division, San Diego, Calif., GDC-ASP-80-012, Feb. 1981.

[2]Hair, L.M., Engel, C.D., and Sulyma, P.R., "Low L/D Aerobrake Test at Mach 10," AIAA Paper 83-1509, 18th Thermophysics Conference, Monteral, Canada, June 1983.

[3]Adcock, C.L., "Aerobraking Design Concept," General Dynamics Convair Division, San Diego, Calif., GDC-ERR-83-411, Dec. 1983.

[4]Dunbar, L.E., Courtney, J.F., McMillen, L.D., "Heating Augmentation in Erosive Hypersonic Environments," AIAA Journal, Vol. 13, July 1975, pp. 908-912.

[5]Fay, J.A. and Riddell, F.R., "Theory of Stagnation Point Heat Transfer in Dissociatied Air," Journal of the Aeronautical Sciences, Vol. 25, Feb. 1958, pp. 73-85.

[6]Park, C., "Calculation of Radiation from Argon Shock Layers," Journal Quantative Spectroscopy & Radiative Transfer, Vol. 28, Jan. 1982, pp. 29-40.

[7]Boison, J.C. and Curtiss, H.A., "An Experimental Investigation of Blunt Body Stagnation Point Velocity Gradient," ARS Journal, Vol. 29, Feb. 1959, pp. 130-135.

[8]Lees, L., "Laminar Heat Transfer Over Blunt-Nosed Bodies at Hypersonic Flight Speed," Jet Propulsion, Vol. 26, April 1956, pp. 259-269.

[9]Fivel, H.J., "Numerical Flow Field Program for Aerodynamic Heating Analysis," AFFDL-TR-79-3129, Sept. 1979.

Calculation of Nonequilibrium Radiation in the Flight Regimes of Aeroassisted Orbital Transfer Vehicles

Chul Park*
NASA Ames Research Center, Moffett Field, California

Abstract

A computer code has been developed that calculates radiative properties of nonequilibrium air in the low-density regimes expected during the flight of aeroassisted, orbital transfer vehicles. From the given nonequilibrium thermodynamic state variables, the code calculates number densities of internal states and the accompanying emission and absorption characteristics. In addition, the code calculates the number density of the hypothetical gas in radiative equilibrium that produces the same radiation emission as the given nonequilibrium gas. Sample results are shown to demonstrate how the code predicts the nonlinear variation of radiation with density at low densities.

Nomenclature

a_o = Bohr radius = 0.529×10^{-8} cm
A_{if}^{o} = radiative transition rate coefficient (transition probability) for transition from initial (upper) state i to final (lower) state f, s^{-1}
$A_{\infty f}$ = free-bound radiative recombination rate coefficient for transition into final state f, s^{-1}
AOTV = aeroassisted orbital transfer vehicle
B = rotational constant, cm^{-1}
B_{ij} = matrix element in quasisteady-state equation (8)
E = energy level, cm^{-1}

Presented as Paper 84-0306 at AIAA 22nd Aerospace Sciences Meeting, Reno, Nev., Jan. 9-12, 1984.

*Research Scientist. Computational Chemistry and Aerothermodynamics Branch.

E^* = threshold energy = $E_f - E_i$, cm^{-1}
f = emission intensity of a nonequilibrium gas divided by emission intensity of the same gas hypothetically in radiative equilibrium at temperature T_E
F = rotational energy level, cm^{-1}
g = statistical weight
G = vibrational energy level, cm^{-1}
J = rotational quantum number
k = Boltzmann constant
K_{if} = rate coefficient for collisional excitation from initial (lower) state i to final (upper) state f, cm^3/s
$K_{i\infty}$ = rate coefficient for collisional ionization or dissociation from state i, cm^3/s
K_o = K_{if} for transition from one initial rotational-vibrational electronic state to one final rotational-vibrational electronic state, cm^3/s
m_e = electron mass
m_h = heavy-particle mass
n_A = number density of atoms, cm^{-3}
n_c = number density of colliding particles, cm^{-3}
n_e = electron density, cm^{-3}
n_i = number density of ith excited state, cm^{-3}
OTV = orbital transfer vehicle
p_∞ = freestream pressure
$q_{v''v'}$ = Franck-Condon factor between upper state v'' and lower state v'
T = heavy-particle translational and rotational temperature, K
T_e = electron temperature, K
T_{ex} = effective electronic excitation temperature
T_v = vibrational temperature, K
T_E = equilibrium radiation temperature
U = electronic energy level, cm^{-1}
v = vibrational quantum number
V_∞ = flight velocity
x = distance along stagnation streamline
ν = predissocation rate coefficient, s^{-1}
σ = inelastic collision cross section, cm^2

Subscripts and Superscripts

E = equilibrium
i = ith internal state
' = lower state
" = upper state

Introduction

The projected design of aeroassisted orbital transfer vehicles (AOTV) has been evolving rapidly in recent years.[1,2] One such vehicle is an aerobraked orbital transfer vehicle (OTV), which decelerates in the atmosphere by using an aerobrake with a large frontal surface area. Such a vehicle is designed to achieve necessary deceleration in the low-density, high-altitude regime while producing relatively small heat-transfer rates to the aerobrake. Thus, reliable data on heat-transfer rates in low-density environments are necessary for use in the design of such a vehicle.

Convective heat-transfer rates, which can be calculated with confidence, provided the surface catalytic reactivity is known, are predicted to be of the order of 10 W/cm^2 or less.[3] Radiative heat-transfer rates cannot be determined as yet with confidence, although they are estimated to be of the order of 10 W/cm^2 or higher.[3,4] Laboratory research on radiative heat-transfer phenomena has been conducted for the density regimes higher than those anticipated for the AOTVs,[5] and for the body sizes smaller than the aerobrakes.[6] Direct extrapolation of the prototype flight data on radiation, obtained by Apollo 4 and 6 (Ref. 7) and Fire 1 and 2 flights,[8,9] is also difficult because of differences in the density and vehicle size.[4]

To facilitate determination of radiative heat-transfer rates in such an environment, a computer code has been developed recently at Ames Research Center that would approximately calculate radiative properties of air in such a low-density regime. This paper describes the code. In the shock layer over an aerobrake, the chemical state varies considerably along a streamline. Dissociation and ionization proceed with a finite rate, and the heavy-particle translational temperature T, vibrational temperature T_V, and electron temperature T_e vary from the widely different initial values to the common equilibrium value,[4] as shown schematically in Fig. 1. Rotational temperature is believed to be nearly the same as heavy-particle translational temperature. It is known that the strongest radiation emanates from a region within the nonequilibrium region.[5] In order to calculate radiation from the nonequilibrium region, one must first know the chemical state in the region, that is, the nonequilibrium species concentration and the three temperatures noted above. These state variables can be calculated by integrating chemical reaction rate equations simultaneously with mass, momentum, and energy conservation equations.[4] The present work assumes that such calculation is done in a different part of a code and that the state variables are available for use as an input to the present code.

The scope of the present code, named NEQAIR (nonequilibrium air radi tion), is shown schematically in Fig. 2. By solving the f ow conservation equations, the number densities of eac chemical species and the three tempera-

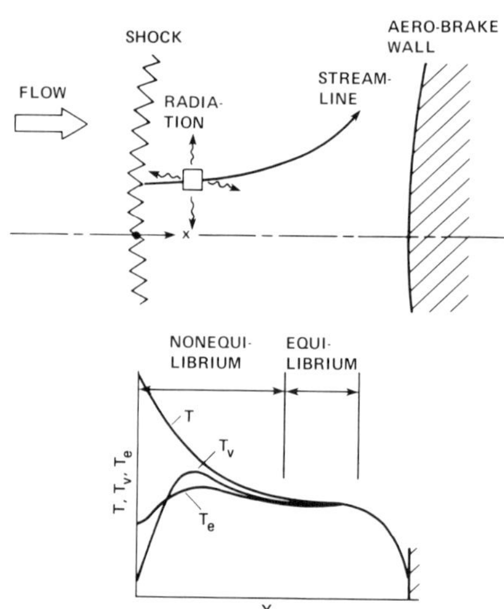

Fig. 1 Schematic of nonequilibrium radiation phenomenon in shock layers.

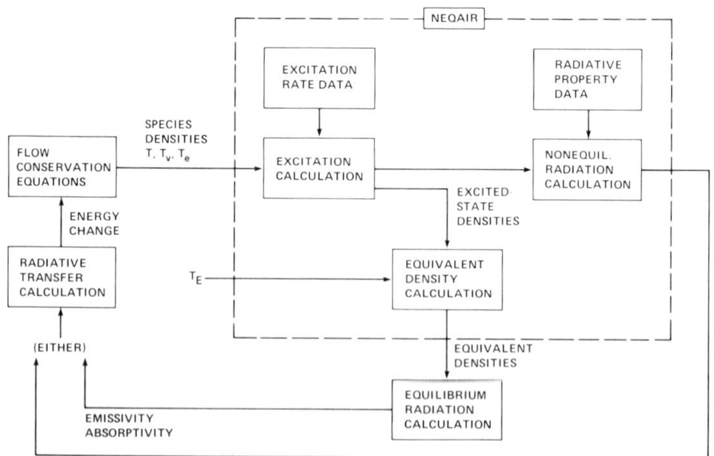

Fig. 2 The scope of the present nonequilibrium air radiation (NEQAIR) code.

tures (T, T_e, and T_V) are obtained outside of the present code. These values are given as inputs to the excitation calculation part of the NEQAIR code, which calculates the number densities of the excited electronic, vibrational, and rotational states of atoms and molecules. The excitation cross-section data necessary for the calculation are included in the NEQAIR package. The excited state number-density values obtained from the calculation are then fed into two different parts of the code. In the first part, the excited state density values are used in calculating radiative properties (i.e., emission and absorption coefficients) in detail, accounting fully for the nonequilibrium nature of the gas. The necessary radiative property data are also a part of the NEQAIR package. In the second option, the excited state densities are used in deriving the equilibrium equivalent species number densities. The equilibrium equivalent number densities are defined as the species number densities of a gas in radiative equilibrium (i.e, emission and absorption coefficients related by the Planck function definable by a unique temperature) that produces the same radiative emission power (but not the same absorptivity) as the given nonequilibrium gas. The temperature of the assumed equilibrium gas T_E can be chosen arbitrarily. The user of the present code has a choice of using either the nonequilibrium radiative properties or the equivalent equilibrium radiative properties in calculating radiative transfer in the given flowfield. The radiative transfer calculation will yield energy changes per unit time owing to the radiative transfer phenomenon, which must be incorporated into the flow conservation equations. In the following, the various components of the code are described.

Atomic Excitation

The spectra of radiation emitted by atoms and molecules in heated air are well known.[5,10] Atomic spectra consist of those from the line radiation of N and O, the free-bound recombination continua, and the free-free bremsstrahlung continua. The molecular radiation consists mostly of the first-negative system of N_2^+, the first-positive and the second-positive systems of N_2, the beta and gamma systems of NO, and the Shumann-Runge system of O_2. In addition, depending on the amount of CO_2, or carbonaceous ablation species, the violet and the red systems of CN may exist. Calculation of excited state populations is necessary only for the states that emit or absorb these radiating systems and the states that affect them. The molecular electronic states that need to be considered are listed in Table 1.

For the line radiation and free-bound continua of the atomic systems, the radiative properties are affected by the populations of all internal electronic states of N and O. The free-free continua of the atoms are unaffected. The electronic states of the atoms are excited and de-excited by collisional and radiative processes (see Fig. 3). A collision by an electron or a heavy particle (i.e., atom, molecule, or their ions) may cause a transition of electronic state of the target atom from an initial lower state i to a final upper state f, provided the relative collision energy exceeds the energy difference between the two states, that is, the threshold energy $E^* = E_f - E_i$. The rate of such a transition is expressed as[11]

$$\text{Collisional transition rate} = K_{if} n_c n_i \tag{1}$$

where n_c is the number density of the colliding species, n_i the number density of the initial state, and K_{if} is the excitation rate coefficient that, in itself, is an integration of the corresponding collision cross sections over the collision energy from the threshold value to infinity.

Of the two colliding particle types, that is, electrons and heavy particles, electrons are much more efficient in causing electronic transitions.[12-17] The cross sections for the collisions of neutral heavy particles are smaller, approximately by the ratio of the two masses, than those of electron collisions, because the amount of energy transmitted from one colliding particle to another is proportional to the mass ratio of the lighter and heavier particles. Experimental data for helium prove this relationship.[18] Collision frequency is proportional to the mean molecular velocity, which is inversely proportional to the square root of mass. As a result, the electron collisions are more efficient than the neutral particle collisions in excitation by the ratio[12-17]

$$\frac{K \text{ for electron}}{K \text{ for neutral particle}} = 0\left(\frac{m_h}{me}\right)^{1.5} \cong 5\times10^6 \tag{2a}$$

where m_h and m_e denote the mass of the neutral particle and mass of electron, respectively. Collisions by ions cause excitation with cross sections that are of the same

Table 1 Electronic states for which population is determined

Species	States	Energy level	Emission	Absorption
N_2^+	$X^2\Sigma_g^+$	0	...	...
	$A^2\Pi_u$	9,016	...	N_2^+ 1-
	$B^2\Sigma_u^+$	25,566	N_2^+ 1-	...
	$D^2\Pi_g$	51,663	...	...
N_2	$X^1\Sigma_g^+$	0	...	...
	$A^3\Sigma_u^+$	49,755	...	N_2 1+
	$B^3\Pi_g$	59,307	N_2 1+	N_2 2+
	$a^1\Pi_g$	68,951	...	...
	$C^3\Pi_u$	88,978	N_2 2+	...
NO	$X^2\Pi$	0	...	NO β,γ
	$A^2\Sigma^+$	43,966	NO γ	...
	$B^2\Pi_r$	45,932	NO β	...
O_2	$X^3\Sigma_g^-$	0	...	O_2 SR
	$a^1\Delta_g$	7,882	...	...
	$b^1\Sigma_g^+$	13,121	...	...
	$B^3\Sigma_u^-$	49,358	O_2 SR	...

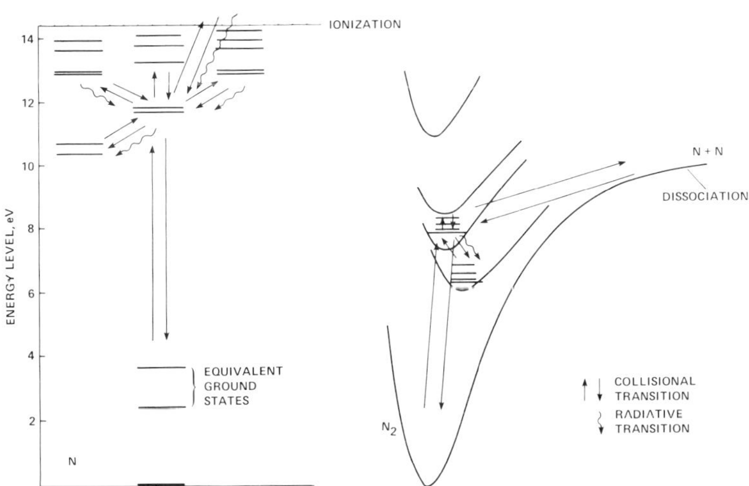

Fig. 3 Schematic of excitation scheme for the nitrogen atom and molecule.

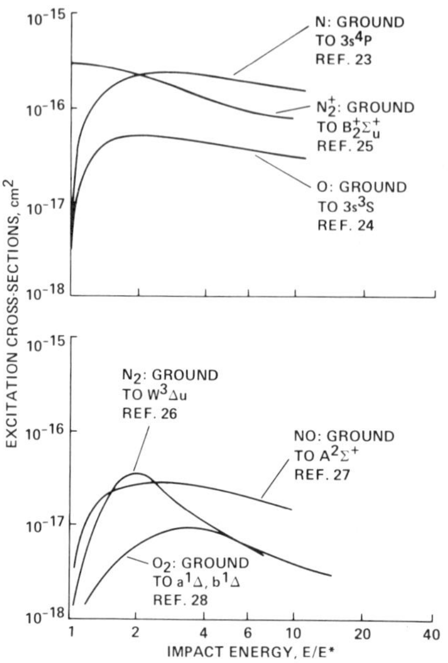

Fig. 4 Examples of measured electron impact excitation cross sections for air species.

order as those of electrons. For the ions, therefore, the ratio becomes

$$\frac{K \text{ for electron}}{K \text{ for ions}} = O\left(\frac{m_h}{m_e}\right)^{0.5} \cong 230 \tag{2b}$$

In the flight regime of the AOTVs, the degree of ionization is expected to be of the order of 1%. At such a level of ionization, electron collisions dominate in collisional excitation. For this reason, heavy-particle collisions are neglected in the collisional excitation of electronic states.

Numerous theoretical and experimental data exist on the collisional excitation and ionization of N, O, N_2, O_2, N_2^+, and NO by electrons.[19-28] Examples of experimental data are shown in Fig. 4, in which cross sections are plotted against collision energy E normalized by threshold value E^*. For N and O, the cross sections are of the order of 10^{-16} cm^2 at their peaks for the transitions shown. As a part of the present work, a complete set of electron impact excitation rate coefficients are generated for N and O by combining existing theories and experimental data. The electronic levels of N and O are grouped into 22 and 19 levels, respectively, for this purpose.

The method used is similar to those used previously.[29-31] The details of this aspect of the work will be published in a separate paper. In brief, the method relies on 1) existing theories, well founded, for the transitions among the highly excited states and among the three low-lying states (see Fig. 3); 2) experimental data for the transitions from the ground state and highly excited states; and 3) for the transitions between an equivalent ground state (see Fig. 3) and an excited state for N (for O, these transitions are forbidden), the assumption that the cross section is the same as that for the transition between the ground state and the excited state (work is in progress to calculate the cross sections from the first principles). The rate coefficients are expressed in the form

$$K_{if} = a\, T^{b} \exp(-E^{*}/kT_{e})$$

where k is the Boltzmann constant. The constants a and b are tabulated for the matrix 22×22 and 19×19 for N and O, respectively. The rate coefficients so determined are believed to be accurate to within a factor of 2 for all transitions, except for those between the equivalent ground states and highly excited states for N, for which the uncertainty is a factor of about 3. The smallest of these rate coefficients is of the order of 10^{-8} cm^3/s for the typical flight regime of AOTVs.

The rates of de-excitation from higher excited states to lower states are related to the excitation rates by the principle of detailed balance. Collisional ionization rate coefficients $K_{j\infty}$ are well known,[32] especially for the highly excited states which affect population distribution most; in the highly excited states, all atoms behave as hydrogen in collisional ionization for which reliable theories and experimental data exist.[19,20] Radiative transition rates include transition probabilities of spontaneous line emission A_{if} and free-bound recombination rate $A_{\infty i}$, both of which are well known.[33,34]

The rate of change of the number density of an electronic state n_i is dictated by the equation[11]

$$\frac{Dn_i}{Dt} = \text{chemical production rate} - \text{chemical removal rate} + \text{diffusion rate}$$

where

$$\text{Chemical production rate} = \sum_{j=1}^{\infty} (K_{ji} n_e + A_{ji}) n_j \qquad (3)$$

$$\text{Chemical removal rate} = \sum_{j=1}^{\infty} (K_{ij} n_e + A_{ij}) n_i \qquad (4)$$

In a shock layer over an AOTV, the scale of change of properties is dictated by the shock standoff distance, which is of the order of 100 cm, and the average flow velocity in the direction toward the wall, which is of the order of 10^5 cm/s. Therefore, the time scale of change of n_i in Dn_i/Dt and in the diffusion rate is of the order of

$$\text{Time scale of change} = 100/10^5 = 10^{-3} \text{ s} \qquad (5)$$

The time scale of change of the two chemical rates is of the order of reciprocal of $K_{if}n_e$. Taking the lowest possible n_e value to be 10^{12} cm^{-3} and K_{if} to be 10^{-8} cm^3/s, the reciprocal is of the order of 10^{-4} s. Thus, the two chemical rates are usually much larger in magnitude than either Dn_i/Dt or the diffusion rate. As a result, the two opposing chemical rates, Eqs. (3) and (4), are approximately equal in magnitude. Expressing this relation, one obtains the so-called quasisteady-state relation,

$$\sum_j (K_{ji} n_e + A_{ji}) n_j - \sum_j (K_{ij} n_e + A_{ij}) n_i = 0 \qquad (6)$$

Equation (6), together with the requirement that the sum of all n_i's equals the given number density of the atom n_A,

$$n_A = \sum_i n_i \qquad (7)$$

forms a set of linear equations for n_i's of the form

$$(B_{ij}) n_i = c n_A + d n_e \qquad (8)$$

where c and d are functions of electron temperature only. The elements of the matrix B_{ij} consist generally of a linear combination of $K_{ij}n_e$ and A_{if}. The solution for Eq. (8) is obtained by the algebraic operation[29,31,35]

$$n_i = (B_{ij})^{-1} (c n_A + d n_e) \qquad (9)$$

A sample solution is presented in Fig. 5 in which the number density n_i divided by statistical weight g_i is plotted on the abscissa against the energy level on the ordinate. The thermodynamic conditions chosen here are a part of the benchmark nonequilibrium conditions given in Table 2, which will be explained later, and correspond to a pressure of about 2×10^{-4} atm. The dotted line in the figure indicates an equilibrium distribution expected at 12,000 K, the given electron temperature. As seen here, the population distribution deviates only slightly from the equilibrium distribution despite the low-pressure environment. This is because the collisional excitation and de-excitation rates are dominant over radiative transition rates. Under this situation, population distribution is insensitive to changes in excitation rate coefficients, or to any possible errors in them.

Molecular Excitation

Since the vibrational temperature T_v and rotational-translational temperature T are assumed to be given, the molecular excitation calculation is limited to that of the electronic states. To calculate electronic excitation for a molecule, the rate coefficient K_{if} must be found for

Table 2 Benchmark state variables used in testing NEQAIR code

	Equilibrium	Nonequilibrium
Translational-rotational temperature T, K	9650	18,000
Vibrational temperature T_v, K	9650	14,000
Electron temperature T_e, K	9650	12,000
Total number density	1.494×10^{17}	2.53×10^{13} -8.01×10^{16}
Molar fractions		
N	0.7305	0.4813
N^+	0.02943	0.03435
O	0.1990	0.1788
O^+	5.315×10^{-3}	6.204×10^{-3}
e^-	3.481×10^{-2}	4.269×10^{-2}
N_2^+	1.767×10^{-5}	6.188×10^{-4}
N_2	9.339×10^{-4}	0.2143
NO	2.598×10^{-5}	0.03032
O_2	3.998×10^{-7}	0.1160
Pressure, atm	0.196	6.21×10^{-5} -0.196
Enthalpy, J/g	≅50,000	≅50,000
Flight velocities, km/s	10	10

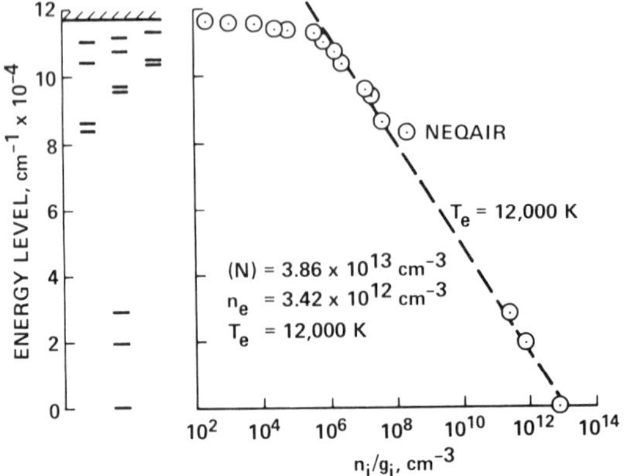

Fig. 5 Typical excited state population distribution for N in nonequilibrium.

the transitions from an electronic state to another that includes all vibrational and rotational combinations. By the definition of a collisional excitation rate coefficient, Eq. (1), the rate coefficient in this case is a sum over all final upper vibrational states v" and rotational states J" and an average over all initial lower vibrational states v' and rotational states j'; that is,

$$K_{if} = \frac{\sum_{v'} \exp\left(-\frac{G'}{kT_v}\right) \sum_{J'} (2J' + 1)\exp\left(-\frac{F'}{kT}\right) \sum_{v''}\sum_{J''} (2J' + 1)K_o}{\sum_{v'} \exp\left(-\frac{G'}{kT_v}\right) \sum_{J'} (2J' + 1) \exp\left(-\frac{F'}{kT}\right)} \tag{10}$$

where G and F are vibrational and rotational energy levels, respectively, and K_o is the rate coefficient for one rotational state-to-rotational state transition. Equation (10) is simplified by the following principles:

1) Rotational selection rule: Electrons are ineffective in causing rotational quantum number changes. As a result, the rotational quantum numbers J' and J" are within 2 of each other.

2) Franck-Condon principle: The cross section K_o can be considered to be a product of Franck-Condon factor and an electronic part, which is nearly independent of either the vibrational or the rotational quantum numbers.

Using these principles, and approximating

$$2J' + 1 \cong 2J'' + 1 \cong 2J' \cong 2J''$$

and replacing summation by an integration, one can reduce Eq. (10) to

$$K_{if} = \frac{\sum_{v''}\sum_{v'} q_{v'v''} \exp\left(-\frac{G'}{kT_v}\right) S}{\sum_{v'} \exp\left(-\frac{G'}{kT_v}\right)\left(\frac{kT}{B'}\right)}$$

$$S \equiv 5.47\times10^{-11} \sqrt{T_e} \exp(-C) \int_0^{\infty} (2J')^2 \exp\left[-\left(\frac{B'' - B'}{kT_e} + \frac{B'}{kT}\right) J'^2\right]\left(D + \beta \frac{B'' - B'}{kT_e} J'^2 \, dJ'\right)$$

$$D \equiv C\beta + \alpha \qquad C \equiv \frac{U'' - U' + G'' - G'}{kT_e} \tag{11}$$

$$\alpha = \int_0^{\infty} \left(\frac{\sigma}{\pi a_0^2}\right) \xi \exp(-\xi) \, d\xi \qquad \beta \equiv \int_0^{\infty} \left(\frac{\sigma}{\pi a_0^2}\right) \exp(-\xi) \, d\xi$$

$$\xi = \frac{(E - E^*)}{kT_e}$$

Here, U is the electronic term energy; B the rotational constant; $q_{v'v''}$ the Franck-Condon factor; σ the electronic part of the excitation cross section; and a_0 is the Bohr radius, $a_0 = 0.529\times10^{-8}$ cm. Since σ is a function mostly of E/E^* and is independent of E^*, α and β remain approximately the same for all vibrational-rotational combinations in a molecular electronic excitation for a given T, T_v, and T_e.

The excitation cross sections have been measured for the air molecules for $T = T_v = 300$ K for almost all transitions in which the initial state is the electronic ground state.[21-28] Examples of such cross sections are given in Fig. 4. As seen here, the cross sections are of the order of 10^{-17} cm^2, except for N_2^+, which has larger cross sections. With the use of Eq. (11), one can extend the room temperature data to arbitrary values of T, T_V, and T_e. For the transitions in which the initial state is not the ground state, and, hence, the cross sections are unknown, the cross sections are assumed to be the same as those known transitions with the same final state. CN is not considered in the present work because nothing is known of its excitation cross sections.

For each electronic state of a molecule, dissociation can occur through collisions. For this collisional dissoci ation process, electrons are less efficient than heavy particles because the mass parity principle [see Eq. (2)] is in favor of heavy particles. Hence, only the heavy-particle collisions are considered for this process. The existing dissociation rate data[36] are utilized in deducing the rate coefficients for collisional dissociation from various electronic states.

The $C^3\Pi_u$ state of N_2, from which the N_2 second-positive radiation emanates, and the $B^3\Sigma_u^-$ state of O_2, from which the O_2 Schumann-Runge radiation originates, need special treatment. They lie above the dissociation limit. Because they are so high, collisional excitation from the lower electronic states is unlikely. Instead, they are likely to be populated by the collisions between the two atoms through a process known as inverse predissociation. Depopulation is by the reverse process of predissociation and by emission of radiation. On applying the principle of detailed balance between the predissociation rate and the inverse predissociation rate, one obtains, for the population of the two states in question,

$$n_i = n_{iE} \frac{\nu}{\nu + A_{if}}$$

where n_{iE} is the equilibrium number density of the two states corresponding to the number density of atoms, and ν is the predissociation rate. The predissociation rate is deduced from the inverse predissociation rate, which is calculated by multiplying the elastic collision frequency by the Boltzmann factor based on the energy difference between these states and the atomic states. The procedure implicitly assumes that predissociation probability is unity. The elastic collision frequency is obtained by assuming the collision radius to be 1 Bohr.

To determine the population of the rest of the electronic states (see Table 1), the quasisteady-state equation (9) is solved. The transition probability A_{if} values for the molecules are obtained from the literature.[10] A sample solution is presented in Fig. 6. The thermodynamic conditions of this solution are the same benchmark conditions given in Table 2 for a pressure of 2×10^{-4} atm, as they were in the solution shown in Fig. 5. In Fig. 6, the vibrational energy levels are spaced deliberately sparse for the sake of clarity. As the figure shows, all rotational states belonging to the same vibrational state lie on straight lines with a common slope corresponding to the given rotational temperature T = 18,000 K. The roots of the rotational distribution (J = 0) of different vibrational quantum numbers lie on a straight line with the slope of the given vibrational temperature T_V = 14,000 K; however, the vibrational-rotational root (J = 0, v = 0) of different electronic states does not lie on any straight line. Instead, the slopes of the lines joining these latter roots range from 7300 K to 9231 K, a result of nonequilibrium excitation. The populations of the $B^3\Pi_g$ and $C^3\Pi_u$ states are depressed by the radiative depopulation accompanied by the emission of the first-positive and second-positive bands.

Radiation Calculation

With the populations of the internal states of atoms and molecules so determined, the radiation calculation is straightforward. Except for the free-free continuum, all radiation mechanisms are affected by the nonequilibrium

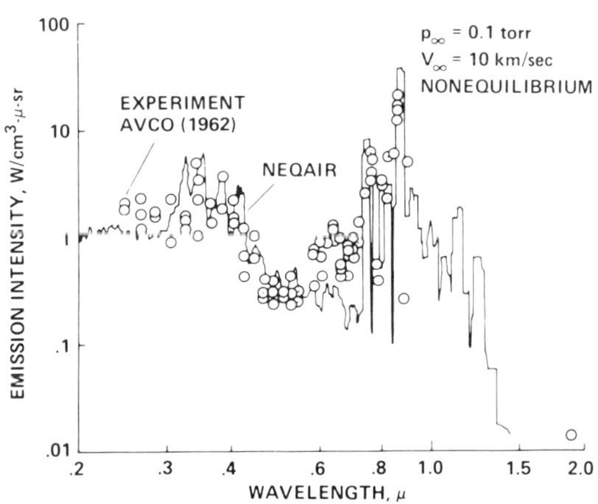

Fig. 6 Partial representation of excited state population distribution for N_2 in nonequilibrium.

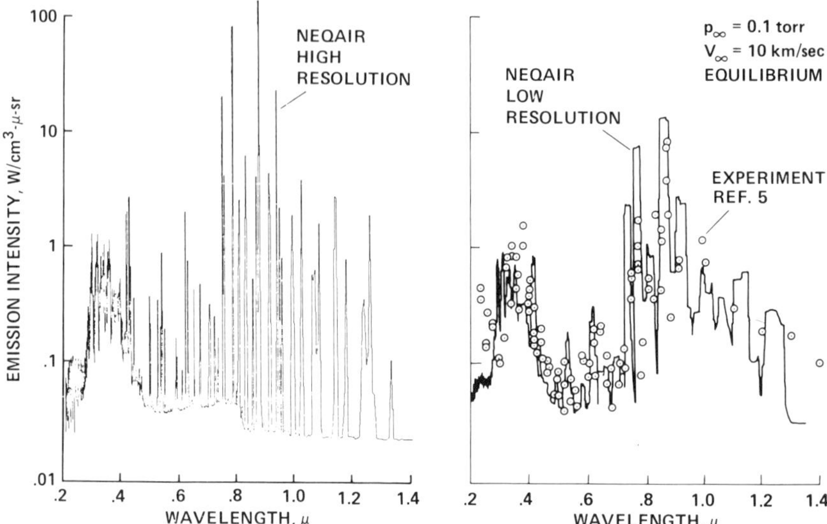

Fig. 7 Comparison of a NEQAIR result and experimental data for an equilibrium region behind a shock wave; species mole fractions and temperature are given in Table 2.

population distribution. Emission and absorption coefficients are not related to each other by a Planck temperature in the nonequilibrium environment. Instead, they are independent summations of all contributions from each emitting and absorbing mechanism; the latter mechanisms are related to each other by a Planck temperature definable only between them. These contributions to emission and absorption coefficients can be calculated according to the following: The emission coefficient contribution is calculated using the number density of the emitting state and the associated emission strength factor; the absorption coefficient contribution is calculated from the number density of the absorbing state and the associated absorption strength factor. This requires only a minor modification of an existing line-by-line radiation code[37] in which contributions from various radiating mechanisms are summed one at a time.

Examples of radiation calculation are shown in Figs. 7-9. In Fig. 7, emission intensity is shown for the wavelength range of 0.2 to 1.4 μm for a thermodynamic condition produced in a shock tube at a preshock pressure of 0.1 Torr and shock speed of 10 km/s. Experimental data are available for this case for both the equilibrium region far behind the shock wave and for the nonequilibrium zone between the shock wave and the equilibrium region.[5] Thermodynamic conditions calculated for the equilibrium region are listed in Table 2. Figure 7 shows both high-resolution and low-resolution computed spectra and compares them with experimental data that had a low spectral resolution.[5] There is a reasonable agreement between the calculated and the experimental data. A small discrepancy between theory and experiment can be attributed partly to the deviation of the experimental condition from true equilibrium. At the pressure tested, radiative depopulation of electronic states of molecules is fast enough to cause deviation from a true equilibrium composition. Concentrations of N_2, O_2, and NO in the experiment were probably many times those calculated assuming equilibrium. There is close agreement between theory and experiment for the integrated radiation power for this equilibrium case. They agree closely also with the prediction in Ref. 38.

In Fig. 8, experimental spectra[5] at the peak of nonequilibrium radiation are compared with the calculated spectra. Thermodynamic conditions of the nonequilibrium flow are estimated from the known general trend in finite-rate air chemistry[4] and are listed in Table 2. Agreement is generally good except for the region around 0.6 μm, where calculation underpredicts by a factor of 3. This can be attributed to the chemiluminescence of NO in the presence of atomic oxygen, which is not considered in the present work. More work is needed to resolve this discrepancy. The integrated radiation power is approximately

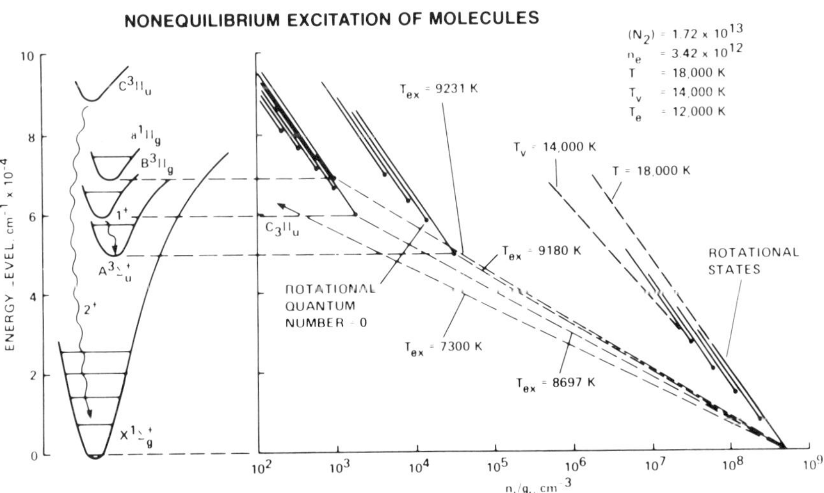

Fig. 8 Comparison of a NEQAIR result and experimental data for a nonequilibrium region behind a shock wave; species mole fractions and temperatures are given in Table 2.

80% of the measured value for this nonequilibrium case. About 60% of the radiation is due to atomic radiation.

In Fig. 9, global radiation power emission under the nonequilibrium conditions is plotted against the postshock total number density. The three temperatures and species mole fractions are fixed, as given in Table 2. Power emitted above 0.25 and 0.35 μm differs only slightly. The emitted powers follow a straight line representing a linear, that is, binary, scaling relationship between density and radiation, down to the density of about $3\times10^{-14}cm^{-3}$. Below this density, emitted power decays faster than the decrease in density. This can be attributed to the fact that particle collisions are not sufficiently frequent to maintain the population of the excited states against the radiative removal rate, which is a phenomenon referred to previously as the collision-limiting phenomenon.[7-9] Referring to the altitude scale at the top of Fig. 9, the collision-limiting phenomenon is seen to be prominent at altitudes above about 100 km for this case. It should be noted that this is purely a benchmark calculation. In a real situation, the three tempera-

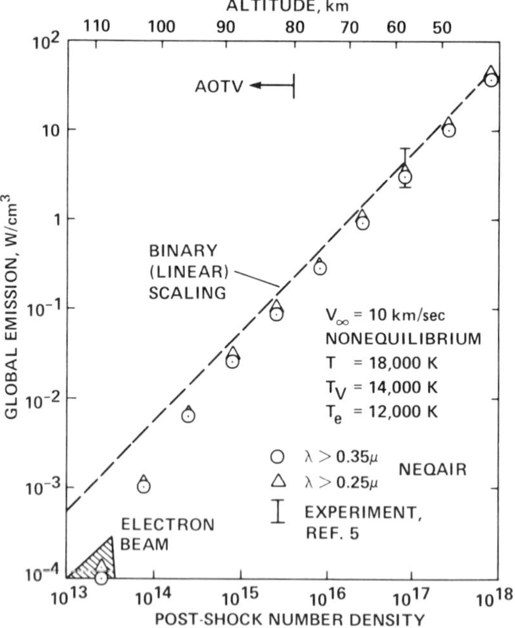

Fig. 9 Departure of global emitted power from binary-scaling (linear) relationship owing to the collision-limiting phenomenon; species mole fractions are assumed to be fixed, as given in Table 2.

tures and species mole fractions in the nonequilibrium zone will vary as altitude changes. Both the point of onset of collision-limiting and absolute radiation intensity at that point may be substantially different from what is shown in Fig. 9.

Equilibrium Equivalent Densities

Radiative properties of a gas can be calculated with methods that are simpler than the present method if the gas is in radiative equilibrium, that is, if there exists a single Planck temperature. Here, an attempt is made to find the hypothetical density of a gas in radiative equilibrium that would produce the same emissive power as the given nonequilibrium gas. Since emission and absorption in a nonequilibrium gas are not related by any unique Planck temperature, the absorptive property of the nonequilibrium gas will not be reproduced by the hypothetical equilibrium gas simultaneously with the emissive property. The number density of such equilibrium gas, which will be defined as equilibrium equivalent density, is dependent on the radiating mechanisms and, hence, will be different among various radiating mechanisms. The equilibrium equivalent densities for the various emitting mechanisms are determined in the NEQAIR code through the following procedures:

1) Calculating the nonequilibrium emitted power for each radiating mechanism for the given nonequilibrium condition using the method described above.

2) Calculating the equilibrium emitted power for each radiating mechanism for the same species densities but at the specified equilibrium radiation temperature T_E.

3) Dividing the nonequilibrium emitted power by the equilibrium emitted power to obtain the ratio, termed here equilibrium equivalent density factor f, by which the density must be multiplied in order that the equilibrium power equals the nonequilibrium power.

4) Multiplying the gas density by the equivalent density factor f.

Steps 1 and 2 are much simpler than the line-by-line calculation procedure described in the previous paragraphs, because line profiles are not generated. Moreover, since only the ratios are calculated, the absolute emission strength factors can be taken to be unity for this calculation. For this reason, the emission strength parameters, such as transition probabilities and transition moments, are not needed.

Figure 10 shows an example of this calculation. Here, the equivalent density factors f determined in step 3 above are plotted against the total number density for selected radiating mechanisms. The species mole fractions and T, T_v, and T_e are fixed, as listed in Table 2, as they were in Fig. 9, and the equilibrium radiation temperature T_E is assumed to be 9650 K. At high densities, the density factors are large because the populations of the internal states approach the equilibrium values at the given electron temperature of 12,000 K. As the total number density decreases, the density factors decrease because of radiative depopulation of the excited states; that is, the collision-limiting phenomenon. The decrease is most rapid with the O_2 Schumann-Runge system, and the slowest with the N_2^+ first-negative system. This is because radiative depopulation rate is large for the O_2 Schumann-Runge system, and collisional excitation rates are large for N_2^+ molecule (see Fig. 4). At densities above 10^{14} cm^{-3}, most of the radiating mechanisms have a density factor value greater than 1, meaning that the total emitted power is probably greater than the equilibrium radiation at 9650 K. The decrease in the density factor values at densities below 10^{14} cm^{-3} corroborates the departure from the binary-scaling relationship seen in Fig. 9.

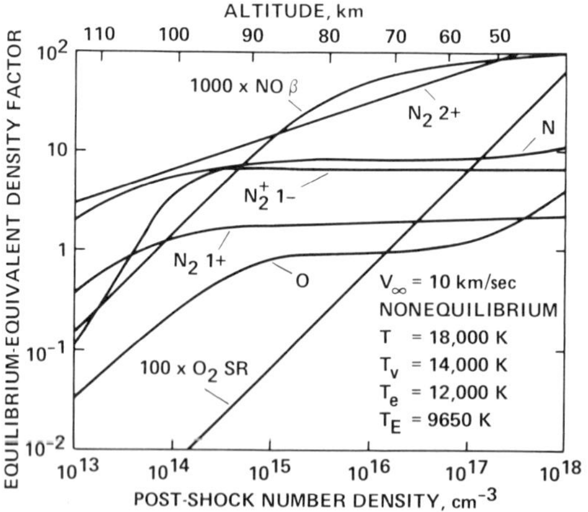

Fig. 10 Equivalent density factors for various radiating mechanisms; species mole fractions are assumed to be fixed, as given in Table 2.

Conclusions

A computer code, named NEQAIR (nonequilibrium air radiation), is developed that calculates radiative properties from nonequilibrium air. The code requires heavy-particle temperature, vibrational temperature, electron temperature, and species mole fractions as inputs. The code calculates number densities of internal states of atoms and molecules using the quasisteady-state assumption. A detailed line-by-line calculation is performed to calculate emission and absorption characteristics independently, using the calculated, nonequilibrium, excited state density values. The code also calculates the number densities of the hypothetical gas in radiative equilibrium that produces the same radiation emission as the given nonequilibrium gas. All needed cross-section and radiative property data are included in the package. Sample results are shown which demonstrate how the code predicts the departure of radiative power emission from the binary scaling relationship at low densities.

Acknowledgments

The author wishes to express his sincere thanks to his colleague S. R. Langhoff for his assistance in computing Franck-Condon factors, and to G. E. Hahne for his help in computing continuum intensities.

References

[1]Walberg, G. D., "A Review of Aeroassisted Orbit Transfer," AIAA Paper 82-1378, AIAA 9th Atmospheric Flight Mechanics Conference, San Diego, Calif., Aug. 1982.

[2]Howe, J. T., "Introductory Aerothermodynamics of Advanced Space Transportation Systems," AIAA Paper 83-0406, AIAA 21st Aerospace Sciences Meeting, Reno, Nev., Jan. 1983.

[3]Menees, G. P., "Thermal-Protection Requirements for Near-Earth Aeroassisted Orbital-Transfer Vehicle Missions," AIAA Paper 83-1513, AIAA 18th Thermophysics Conference, Montreal, Canada, June 1983.

[4]Park, C., "Radiation Enhancement by Nonequilibrium in Earth's Atmosphere," AIAA Paper 83-0410, AIAA 21st Aerospace Sciences Meeting, Reno, Nevada, Jan. 1983.

[5]Allen, R. A., Rose, P. H., and Camm, J. C., "Nonequilibrium and Equilibrium Radiation at Super-Satellite Re-Entry Velocities," AVCO-Everett Research Laboratory, Everett, Mass., Research Report 156, Sept. 1962.

[6]Page, W. A. and Arnold, J. O., "Shock Layer Radiation of Blunt Bodies at Entry Velocities," NASA TR R-193, 1964.

[7]Lee, D. B. and Goodrich, W. D., "The Aerothermodynamic Environment of the Apollo Command Module During Super-Orbital Entry," NASA TN D-6792, 1972.

[8]Cauchon, D. L., "Project Fire Flight 1 Radiative Heating Experiment," NASA TM X-1222, 1966.

[9]Cauchon, D. L., "Radiative Heating Results From the Fire 2 Flight Experiment at a Reentry Velocity of 11.4 Kilometers Per Second," NASA TM X-1402, 1967.

[10]Allen, R. A., "Air Radiation Tables: Spectral Distribution Functions for Molecular Band Systems," AVCO-Everett Research Laboratory, Everett, Mass., Research Report 236, April 1966.

[11]Hansen, C. F., "Rate Processes in Gas Phase," NASA RP 1090, May 1983.

[12]Drawin, H. W., "Influence of Atom-Atom Collisions on the Collisional-Radiative Ionization and Recombination Coefficients of Hydrogen Plasmas," Zeitschrift fur Physik, Vol. 225, No. 5, July 1969, pp. 483-493.

[13]Drawin, H. W., "Influence of Atom-Atom Collisions on the Population Densities of Excited Atomic Levels," Zeitschrift fur Naturforschung, Vol. 25A, No. 7, July 1969, pp. 145-147.

[14]Milford, S. N., "Inelastic Atomic Collisions in Plasmas," AIAA Paper 64-54, AIAA 1st Aerospace Sciences Meeting, New York, New York, Jan. 1964.

[15]Gryzinski, M., "Two-Particle Collisions: I. General Relations for Collisions in the Laboratory System," Physical Review, Vol. 138, No. 2A, April 1965, pp. 305-321.

[16]Gryzinski, M., "Two-Particle Collisions: II. Coulomb Collisions in the Laboratory System of Coordinates," Physical Review, Vol. 138, No. 2A, April 1965, pp. 322-335.

[17]Gryzinski, M., "Classical Theory of Atomic Collisions: I. Theory of Inelastic Collisions," Physical Review, Vol. 138, No. 2A, April 1965, pp. 336-358.

[18]Utterback, N. G., "Ionization Cross Sections for Neutral Helium Atoms on Helium, Neon and Nitrogen," Proceedings of the Third International Conference on the Physics of Electronic and Atomic Collisions, North Holland Press, Amsterdam, 1963.

[19]Moiseiwitsch, B. L., "Electron Impact Excitation of Atoms," Review of Modern Physics, Vol. 40, No. 2, April 1968, pp. 238-353.

[20]Gilmore, F. R., Bauer, E., and McGowan, J. W., "A Review of Atomic and Molecular Excitation Mechanisms in Nonequilibrium Gases

Up to 20,000°K," Journal of Quantitative Spectroscopy and Radiative Transfer, Vol. 9, No. 1, Jan. 1969, pp. 157-183.

[21]Ali, A. W., "Excitation and Ionization Cross-Sections for Electron Beam and Microwave Energy Deposition in Air," Naval Research Laboratory, Washington, D.C., Memorandum Report 4598, Aug. 1981.

[22]Slinker, S. and Ali, A. W., "Electron Excitation and Ionization Rate Coefficients for N_2, O_2, NO, N and O," Naval Research Laboratory, Washington, D.C., Memorandum Report 4756, Feb. 1982.

[23]Stone, E. J. and Zipf, E. C., "Excitation of Atomic Nitrogen by Electron Impact," Journal of Chemical Physics, Vol. 58, No. 10, May 1973, pp. 4278-4284.

[24]Stone, E. J. and Zipf, E. C., "Electron-Impact Excitation of the ^3S and ^5S States of Atomic Oxygen," Journal of Chemical Physics, Vol. 60, No. 11, June 1974, pp. 4237-4243.

[25]Crandall, D. H., Kauppila, W. E., Phaneuf, R. A., Taylor, P. O., and Dunn, G. H., "Absolute Cross Sections for Electron-Impact Excitation of N_2^+," Physical Review, Series A, Vol. 9, No. 6, June 1974, pp. 2545-2551.

[26]Cartwright, D. C., Trajmar, S., Chutjian, A., and Williams, W., "Electron Impact Excitation of the Electronic States of N_2: II. Integral Cross-Sections at Incident Energies from 10 to 50 ev," Physical Review, Series A, Vol. 16, No. 3, Sept. 1977, pp. 1041-1051.

[27]Imami, M. and Borst, W. L., "Electron Impact Excitation of the Gamma Bands of Nitric Oxide," Journal of Chemical Physics, Vol. 63, No. 8, Oct. 1975, pp. 3602-3605.

[28]Wakiya, K., "Differential and Integral Cross Sections for the Electron Impact Excitation of O_2: II. Optically Forbidden Transitions from the Ground State," Journal of Physics, Series B: Atomic and Molecular Physics, Vol. 11, No. 22, Nov. 1978, pp. 3931-3938.

[29]Park, C., "Spectral Line Intensities in a Nonequilibrium Nitrogen Plasma," Journal of Quantitative Spectroscopy and Radiative Transfer, Vol. 8, No. 10, Oct. 1968, pp. 1633-1653.

[30]Park, C., "Electron Impact Excitation Rate Coefficients for Hydrogen, Helium and Alkali Atoms," Journal of Quantitative Spectroscopy and Radiative Transfer, Vol. 11, No. 1, Jan. 1971, pp. 7-36.

[31]Park, C., "Comparison of Electron and Electronic Temperatures in Recombining Nozzle Flow of Ionized Nitrogen-Hydrogen Mixture, Part 1, Theory," Journal of Plasma Physics, Vol. 9, Pt. 2, July 1973, pp. 187-215.

[32]Lotz, W., "Electron-Impact Ionization Cross-Sections and Ionization Rate Coefficients for Atoms and Ions from Hydrogen to

Calcium," Zeitschrift fur Physik, Vol. 216, No. 3, Oct. 1968, pp. 241-247.

[33]Wiese, W. L., Smith, M. W., and Glennon, B. M., "Atomic Transition Probabilities, Vol. I. Hydrogen Through Neon," National Bureau of Standards, Washington, D.C., National Standard Reference Data Series — National Bureau of Standards 4, May 1966.

[34]Peach, G., "Continuous Absorption Coefficients for Non-Hydrogenic Atoms," Memoirs of the Royal Astronomical Society, Vol. 73, Pt. 1, 1970, pp. 1-123.

[35]Park, C., "Calculation of Radiative Properties of Nonequilibrium Hydrogen Plasma," Journal of Quantitative Spectroscopy and Radiative Transfer, Vol. 22, No. 1, July 1979, pp. 101-112.

[36]Park, C. and Menees, G. P., "Odd Nitrogen Production by Meteoroids," Journal of Geophysical Research, Vol. 83, No. 8, Aug. 1978, pp. 4029-4035.

[37]Arnold, J. O., Cooper, D. M., Park, C., and Prakash, S. G., "Line-by-Line Transport Calculations for Jupiter Entry Probes," Progress in Astronautics and Aeronautics, Vol. 69, Entry Heating and Thermal Protection, edited by W. G. Olstad, AIAA, New York, 1980, pp. 52-82.

[38]Nardone, M. C., Breene, R. G., Zeldin, S. S., and Riethof, T. R., "Radiance of Species in High Temperature Air," General Electric Co., Missile and Space Division, Philadelphia, Pa., R63SD3, June 1963.

Air Radiation Revisited

Kenneth Sutton*
NASA Langley Research Center, Hampton, Virginia

Abstract

The experimental data for air radiation from ground-based experiments and from the Fire II and Apollo 4 flights have been reviewed and compared with a single prediction method. The ground-based data agree generally with the trends predicted by the present calculations, but differences of a factor of 2 are noted. The flight data from Fire II and Apollo 4 are in good agreement with the present calculations, with differences of less than 50%.

Nomenclature

D	= diameter, m
h	= Planck constant = 4.14×10^{-15} eV-s
L	= gas path length, m
P	= pressure, atm
P_1	= initial shock tube pressure, Torr
q_c	= cold-wall, convective heating rate, W/m^2
R_e	= equivalent hemisphere radius, m
R_n	= hemisphere nose radius, m
V_∞	= freestream velocity, m/s
ρ_∞	= freestream density, kg/m^3
ν	= frequency, s^{-1}

Presented as Paper 84-1733 at AIAA 19th Thermophysics Conference, Snowmass, Colorado, June 25-28, 1984.

*Assistant Head, Aerothermodynamics Branch, Space Systems Division.

Introduction

There is a renewed interest in the radiative heating to a vehicle entering the Earth's atmosphere at high velocities. This interest is due to the potential use of aeroassisted orbit transfer vehicles (AOTV's) to transfer payloads between high and low Earth orbits.[1,2] Upon return from a high Earth orbit (geosynchronous in most cases), an AOTV would dip in the Earth's atmosphere to use aerodynamic drag for velocity depletion before rendezvous with the Shuttle Orbiter or possibly a space station. An AOTV will enter the atmosphere at a high velocity, 10-11 km/s, and will require a large frontal area for sufficient drag to meet the required velocity depletion. Both high velocities and large frontal area promote increased radiative heating.

Most of the studies on air radiation occurred in the 1960's in preparation for the Apollo mission. During this era, both ground-based and flight experiments were being conducted, and theoretical radiation models were being developed. This large research effort was prompted by early predictions that showed radiative heating to be a major problem for the Apollo vehicle during re-entry. In recent years, little attention has been devoted to air radiation during re-entry because radiative heating was not dominant for any Earth mission; but, significant advances were made in radiative flowfields over planetary entry vehicles in support of the Pioneer Venus probes and the Galileo probe (Jupiter entry) for which radiation is important.

With the interest in an AOTV as a future mission for Earth application, now is a good time to revisit the air radiation data. The present paper presents the results of a study to compare the ground-based and flight experimental data with a single theoretical method and assess our capability to predict air radiation for a future mission. The primary interest is the radiation for air in chemical equilibrium. Some recent papers[2-4] have discussed the importance of air radiation for finite-rate (nonequilibrium) chemistry in an AOTV application because the vehicle will decelerate at a high altitude. However, the demarcation on the effects of air radiation between equilibrium and nonequilibrium chemistry is not clearly defined; our understanding of radiation with nonequilibrium chemistry is not as advanced as for equilibrium chemistry; and basic information on the properties of air radiation can be learned from an equilibrium analysis. Thus, an evaluation of the experimental data for the radiation from air in chemical equilibrium provides a good foundation to begin our development of engineering methods to predict the radiation to an AOTV. The present paper is the first to provide a collection of the large body of experimental data, both ground-based and flight, for air radiation and to compare these data with a single prediction method.

Flowfield Solution

The present predictions of the radiative heating rates or intensities to the test models or flight vehicles are made with a radiatively coupled solution of the inviscid flow equations at the stagnation point of a hemisphere by the method of Ref. 5. For those configurations that are not hemispheres, hemispherical radii are selected that give the equivalent stagnation-point shock standoff distances for an adiabatic solution. Basically, this was done by equating the stagnation-point velocity gradient of the configuration to that of an equivalent hemisphere by the method of Ref. 6. The radiation model coupled to the flowfield solution is that of Nicolet[7,8] and includes the atomic line, continuum, and molecular band transitions. The radiation frequency is expressed in terms of photon energy, $h\nu$, in electron-volt units. Most of the experimental data were measured only over an interval of the total frequency range of air due to the required use of windows in the instrumentation systems. The present calculations are made for the entire frequency range (0-16eV) to correctly account for radiation cooling, and then the results for the required frequency interval are extracted for comparison with the experimental data. For comparisons with the experimental data that measured the radiation across a gas layer rather than in front of a test model, calculations were made for a constant property layer using only the radiation code.[7,8] All calculations were made for the gas in chemical equilibrium.

Ground-Based Experimental Data

Comparisons have been made with the experimental data for air radiation given in Refs. 9-20. These data can be grouped in four categories: 1) measurements of the radiation behind the shock of an instrumented model in a shock tube,[9-13] 2) measurements of the radiation behind the shock of a model launched by a light-gas gun using external instrumentation,[14] 3) measurements of the radiation behind the reflected shock in a shock tube with the instrumentation mounted on the tube wall,[15-17] and 4) measurements of the radiation across a constricted arc.[18-20] Most of the data were obtained behind quartz or sapphire windows at photon energies less than 7 eV. Limited data for the radiation over the entire frequency range for air were obtained with windowless systems. For the window data, the effects of window transmission were included in the data reduction; thus, the data denotes the radiation in front of the window. An exception is the data from Refs. 16 and 17. The transmission characteristics, as given in Ref. 17, had to be included in the present calculations.

The ranges of temperature and pressure representative of the test conditions for the various experiments are given in Fig. 1. Note there were no duplications of test conditions between the various experiments except for the constricted arc data of Krey and Morris[18] and Schreiber et al.[19-20] Additionally, there are differences in model configurations or measurement locations. Thus, a direct comparison among the experimental data cannot be made. The ground-based data are at higher pressures and temperatures than the flight data of Fire II and Apollo 4.

Shock Tube Data

Nerem and Stickford[9] measured the radiation to the stagnation point of hemispherical models with radii of

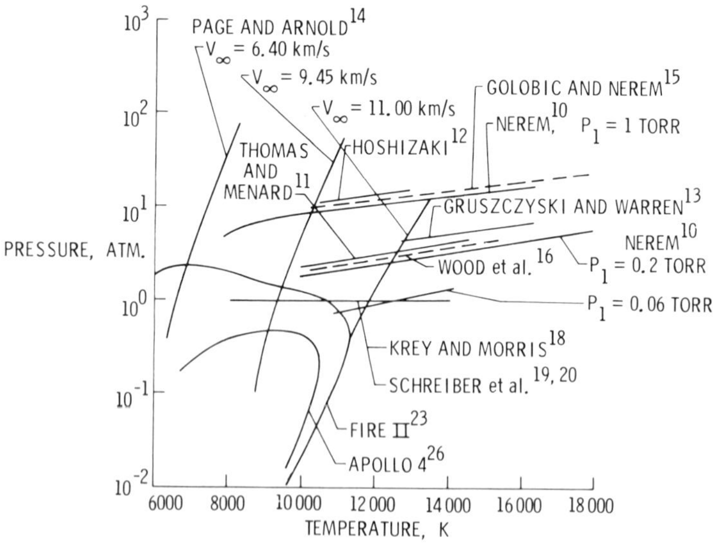

Fig. 1 Comparison of thermodynamic conditions for experimental data.

2.54 and 1.27 cm in a shock tube for a range of incident shock velocities at initial pressures of 1, 0.2, and 0.06 Torr. The data were measured behind a sapphire window. For the present study, these data were taken from the tabulations given in the report by Nerem.[10] Because of scatter in the data, Nerem averaged the results clustered around velocity points, and these averaged data are those for which comparisons are generally presented in the literature. Comparisons between the present calculations and Nerem's averaged data are presented in Fig. 2. There is good agreement between the calculations and the data

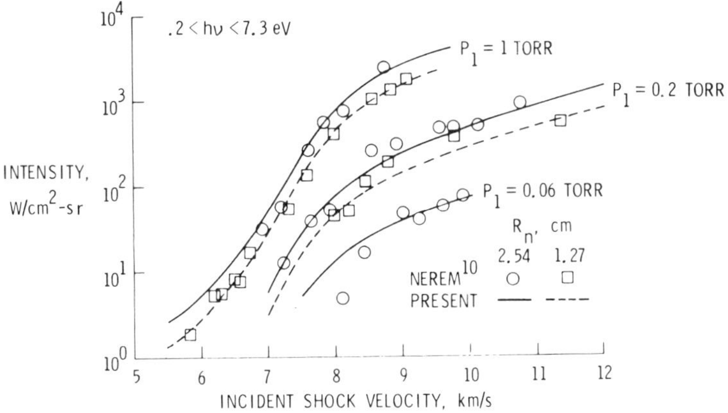

Fig. 2 Comparisons with shock tube data of Nerem (Ref.10).

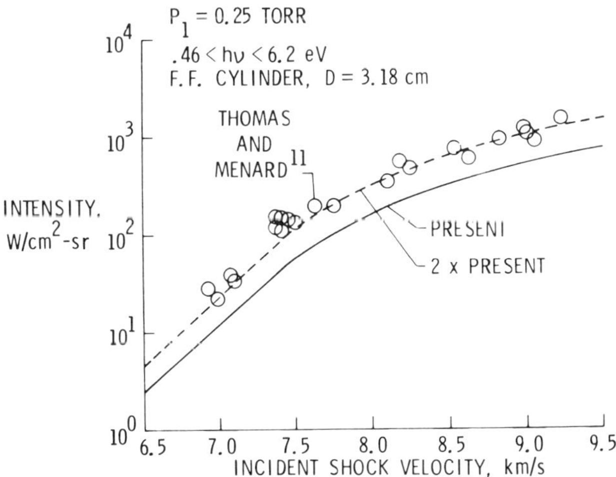

Fig. 3 Comparisons with shock tube data of Thomas and Menard (Ref. 11).

in both magnitude and trend. For the 0.06-Torr comparison, the data deviate from the present calculations at the lower velocities. However, Nerem and Stickford[9] state that their data are questionable for this pressure because of partial nonequilibrium chemistry effects. Comparisons were made with all of Nerem's data and also with the data presented in the form of heating rate rather than intensity. In general, the results presented in Fig. 2 are representative of the comparisons between the present calculations and Nerem's data.

Thomas and Menard[11] measured the radiation to the stagnation point of a 3.18-cm-diam flat-faced cylinder (R_e = 5.58 cm) in a shock tube for a range of incident shock velocities at an initial pressure of 0.25 Torr. The data were measured behind a Pyrex window. Comparisons between the present calculations and these data are presented in Fig. 3. The trend of the data agree with the present calculations. However, the values of the experimental data are approximately a factor of 2 greater than the present calculations, as shown by the dashed curve in Fig. 3. Thomas and Menard[11] present results and discuss the importance of the N^- continuum. Their calculated results go through the data points. There is an option in the radiation code[7] used in the present calculations for including the N^- continuum. However, the present author does not normally include the N^- continuum, because of controversy over the magnitude for the cross section. Present calculations that did include the N^- continuum give results that are approximately 25% greater than the calculated results shown in Fig. 3.

Hoshizaki[12] measured the radiation to the stagnation point of a 2.87-cm-diam flat-faced cylinder in a shock tube over a narrow range of incident shock velocities at an initial pressure of 1.16 Torr. The data were measured behind a quartz window. Calculations were made for equivalent hemispherical radii of 5.04 and 7.56 cm. A comparison of these results with Hoshizaki's data is presented in Fig. 4. The smaller radius is that for the present method of defining an equivalent radius; whereas, the larger radius was selected to agree with the shock-layer thickness as given by Hoshizaki. The difference between the two calculations is 40%. The present calculations agree with the trend of the experimental data with the results for the two calculations bracketing the data at the highest velocities. Hoshizaki's shock-layer thickness is approximately 30% greater than the thicknesses given by Thomas and Menard[11] and Gruszczynski and Warren[13] for equivalent diameters of a flat-faced cylinder.

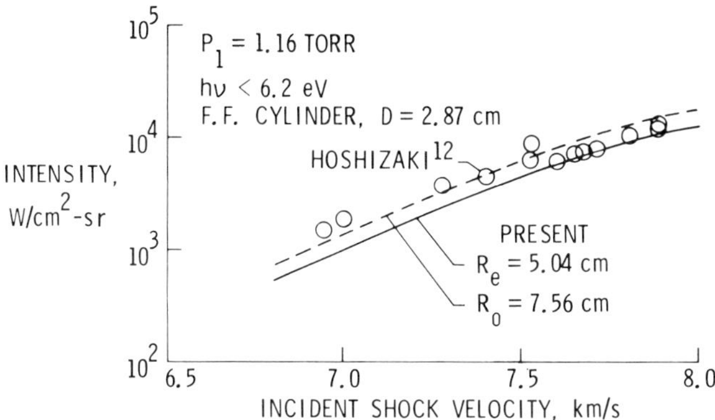

Fig. 4 Comparisons with shock tube data of Hoshizaki (Ref. 12).

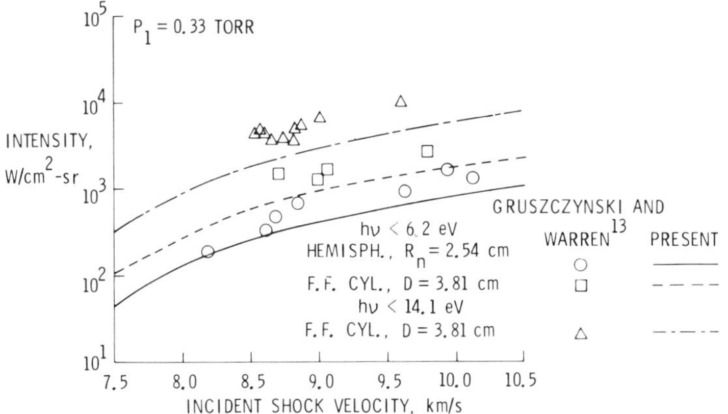

Fig. 5 Comparisons with shock tube data of Gruszczynski and Warren (Ref. 13).

Gruszczynski and Warren[13] measured the radiation to the stagnation point of a 3.81-cm-diam flat-faced cylinder (R_e = 6.70 cm), and a 2.54-cm-radius hemisphere in a shock tube at an initial pressure of 0.33 Torr for a range of incident shock velocities. Data were measured behind a quartz window for both models; and, also, a windowless technique was used in the flat-faced cylinder. Comparisons between the present calculations and the data are presented in Fig. 5. As shown by the comparison, the experimental data are roughly 50% greater than the present calculations. The differences between the sets of data do agree with the differences predicted by the present calculations. For example, the increase in radiation for the total spectrum over that at photon energies less than 6.2 eV (see flat-faced cylinder results), as shown by the experimental data, is approximately the same as that calculated.

A comparison between all the experimental data taken in shock tubes with instrumented models using windows and the present calculations is presented in Fig. 6. The calculations are for a constant property layer of 0.5 cm at 5 atm for the spectral interval of 0.3 to 6.2 eV. These selected conditions lie in the middle of the conditions for the different experiments. The experimental data were translated to these conditions based on the present calculated trends in path length and pressure. As shown, these data are bounded by the data of Nerem[10] and Thomas and Menard.[11] The data agree with the general trend of the calculations, with the data of Nerem[10] being in best agreement.

Light-Gas Gun Data

Page and Arnold[14] launched models in a light-gas gun facility over a range of freestream velocities and densities. Radiometers viewed the models at right angles to the flight path through slits in the facility walls. The data from these observations were reduced to represent the radiative heating rates to the stagnation point of the models. The models were made of various materials. However, only the data for models made of aluminum and polyethylene are considered in the present study because of increased ablation effects for the other materials.[14] The models were truncated hemispheres with nose radii of 0.51 and 1.56 cm (R_e = 0.49 and 1.25 cm). Page and

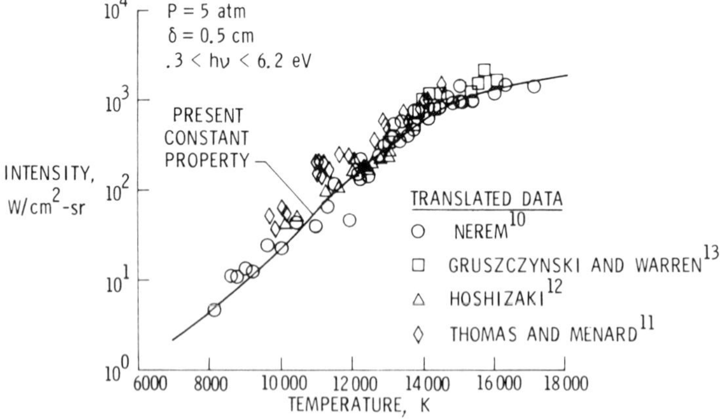

Fig. 6 Comparison of the present constant-property results with the translated shock tube data.

Arnold reduced the data for the various velocities to three specified velocities. Comparisons of these data with the present calculations are presented in Fig. 7. There is scatter in the experimental data, but the data do cluster about the present calculations at the higher freestream densities, where the shock-layer air was in chemical equilibrium. At the lower densities, the shock-layer air was in chemical nonequilibrium, and the present calculations would not be expected to agree with the data.

Reflected Shock Data

Golobic and Nerem[15] measured the radiative heating rate to the end wall of a shock tube behind a reflected shock at an initial pressure of 1 Torr for a range of incident shock velocities. The measurements were made behind a sapphire window. The results were presented for a time of 10 μs after shock reflections. The present calculations were made for constant-property layers at the temperatures and pressures behind the reflected shock. The layer thicknesses were calculated from the time given (10 μs) and the reflected shock velocities. The thickness ranged from 0.9 to 1.4 cm, increasing with incident shock velocity. Comparisons between the present calculations and the data of Golobic and Nerem[15] are presented in Fig. 8. There is considerable scatter in

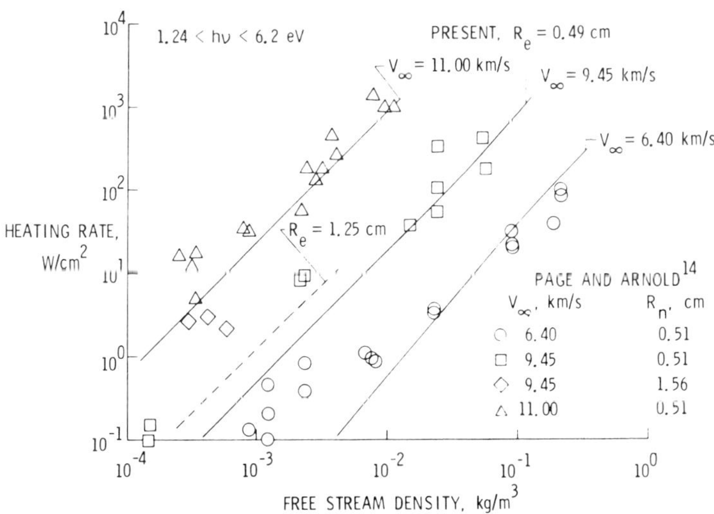

Fig. 7 Comparisons with light-gas gun data of Page and Arnold (Ref. 14).

the experimental data, with the results from the present, constant-property calculations (solid curve) lying at the upper edge of the data. Golobic and Nerem,[15] using a theoretical calculation for the reflected shock flow and a two-step absorption coefficient model, showed the importance of radiative cooling on the radiation behind the reflected shock; however, their calculated results did not go through the bulk of their experimental data. Knott et al.[21] repeated the calculations using a five-step absorption coefficient model, although for nitrogen only, and showed good agreement with the data. In the present study, two methods were used to simulate the effects of radiative cooling for comparison with the data of Golobic and Nerem. In method I, the present flowfield solution was used with hemispherical radii selected to give shock-layer thickness equal to the layer thickness behind the reflected shock at 10 μs. In method II, the constant-property results were reduced by the same ratio as given by Knott et al. The results from these two methods are presented in Fig. 7 by the dashed curves. The present, simulated-cooling results are in better agreement with the data of Golobic and Nerem than the constant-property results.

Wood et al.[16] measured the radiation across the end of a shock tube behind a reflected shock with a plate arrangement to obtain gas lengths of 3, 6, 8, and 12 cm. The measurements were made at an initial pressure of 0.2 Torr for a range of incident shock velocities. Both the

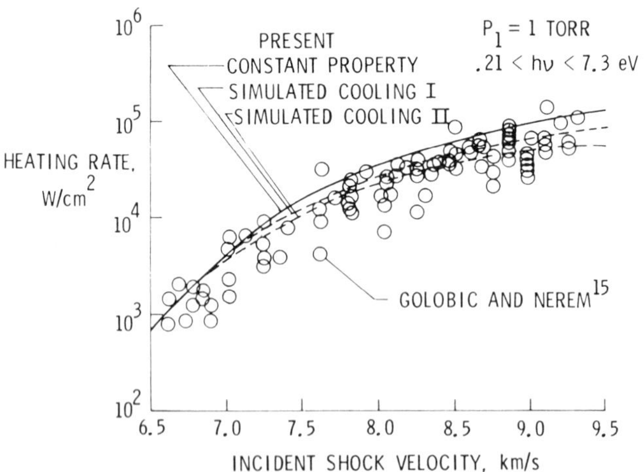

Fig. 8 Comparisons with reflected shock data of Golobic and Nerem (Ref. 15).

total radiation (windowless gage) and the radiation behind a quartz window were measured. The present calculations were made for a constant-property layer for the conditions behind the reflected shock. Wood et al. scaled their results, based on theoretical calculations, to account for the various path lengths where the intensity data for the quartz window were scaled to the 0.8 power of path length and the windowless data to the 0.6 power. The results of the present calculations for a path length of 8 cm are compared with the data of Wood et al. in Figs. 9 and 10 for the quartz window and windowless gage, respectively. The present results scaled for the other path lengths agree with the 8-cm results within 10% and are not presented. The experimental data for the quartz window, Fig. 9, are in general agreement with the present calculations at the lower velocities but lie approximately 30% lower at the highest velocities. The windowless data, Fig. 10, are approximately 30% to 50% lower than the present calculations over the entire velocity range. Also presented in Fig. 10 are present results neglecting line radiation at photon energies greater than 3.71 eV. The line radiation in the vacuum ultraviolet region is easily absorbed, and this comparison was made in case the wall boundary layer or the windowless cavity absorbed some radiation during the test.

Wood et al.[16] did not identify the data according to path length; however, in an earlier paper,[17] the data for

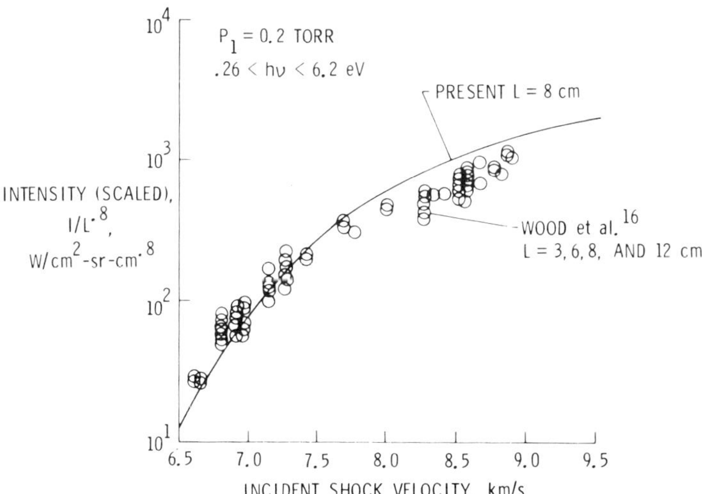

Fig. 9 Comparison with the reflected shock data of Wood et al. (Ref. 16) for a quartz window.

the 3- and 8-cm path lengths were identified. Present results are compared with these data in Figs. 11 and 12 for the 3- and 8-cm data, respectively. At the lower velocities, the experimental data for the quartz window and windowless gage agree and lie within the present results. There is even a trend for the data to come together at the highest velocities.

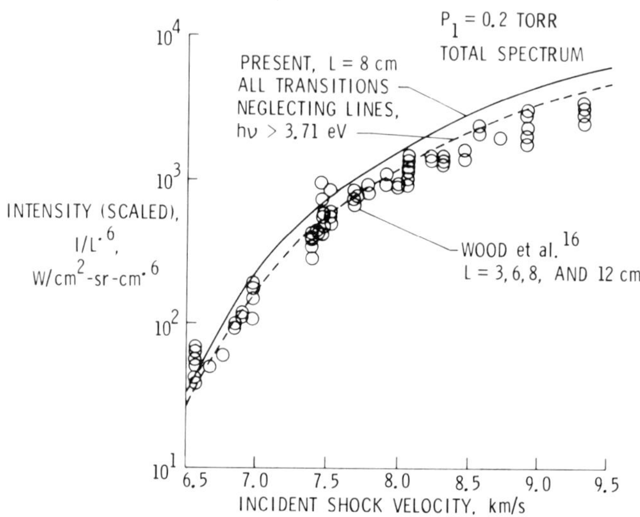

Fig. 10 Comparisons with reflected shock data of Wood et al. (Ref. 16) for a windowless gage.

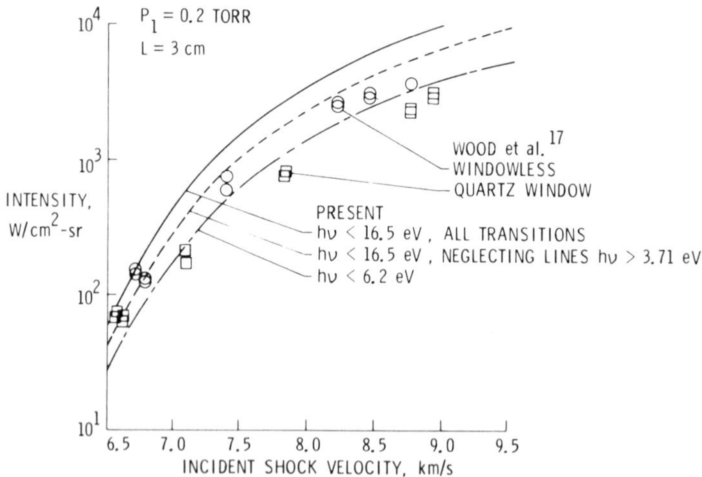

Fig. 11 Comparisons with reflected shock data, L = 3 cm, of Wood et al. (Ref. 17).

Constricted Arc Data

Krey and Morris[18] and Schreiber et al.[19,20] measured the radiation across a constricted arc at a pressure of 1 atm. Krey and Morris[18] measured the radiation for nitrogen and oxygen and Schreiber et al. measured the radiation for nitrogen[19] and air[20] behind quartz windows. Schreiber et al.[19] also presented data for the total radiation for nitrogen. The data were presented in the form of emission coefficients. The present results were calculated with the radiation code by dividing the intensity by the gas path length in which the path length was reduced until the gas was optically thin over the entire frequency range of interest. The calculated emission coefficients were required to be constant for at least a factor of 10 difference in path length. The present results are compared with the data for air and nitrogen behind quartz windows in Fig. 13. The present results for air and nitrogen were essentially the same and are shown by a single curve. As shown in Fig. 13, the present results are in excellent agreement with the data at the higher temperatures and lie between the data of Krey and Morris and Schreiber et al. at the lower temperatures. The present results for oxygen are compared with the data of Krey and Morris in Fig. 14. As shown, the experimental data are slightly lower than the present calculations.

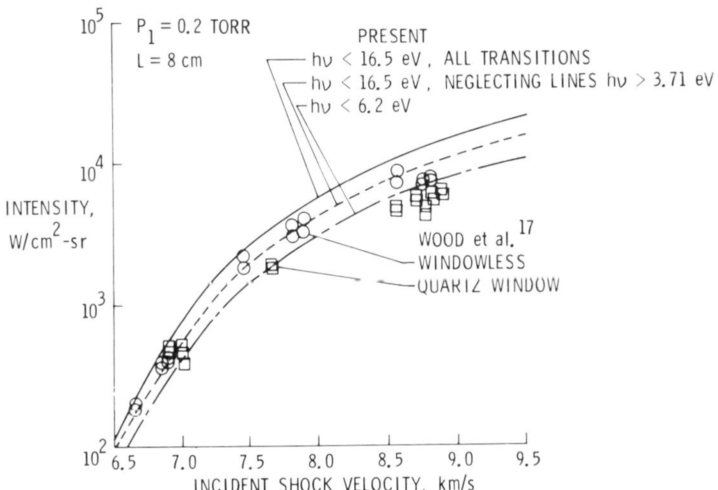

Fig. 12 Comparisons with reflected shock data, L = 8 cm, of Wood et al. (Ref. 17).

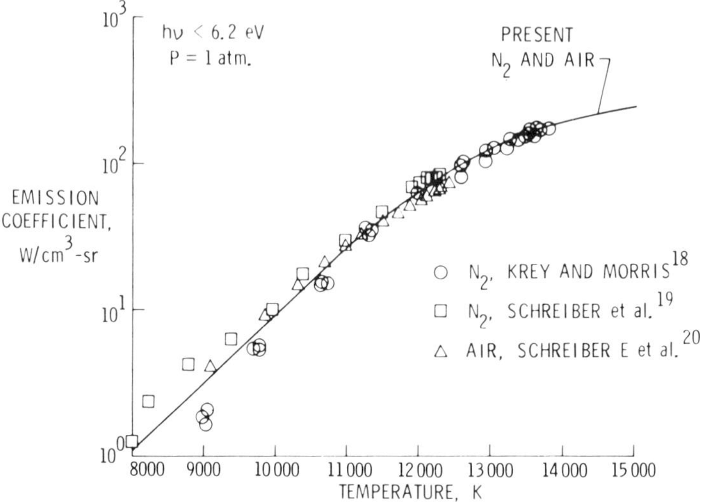

Fig. 13 Comparison with constricted arc data for air and nitrogen behind a quartz window.

Schreiber et al.[19] presented data for the total spectrum of nitrogen as measured by an optical technique and a Hall probe. The results from the present calculations for the total spectrum, except neglecting line radiation at photon energies greater than 3.71 eV, are compared with the experimental data in Fig. 15. The present results lie between the data from the two different measurement techniques. Present results that include high-frequency line radiation are a factor of 10 greater than that shown by the curve in Fig. 15. The data reduction by Schreiber et al. is based on the radiation being optically thin, and they discuss that absorption of the line radiation in the vacuum uv spectra could have impacted their results.[19] The present calculations show the gas to be optically thin up to path length of 0.02 cm for photon energies below 6.2 eV and up to 0.0005 cm for the total spectrum neglecting line radiation greater than 3.71 eV; but, a path length down to 1×10^{-6} cm was required for the lines located at photon energies greater than 3.71 eV.

Flight Experimental Data

Flight experimental data[22-26] have been obtained for the intensity of air radiation by project Fire (Fire I and Fire II) and during the testing phase of the Apollo

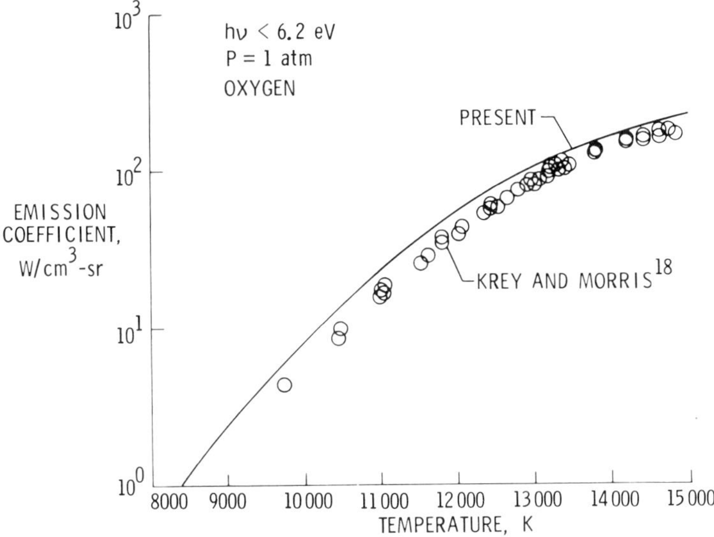

Fig. 14 Comparison with constricted arc data for oxygen behind a quartz window.

program (Apollo 4 and Apollo 6). The present predictions are compared only with the data from Fire II and Apollo 4 because unexpected body motions occurred for Fire I, and a malfunction occurred in the radiometer for Apollo 6. These flight data were measured behind quartz windows and included radiation up to 6.2 eV.

Fire II Data

The Fire II vehicle had a layered heatshield composed of three beryllium layers, each backed by phenolic asbestos. Heatshields 1 and 2 were ejected during the entry to expose a clean surface for the next data period. Radiometers were located at several locations, but the present analysis considers only the centerline location. The vehicle entered at a zero angle of attack; thus, the centerline location was the flow stagnation point. Two types of radiometers were used. A total radiometer (thermopile) measured the intensity in the 0.2-6.2 eV range (limited by the quartz window). A spectral radiometer measured the spectral radiation, which was then integrated to provide the total intensity in the 2-4 eV interval. The forebody configuration was a truncated hemispherical shape with a small corner radius. The dimensions of the heatshields were smaller from heatshield 1 to 3. For the present study, equivalent hemispherical nose radii of 0.75, 0.66, and 0.60 m were used for heatshields 1, 2, and 3, respectively.

Comparisons of the present results with the Fire II data[23] throughout the entry are presented in Fig. 16 for both the 0.2-6.2 and 2-4 eV intervals. The Fire II data represented by the solid curve, 0.2-6.2 eV, are the data from the prime data periods for each heatshield; that is, data obtained at times when the outer-surface temperature of the quartz window was calculated to still be below the melting point. The prime data periods for heatshields 2 and 3 were less than 0.5 s. The dashed curve is the data taken outside the prime data period. The prime data periods also apply to the data for the 2-4 eV interval. The error band for the experimental data was estimated to be $\pm 20\%$ (Ref. 23). An explanation of the anomalies seen in the data is given in Ref. 23 in terms of melting of the beryllium layers, exposure of the phenolic asbestos, and ejection of the heatshields.

The present calculations are in good agreement with the Fire II data for both spectral intervals throughout the entry, as shown by the comparisons in Fig. 16. For the 0.2-6.2 eV interval, the present calculations agree almost exactly with the data during the third prime data period and are 35% lower during the second prime data period. However, the present calculations assume the air

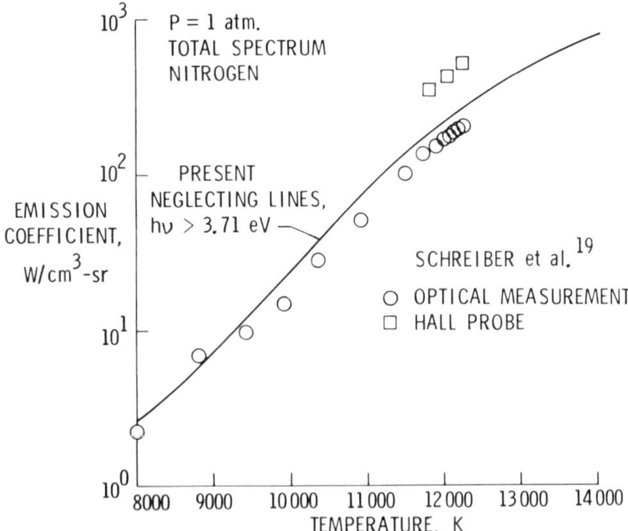

Fig. 15 Comparison with the constricted arc data for the total spectrum of nitrogen.

in the shock layer to be in chemical equilibrium. For entry times up to approximately 1637 s, the Fire II vehicle was at a high altitude, and the shocked air was most likely in chemical nonequilibrium. In fact, a flowfield calculation made in the present study at 1632 s showed the air in the shock layer to be entirely in nonequilibrium. Thus, the good agreement before approximately 1637 s is probably fortuitous, and the present results are presented as a point of reference. After 1637 s, the shock-layer air during the flight should be close to chemical equilibrium and the comparisons presented are valid.

Additional data on air radiation can be inferred from data obtained on the Fire II vehicle using the beryllium layers as total calorimeters that measured the total heating rate; that is, the sum of the convective heating rate and the absorbed radiative heating rate.[24] In the present study, the absorbed radiative heating was calculated with the flowfield solution using the spectral dependency of the absorptance for beryllium as given in Ref. 24. The cold-wall convective heating rates were calculated by

$$q_c \sqrt{R_e} = 1.76 \times 10^{-4} \quad \rho_\infty^{1/2} \; V_\infty^3 \tag{1}$$

where the heat-transfer coefficient is based on Ref. 27. Hot-wall corrections were made to the cold-wall values, although the corrections were less than 5%.

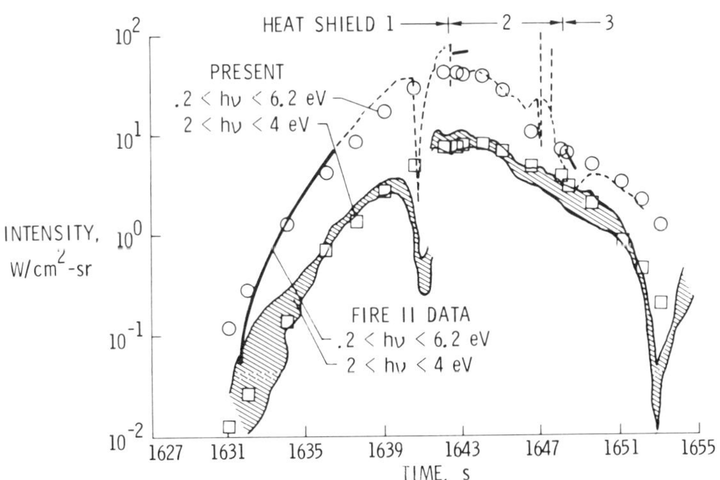

Fig. 16 Comparisons with the radiation data from Fire II (Ref. 23).

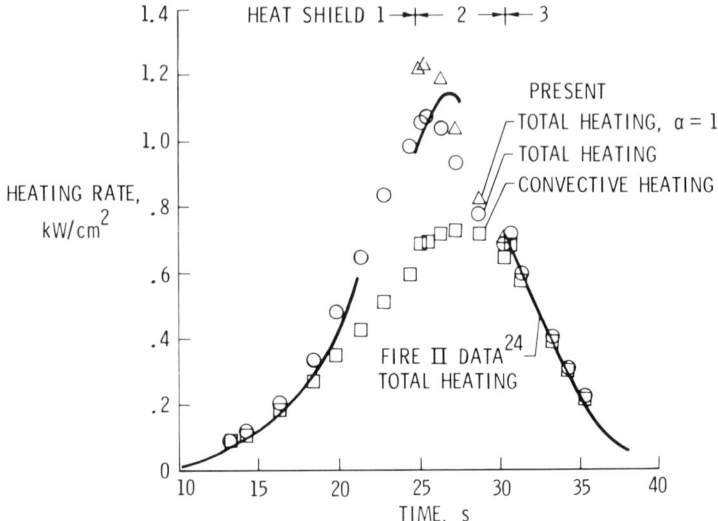

Fig. 17 Comparisons with the total heating data from Fire II (Ref. 24).

The results from the present calculations are compared with the total heating rate data from Fire II (Ref. 24) in Fig. 17. (Note, the time scale shown is the elapsed time from an altitude of 121.92 km, and 1617.75 s must be added to be equal to the flight time presented in Fig. 16.) The present calculations for the convective heating and the total heating are presented throughout the entry. Also, the present calculations, assuming the beryllium absorptance to be a constant value of 1 are presented for the time period of the second heatshield. The present results are in excellent agreement with the data for heatshields 1 and 3. However, the present results show that these times are dominated by convective heating with some significant radiation beginning at 20 s for heatshield 1. The present convective heating rates are in good agreement with the calculated values for Fire II presented in Ref. 24, which were based on the method by Cohen.[28] For the times of peak heating, heatshield 2, the present results are in good agreement with the Fire II data. However, the present maximum value occurs at an earlier time and then decreases earlier than shown by the flight data. The present maximum value was 1070 W/cm^2 as compared to 1140 W/cm^2 for the flight data. The estimated accuracy for the flight data is $\pm$50 W/cm^2 at the peak value.[24] The present calculations assuming a constant value, α = 1, for the absorptance of beryllium were made to determine if the differences between the calculations and data could be explained on the basis of a change in absorptance as the melt temperature was approached. As shown by the results in Fig. 16, these calculations still show an earlier fall-off in the calculated results after peak heating than the data. At peak heating, the present calculations show that 35% of the total heating is due to absorbed radiation. Also, 73% of the absorbed radiation comes from the spectra at photon energies greater than 6.2 eV. While the measured radiation data from Fire II, $h\nu < 6.2$ eV, provided a means to compare theoretical results at the lower end of the spectrum, the total heating data provide some results to infer a comparison with data for the vacuum uv region.

Apollo 4 Data

The Apollo 4 data were measured with a radiometer located off the centerline at a $s/r_n = 0.73$ location. However, the vehicle entered at an angle of attack of 24 deg, and, as discussed in Ref. 26, the radiometer would be located near the flow stagnation point. The present calculations were made for a 3-m nose radius, which provides an equivalent shock standoff distance for the radiometer location.[26] A comparison of the present calculations with the Apollo 4 data[26] is presented in Fig. 18. The present predictions are in excellent agreement with the flight data for the major portion of the radiative heat pulse. The shocked flow should have been in chemical equilibrium except possibly for the very

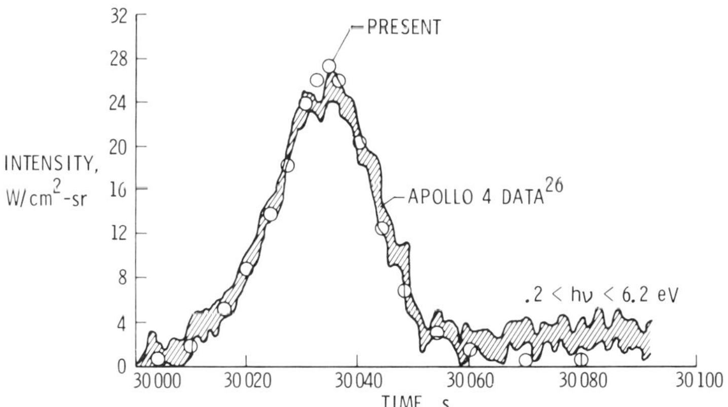

Fig. 18 Comparisons with the radiation data from Apollo 4 (Ref. 26).

earliest times, less than 30,020 s. At times greater than approximately 30,060 s, Ried et al.[26] presented data to indicate the possibility of instrument drift.

Discussion

The purpose of the present study was to assess our capability to predict air radiation by comparing the available ground-based and flight data with one theoretical method. The radiation model[7] used is considered to be a good model that can realistically be used in engineering applications and coupled to flowfield solutions. Also, one computer code was used to calculate the shock tube conditions using the incident shock velocities and initial pressures from the different experiments. No changes were made in the flowfield solutions between calculations for the models in the ground-based experiments and the flight vehicles. Thus, the assessment of the ground-based data and the flight data was made on a common base.

Additional calculations, beyond those presented, were made in an attempt to examine some of the differences noted in the various experiments. However, the results presented are representative of the overall comparison with the data. Other investigators, using constant-property layers, have compared their calculations with some of the ground-based data. (See, for example, Refs. 16, 17, 29, and 30.) These comparisons are comparable to the present comparisons as to either general agreement with the data or being predominantly greater or smaller for the same data set.

An overall assessment for the ground-based data is as follows. The data agree generally with the trends predicted by the present calculations, but differences as great as a factor of 2 are noted. There is scatter in much of the data that can be comparable to the differences between the data and the calculated results.

The flight data from Fire II and Apollo 4 are in better agreement with the present calculations than the ground-based data. The overall assessment is that the flight data deviate from the calculations by less than 50%. An interesting point is raised in the good comparisons with the Fire II data for the highest altitudes where the flowfield was most likely in chemical nonequilibrium. It is generally considered that radiation from air in chemical nonequilibrium is greater than that for equilibrium. The general continuity of the Fire II data during the entry and the good comparisons of the data at the lower altitudes where the present calculations are valid indicate the Fire II data are of good quality. Thus, the Fire II data at the earliest entry times should provide a good data base for comparing any theoretical model developed to predict radiation from air in chemical nonequilibrium or low-density effects for an AOTV mission.

Concluding Remarks

The experimental data for air radiation from ground-based experiments and from the flights of Fire II and Apollo 4 have been reviewed and compared with a radiatively coupled, flowfield solution. This comparison with a single prediction method allows one to more easily assess the accuracy of the large body of experimental data from various sources and judge our ability to predict air radiation for future Earth entry missions such as an aeroassisted orbital transfer vehicle.

There is considerable scatter in the experimental data from ground-based experiments. These data agree generally with the trends predicted by the present calculations, but differences as great as a factor of 2 are noted. The flight data from Fire II and Apollo 4 are in good agreement with the present calculations with differences of less than 50%. Furthermore, this good agreement with the flight data suggests that the Fire II data at the earliest entry times should provide a good data base for comparing any theoretical model developed to predict radiation from air in chemical nonequilibrium or low-density effects for an AOTV mission.

References

[1]Walberg, G. D., "Aeroassisted Orbit Transfer--Window Opens on Missions," Astronautics and Aeronautics, Vol. 21, Nov. 1983, pp. 36-43.

[2]Howe, J. T., "Introductory Aerothermodynamics of Advanced Space Transportation Systems," AIAA Paper 83-0406, AIAA 21st Aerospace Sciences Meeting, Reno, Nevada, Jan. 1983.

[3]Park, C., "Radiation Enhancement by Nonequilibrium in Earth's Atmosphere," AIAA Paper 83-0410, AIAA 21st Aerospace Sciences Meeting, Reno, Nevada, Jan. 1983.

[4]Park, C., "Calculation of Nonequilibrium Radiation in AOTV Flight Regimes," AIAA Paper 84-0306, AIAA 22nd Aerospace Sciences Meeting, Reno, Nevada, Jan. 1984.

[5]Sutton, K., "Characteristics of Coupled Nongray Radiating Gas Flows With Ablation Products Effects About Blunt Bodies During Planetary Entries," Ph.D. Thesis, North Carolina State University, Raleigh, North Carolina, 1973.

[6]Zoby, E. V. and Sullivan, E. M., "Effects of Corner Radius on Stagnation-Point Velocity Gradients on Blunt Axisymmetric Bodies," NASA TM X-1067, March 1965.

[7]Nicolet, W. E., "Advanced Methods for Calculating Radiation Transport in Ablation-Product Contaminated Boundary Layers," NASA CR-1656, Sept. 1970.

[8]Nicolet, W. E., "User's Manual for the Generalized Radiation Transfer Code (RAD/EQUIL)," NASA CR-116353, Oct. 1969.

[9]Nerem, R. M. and Stickford, G. H., "Shock-Tube Studies of Equilibrium Air Radiation," AIAA Journal, Vol. 3, June 1965, pp. 1011-1018.

[10]Nerem, R. M., "Stagnation Point Heat Transfer in High Enthalpy Gas Flows," Air Force Systems Command, Wright-Patterson AFB, Ohio, FDL-TDR-64-41, Part II, March 1964.

[11]Thomas, G. M. and Menard, W. A., "Measurements of the Continuum and Atomic Line Radiation from High-Temperature Air," AIAA Journal, Vol. 5, Dec. 1967, pp. 2214-2223.

[12]Hoshizaki, M., "Equilibrium Total Radiation Measurements in Air at Superorbital Entry Velocities," Lockheed Missiles and Space Company, Sunnyvale, California, Rep. 6-90-63-97, Oct. 1963.

[13]Gruszczynski, J. S. and Warren, W. R. Jr., "Study of Equilibrium Air Total Radiation," AIAA Journal, Vol. 5, March 1967, pp. 517-525.

[14]Page, W. A. and Arnold, J. O., "Shock-Layer Radiation of Blunt Bodies at Reentry Velocities," NASA TR-R-193, April 1964.

[15]Golobic, R. A. and Nerem, R. M., "Shock-Tube Measurements of End-Wall Radiative Heat Transfer in Air," AIAA Journal, Vol. 6, Sept. 1968, pp. 1741-1793.

[16]Wood, A. D., Hoshizaki, H., Andrews, J. C., and Wilson, K. H., "Measurements of the Total Radiant Intensity of Air," AIAA Journal, Vol. 7, Jan. 1969, pp. 130-139.

[17]Wood, A. D., Hoshizaki, H., Andrews, J. C., and Wilson, K. H., "Measurements of the Total Radiant Intensity of Air, AIAA Paper 67-311, AIAA Thermophysics Specialist Conference, New Orleans, Louisiana, April 1967.

[18]Krey, R. U. and Morris, J. C., "Experimental Total and Total Line Radiation of Nitrogen, Oxygen, and Argon Plasmas," The Physics of Fluids, Vol. 13, June 1970, pp. 1483-1487.

[19]Schreiber, P. W., Hunter, A. M., II, and Benedetto, K. R., "Measurement of Nitrogen Plasma Transport Properties," AIAA Journal, Vol. 10, May 1972, pp. 670-674.

[20]Schreiber, P. W., Hunter, A. M. II, and Benedetto, K. R., "Electrical Conductivity and Total Emission Coefficient of Air Plasma," AIAA Journal, Vol. 11, June 1973, pp. 815-821.

[21]Knott, P. R., Carlson, L. A., and Nerem, R. M., "A Further Note on Shock-Tube Measurements of End-Wall Radiative Heat Transfer in Air," AIAA Journal, Vol. 7, Nov. 1969, pp. 2170-2172.

[22]Cauchon, D. L., "Project Fire Flight 1 - Radiative Heating Experiment," NASA TM X-1222, April 1966.

[23]Cauchon, D. L., "Radiative Heating Results from the Fire II Flight Experiment at a Reentry Velocity of 11.4 Kilometers per Second," NASA TM X-1402, July 1967.

[24]Cornette, E. S., "Forebody Temperatures and Calorimeter Heating Rates Measured During Project Fire II Reentry at 11.35 Kilometers per Second," NASA TM X-1305, Nov. 1966.

[25]Lee, D. B. and Goodrich, W. D., "The Aerothermodynamics Environment of the Apollo Command Module During Superorbital Entry," NASA TN D-6792, April 1972.

[26]Ried, R. C. Jr., Rochelle, W. C., and Milhoan, J. D., "Radiative Heating to the Apollo Command·Module: Engineering Prediction and Flight Measurement," NASA TM X-58091, April 1972.

[27]Sutton, K. and Graves, R. A., Jr., "A General Stagnation-Point Convective-Heating Equation for Arbitrary Gas Mixtures," NASA TR R-376, Nov. 1971.

[28]Cohen, N. B., "Boundary-Layer Similar Solutions and Correlation Equations for Laminar Heat-Transfer Distribution in Equilibrium Air at Velocities up to 41,100 Feet per Second," NASA TR R-118, 1961.

[29]Page, W. A., Compton, P. L., Borucki, W. J., Ciffone, D. L., and Cooper, D. M., "Radiative Transport in Inviscid Nonadiabatic Stagnation-Region Shock Layers," AIAA Paper 68-784, AIAA 3rd Thermophysics Conference, Los Angeles, California, June 1968.

[30]Wilson, K. M., and Greif, R., "Radiation Transport in Atomic Plasmas," Journal of Quantitative Spectroscopy and Radiative Transfer, Vol. 8., April 1968, pp. 1061-1086.

Chapter IV. Surface Effects

Temperature-Dependent Reaction Rate Expressions for Oxygen Recombination

E.V. Zoby,* R.N. Gupta,† and A.L. Simmonds‡
NASA Langley Research Center, Hampton, Virginia

Abstract

A temperature-dependent oxygen surface reaction rate coefficient has been determined from experimental STS-2 heating and wall temperature data at altitudes of 77.91, 74.98, and 71.29 km. The coefficient is presented in an Arrhenius form and is shown to be less temperature dependent than previous results. Finite rate viscous shock-layer heating rates based on this present expression have been compared with predicted heating rates using the previous rate coefficients and with experimental heating data obtained over an extensive range of STS-2 and STS-3 entry conditions. A substantial improvement is obtained in comparison of experimental data and predicted heating rates using the present oxygen reaction rate expression.

Nomenclature

CSE = Catalytic Surface Experiment
K = Boltzmann constant
k_w = surface reaction rate coefficient
L = Shuttle axial length to hinge line
m = mass of an atom
M = Mach number
q = heating rate

AIAA Paper 84-0224 presented at 22nd Aerospace Sciences Meeting, Reno, Nev., Jan. 9-12, 1984.

*Aero-Space Technologist, Aerothermodynamics Branch, Space Systems Division.

†Old Dominion University.

‡Mathematician, Aerothermodynamics Branch, Space Systems Division.

T = temperature
TPS = thermal protection system
u = velocity
VSL = viscous shock layer
X = axial measurement
α = angle of attack
γ = energy-transfer recombination coefficient
ρ = density

Subscripts

∞ = freestream conditions
N = nitrogen
O = oxygen
w = wall conditions

Introduction

The experimental wall temperature measurements and resulting heat-transfer rates obtained during the first flights of the Space Shuttle have been demonstrated[1-4] to be lower than predicted equilibrium values at least over the first 40 percent of the shuttle length and for much of the altitude range of interest. The flight data from the Catalytic Surface Experiment[2] (CSE), which is a Space Shuttle Orbiter experiment by NASA Ames Research Center, have verified that the lower rates can be attributed primarily to the fairly noncatalytic nature of the Shuttle thermal protection system (TPS) and not to unknowns in freestream or flowfield quantities, and that some degree of nonequilibrium flow persists to altitudes as low as 50 km. For the design of future transportation systems, the need to develop confidence in predicting the thermal environment about arbitrary geometries in a finite reacting flowfield coupled with a nonfully catalytic TPS is evident.

Advancement in flowfield computational capabilities, e.g., three-dimensional applications,[5] is obviously an important factor in improving our confidence level. However, the need for determining the best values for flowfield and surface properties is no less important. For example, the heating rate in a dissociated nonequilibrium flow environment is a function of the surface reaction rate coefficient[6] (or energy-transfer recombination coefficient). Both temperature-dependent[1] and constant[2] values of the coefficients for oxygen and nitrogen recombination have been used to calculate laminar heat fluxes to the Space Shuttle. The reaction rates were inferred from ground-test arcjet heat-transfer measurements.[7,8] The heating rates, or surface temperatures, that have been calculated with the existing reaction rates yield only a fair comparison[1-3] with the Shuttle experimental windward-ray thermal data. In fact, at altitudes lower than approximately 65 km, the calculated heating rates are as much as 30 to 40% lower than the experimental data. As noted in a recent investigation,[9] no set of experimental reaction rates or those assumed[3] for sensitivity studies could consistently produce the "best" comparison.

The purpose of this investigation is to determine a temperature-dependent surface reaction rate expression that results in improved comparisons with the Shuttle laminar heating data. Oxygen recombination has been defined[2] as the most important surface reaction for the Shuttle thermal environment. Therefore, oxygen reaction rate values that give the best fit to experimental flight heating rates at STS-2 entry conditions for altitudes of 77.91, 74.98, and 71.29 km are correlated as a function of the surface temperature in an Arrhenius form. For this study, the heating data are selected from only the windward-ray measurements, since local flow conditions in this region are easier to model. The validity of the expression is demonstrated by incorporating the expression in heating calculations for comparisons with Shuttle heating data over an extensive range of STS-2 and of STS-3[10] entry conditions.

Analysis

This section presents the procedures used to develop the present expression for the oxygen surface reaction rate coefficient, as well as the resulting equation. The coefficients are incorporated in a detailed finite reacting flowfield method to compute laminar heating rates for nonfully catalytic conditions. A brief outline of the flowfield method is presented. Also, the relative merits of the method compared to other computational methods used for application to the shuttle environment are discussed.

Flowfield Method

The flowfield method is an axisymmetric, 0-deg angle-of-attack, finite reacting, viscous shock-layer (VSL) code.[11] The code has been modified[3] recently to include surface rate controlled oxygen and nitrogen reactions. The VSL equations are applied to the windward symmetry plane by using the concept of an equivalent axisymmetric body at 0-deg angle of attack.[12,13] For 25- to 45-deg angle of attack, a hyperbola with a computed nose radius and body half-angle is used[3,9,12-14] to model the coordinates of the Shuttle windward symmetry plane. Confidence in using the equivalent body concept in detailed or approximate two-dimensional methods to compute Shuttle flow condition at large angles of attack is based on the consistently good comparison with ground[13] and flight[14] Shuttle heat-transfer data over a wide range of flow conditions. The primary reason for the good comparison is that the vehicle has a rather wide flat-bottom surface with a smoothly decreasing slope, and the equivalent body models adequately the longitudinal and circumferential cross sections.

The VSL equations, solution technique, thermodynamic and transport properties, chemical kinetics, and surface catalycity conditions are documented[3,9,11] sufficiently and are not repeated herein. The properties are based on detailed calculations. While the absolute accuracy of these properties is not a subject of the present investigation, errors in the properties could impact the results of investigations similar to the present one.

The nature of the present study is, obviously, dependent on the numerical method used in the analysis, and a 10-30% difference in the heating predicted by the different detailed numerical methods used for the Shuttle analysis has been reported in a recent investigation.[4] While the capabilities of the detailed methods represent significant advances over the heating methods used for Shuttle design, such discrepancies could readily impact the present conclusions. The VSL code has been selected for this study based on the previously mentioned features of the code as well as on the following points.

In comparison to other techniques, the VSL method provides a direct means of computing heat fluxes as well as interactions between inviscid and viscous flow regions due to heat transfer, entropy layer swallowing, and mass injection. Viscous-inviscid matching methods using either detailed[2] or approximate[13] perfect gas or equilibrium inviscid flows and heat-transfer calculations account for variable entropy effects on surface heating with approximate techniques.[12,13,15] However, a recent investigation[4] has shown that the results of a nonequilibrium inviscid-viscous method overpredicts Shuttle experimental laminar heating rates for $X < 0.25$ L. For this region, a better comparison is obtained with the VSL results. The problem with the matching procedure is attributed[4] possibly to a too-rapid swallowing of the entropy layer. Also, many methods require an initial well-defined shock shape to avoid possible convergence problems. The present VSL technique generates an approximate initial shock, with body and shock angle equal, and iterates the shock to satisfactory convergence. Such a procedure is reasonable for wide-angle bodies. Also, the computational time of the VSL code is much less relative to the other detailed codes. Thus not only can confidence be placed in the results of the detailed VSL code, but also the code is very satisfactory for application to the present investigation.

Reaction Rate Coefficients

A recent investigation has presented a temperature-dependent energy-transfer recombination expression in an Arrhenius form. Such an equation is written as

$$\gamma_i = Ae^{-B/T} \tag{1}$$

The unknowns A and B have been determined[7] for both oxygen and nitrogen surface recombination such that

$$\gamma_O = 16.0\ e^{-10271./T_w} \tag{2}$$

and

$$\gamma_N = 0.071\ e^{-2219./T_w} \tag{3}$$

The reaction rate coefficient is related[6] to the recombination coefficient by

$$k_{w,i} = \sqrt{KT_w/2\pi m}\ \gamma_i \tag{4}$$

Temperature-dependent oxygen and nitrogen surface reaction rate coefficients obtained by substituting Eqs. (2) and (3) into Eq. (4), respectively, or constant $k_{w,O}$ and $k_{w,N}$ values have been incorporated[3] in the VSL code, and, as previously noted, the resulting heating predictions[3] are only in fair agreement with the STS-2 experimental data. Similar conclusions can be drawn from other analyses.[1,2] Obviously, some of the discrepancy can be attributed to the numerical methods. However, the values of A and B used in Eqs. (2) and (3) were inferred from arcjet experimental heat-transfer data,[7,8] and

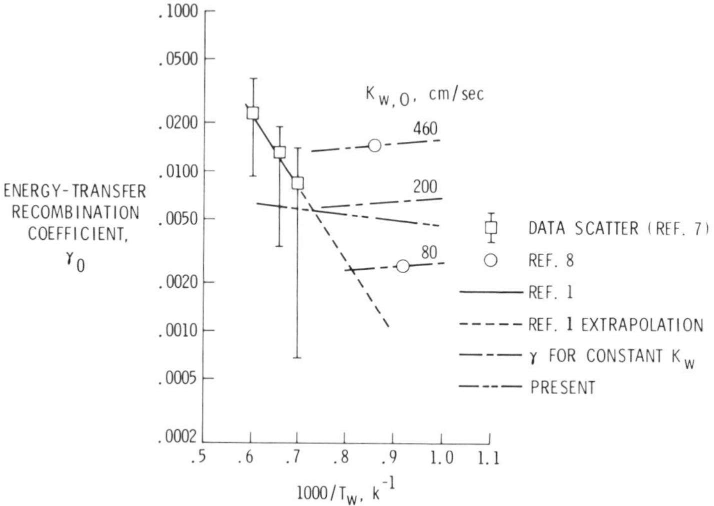

Fig. 1 Oxygen energy-transfer recombination coefficient for Shuttle TPS.

large data scatter in the recombination values are shown (see Fig. 1 for oxygen). Also, the data for oxygen were measured at wall temperature levels ($T_w > 1400$ K) in excess of the Shuttle wall temperature data range of approximately 800 to 1400 K. Thus, the relation inferred from experimental ground-test measurements must be extrapolated to the appropriate flight wall temperature conditions. As observed in Ref. 9, the comparisons of predicted heating rates based on the extrapolated results and experimental data imply that improved comparisons may result if the oxygen reaction rate expression exhibited less temperature dependence. Also, no single expression or constant value for the coefficient yields predicted heating rates that result in a consistently "best" comparison. In Fig. 1, a constant $k_{w,O}$ value is shown to correspond to increasing values of the recombination parameter with decreasing temperature, which is not a physically realistic trend. These values were assumed in a recent investigation for a parametric study on the effect of varying the reaction rate coefficient in heating calculations and were not intended to specifically characterize the material property.

This investigation also determines the unknowns in Eq. (1) for use in Eq. (4) but bases the correlation on free-flight data. STS-2 entry conditions, as well as surface heating and temperature distributions for altitudes of 77.91, 74.98, and 71.29 km, are used to determine an oxygen reaction rate expression. Since oxygen recombination is the most important[2] surface reaction at Shuttle entry conditions, nitrogen recombination rates given by Eq. (3) are assumed to be valid for the major part of this investigation. Obviously, a correlation determined from the local environment where it will be used should be more applicable than an extrapolated correlation. The limitation of such an approach is that either detailed calculations or appropriate experimental data must be available to properly determine and assess the applicability of the correlation.

Initially, heat-transfer rates are computed at the selected STS-2 freestream conditions for body locations corresponding to the actual measurement stations and for a range of constant $k_{w,O}$ values. Similar calculated results have been documented[9] recently. The STS-2 experimental laminar heating rates are included with the results. The comparison between the computed and experimental values permits the selection of the $k_{w,O}$ values that yield "best agreement" with the actual data. The $k_{w,O}$ values are then plotted as a function of the measured surface temperatures and a "best-fit" coefficient is determined. The required form of the equation is obtained by substituting Eq. (1) into Eq. (4), which for oxygen reduces to

$$k_{w,O} = 909.5A\sqrt{T_w}\; e^{-B/T_w} \tag{5}$$

A "best-fit" expression determined by comparison with the selected STS-2 heating data is obtained as

$$k_{w,O} = 8.557\sqrt{T_w}\; e^{-658.9/T_w} \tag{6}$$

or, in terms of the energy-transfer recombination coefficient,

$$\gamma_O = 0.00941\; e^{-658.9/T_w} \tag{7}$$

Results and Discussion

In this section of the paper, the results of the present oxygen reaction rate coefficient are compared with existing results. Also, heating rates computed with the present $k_{w,O}$ expression incorporated in the finite reacting VSL code are compared with Shuttle laminar heating data. The comparisons of predicted and experimental heating data are limited to STS-2 and STS-3 entry conditions. The onboard recorder malfunctioned during the STS-1 and STS-4 entries and data measured at freestream Mach numbers greater than approximately 13 were lost. Most of the windward symmetry plane instrumented tiles on STS-5 were specially treated for the CSE experiment. A discussion of the sensitivity and applicability of the present $k_{w,O}$ expression, as well as the present assumption of the applicability of an existing expression for $k_{w,N}$, is presented.

Reaction Rate Coefficients

Values of the oxygen and nitrogen reaction rates used in this study are shown as a function of temperature in Fig. 2. The $k_{w,N}$ values are calculated from the expression obtained by substituting Eq. (3) into Eq. (4). The two sets of $k_{w,O}$ values, referred to as existing and present, are calculated from the expressions obtained by substituting Eq. (2) into Eq. (4) and from Eq. (6), respectively. Results based on Eq. (7) for the related oxygen energy-transfer recombination parameter are presented in Fig. 1. The present results are shown to yield much less temperature dependence than the existing predicted results and to lie generally within the

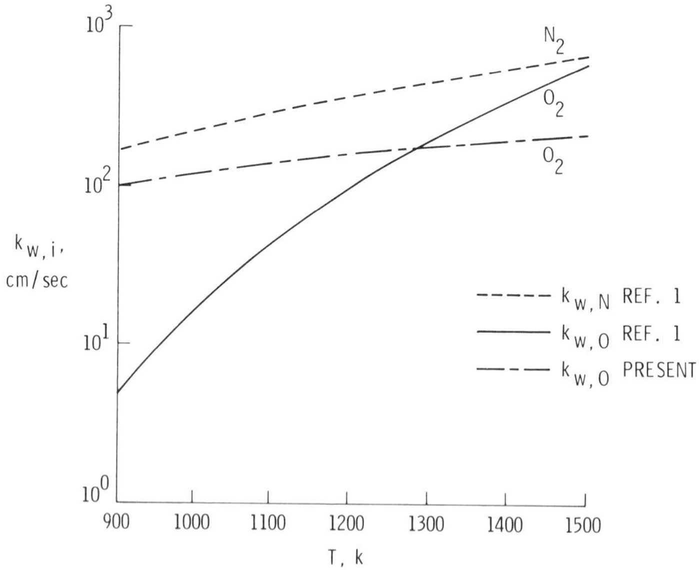

Fig. 2 Temperature-dependent surface reaction rates.

experimental data scatter. The present calculated results shown in Fig. 2 for $k_{w,O}$ are greater than the existing values[1] for temperatures less than approximately 1280 K. At 900 K, a factor of 20 in the results of the two calculations is noted. For temperatures greater than 1280 K, the existing $k_{w,O}$ data are greater than the present results with a factor of approximately 3 obtained at 1500 K. Note that a factor of 2.5 in $k_{w,O}$ data has been reported[9] to yield 12% differences in heating predictions. Thus, large differences in $k_{w,O}$ values do not translate necessarily to similar differences in heating rates. This lack of sensitivity of $k_{w,O}$ to the surface heating can impact the definitiveness of the present study since the current approach for determining the "best-fit" expression is somewhat subjective. In addition, the present results are determined for selected STS-2 entry conditions that are representative of the present Shuttle entries for which the wall temperature range is approximately 800 to 1400 K. There are no higher flight temperature data to assess the applicability of the current values to higher velocity entry conditions. Also at these higher velocity entry conditions, nitrogen dissociation should also have an appreciable impact on nonfully catalytic heat-transfer predictions due to a greater degree of nitrogen dissociation at the higher entry velocities and to recognizing that nitrogen dissociation energy is approximately twice as large as oxygen. As noted previously,[4,16] nitrogen recombination may impact heating levels in the Shuttle nose region. An interesting observation is obtained from the comparison of predicted and measured heating rates used to develop the present expressions. The predicted heating rates based on either the present or existing values for $k_{w,O}$ overestimate the actual measured data at X/L = 0.025 by approximately 10%. Also, the differences in the two predicted values at this station are less than 5%. These results could mean possibly that, similar to the existing oxygen expression, the existing nitrogen recombination expression [Eq. (3)] exhibits too great a temperature sensitivity and instead should yield lower values at the higher temperature range (1200 to 1500 K). A preliminary study of these effects has been conducted and the results suggest that a nitrogen expression with less temperature dependence yields heating rates that are in better agreement with the limited Shuttle data. The impact of these nitrogen reaction rate values on the downstream heating predictions, as well as for the entire Shuttle length at altitudes lower than 68 km, is insignificant. However, owing to the limited Shuttle data and to assuming the applicability of the present $k_{w,O}$ expression to temperatures greater than 1400 K, further analysis is not considered to be fruitful at present.

Heating Rate Comparison

Predicted heating rates are compared with STS-2 and STS-3 laminar heating data in Figs. 3a)-l). The STS-2 data are presented in Figs. 3a)-g), and the STS-3 results are given in the remaining figures. The data represent measurements over an altitude range from approximately 78 to 53 km. This altitude range represents an extensive variation in flow properties and should represent an adequate test to validate the applicability of the present oxygen surface reaction rate expression. The results that are shown on each figure are the experimental data and VSL predictions based on an equilibrium flow assumption and a finite catalytic wall condition using $k_{w,O}$ based on Eq. (2) and on the present results given by Eq. (6).

The data measured at $X < 0.4\ L$ and for the higher altitude conditions demonstrate the greatest departure from the equilibrium state. While the oxygen reaction rate coefficient was correlated only at STS-2 entry altitudes of 77.91, 74.98, and 71.29 km (Figs. 3a, 3b, and 3d, respectively), the present expression incorporated in the VSL code yields levels and trends of predicted heating rates that are in generally good agreement with the experimental data for the range of STS-2 and STS-3 free-stream conditions. The finite catalytic wall heating rates using $k_{w,O}$ based on Eq. (6) result in improved agreement with experimental data compared to calculated heating rates based on an existing $k_{w,O}$ expression,[1] especially at altitudes lower than 65 km. At these altitude conditions, the predicted heating rates based on the existing oxygen reaction rate expression are 30 to 40% lower than the experimental data. However, predicted heating rates using the present reaction rate expression are generally within 10% of the experimental rates.

An interesting and encouraging result of the present investigation is noted in comparison of the experimental heating data and the current predicted values with decreasing altitudes. The data and the finite rate predicted results approach the equilibrium heating levels in a very similar manner. For consistency, the equilibrium code uses the same definition[17] for viscosity as the finite reacting code. A recent investigation[18] has shown that heating calculations based on this viscosity model

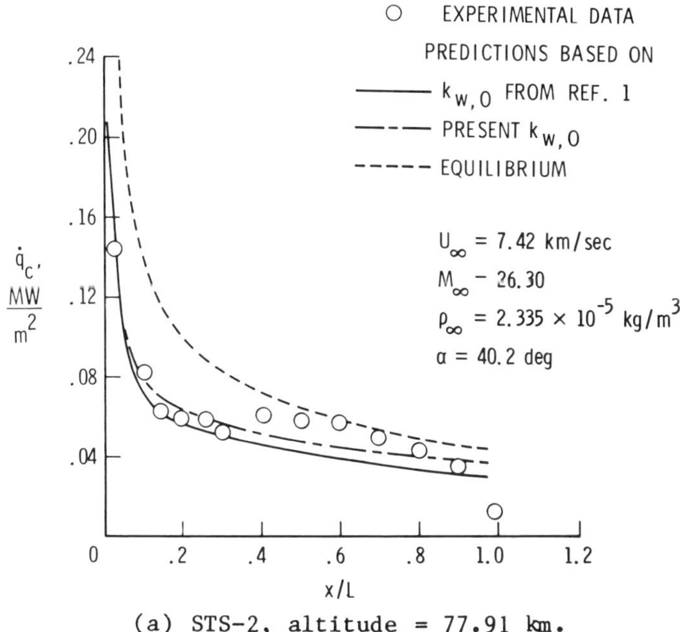

(a) STS-2, altitude = 77.91 km.

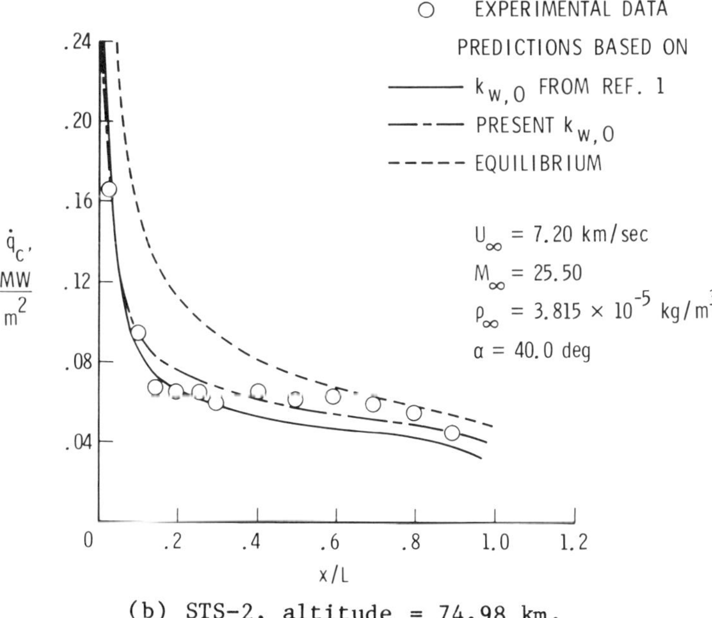

(b) STS-2, altitude = 74.98 km.

Fig. 3 Comparison of predicted and experimental heating rates for different surface conditions.

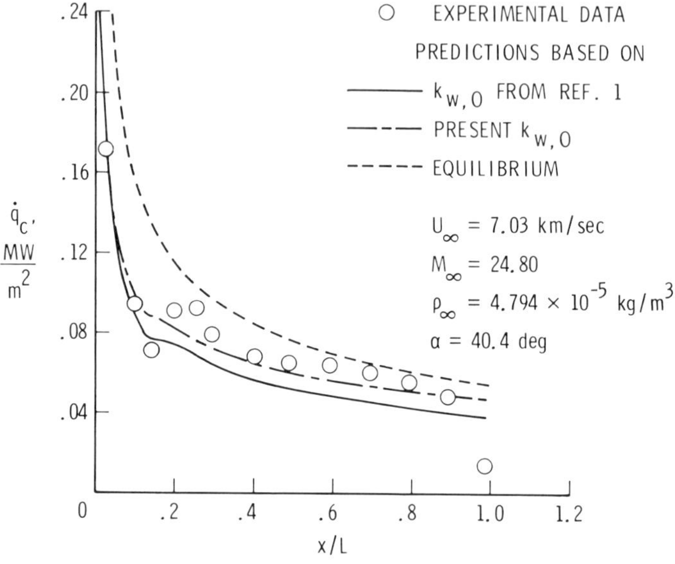

(c) STS-2, altitude = 73.33 km.

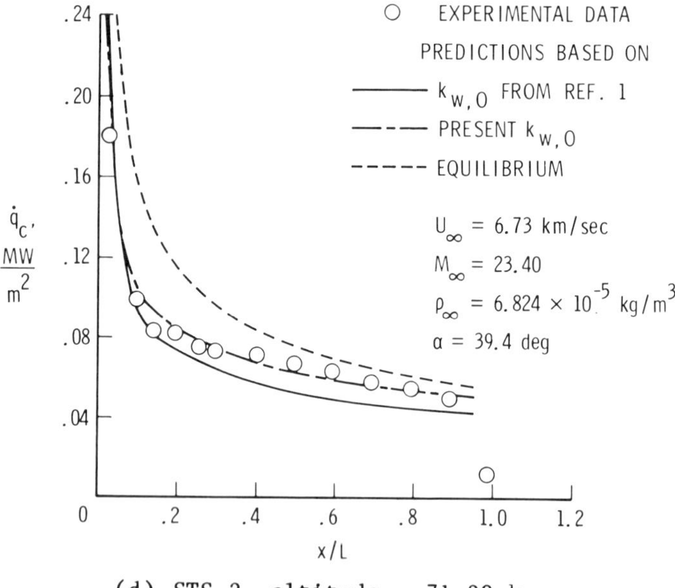

(d) STS-2, altitude = 71.29 km.

Fig. 3 (continued) Comparison of predicted and experimental heating rates for different surface conditions.

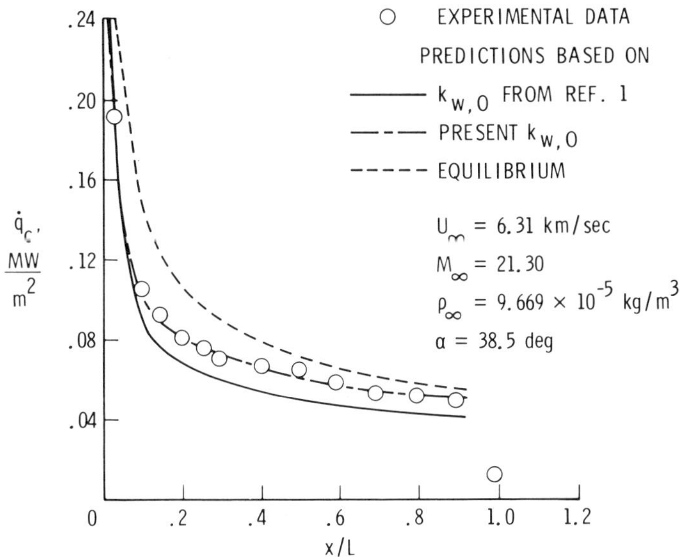

(e) STS-2, altitude = 68.67 km.

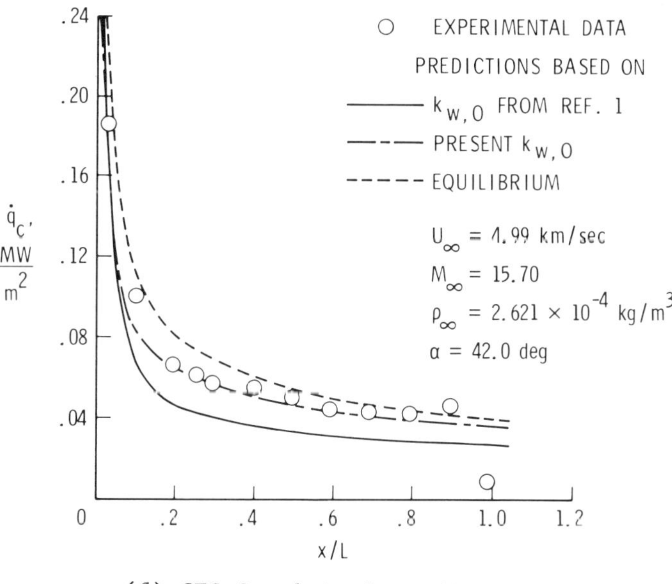

(f) STS-2, altitude = 60.56 km.

Fig. 3 (continued) Comparison of predicted and experimental heating rates for different surface conditions.

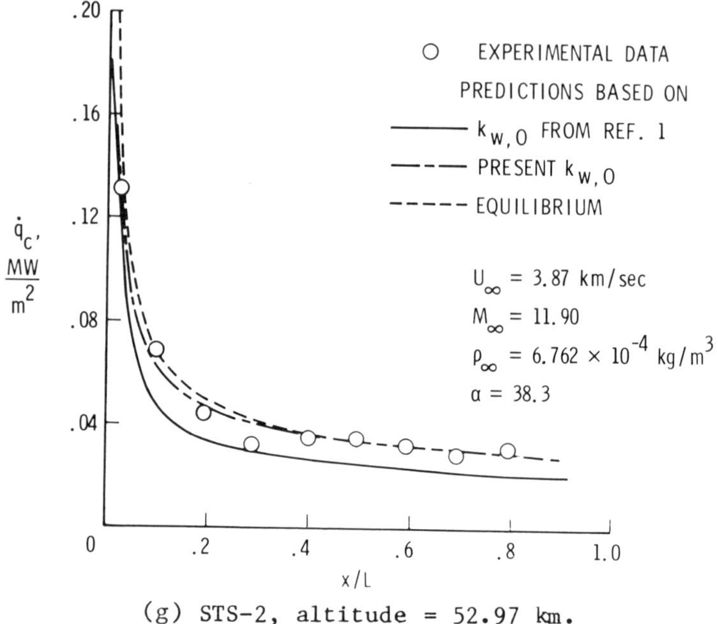

(g) STS-2, altitude = 52.97 km.

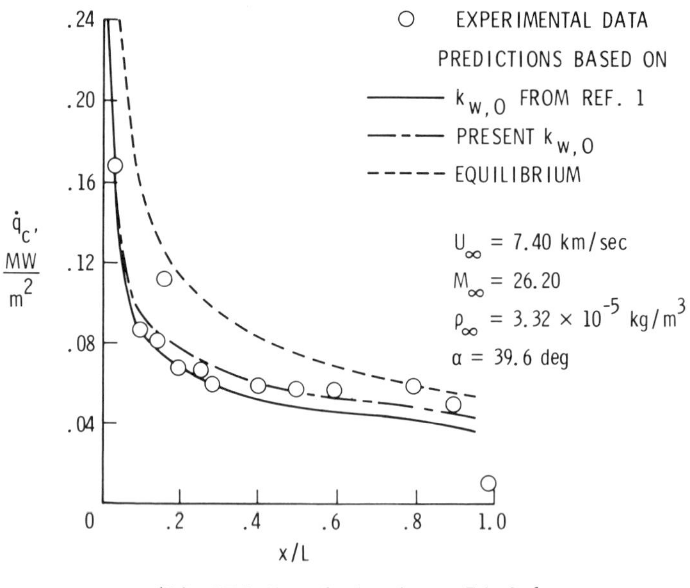

(h) STS-3, altitude = 76.2 km.

Fig. 3 (continued) Comparison of predicted and experimental heating rates for different surface conditions.

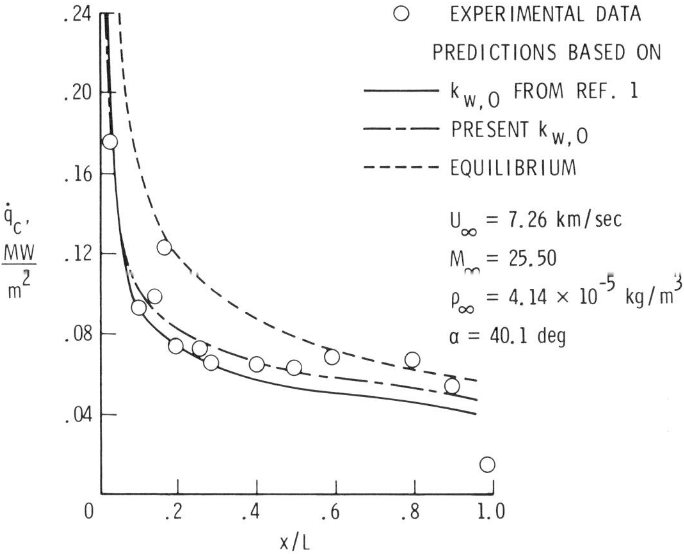

(i) STS-3, altitude = 74.7 km.

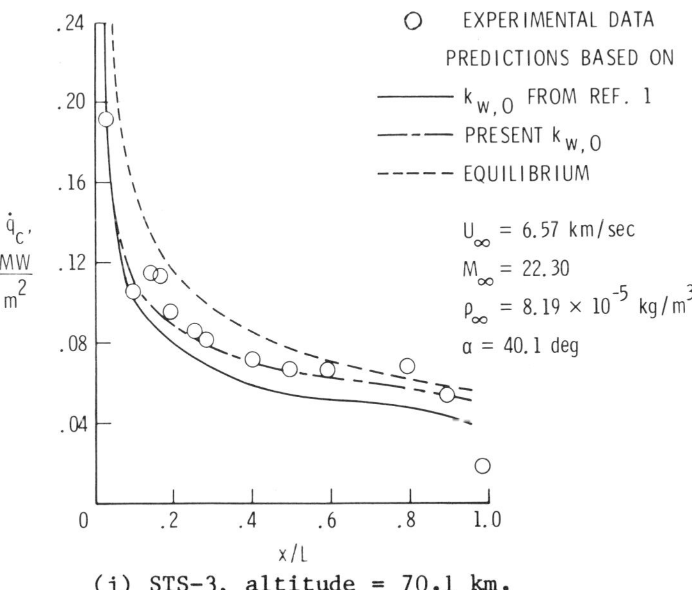

(j) STS-3, altitude = 70.1 km.

Fig. 3 (continued) Comparison of predicted and experimental heating rates for different surface conditions.

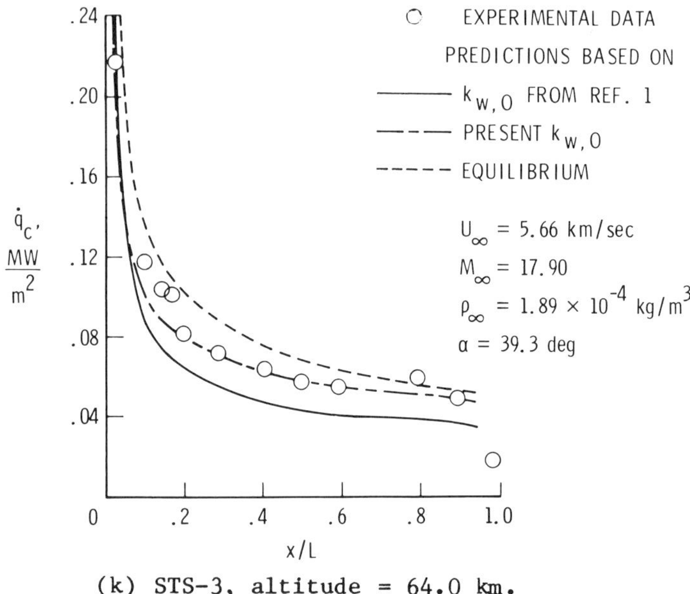

(k) STS-3, altitude = 64.0 km.

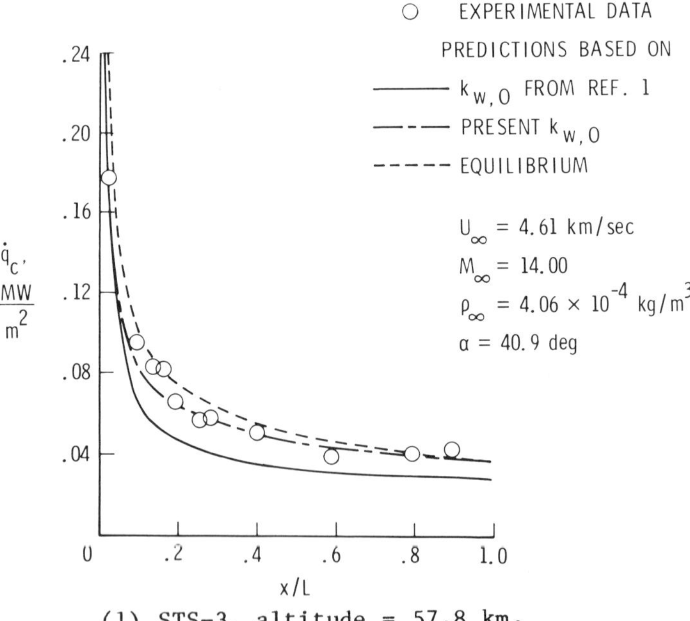

(l) STS-3, altitude = 57.8 km.

Fig. 3 (continued) Comparison of predicted and experimental heating rates for different surface conditions.

are approximately 10% greater than heating rates computed with a transport model similar to Hansen[19] or Sutherland.

A discussion of possible surface contamination from the deposition of melted acoustic sensors increasing the surface catalycity and consequently the temperatures and resulting heating rates has been documented.[1,10,14] However, this increase in predicted and experimental heating rates appears to be restricted primarily to $0.2 < X/L < 0.4$ on the Shuttle Orbiter (Fig. 3c) and to impact the heating levels over a limited altitude range. While such an occurrence could alter the TPS characteristics which the present expressions cannot treat, the present expression for $k_{w,O}$ is temperature dependent and appears to account for the major thermal effects over this region. Note that the present correlations are determined from three selected STS-2 entry conditions. While surface contamination occurred during the STS-3 entry also, the predicted heating rates using the present rate expressions are in good agreement with the STS-3 measured heating rates also.

Concluding Remarks

Finite rate viscous shock-layer (VSL) laminar heat-transfer predictions using existing oxygen and nitrogen reaction rate coefficients have been shown to be in only fair agreement with STS-2 windward symmetry plane experimental data. The existing reaction rates were inferred from arcjet heat-transfer measurements. These reaction rates, as a function of temperature, exhibited large data scatter. Also oxygen recombination which represents the primary surface reaction at Shuttle entry conditions was determined at surface temperatures in excess of Shuttle values. An extrapolation of the results was required. In contrast, the present investigation uses experimental flight heat-transfer data to determine an oxygen reaction rate relation. An existing nitrogen reaction rate expression is assumed to be valid for this study.

The oxygen reaction rate expression is determined at STS-2 freestream conditions and surface measurements corresponding to altitudes of 77.91, 74.98, and 71.29 km. Constant values for the oxygen reaction rates that are incorporated in the VSL code and yield the "best" heating comparison with experimental data at a given altitude and body station are correlated with an Arrehenius model as a function of temperature. The temperature-dependent oxygen reaction rate expression that is selected as the "best-fit" correlation is determined based on the resulting comparison with the heating data. The current reaction rate expression is shown to be less temperature dependent than previous results. A comparison of predicted heating rates based on the present results with experimental data for a range of STS-2 and STS-3 entry conditions from approximately 78 to 53 km demonstrates considerable improvement over previous comparisons. The altitude range represents an extensive variation in the flow properties, and the resulting good comparison of predicted and experimental heating rates should provide sufficient validity for the present oxygen surface reaction rate equation. In addition, the trend of the experimental data and finite rate predictions approach the equilibrium heating levels in a very similar manner.

References

[1]Scott, C. D. and Derry, S. H., "Catalytic Recombination and the Space Shuttle Heating," AIAA Progress in Astronautics and Aeronautics: Entry Vehicle Heating and Thermal Protection Systems: Space Shuttle Solar Starprobe, Jupiter Galileo Probe, Vol. 85, edited by P. E. Bauer and H. E. Collicott, AIAA, New York, 1983, pp. 123-149.

[2]Rakich, J. V., Stewart, D. A., and Lanfranco, M. J., "Results of a Flight Experiment on the Catalytic Efficiency of the Space Shuttle Heat Shield," AIAA Progress in Astronautics and Aeronautics: Entry Vehicle Heating and Thermal Protection Systems: Space Shuttle Solar Starprobe, Jupiter Galileo Probe, Vol. 85, edited by P. E. Bauer and H. E. Collicott, AIAA, New York, 1983, pp. 97-123.

[3]Shinn, J. L., Moss, J. N., and Simmonds, A. L., "Viscous-Shock-Layer Heating Analysis for the Shuttle Windward Plane with Surface Finite Catalytic Recombination Rates," AIAA Progress in Astronautics and Aeronautics: Entry Vehicle Heating and Thermal Protection Systems: Space Shuttle Solar Starprobe, Jupiter Galileo Probe, Vol. 85, edited by P. E. Bauer and H. E. Collicott, AIAA, New York, 1983, pp. 149-181.

[4]Scott, C. D., "Effects of Nonequilibrium and Catalysis on Shuttle Heat Transfer," AIAA Paper 83-1485, 18th Thermophysics Conference, Montreal, Canada, June 1983.

[5]Kim, M. D., Swaminathan, S., and Lewis, C. H., "Three-Dimensional Nonequilibrium Viscous Shock-Layer Flow over the Space Shuttle Orbiter," AIAA Paper 83-0487, 21st Aerospace Sciences Meeting, Reno, Nevada, Jan. 1983.

[6]Goulard, R., "On Catalytic Recombination Rates in Hypersonic Stagnation Heat Transfer," Jet Propulsion, Vol. 28, Nov. 1958, pp. 737-745.

[7]Scott, C. D., "Catalytic Recombination of Nitrogen and Oxygen on High-Temperature Reusable Surface Insulation," AIAA Progress in Astronautics and Aeronautics: Aerothermodynamics and Planetary Entry, Vol. 77, edited by A. L. Crosbie, AIAA, New York, 1981, pp. 192-213.

[8]Stewart, D. A., Rakich, J. V., and Lanfranco, M. J., "Catalytic Surface Effects Experiment on the Space Shuttle," AIAA Progress in Astronautics and Aeronautics: Thermophysics of Atmospheric Entry, Vol. 82, edited by T. E. Horton, AIAA, New York, 1982, pp. 248-273.

[9]Gupta, R. N., Moss, J. N., Simmonds, A. L., Shinn, J. L., and Zoby, E. V., "Space Shuttle Heating Analysis with Variation in Angle of Attack and Surface Condition," Journal of Spacecraft and Rockets, Vol. 21, March-April 1984, pp. 217-219.

[10]Throckmorton, D. A., Hamilton, H. H., and Zoby, E. V., "Preliminary Analysis of STS-3 Entry Heat-Transfer Data for the Orbiter Windward Centerline," NASA TM 84500, June 1982.

[11]Moss, J. N., "Reacting Viscous-Shock-Layer Solutions with Multicomponent Diffusion and Mass Injection," NASA TR-R-411, June 1974.

[12]Adams, J. C. Jr., Martindale, W. R., and Mayne, A. W. Jr., "Real-Gas Effects on Hypersonic Laminar Boundary-Layer Parameters Including Effects of Entropy-Layer Swallowing," AEDC TR-75-2, Dec. 1975.

[13]Zoby, E. V., "Approximate Heating Analysis for the Windward Symmetry Plane of Shuttle-Like Bodies at Large Angle of Attack," AIAA Progress in Astronautics and Aeronautics: Thermophysics of Atmospheric Entry, Vol. 82, edited by T. E. Horton, AIAA, New York, 1982, pp. 229-247.

[14]Zoby, E. V., "Analysis of STS-2 Experimental Heating Rates and Transition Data," Journal of Spacecraft and Rockets, Vol. 20, May-June 1983, pp. 232-237.

[15]Goodrich, W. D., Li, C. P., Houston, C. K., Chiu, P., B., and Olmedo, L., "Numerical Computations of Orbiter Flowfields and Laminar Heating Rates," Journal of Spacecraft and Rockets, Vol. 14, May 1977, pp. 257-264.

[16]Scott, C. D., "Catalytic Recombination of Nitrogen and Oxygen on Iron-Cobalt-Chromia Spinel," AIAA Paper 83-0585, 21st Aerospace Sciences Meeting, Reno, Nevada, Jan. 1983.

[17]Armaly, B. F. and Sutton, K., "Viscosity of Multicomponent Partially Ionized Gas Mixtures," AIAA Progress in Astronautics and Aeronautics: Aerothermodynamics and Planetary Entry, Vol. 77, edited by A. L. Crosbie, AIAA, New York, 1981, pp. 335-351.

[18]Zoby, E. V. and Moss, J. M., "Thermodynamic Equilibrium-Air Correlations for Flowfield Applications," AIAA Journal, Vol. 20, June 1982, pp. 849-854.

[19]Hansen, C. F., "Approximations for the Thermodynamic and Transport Properties of High-Temperature Air," NASA TR R-50, 1959.

Surface-Slip Equations for Low Reynolds Number Multicomponent Air Flow

Roop N. Gupta*
Old Dominion University, Norfolk, Virginia
Carl D. Scott†
NASA Johnson Space Center, Houston, Texas
and
James N. Moss‡
NASA Langley Research Center, Hampton, Virginia

Abstract

The surface-slip equations have been obtained from the closed form solutions of the mass, momentum, and energy flux equations using the Chapman-Enskog velocity distribution function. This function represents a solution of the Boltzmann equation in the Navier-Stokes approximation. The obtained expressions provide jump (or slip) in the wall values of species concentraion, pressure, velocity, and temperature for the low Reynolds number high-altitude regime of a space vehicle. The analysis includes multicomponent diffusion with finite-rate surface catalytic recombination. A consistent set of equations is provided for multicomponent, binary, and single-species mixtures. Expression is also provided for finite-rate surface catalytic recombination without slip for a multicomponent mixture.

Presented as Paper 84-1732 at AIAA 19th Thermophysics Conference, Snowmass, Colorado, June 25-28, 1984.

*Research Professor of Mechanical Engineering and Mechanics.

†Research Engineer.

‡Research Leader, Aerothermodynamics Branch, Space Systems Division.

Nomenclature

A_k^i, B_{kl}^i, C_k^i = coefficients of the velocity distribution function for ith species
a_{i0}, a_{i1}, (j) b_{i0}, C_{i0} = distribution function coefficients defined in Ref. 9 for ith species
C_i = mass function of species i
C_p = frozen specific heat at constant pressure of the mixture $= \sum_{i=1}^{NS} C_i C_{pi}$
D_{ij} = multicomponent diffusion coefficient for species pair i and j
$\mathcal{D}_{ij}$ = binary diffusion coefficient
d_j^k = diffusion vector of jth species
E = energy flux from translational energy
F = general flux of property such as mass momentum, or energy
f^i = distribution function of ith species
$f^{(0)i}$ = equilibrium (or Maxwellian) distribution function of ith species
K = thermal conductivity of mixture
k = Boltzmann constant
k_{wi} = wall catalytic recombination rate constant for species i
Le = binary Lewis number $= \rho C_p \mathcal{D}_{12}/K$
M = mass flux
m_i = mass of ith species
$\bar{\bar{m}}$ = mass of a mixture molecule
NS = number of chemical species
n_i = number density of species i
$\bar{\bar{n}}$ = total number density $= \sum_{i=1}^{NS} n_i$
P = momentum flux
Pr = Prandtl number
p = pressure $= \sum_{i=1}^{NS} p_i$
p_i = partial pressure of species i
q_w = wall heat-transfer rate
$\mathcal{R}$ = universal gas constant
T = temperature

$\vec{V}_i$	= thermal (or peculiar) velocity of the ith species
V_k^i	= k-component of thermal velocity of ith species
$\bar{V}_k^i$	= k-component of diffusion velocity
v_{0k}	= bulk or mass averaged velocity
W_i	= molecular (or atomic) weight of species i
$\bar{\bar{W}}$	= molecular weight of mixture
x	= coordinate parallel to body (Fig. 1)
y	= coordinate perpendicular to body (Fig. 1)
z	= coordinate parallel to body (Fig. 1)
γ_i	= recombination coefficient (fraction of striking atoms that recombine at surface)
λ	= mean free path
μ	= viscosity of mixture
$\Phi^i(\vec{V})$	= perturbation part of distribution function
$\phi^i(\vec{V})$	= general property of the ith species such as mass, momentum, and energy
ρ	= density
σ	= Stephan-Boltzmann constant
ψ_i	= source term defined through Eqs. (23a-c)
θ, θ_i	= accommodation coefficient (or fraction of incident particles that stick)

Affixes (Note: employed both as subscripts and superscripts)

i,j	= species indices
s	= edge of Knudsen layer
w	= wall
$\downarrow$	= incident flux
$\uparrow$	= specularly reflected flux

Subscripts

A	= atom
M	= molecule
k,l	= coordinate indices
q	= species index
x,y,z	= component directions

Introduction

Under low Reynolds number high-altitude flight conditions, the density at the surface of a space vehicle is sufficiently low so that the velocity, temperature, pressure, and species concentration at the wall are no longer the same as those of the gas adjacent to the wall. This difference in the values of various quantities is known as "surface slip" or "wall jump," and the equations describing this phenomena are known as surface-slip or wall-jump equations. These equations play an important role in the accurate prediction of aerothermal environment of space vehicles[1] entering the Earth's atmosphere. In this paper, these equations have been obtained for the case where catalytic recombination reactions occur on the surface.

With the decreased density at higher altitudes, the continuum flow equations that describe the flow adequately under high Reynolds number flow conditions are no longer adequate near the wall. The flow in a region next to the wall having a thickness on the order of a local mean free path (the Knudsen layer) cannot be analyzed through the continuum description[2] because there are insufficient collisions for this description to be valid. The velocity distribution function for such a case may deviate significantly from equilibrium, and a rigorous treatment of the Knudsen layer would require the solution of the Boltzmann equation. Such a solution, with appropriate kinetic boundary conditions at the wall, would then provide boundary conditions at the top of the Knudsen layer for solution of the Navier-Stokes equations in the bulk outer flow. The gas-surface interaction may be accounted for through the use of accommodation coefficients.

The mathematical difficulties associated with the full Boltzmann equation generally preclude direct approaches that would lead to exact analytical solutions. Of the several indirect approaches that have been proposed, the Chapman-Enskog method has been utilized by Patterson[2] and Shidlovskiy[3] to obtain slip expressions for the Knudsen layer that are appropriate for a single-species gas that is close to equilibrium. This solution is of the form of an equilibrium velocity distribution function times a power series in velocity space, which characterizes the deviation from the equilibrium state. These results are applicable to a single-species gas. Davis[4] hads included first-order slip effects for the analysis of a binary-mixture flow past a blunt body. However, certain assumptions contained in the slip expressions for temperature, pressure, and species concentration may not generally be applicable.

Following the approach of Shidlovskiy,[3] Scott[5] first obtained the wall boundary equations for a multicomponent mixture with diffusion and wall-catalyzed atom recombination under slip (low-density) flow conditions. The jump in the wall values was treated as a discontinuity across a thin Knudsen layer (of thickness on the order of one mean free path). The Chapman-Enskog type distribution function (which accounts for diffusion) with first-order accuracy was employed. Hendricks,[6] employing Scott's formulation, obtained simplified expressions for engineering applications with some corrections to Scott's expressions.[5] Hendricks' expressions,[6] however, contained some gross errors. Nevertheless, he was able to demonstrate, with the help of some simple calculations made with single and multicomponent species surface-slip conditions, the importance of including the multicomponent species in the slip conditions.

With the recently increased interest in space vehicles[1,7] operating in the nonequilibrium, low-density (or low Reynolds number) regime, the accurate prediction of the aerothermal environment of such vehicles requires the use of appropriate boundary conditions. This is quite important because most of the existing ground-based experimental facilities fail to simulate the typical nonequilibrium, low-density re-entry conditions. Thus, more reliance is placed on the theoretical predictions. With this and the shortcomings of the earlier analyses in view, it became necessary to reanalyze the boundary equations. In the present work, the wall-jump equations have been obtained for a multicomponent, nonequilibrium gas mixture by employing the approaches of Refs. 3 and 5. Appropriate expressions have been obtained for the slip (or jump) values of temperature, velocity, pressure, and species concentration (with or without full surface accommodation) in a form which can readily be employed for flowfield computations. The present analysis provides a consistent formulation for the slip equations for a multicomponent, binary, and single-species mixture. Further, wall boundary equations for the species concentrations have also been obtained for a gas flow with multicomponent diffusion and finite surface catalycity in absence of the slip.

The catalytic model treated here assumes simply that there is full accommodation of the energy of dissociation. Also, internal degrees of freedom of molecules (or atoms) are assumed to be frozen in the jump relations. The slip boundary equations, allowing for the change in internal energy of the reflecting molecules, are contained in Ref. 8.

Analysis

Interaction Model for the Gas-Solid Interface

The slip conditions are taken to exist across a thin Knudsen layer, which is on the order of one mean-free path in thickness as sketched in Fig. 1. The analysis outlined here follows the approach of Refs. 3 and 5 and contains the following assumptions:

1) Variations of the velocity distribution function through the Knudsen layer are small. This implies constant fluxes of mass, momentum, and energy across the Knudsen layer.

2) The energy and momentum accommodation coefficients are considered to have the same value.

3) The change in internal energy of the reflecting molecules, such as rotational, vibrational, and electronic excitation, is neglected.

The interaction model at the gas-solid interface, with the various fluxes sketched in Fig. 1, can be mathematically stated for dissociated air as

For a recombining atom:

$$F_A = F_A^{\downarrow} + (1-\theta_A)\, F_A^{\uparrow} + (\theta_A-\gamma_A)\, F_A^{w} \;; \quad A = O,N \tag{1}$$

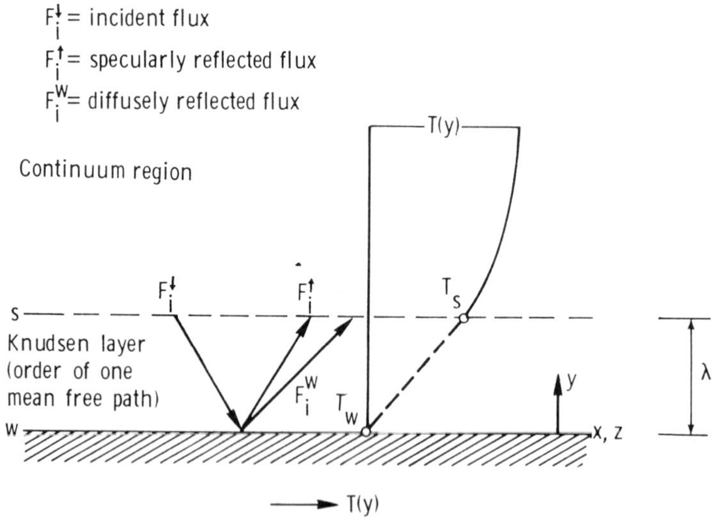

Fig. 1 The Knudsen layer showing general fluxes and coordinate axes. The temperature as a function of normal distance is schematically overlayed.

For a molecule gaining from the corresponding atom recombination:

$$F_M = F_M^{\downarrow} + (1-\theta_M)\ F_M^{\uparrow} + \theta_M F_M^W + \gamma_A F_A^W;\ M = O_2,\ N_2 \qquad (2)$$

For all other atoms and molecules (surface is assumed to be noncatalytic with respect to them):

$$F_i = F_i^{\downarrow} + (1-\theta_i)\ F_i^{\uparrow} + \theta_i F_i^{W} \qquad (3)$$

where F_i denotes a convective property such as mass, momentum, or energy.

Summing over all the species, we obtain from Eqs. (1), (2), and (3), the following expression for the net flux of momentum or energy:

$$\sum_{i=1}^{NS} F_i = \sum_{i=1}^{NS} F_i^{\downarrow} + \sum_{i=1}^{NS} (1-\theta_i)\ F_i^{\uparrow} + \sum_{i=1}^{NS} \theta_i F_i^{W} \qquad (4)$$

That is, the net flux at the outer edge of the Knudsen layer equals the incident flux, plus the specularly reflected flux (incident minus the fraction that sticks) at the wall, in addition to the diffusely reflected flux (those that accommodate to the wall) from the wall.

Each species is treated separately in the mass balance equations. Therefore, Eqs. (1), (2), or (3) are employed depending on the species being considered. In Eq. (1), the diffusely reflected flux consists of those atoms that are accommodated to the wall minus those that recombine. For the molecules in Eq. (2), the diffusely reflected term is present along with the source term resulting from the appropriate atoms recombining on the surface.

Various Fluxes and the Distribution Functions

The interaction model of Eqs. (1-4) is employed to obtain the slip boundary equations at the gas-solid interface. Through these equations, the net fluxes of mass, momentum, and energy at the outer edge of the Knudsen layer are equated to the difference between the incident and reflected fluxes at the wall. These fluxes, assumed to be constant across the Knudsen layer, are obtained from moments of the distribution function. For a convected property $\phi^i(\vec{V})$ such as mass, momentum, or energy for the ith species, the net flux of that property normal to the wall at the outer edge of the Knudsen layer, for example, is

$$F_i = \int_{-\infty}^{\infty}\int_{-\infty}^{\infty}\int_{-\infty}^{\infty} V_y^i \, \phi^i(\vec{V}) \, f_s^i(\vec{V}) \, d^3V^i \tag{5}$$

where V_y^i is the normal component of the molecular (or thermal) velocity and f_s^i is the velocity distribution function at the outer edge of the Knudsen layer.

Similar integrals are obtained for the incident and reflected fluxes by integration over appropriate half-spaces in molecular velocity.

Incident flux:

$$F_i^{\downarrow} = \int_{-\infty}^{\infty}\int_{-\infty}^{0}\int_{-\infty}^{\infty} V_y^i \, \phi^i(\vec{V}) \, f_s^i(\vec{V}) \, d^3V^i \tag{6}$$

Specularly reflected flux:

$$F_i^{\uparrow} = \int_{-\infty}^{\infty}\int_{0}^{\infty}\int_{-\infty}^{\infty} V_y^i \, \phi^i(\vec{V}) \, f_s^i(V_x, - V_y, V_z) \, d^3V^i \tag{7}$$

Diffusely reflected flux:

$$F_i^{w} = \int_{-\infty}^{\infty}\int_{0}^{\infty}\int_{-\infty}^{\infty} V_y^i \, \phi^i(\vec{V}) \, f_w^i(\vec{V}) \, d^3V^i \tag{8}$$

where f_w^i is the Maxwellian velocity distribution function at the temperature of the wall.

The velocity distribution functions used in integrals contained in relations (5) through (8) are those for a nonuniform multicomponent mixture perturbed out of equilibrium

$$f^i(\vec{V}) = f^{(0)i}(\vec{V}) \, [1+\Phi^i(\vec{V})] \tag{9}$$

where $f^{(0)i}(\vec{V})$ is the Maxwellian distribution function for the ith species defined as[3]

$$f^{(0)i}(\vec{V}) = \frac{n_i m_i^{3/2}}{(2\pi kT)^{3/2}} \exp\left\{- \left(\frac{m_i}{2kT}\right) V_i^{2}\right\} \tag{10}$$

and

$$\Phi^i(\vec{V}) = - A_k^i \frac{\partial(\ln T)}{\partial x_k} - B_{kl}^i \frac{\partial v_{0k}}{\partial x_l} + n \sum_{j=1}^{NS} C_k^{i(j)} d_k^j \tag{11}$$

Horo k and l aro tho ooordinate indices for three coordinate directions using the summation convention for repeated indices. The coefficients A_k^i, B_{kl}^i, and $C_k^{i(j)}$ given in Refs. 5 and 8 are taken from Ref. 9.

Slip Equations

By employing expressions for the various fluxes§ of mass, momentum, and energy, as defined by Eqs. (5-8) in Eqs. (1-4), the following slip equations are obtained.[8] These equations provide the slip values at the top of the Knudsen layer in terms of the wall qualities and gradients at the edge of the Knudsen layer. For the present equations, all accommodation coefficients θ_i are assumed equal to θ and the effects of thermal diffusion are neglected.

Number density (or concentration) slip (obtained from mass flux balance):

$$\frac{n_i^s}{n_i^w} = \frac{2\left[1 + \frac{1}{2}\frac{M_{iy}}{n_i^w m_i}\sqrt{\frac{2\pi m_i}{kT_w}}\right]}{\left(\frac{P_{iy}}{p_i^s} + 1\right)}\sqrt{\frac{T_w}{T_s}} \tag{12}$$

Pressure slip (obtained from the flux balance of normal component of momentum):

$$\frac{p^s}{p^w} = \left\{\frac{\theta}{2} - \frac{\theta}{3\bar{\bar{n}}_w kT_w}\left[\mu\left(\frac{\partial v_{ox}}{\partial x} + \frac{\partial v_{oz}}{\partial z} - 2\frac{\partial v_{0y}}{\partial y}\right)\right]_s\right.$$

(continued)

§The explicit expressions for the fluxes of mass M_{ik}, normal and tangential momentum P_{iy} and τ_{xyi}, and energy E_{ik} are contained in Appendix A. Appendix B provides the mass flux expression in a form similar to that of Ref. 10.

$$+ \frac{2}{5}\left(\frac{2-\theta}{\sqrt{\pi}}\right)\frac{1}{\bar{\bar{n}}_w kT_w}\left[\frac{1}{\bar{\bar{n}}}\frac{\partial T}{\partial y}\sum_{i=1}^{NS} n_i K_i \sqrt{\frac{m_i}{2kT}}\right]_s\Bigg\}$$

$$\Bigg/\Bigg\{\frac{\theta}{2} + 2\left(\frac{2-\theta}{\sqrt{\pi}}\right)\sum_{i=1}^{NS}\sqrt{\frac{m_i}{2kT_s}}\sum_{\substack{j=1\\ j\neq i}}^{NS} D_{ij}^s$$

$$\times\left[\frac{\partial C_j}{\partial y} - C_j\sum_{q=1}^{NS}\left(\frac{\bar{\bar{m}}}{m_q}\frac{\partial C_q}{\partial y}\right)\right]_s\Bigg\} \quad (13)$$

where

$$K_i = -\frac{5}{4}k\,\bar{\bar{n}}\sqrt{\frac{2kT}{m_i}}\;a_{i1}$$

is consistent with the mixture approximation

$$K = \sum_{i=1}^{NS}\frac{n_i}{\bar{\bar{n}}}K_i$$

Velocity slip (obtained from the flux balance of tangential component of momentum):

$$v_{0x}^s = \Bigg\{\sqrt{\frac{\pi}{2}}\left(\frac{2-\theta}{\theta}\right)\left[\frac{\mu}{\sqrt{kT}}\left(\frac{\partial v_{0x}}{\partial y} + \frac{\partial v_{0y}}{\partial x}\right)\right]_s$$

$$+ \frac{1}{5}\left[\frac{1}{\bar{\bar{n}}kT}\frac{\partial T}{\partial x}\sum_{i=1}^{NS} n_i\sqrt{m_i}\,K_i\right]_s$$

$$- \bar{\bar{n}}_s\sum_{i=1}^{NS}\sqrt{m_i}\sum_{\substack{j=1\\ j\neq i}}^{NS} D_{ij}^s\left[\frac{\partial C_j}{\partial x} - C_j\sum_{q=1}^{NS}\right.$$

$$\left.\times\left(\frac{\bar{\bar{m}}}{m_q}\frac{\partial C_q}{\partial x}\right)\right]_s\Bigg\}\Bigg/\sum_{i=1}^{NS} n_i^s\sqrt{m_i} \quad (14)$$

$$v_{0z}^s = \Bigg\{\sqrt{\frac{\pi}{2}}\left(\frac{2-\theta}{\theta}\right)\left[\frac{\mu}{\sqrt{kT}}\left(\frac{\partial v_{0z}}{\partial y} + \frac{\partial v_{0y}}{\partial z}\right)\right]_s$$

$$+ \frac{1}{5}\left[\frac{1}{\bar{\bar{n}}kT}\frac{\partial T}{\partial Z}\sum_{i=1}^{NS} n_i\sqrt{m_i}\,K_i\right]_s$$ (continued)

$$- \bar{\bar{n}}_s \sum_{i=1}^{NS} \sqrt{m_i} \sum_{\substack{j=1 \\ j\neq i}}^{NS} D_{ij}^s \left[\frac{\partial C_i}{\partial z} - C_j \sum_{q=1}^{NS} \right.$$

$$\left. \times \left(\frac{\bar{\bar{m}}}{m_q} \frac{\partial C_q}{\partial z} \right) \right]_s \Bigg\} \Bigg/ \sum_{i=1}^{NS} n_i^s \sqrt{m_i} \qquad (15)$$

Temperature slip (obtained from the flux balance of energy):

$$\frac{T_s}{T_w} = \left\{ - \frac{\sqrt{\pi}}{\bar{\bar{n}}_s} \sum_{i=1}^{NS} \frac{M_{iy}}{m_i} + \frac{1}{2} \sum_{i=1}^{NS} \sqrt{\frac{2kT_s}{m_i}} \left(\frac{\bar{\bar{m}}}{m_i} \right) C_i^s \right.$$

$$\left. \times \left(\frac{P_{iy}}{p_i^s} + 1 \right) \right\} \Bigg/ \left\{ - \sqrt{\pi} \left(\frac{2-\theta}{\theta} \right) \left[\frac{1}{2} \frac{K}{p^s} \frac{\partial T}{\partial y} \right. \right.$$

$$\left. - \frac{5}{4} \sum_{i=1}^{NS} \sum_{\substack{j=1 \\ j\neq i}}^{NS} D_{ij}^s \left(\frac{\partial C_j}{\partial y} - C_j \sum_{q=1}^{NS} \frac{\bar{\bar{m}}}{m_q} \frac{\partial C_q}{\partial y} \right)_s \right]$$

$$\left. + \sum_{i=1}^{NS} \sqrt{\frac{2kT_s}{m_i}} \left(\frac{\bar{\bar{m}}_s}{m_i} \right) C_i^s \left[1 + \frac{3}{4} \left(\frac{P_{iy}}{p_i^s} - 1 \right) \right] \right\} \qquad (16)$$

where the mass fraction C_i and the mixture molecule mass $\bar{\bar{m}}$ are defined, respectively, as

$$C_i = n_i m_i / \rho \qquad \text{and} \qquad \bar{\bar{m}} = \left(\sum_{i=1}^{NS} C_j / m_j \right)^{-1} \qquad (17)$$

For a first-order recombination at the surface, the following relation between the atom mass flux M_{Ay} and the wall number density n_A^w may be employed

$$M_{Ay} = - k_{wA} \left(n_A^w m_A \right) \qquad (18)$$

where the minus sign indicates that the flux is in the direction opposite to the outward normal, and the expression for rate constant K_{wA} with diffusion and slip is[5]

$$k_{wA} = \gamma_A \sqrt{\frac{kT_w}{2\pi m_A}} \qquad (19)$$

Here, γ_A is the recombination coefficient.

Thus, employing Eq. (18), the net mass flux M_{iy} appearing in Eqs. (12) and (16) may be defined as

$$M_{Ay} = - k_{wA} n_A^w m_A ; \quad A = O, N \tag{20a}$$

$$M_{My} = k_{wA} n_A^w m_A ; \quad \begin{cases} A = O \text{ for } M = O_2 ; \\ A = N \text{ for } M = N_2 \end{cases} \tag{20b}$$

For all other atoms and molecules,

$$M_{iy} = 0 \tag{20c}$$

Equation (12), with M_{iy} defined by Eqs. (20a-c), gives the number density ratio n_i^s/n_i^w. However, to obtain n_i^w from this ratio, an expression for n_i^s is required.

The net mass flux of O and N atoms to the surface M_{Ay} is also equal to the rate of consumption of these atoms at the wall from surface recombination

$$M_{Ay} = \gamma_A M_{Ay}^{\downarrow} \qquad A = O, N \tag{21a}$$

and the corresponding net mass flux of O_2 and N_2 molecules will be

$$M_{My} = - \gamma_A M_{Ay}^{\downarrow} ; \quad \begin{cases} A = O \text{ for } M = O_2 ; \\ A = N \text{ for } M = N_2 \end{cases} \tag{21b}$$

For all other species, the net mass flux to the surface may be assumed to be zero. Thus,

$$M_{iy} = 0 \tag{21c}$$

Substituting values of the net and incident fluxes M_{iy}, $M_{iy}^{\downarrow}$ from Appendix A in Eq. (21), the following expression is obtained

$$\sum_{\substack{j=1 \\ j \neq i}}^{NS} D_{ij}^s \left[\frac{\partial C_j}{\partial y} - C_j \sum_{q=1}^{NS} \left(\frac{\bar{\bar{m}}}{m_q} \frac{\partial C_q}{\partial y} \right) \right]_s = \frac{\psi_i^{\,s}}{m_i} \tag{22}$$

which may be used to obtain n_i^s. Here, ψ_i^s is the source term defined as

For O and N atoms:

$$\psi_A^s = - \frac{\gamma_A}{(2-\gamma_A)} \frac{1}{\sqrt{\pi}} \frac{m_A n_A^s}{2\bar{\bar{n}}_s} \sqrt{\frac{2\ kT_s}{m_A}} \left(\frac{P_{Ay}}{p_A^s} + 1 \right) \qquad (23a)$$

For O_2 and N_2 molecules:

$$\psi_M^s = - \psi_A^s \quad ; \quad \begin{cases} A = O \text{ for } M = O_2; \\ A = N \text{ for } M = N_2 \end{cases} \qquad (23b)$$

For all other species:

$$\psi_i^s = 0 \qquad (23c)$$

If no assumption is made about D_{ij}, Eq. (22) would give an expression for $(\partial C_i/\partial y)_s$ for all the species:

$$\left(\frac{\partial C_i}{\partial y}\right)_s = \left(\frac{m_i}{\bar{\bar{m}}_s}\right) \left\{ \sum_{j \neq i}^{NS} D_{ij}^s \left[\frac{\partial C_j}{\partial y} - C_j \sum_{q \neq i}^{NS} \left(\frac{\bar{\bar{m}}}{m_q} \frac{\partial C_q}{\partial y} \right) \right]_s - \frac{\Psi_i^s}{m_i} \right\} \Big/ \sum_{j \neq i}^{NS} (D_{ij} C_j)_s$$

The source term Ψ_i in this expression, however, may be simplified to yield an expression for C_A^s (or n_A^s) for the recombining atoms only:

$$C_A^s = \frac{m_A n_A^s}{\rho_s} = - 2\sqrt{\pi} \left(\frac{2-\gamma_A}{\gamma_A} \right) \left(\frac{m_A}{\bar{\bar{m}}_s} \right) \sqrt{\frac{m_A}{2kT_s}} \sum_{\substack{j=1 \\ j \neq A}}^{NS} D_{Aj}^s$$

$$\times \left[\frac{\partial C_j}{\partial y} - C_j \sum_{q=1}^{NS} \left(\frac{\bar{\bar{m}}}{m_q} \frac{\partial C_q}{\partial y} \right) \right]_s \Big/ \left(\frac{P_{Ay}}{p_A^s} + 1 \right)$$

Simplified Expressions for a Multicomponent Mixture

Equations (9-13) can be simplified[8] considerably for use with the flowfield calculations if one makes the following assumptions:

1) The multicomponent diffusion coefficients D_{ij} are the same for all pairs of species i and j. This assumption¶ implies that $D_{ij} = D_{12}$, and D_{12} is the same as the binary diffusion coefficient $\mathcal{D}_{12}$ or $\mathcal{D}_{AM}$ if all the species are grouped into atoms and molecules only.

2) The normal momentum flux to the pressure ratio, P_{iy}/p_i, is same for all species and equal to that of the mixture. Also, the tangential and normal shear stresses τ_{xyi}, τ_{yyi} for species i are the same as that for the mixture.

These assumptions retain the major effects of multicomponent fluxes on various slip quantities and provide considerable savings in computational effort required for the analysis of a flow problem.[11] With these simplifications, Eqs. (9-13) yield:

Concentration slip:

$$\frac{n_i^s}{n_i^w} = 2\left[1 + \frac{1}{2}\,\frac{M_{iy}}{n_i^w\, m_i}\sqrt{\frac{2\pi m_i}{kT_w}}\right]\sqrt{\frac{T_w}{T_s}} \Bigg/ \left(\frac{P_y}{p^s} + 1\right) \qquad (24)$$

Pressure slip:

$$\frac{p^s}{p^w} = \left\{\frac{\theta}{2} - \frac{\theta}{3\bar{\bar{n}}_w kT_w}\left[\mu\left(\frac{\partial v_{0x}}{\partial x} + \frac{\partial v_{0z}}{\partial z} - 2\frac{\partial v_{0y}}{\partial y}\right)\right]_s + \frac{2}{5}\left(\frac{2-\theta}{\sqrt{\pi}}\right)\frac{1}{\bar{\bar{n}}_w kT_w} \times\left[\frac{1}{\sqrt{2kT}}\frac{\partial T}{\partial y}\sum_{i=1}^{NS}\frac{n_i K_i}{\bar{\bar{n}}}\sqrt{m_i}\right]_s\right\}$$

(continued)

¶This is a somewhat stronger assumption. Because D_{ij} are concentration dependent whereas $\mathcal{D}_{ij}$ are virtually independent of composition.[9] However, for an air mixture consisting predominantly of nitrogen molecules and oxygen atoms, i.e. prior to an appreciable dissociation of N_2, the simple and multicomponent transport gave virtually the same results.[12]

$$\Bigg/ \left\{ \frac{\theta}{2} - \frac{2(2-\theta)}{\sqrt{\pi}} \frac{\mathcal{D}_{12}^{s}}{\sqrt{2kT_s}} \sum_{i=1}^{NS} \sqrt{m_i} \right.$$

$$\left. \times \left[\frac{\partial C_i}{\partial y} + (1-C_i) \sum_{q=1}^{NS} \left(\frac{\bar{\bar{m}}}{m_q} \frac{\partial C_q}{\partial y} \right) \right]_s \right\} \qquad (25)$$

Velocity slip:

$$v_{0x}^{s} = \left\{ \sqrt{\frac{\pi}{2}} \frac{(2-\theta)}{\theta} \left[\frac{\mu}{\sqrt{kT}} \left(\frac{\partial v_{0x}}{\partial y} + \frac{\partial v_{0y}}{\partial x} \right) \right]_s \right.$$

$$+ \frac{1}{5} \left[\frac{1}{kT} \frac{\partial T}{\partial x} \sum_{i=1}^{NS} \frac{n_i K_i}{\bar{\bar{n}}} \sqrt{m_i} \right]_s + \bar{\bar{n}}_s \mathcal{D}_{12}^{s} \sum_{i}^{NS} \sqrt{m_i}$$

$$\left. \times \left[\frac{\partial C_i}{\partial x} + (1-C_i) \sum_{q=1}^{NS} \left(\frac{\bar{\bar{m}}}{m_q} \frac{\partial C_i}{\partial x} \right) \right]_s \right\} \Bigg/ \sum_{i=1}^{NS} n_i^{s} \sqrt{m_i} \qquad (26)$$

and

$$v_{0z}^{s} = \left\{ \sqrt{\frac{\pi}{2}} \left(\frac{2-\theta}{\theta} \right) \left[\frac{\mu}{\sqrt{kT}} \left(\frac{\partial v_{0z}}{\partial y} + \frac{\partial v_{0y}}{\partial z} \right) \right]_s \right.$$

$$+ \frac{1}{5} \left[\frac{1}{KT} \frac{\partial T}{\partial Z} \sum_{i=1}^{NS} \frac{n_i K_i}{\bar{\bar{n}}} \sqrt{m_i} \right]_s$$

$$+ \bar{\bar{n}}_s \mathcal{D}_{12}^{s} \sum_{i} \sqrt{m_i} \left[\frac{\partial C_i}{\partial z} + (1-C_i) \sum_{q=1}^{NS} \right.$$

$$\left. \left. \times \left(\frac{\bar{\bar{m}}}{m_q} \frac{\partial C_q}{\partial z} \right) \right]_s \right\} \Bigg/ \sum_{i=1}^{NS} n_i^{s} \sqrt{m_i} \qquad (27)$$

Temperature slip:

$$\frac{T_s}{T_w} = \left\{ - \frac{\sqrt{\pi}}{\bar{\bar{n}}_s} \sum_{i=1}^{NS} \frac{M_{iy}}{m_i} + \frac{1}{2} \sum_{i=1}^{NS} \sqrt{\frac{2kT_s}{m_i}} \left(\frac{\bar{\bar{m}}_s}{m_i} \right) \right.$$

$$\left. \times C_i^{s} \left(\frac{p_y}{p^s} + 1 \right) \right\} \qquad \text{(continued)}$$

$$/\left\{- \sqrt{\pi}\left(\frac{2-\theta}{\theta}\right)\left[\frac{1}{2}\frac{K}{p}\frac{\partial T}{\partial y} - \frac{5}{4}\sum_{i=1}^{NS}\left(\frac{M_{iy}}{\bar{\bar{n}}\, m_i}\right)\right]_s + \sum_{i=1}^{NS}\sqrt{\frac{2kT_s}{m_i}}\left(\frac{\bar{\bar{m}}_s}{m_i}\right) c_i^s\left[1 + \frac{3}{4}\left(\frac{p_y}{p^s} - 1\right)\right]\right\} \tag{28}$$

Equation (22) may also be simplified to yield an explicit expression for n_i^s:

$$n_i^s = \frac{\rho_s}{m_i}\left[1 + \frac{\left(\frac{\partial C_i}{\partial y}\right)_s + \frac{1}{m_i}\left(\frac{\psi_i}{\mathcal{D}_{12}}\right)_s}{\sum_{q=1}^{NS}\left(\frac{\bar{\bar{m}}}{m_q}\frac{\partial C_q}{\partial y}\right)_s}\right] \tag{29}$$

where

$$\psi_A^s = -\frac{\gamma_A}{(2-\gamma_A)}\frac{1}{\sqrt{\pi}}\frac{m_A n_A^s}{2\bar{\bar{n}}_s}\sqrt{\frac{2kT_s}{m_A}}\left(\frac{p_{Ay}}{p_A^s} + 1\right); \; A = O,N \tag{30a}$$

$$\psi_M^s = -\psi_A^s \;; \quad \begin{cases} A = O \text{ for } M = O_2; \\ A = N \text{ for } M = N_2 \end{cases} \tag{30b}$$

and

$$\psi_i^s = 0 \text{ for all other species} \tag{30c}$$

It is suggested here that the concentration for the major specie (e.g., nitrogen) be obtained by requiring the sum of concentrations of all the species to equal unity. It may be mentioned here that the mass of the ith species m_i is related to the molecular (or atomic) weight W_i through the relation

$$\frac{\bar{\bar{m}}_i}{W_i} = \frac{k}{\mathcal{R}} \tag{31}$$

where k is the Boltzmann constant and $\mathcal{R}$ is the universal gas constant.

Concentration Slip Boundary Condition for a Fully Catalytic and a Noncatalytic Surface

For a fully catalytic surface, the recombination coefficient $\gamma_A = 1$ and use of Eq. (30) in Eq. (29) give the appropriate slip value for concentration n_i^s.

For a noncatalytic surface ($\gamma_A = 0$), Eq. (30) gives $\psi_i^s = 0$ for all the species. For this case, Eq. (29) becomes

$$n_i^s = \frac{\rho_s}{m_i}\left[1 + \frac{\left(\frac{\partial C_i}{\partial y}\right)_s}{\sum_{q=1}^{NS}\left(\frac{\bar{\bar{m}}}{m_q}\frac{\partial C_q}{\partial y}\right)_s}\right] \tag{32a}$$

which may also be written as

$$\left(\frac{\partial C_i}{\partial y}\right)_s = -(1-C_i^s)\sum_{q=1}^{NS}\left(\frac{\bar{\bar{m}}}{m_q}\frac{\partial C_q}{\partial y}\right)_s \tag{32b}$$

Summing the above equation over all the species gives

$$\sum_{q=1}^{NS}\left(\frac{\bar{\bar{m}}}{m_q}\frac{\partial C_q}{\partial y}\right)_s = 0 \tag{33}$$

for a noncatalytic surface. Therefore, Eq. (32a) is not an appropriate boundary condition. However, employing Eq. (33) with Eq. (32b) yields:

$$\left(\frac{\partial C_i}{\partial y}\right)_s = 0 \tag{34}$$

which may be used as the boundary condition for a noncatalytic surface with a multicomponent gas mixture.

Simplified Expressions for a Binary Mixture

For a binary mixture consisting of atoms and molecules,[4] Eqs. (24) through (30) can be simplified further. By neglecting higher-order shear (i.e., $P_y/p^s \simeq 1$), these equations are obtained as

Concentration slip:

$$\frac{n_A^s}{n_A^w} = \left(\frac{2-\gamma_A}{2}\right)\sqrt{\frac{T_w}{T_s}} \tag{35}$$

Pressure slip:

$$\frac{p^s}{p^w} = \left\{1 + \frac{4}{5\sqrt{\pi}}\left(\frac{2-\theta}{\theta}\right)\frac{1}{\bar{\bar{n}}_w k T_w}\left(\frac{\bar{\bar{m}}}{\sqrt{2kT}}\frac{\partial T}{\partial y}\right)_s \right.$$

$$\left.\left[\frac{C_A^s K_A}{\sqrt{m_A}} + \frac{(1-C_A^s)K_M}{\sqrt{m_M}}\right]\right\} \Big/ \left\{1 - \frac{4}{\sqrt{\pi}}\left(\frac{2-\theta}{\theta}\right)\frac{\mathcal{D}_{AM}^s}{\sqrt{2kT_s}}\right.$$

$$\left.\times\ \bar{\bar{m}}_s\left(\frac{\sqrt{m_M} - \sqrt{m_A}}{\sqrt{m_A m_M}}\right)\left(\frac{\partial C_A}{\partial y}\right)_s\right\} \tag{36}$$

Velocity slip:

$$v_{ox}^s = \left\{\sqrt{\frac{\pi}{2}}\left(\frac{2-\theta}{\theta}\right)\left[\frac{\mu}{\sqrt{kT}}\frac{\partial v_{0x}}{\partial y}\right]_s\right\}$$

$$\Big/\left\{(\sqrt{m_A} - \sqrt{m_M})\, n_A^s + \bar{\bar{n}}_s \sqrt{m_M}\right\} \tag{37}$$

and

$$v_{oz}^s = \left\{\sqrt{\frac{\pi}{2}}\left(\frac{2-\theta}{\theta}\right)\left[\frac{\mu}{\sqrt{kT}}\frac{\partial v_{0z}}{\partial y}\right]_s\right\}$$

$$\Big/\left\{(\sqrt{m_A} - \sqrt{m_M})\, n_A^s + \bar{\bar{n}}_s \sqrt{m_M}\right\} \tag{38}$$

Temperature slip:

$$\frac{T_s}{T_w} = \left\{\sqrt{\frac{2kT_s}{m_A}}\left(\frac{\bar{\bar{m}}_s}{m_A}\right)\left[\left\{1 - \left(\frac{m_A}{m_M}\right)^{3/2}\right\} c_A\right.\right.$$

$$\left.\left. + \left(\frac{m_A}{m_M}\right)^{3/2}\right] + \sqrt{\frac{kT_w}{2m_A}}\left(\frac{\gamma_A n_A^w}{\bar{\bar{n}}_s}\right)\left(1 - \frac{m_A}{m_M}\right)\right\}$$

(continued)

$$\Big/ \Bigg\{ - \sqrt{\pi} \left(\frac{2-\theta}{\theta} \right) \left[\frac{1}{2} \frac{K}{p} \frac{\partial T}{\partial y} + \frac{5}{4\sqrt{\pi}} \left(1 - \frac{m_A}{m_M} \right) \right.$$

$$\left. \sqrt{\frac{2kT_w}{m_A}} \left(\frac{\gamma_A n_A^w}{2\bar{\bar{n}}_s} \right) \right] + \sqrt{\frac{2kT_s}{m_A}} \left(\frac{\bar{\bar{m}}_s}{m_A} \right)$$

$$\times \left[\left\{ 1 - \left(\frac{m_A}{m_M} \right)^{3/2} \right\} C_A^s + \left(\frac{m_A}{m_M} \right)^{3/2} \right] \Bigg\} \qquad (39)$$

where n_A^w is obtained from n_A^s by using Eq. (35).

Expression for n_A^s:

$$n_A^s = \frac{\rho_s}{m_A} \left[\left(\frac{1}{\bar{\bar{m}}} + \frac{1}{m_A} - \frac{1}{m_M} \right) \frac{\partial C_A}{\partial y} - \frac{2\gamma_A}{(2-\gamma_A)} \right.$$

$$\left. \times \sqrt{\frac{kT_w}{2\pi m_A}} \left(\frac{n_A}{\rho \mathcal{D}_{AM}} \right) \right]_s$$

$$\Big/ \left[\left(\frac{1}{m_A} - \frac{1}{m_M} \right) \left(\frac{\partial C_A}{\partial y} \right)_s \right] \qquad (40a)$$

which may also be rewritten as

$$\frac{n_A^s m_A}{\rho_s} = C_A^s = \left(\frac{2-\gamma_A}{2\gamma_A} \right) \sqrt{\frac{2\pi m_A}{kT_s}} \; \mathcal{D}_{AM}^s$$

$$\times \left(\frac{\partial C_A}{\partial y} \right)_s \qquad (40b)$$

For a fully catalytic surface (γ_A=1), Eq. (40b) gives appropriate value for the concentration slip. The relevant boundary condition for a noncatalytic surface (γ_A=0) will be

$$\left(\frac{\partial C_A}{\partial y} \right)_s = 0 \qquad (41)$$

as can readily be seen from Eq. (40b).

By neglecting the diffusion term in Eqs. (36) and (39) and by making further simplifying assumptions, Eqs. (36), (37), (39), and (40b) may be reduced to those obtained in Ref. 4 and employed in Ref. 7.

Simplified Expressions for a Single-Species Mixture

Equations (35-39) may also be simplified, for a single-species mixture, to those obtained in Ref. 3 if higher-order shear terms are retained (i.e. P_y/ps is not taken as unity). For small slip values, the simplified equations may be written as[8] (with γ_A=0 for a single-species mixture).

Density slip:

$$\frac{\rho_s}{\rho_w} = \frac{\bar{\bar{m}}\,\bar{\bar{n}}^s}{\bar{\bar{m}}\,\bar{\bar{n}}^w} = \sqrt{\frac{T_w}{T_s}}\left\{1 - \frac{5}{24}\sqrt{\frac{\pi}{2}}\,\frac{\lambda_s}{\sqrt{RT_s}} \times \left(\frac{\partial v_{0x}}{\partial x} + \frac{\partial v_{0z}}{\partial z} - 2\,\frac{\partial v_{0y}}{\partial y}\right)_s\right\} \tag{42}$$

Pressure slip:

$$\frac{p^s}{p_w} = 1 + \left(\frac{2-\theta}{\theta}\right)\frac{15}{16}\left(\frac{\lambda}{T}\frac{\partial T}{\partial y}\right)_s - \frac{5}{12}\sqrt{\frac{\pi}{2}}\,\frac{\lambda_s}{\sqrt{RT_s}} \times \left(\frac{\partial v_{0x}}{\partial x} + \frac{\partial v_{0z}}{\partial z} - 2\,\frac{\partial v_{0y}}{\partial y}\right)_s \tag{43}$$

Velocity slip:

$$v_{0x}^s = \left(\frac{2-\theta}{\theta}\right)\frac{5\pi}{16}\,\lambda_s\left(\frac{\partial v_{0x}}{\partial y} + \frac{\partial v_{0y}}{\partial x}\right)_s + \frac{15}{32}\sqrt{\frac{\pi}{2}\,R\,T_s}\left(\frac{\lambda}{T}\frac{\partial T}{\partial x}\right)_s \tag{44}$$

and

$$v_{0z}^s = \left(\frac{2-\theta}{\theta}\right)\frac{5\pi}{16}\,\lambda_s\left(\frac{\partial v_{0z}}{\partial y} + \frac{\partial v_{0y}}{\partial z}\right)_s \quad \text{(continued)}$$

$$+ \frac{15}{32} \sqrt{\frac{\pi}{2} R \, T_s} \left(\frac{\lambda}{T} \frac{\partial T}{\partial z} \right)_s \tag{45}$$

Temperature slip:

$$\frac{T_s}{T_w} = 1 + \left(\frac{2-\theta}{\theta} \right) \frac{75\pi}{128} \left(\frac{\lambda}{T} \frac{\partial T}{\partial y} \right)_s$$

$$- \frac{5}{48} \sqrt{\frac{\pi}{2}} \frac{\lambda_s}{\sqrt{RT_s}} \left(\frac{\partial v_{0x}}{\partial x} + \frac{\partial v_{0z}}{\partial z} - 2 \frac{\partial v_{0y}}{\partial y} \right)_s \tag{46}$$

No-Slip Boundary Condition for the Species Concentration

Multicomponent Mixture. The no-slip boundary condition may be obtained from Eqs. (29) and (30). In absence of slip, the Knudsen layer thickness shrinks to almost zero, then the values at the top of the Knudsen layer become the wall values (see Fig. 1):

$$n_i^w = \frac{\rho_w}{m_i} \left[1 + \frac{\left(\frac{\partial C_i}{\partial y} \right)_w + \frac{1}{m_i} \left(\frac{\psi_i}{\mathcal{D}_{12}} \right)_w}{\sum_{q=1}^{NS} \left(\frac{\bar{\bar{m}}}{m_q} \frac{\partial C_q}{\partial y} \right)_w} \right] \tag{47}$$

with

$$\psi_A^w = - \frac{2\gamma_A}{(2-\gamma_A)} \frac{1}{\sqrt{\pi}} \frac{m_A n_A^w}{2\bar{\bar{n}}_w} \sqrt{\frac{2kT_w}{m_A}} \; ; \quad A = 0, N \tag{48a}$$

$$\psi_M^w = - \psi_A^w \; ; \qquad \begin{cases} A = 0 \text{ for } M = O_2; \\ A = N \text{ for } M = N_2 \end{cases} \tag{48b}$$

and

$$\psi_i^w = 0 \text{ for all other species} \tag{48c}$$

where we have neglected the higher-order shear (i.e., $P_y/p^w \simeq 1$). It is suggested here again that the concentration for the major specie (e.g., nitrogen) be obtained by requiring the sum of concentrations of all the species to equal unity.

For the recombining O and N atoms, Eq. (47) may further be simplified to

$$n_A^w = \frac{1}{k_{wA}} \left[\bar{\bar{n}} \mathcal{D}_{12} \left\{ \frac{\partial C_A}{\partial y} + (1-C_A) \sum_{q=1}^{NS} \left(\frac{\bar{\bar{m}}}{m_q} \frac{\partial C_q}{\partial y} \right) \right\} \right]_w \tag{49}$$

The recombination rate constant k_{wA} in Eq. (49) has been defined as[5]

$$k_{wA} = \left(\frac{2\gamma_A}{2-\gamma_A} \right) \sqrt{\frac{kT_w}{2\pi m_A}} \tag{50}$$

without slip and higher-order shear.

For a noncatalytic wall (with $\gamma_A = 0$), Eq. (34) gives

$$\left(\frac{\partial C_i}{\partial y} \right)_w = 0 \tag{51}$$

for all the species of a multicomponent mixture.

Binary Mixture. For a two-species mixture of atoms and molecules, Eq. (40b) gives

$$C_A^w = \frac{1}{k_{wA}} \times \left(\mathcal{D}_{AM} \frac{\partial C_A}{\partial y} \right)_w \tag{52}$$

for a surface with finite catalycity and

$$\left(\frac{\partial C_A}{\partial y} \right)_w = 0 \tag{53}$$

for a noncatalytic surface. Eqs. (52) and (53) are obtained in Ref. 12.

Appendices D and E of Ref. 8 give the slip and no-slip boundary conditions (presented here earlier) in dimensionless form for the body-fitted and spherical polar coordinates, respectively.

Concluding Remarks

A consistent formulation for the slip equations for a multicomponent, binary, and single-species mixture (reacting catalytically on a surface) has been provided. The complete expressions give implications of various assumptions (some of them inconsistent) employed by other researchers. In Ref. 8, slip equations have also been obtained in body-fitted and spherical polar coordinates in a form which can readily be employed in the flowfield calculations.

The equation for pressure slip is not required as a boundary condition. It is needed for obtaining the wall pressure only. Similarly, when the surface temperature is not specified and the surface is adiabatic, the temperature slip equation is needed to obtain the surface temperature T_w. The slip temperature T_s is obtained in this case by equating the heat-transfer rate to zero, i.e.,

$$q_w = \left(K \frac{\partial T}{\partial y} - \sum_{i=1}^{NS} j_i \, h_i + \mu \, v_{0x} \frac{\partial v_{0x}}{\partial y} \right)_s = 0 \tag{54}$$

where the expression for mass flux j_i is provided in Appendix B, and the higher-order terms have been neglected. A study to access the importance of various terms in the general boundary equations provided here is under way.

In conclusion, the boundary slip expressions obtained here are closed form solutions of the mass, momentum, and energy flux equations using the Chapman-Enskog velocity distribution function. This function represents a solution of the Boltzmann equation in the Navier-Stokes approximation.

Appendix A: Flux Expressions

The explicit expressions for the fluxes of mass, momentum, and energy for the ith species are (after neglecting thermal diffusion):

$$M_{ik} = \bar{\bar{n}}_s m_i \sum_{\substack{j=1 \\ j \neq i}}^{NS} \Bigg\{ D_{ij} \left[\frac{\partial C_j}{\partial x_k} - C_j \sum_{q=1}^{NS} \right.$$

(continued)

$$\times\left(\frac{\bar{\bar{m}}}{m_q}\frac{\partial C_q}{\partial x_k}\right)\Big]\Big\}_s \quad ; \quad k = x,\ y,\ z \tag{A1}$$

$$M^{\downarrow}_{ik} = -\frac{m_i\, n^s_i}{2\sqrt{\pi}}\sqrt{\frac{2kT_s}{m_i}}\left[\frac{1}{2}\left(\frac{P_{iy}}{p^s_i} + 1\right)\right] + \frac{1}{2}\, M_{ik} \tag{A2}$$

$$P_{iy} = p^s_i\left(1 + \tau_{yyi}\right)_s = p^s_i\Big[1 + \frac{b_{i0}}{3}$$

$$\times\left(\frac{\partial v_{0x}}{\partial x} + \frac{\partial v_{0z}}{\partial z} - 2\,\frac{\partial v_{0y}}{\partial y}\right)\Big]_s \tag{A3}$$

$$\tau_{xyi} = -\, p^s_i\,\frac{b_{0i}}{2}\left(\frac{\partial v_{0x}}{\partial x} + \frac{\partial v_{0y}}{\partial x}\right)_s \tag{A4}$$

$$E_{ik} = \frac{5}{8}\, m_i\, n_i{}^s\left(\frac{2kT_s}{m_i}\right)^{3/2} a_{i1}\,\frac{\partial \ln T}{\partial x_k}$$

$$+\,\frac{5}{2}\,\bar{\bar{n}}_s k\, T_s \sum_{\substack{j=1\\ j\neq i}}^{NS}\Big\{ D_{ij}\Big[\frac{\partial C_i}{\partial x_k}$$

$$-\, C_j \sum_{q=1}^{NS}\frac{\bar{\bar{m}}}{m_q}\frac{\partial C_q}{\partial x_k}\Big]\Big\}_s \quad ; \quad k = x,y,z \tag{A5}$$

Where the Sonine expansion coefficients a_{i1} and b_{i0} are related to the thermal conductivity and viscosity, respectively, of the mixture through the relations[9]

$$K = -\frac{5}{4}\, k \sum_{i=1}^{NS} n_i\sqrt{\frac{2kT}{m_i}}\; a_{i1}$$

$$\mu = \frac{1}{2}\, kT \sum_{i=1}^{NS} n_i\, b_{i0} \tag{A6}$$

Appendix B: Alternate Form for Mass Flux Expressions

An alternate form for the mass flux Eq. (A1), with the simplifications for a multicomponent mixture described earlier in the text, may be written as[8]:

$$j_i = M_{iy} = -\frac{\mu}{Pr}\left(Le_i \frac{\partial C_i}{\partial y} + \sum_{\substack{j=1 \\ j \neq i}}^{NS} \overline{\overline{\Delta b}}_{ij} \frac{\partial C_i}{\partial y} \right) \tag{B1}$$

where

$$Le_i = Le\left[\frac{W_i}{\overline{\overline{W}}} + (1-C_i)\left(\frac{\partial C_i}{\partial y} \right)^{-1} \sum_{j=1}^{NS} \times \left(\frac{W_i}{W_j} \frac{\partial C_i}{\partial y} \right) \right] \tag{B2}$$

$$\overline{\overline{\Delta b}}_{ij} = Le_i - Le\left[\frac{W_i}{\overline{\overline{W}}} + \left(1 - \frac{W_i}{W_j} \right)(1-C_i) \right] \tag{B3}$$

$$Le = \rho C_p \mathcal{D}_{12}/K = \rho Pr\, \mathcal{D}_{12}/\mu \tag{B4}$$

and $\mathcal{D}_{12}$ is the binary diffusion coefficient.

The form of Eq. (B1) is more appealing in the flowfield calculations.[11] Equation (B2) differs from the corresponding expression for Le_i given in Ref. 10 for the reasons explained in Ref. 8.

Acknowledgments

The authors wish to acknowledge many useful discussions (pertaining to the mass flux expressions contained in Appendix B) with Dr. F. G. Blottner of Sandia National Laboratories, Albuquerque, New Mexico. This work was sponsored by NASA Langley Research Center, Hampton, Virginia, under Grant NAG-1-346.

References

[1]Walberg, G. D., "Aeroassisted Orbit Transfer-Window Opens on Missions," Astronautics and Aeronautics, Nov. 1983, pp. 36-43.

[2]Patterson, G. N., Molecular Flow of Gases, John Wiley and Sons, New York, 1956.

[3]Shidlovskiy, V. P., Introduction to the Dynamics of Rarefied Gases, edited and translated by J. A. Laurmann, Americal Elsevier Publishing Co., New York, 1976.

[4]Davis, R. T., "Hypersonic Flow of a Chemically Reacting Binary Mixture Past a Blunt Body," AIAA Paper 70-805, AIAA 3rd Fluid and Plasma Dynamics Conference, Los Angeles, California, July 1970.

[5]Scott, C. D., "Wall Boundary Equations with Slip and Catalysis for Multicomponent, Nonequilibrium Gas Flows," NASA TM X-58111, Dec. 1973.

[6]Hendricks, W. L., "Slip Conditions with Wall Catalysis and Radiation for Multicomponent, Nonequilibrium Gas Flows," NASA TM X-64942, June 1974.

[7]Shinn, J. L. and Simmonds, A. L., "Comparison of Viscous-Shock-Layer Heating Analysis with Shuttle Flight Data in Slip Flow Regime," AIAA Paper 84-0226, Jan. 1984.

[8]Gupta, R. N., Scott, C. D., and Moss, J. N., "Surface-Slip Equations for Multicomponent, Nonequilibrium Gas Flows," NASA TM 85820, Dec. 1984.

[9]Hirshfelder, J. O., Curtiss, C. F., and Bird, R., Molecular Theory of Gases and Liquids, John Wiley and Sons, New York, 1954, pp. 468-490.

[10]Blotter, F. G., "Finite Difference Methods of Solution of the Boundary-Layer Equations," AIAA Journal, Vol. 8, No. 2, Feb. 1970, pp. 193-205.

[11]Scott, C. D., "An Experimental and Analytical Study of Slip and Catalytic Boundary Conditions Applied to Spheres in Low Reynolds Number Arc Jet Flows," Proceedings of the 9th International Symposium on Rarefied Gas Dynamics, Gottingen, July 15-20, 1974, pp. D.14-1, 11.

[12]Shinn, J. L., Moss, J. N., and Simmonds, A. L., "Viscous Shock-Layer Heatings Analysis for the Shuttle Windward-Symmetry Plane with Surface Finite Catalytic Recombination Rates," AIAA Paper 82-0842, AIAA/ASME 3rd Joing Thermo-physics, Fluids, Plasma and Heat-Transfer Conference, St. Louis, Missouri, June 1982.

Comparison of Viscous Shock-Layer Heating Analysis with Shuttle Flight Data in Slip Flow Regime

Judy L. Shinn* and Ann L. Simmonds†
NASA Langley Research Center, Hampton, Virginia

Abstract

Comparison of Space Transportation System Shuttle flight 2 heating data along the windward centerline has been made with two-dimensional nonequilibrium viscous shock-layer solutions obtained with shock slip and wall slip conditions at an altitude range of 90 to 110 km. The shock slip condition used is the modified Rankine-Hugoniot relations of Cheng as used by Davis, and the wall slip conditions are based on the first-order consideration derived from kinetic theory as given by Scott and Hendricks. The results indicate that the calculated heating distributions with slip boundary conditions agree better with the flight data than those without slip conditions. The agreement improves when the accommodation coefficient or freestream density is decreased to one-half, suggesting the possibility of less than full accommodation for the tile surface and (or) an overestimate of freestream density using the Jacchia-Roberts model. Heating reduction due to the slip effect becomes very pronounced as the flow becomes more rarefied, and the effect is more significant for the stagnation region than the aft region of the vehicle.

Nomenclature

C_i - mass fraction of species i = ρ_i/ρ

C_p = frozen specific heat of mixture = $\sum_i C_i C_{p,i}$

Presented as Paper 84-0226 at AIAA 22nd Aerospace Sciences Meeting, Jan. 9-12, 1984, Reno, Nev.

*Research Engineer, Aerothermodynamics Branch, Space Systems Division.

†Mathematician, Aerothermodynamics Branch, Space Systems Division.

$C_{p,i}$ = specific heat of species i = $C^*_{p,i}/C^*_{p,\infty}$
D_{ij} = binary diffusion coefficients
h = enthalpy of mixture = $\sum_i C_i h_i$
h_i = enthalpy of species i = h^*_i/U^{*2}_∞
k^*_w = surface catalytic recombination rate
K = thermal conductivity of mixture = $K^*/\mu^*_{ref}\ C^*_{p,\infty}$
M^*_i = molecular weight of species i
$\bar{M}^*$ = molecular weight of mixture
n = coordinate measured normal to the body = n^*/R^*_N
N_s = number of reacting species
N_{Le} = Lewis number = $\rho^* D^*_{ij} C^*_p/K^*$
N_{Pr} = Prandtl number = $\mu^* C^*_p/K^*$
N_{Re_∞} = Reynolds number = $\rho^*_\infty\ U^*_\infty\ R^*_N/\mu^*_\infty$
p = pressure = $p^*/\rho^*_\infty U^{*2}_\infty$
q_w = wall heat-transfer rate = $q^*_w/\rho^*_\infty\ U^{*3}_\infty$
r = radius measured from axis of symmetry to a point on the body surface = r^*/R^*_N
R^* = universal gas constant
R^*_N = nose radius
s = coordinate measured along the body surface = s^*/R^*_N
STS-2 = Space Transportation System Shuttle flight 2
T = temperature = T^*/T^*_{ref}
T^*_{ref} = reference temperature = $U^{*2}_\infty/C^*_{p\infty}$
u = velocity component tangent to body surface = u^*/U^*_∞
U^*_∞ = freestream velocity
v = velocity component normal to body surface = v^*/U^*_∞
VSL = viscous shock layer
$\dot{w}_i$ = mass rate of formation of species i = $\dot{w}^*_i\ R^*_N/\rho^*_\infty U^*_\infty$
x/L = nondimensionalized orbiter axial length
α = shock angle defined in Fig. 1
γ_i = catalytic recombination coefficient of species i

ε = Reynolds number parameter = $(\mu^*_{ref}/\rho^*_\infty U^*_\infty R^*_N)^{1/2}$
θ = body angle defined in Fig. 1
κ = body curvature = $\kappa^* R^*_N$
λ = molecular mean free path
σ = accommodation coefficient
μ = viscosity of mixture = μ^*/μ^*_{ref}
μ_{ref} = reference viscosity = $\mu^*(T^*_{ref})$
ρ = density of mixture = ρ^*/ρ^*_∞

Subscripts and Superscripts

i = ith species
w = wall value
∞ = freestream condition
$*$ = dimensional quantity
' = shock oriented velocity components (see Fig. 1)

Introduction

Research emphasis on the hypersonic, low Reynolds number flow regime has increased recently owing to various future space programs such as space stations and orbital transfer vehicles.[1-2] Several studies[3-7] in the past have indicated that the continuum flow analysis can be extended to low Reynolds number flow if slip effects are properly accounted for. One of the continuum approaches which has been explored in the past is the viscous shock-layer (VSL) method. This method has the advantage of requiring much less computing cost when compared to the time-dependent Navier-Stokes solutions. Davis,[3] who developed the VSL method, obtained favorable comparisons with Little's experimental data[8] for drag coefficients in the low Reynolds number flow regime when slip conditions were included. Anderson and Moss[4] further verified the accuracy of the VSL method at low Reynolds number with slip conditions by comparing it with the steady-state Navier-Stokes solutions. Recently, Chrusciel et al.[9] made a comparison between Monte Carlo predictions and the results of three-dimensional VSL analysis with shock and wall slip boundary conditions for a hypersonic blunt body in the slip flow regime and obtained fairly good agreement. However, none of the aforementioned verifications of the VSL method in slip flow regime was for a chemically reacting gas.

The importance of including chemical reactions and wall catalysis in predicting flowfield and surface heating around a hypersonic vehicle at high altitude has been recognized for some time. This was demonstrated in a previous study[10] in which an analysis of the Space Shuttle heating data at an altitude range of 50 to 90 km for Space Transportation System Shuttle flight 2 (STS-2) was conducted. The windward centerline heating was predicted using a two-dimensional VSL analysis around an "equivalent axisymmetric hyperboloid."[10] The results showed that the flight data were in much closer agreement to the nonequilibrium (chemically reacting) VSL, finite catalytic wall predictions than the equilibrium predictions.

The purpose of the present study is to extend the previous[10] analysis to a higher altitude range where slip effect on heating becomes significant. Comparisons of VSL analysis with STS-2 flight heating data along the windward centerline are made for several trajectory points for the altitude range of 90 to 110 km. Both shock slip and wall slip boundary conditions for chemically reacting gas are incorporated into the existing two-dimensional nonequilibrium VSL code. The shock slip conditions used are the modified Rankine-Hugoniot relations of Cheng as having been used by Davis.[11] The wall slip conditions used are based on the first-order considerations[12-13] derived from kinetic theory. Inputs for surface accommodation coefficient (which appears in wall slip conditions) and freestream density are also varied in an attempt to identify possible causes of discrepancy. Typical shock layer profiles, in this low Reynolds number regime, with and without slip conditions are also compared.

Analysis

Governing Equations

The viscous shock-layer (VSL) equations are obtained from the steady-state Navier-Stokes equations by keeping terms up to second order in ε, the inverse square root of the Reynolds number. Consequently, one set of equations is solved for both the inviscid and viscous region. The nondimensional VSL equations developed in Ref. 14, with the body oriented coordinate system shown in Fig. 1, are global continuity:

$$\frac{\partial}{\partial s}[(r + n\cos\theta)\,\rho u] + \frac{\partial}{\partial n}[(1 + n\kappa)(r + n\cos\theta)\,\rho v] = 0 \qquad (1)$$

s momentum:

$$\rho\left(\frac{u}{1+n\kappa}\frac{\partial u}{\partial s}+v\frac{\partial u}{\partial n}+\frac{uv\kappa}{1+n\kappa}\right)$$

$$+\frac{1}{1+n\kappa}\frac{\partial p}{\partial s}=\varepsilon^2\left\{\frac{\partial}{\partial n}\left[\mu\left(\frac{\partial u}{\partial n}-\frac{u\kappa}{1+n\kappa}\right)\right]\right.$$

$$\left.+\mu\left(\frac{2\kappa}{1+n\kappa}+\frac{\cos\theta}{r+n\cos\theta}\right)\left(\frac{\partial u}{\partial n}-\frac{u\kappa}{1+n\kappa}\right)\right\} \tag{2}$$

n momentum:

$$\rho\left(\frac{u}{1+n\kappa}\frac{\partial v}{\partial s}+v\frac{\partial v}{\partial n}-\frac{u^2\kappa}{1+n\kappa}+\frac{\partial p}{\partial n}\right)=0 \tag{3a}$$

where the thin shock-layer form of Eq. (3a) is

$$\frac{\partial p}{\partial n}=\frac{\rho u^2\kappa}{1+n\kappa} \tag{3b}$$

energy (temperature):

$$\rho C_p\left(\frac{u}{1+n\kappa}\frac{\partial T}{\partial s}+v\frac{\partial T}{\partial n}\right)-\left(\frac{u}{1+n\kappa}\frac{\partial p}{\partial s}+v\frac{\partial p}{\partial n}\right)$$

$$=\varepsilon^2\left[\frac{\partial}{\partial n}\left(K\frac{\partial T}{\partial n}\right)+\left(\frac{\kappa}{1+n\kappa}+\frac{\cos\theta}{r+n\cos\theta}\right)\right.$$

$$\times\left(K\frac{\partial T}{\partial n}\right)+\frac{\mu N_{Le}}{N_{Pr}}\left(\sum_{i=1}^{N_s}C_{p,i}\right)\frac{\partial C_i}{\partial n}\frac{\partial T}{\partial n}$$

$$\left.+\mu\left(\frac{\partial u}{\partial n}-\frac{\kappa u}{1+n\kappa}\right)^2\right]-\sum_{i=1}^{N_s}h_i\dot{w}_i \tag{4}$$

species continuity:

$$\rho\left(\frac{u}{1+n\kappa}\frac{\partial C_i}{\partial s}+v\frac{\partial C_i}{\partial n}\right)=\dot{w}_i+\frac{\varepsilon^2}{(1+n\kappa)(r+n\cos\theta)}$$

$$\times\left[\frac{\partial}{\partial n}\left((1+n\kappa)(r+n\cos\theta)\frac{\mu N_{Le}}{N_{Pr}}\frac{\partial C_i}{\partial n}\right)\right] \tag{5}$$

state:

$$p = \rho T R^*/\bar{M}^* C^*_{p,\infty} \tag{6}$$

Boundary Conditions

In low Reynolds number hypersonic flow, slip phenomena occur near the wall such that the velocity, temperature, pressure, and species concentration of the wall are no longer the same as that of the gas immediately adjacent to the wall. In this flow regime, the mean free path is not small compared to the characteristic length of the body, and the continuum model breaks down in regions of large gradient, such as near the wall. The approach of kinetic theory taken by various researchers[12-13,15-16] for obtaining wall slip expressions are basically the same; that is, use of Chapman-Enskog expansion for the distribution function and assumption of a certain "slip model" within the Knudsen layer. Among these, only Scott[12] and Hendricks[13] have treated for multicomponent nonequilibrium gas with wall catalysis. The "slip model" assumed in this study follows that of Scott and Hendricks, such that the molecules (atoms) specularly reflected from the wall are associated with the slip temperature, while the molecules diffusely reflected are associated with the wall temperature. The accommodation coefficient for species i, σ_i is the fraction of those molecules (atoms) colliding with the surface that stick (and then diffusely reflect), and γ_i is the fraction of those incident atoms that stick and also recombine at the surface. The wall slip boundary conditions so derived can then be simplified to the following:

$$u = \sqrt{\frac{\pi}{2}}\left(\frac{2-\sigma}{\sigma}\right)\varepsilon^2 \frac{\mu}{\sqrt{p\rho}}\left(\frac{\partial u}{\partial n} - \frac{\kappa u}{1+\kappa n}\right) \tag{7}$$

$$T = T_w + \frac{15}{8}\sqrt{\frac{\pi}{2}}\left(\frac{2-\sigma}{\sigma}\right)\varepsilon^2 \frac{\mu}{\sqrt{p\rho}}\left(\frac{\partial T}{\partial n}\right) \tag{8}$$

$$p = p_w + \frac{3}{\sqrt{2\pi}}\left(\frac{2-\sigma}{\sigma}\right)\varepsilon^2 \frac{\mu}{T}\frac{p}{\rho}\left(\frac{\partial T}{\partial n}\right) \tag{9}$$

$$C_i = \frac{N_{Le}\,\mu\,\varepsilon^2}{k_{wi}\,\rho\,N_{Pr}}\left(\frac{2-\gamma_i}{2}\right)\sqrt{\frac{T_w}{T}}\left(\frac{\partial C_i}{\partial n}\right) \tag{10}$$

where the simplification $\sigma=\sigma_i$ for all species i is made, and

$$\gamma_i = \sqrt{\frac{2\pi M_i^\star}{R^\star T^\star}}\ k_{wi}^\star$$

as before.[10]

The total heat transfer to the wall due to conduction, diffusion, and sliding friction[17] are given by

$$q_w = \varepsilon^2 \left(K\frac{\partial T}{\partial n} + \frac{\mu N_{Le}}{N_{Pr}} \sum_{i=1}^{N_s} h_i \frac{\partial C_i}{\partial n} + \mu\, u\frac{\partial u}{\partial n} \right)_w \qquad (11)$$

It is known that as the Reynolds number decreases in hypersonic flow, the thickness of shock wave increases such that the Rankine-Hugoniot discrete shock relations become inadequate. In the present study, the modified Rankine-Hugoniot relations[11] of Cheng are used for shock slip conditions. With u' and v' denoting the velocity components tangent and normal to the shock (see Fig. 1), respectively, the conditions are given by

$$\rho_s V_s' = -\sin\alpha \qquad (12)$$

$$\varepsilon^2 \mu_s \left(\frac{\partial u'}{\partial n}\right)_s + \sin\alpha \times u_s' = \sin\alpha \times \cos\alpha \qquad (13)$$

$$p_s - \sin\alpha \times V_s' = p_\infty + \sin^2\alpha \qquad (14)$$

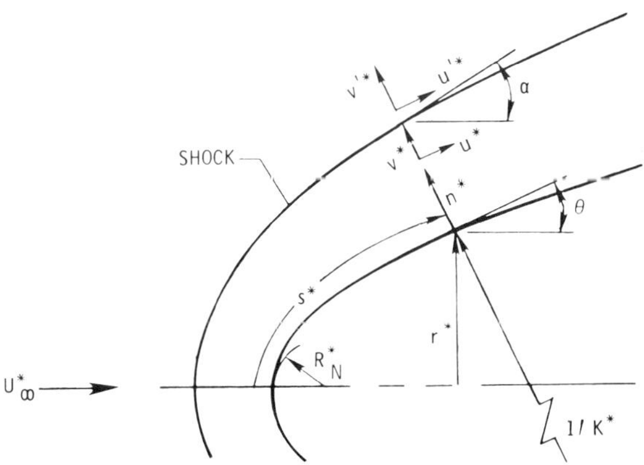

Fig. 1 Coordinate system.

$$\varepsilon^2 K_s \left(\frac{\partial T}{\partial n}\right)_s + \sin\alpha \times \sum_{i=1}^{N_s} C_{i,\infty} h_{i,s} - \frac{\sin\alpha}{2}$$

$$\times \left[(u_s' - \cos\alpha)^2 + \sin\alpha^2 - V_s'^2\right] = \sin\alpha \times \sum_{i=1}^{N_s} C_{i,\infty} h_{i,\infty} \quad (15)$$

$$\varepsilon^2 \left(\frac{\mu_s N_{Le,s}}{N_{Pr,s}}\right) \left(\frac{\partial C_i}{\partial n}\right)_s + \sin\alpha \times C_{i,s} = \sin\alpha \times C_{i,\infty} \quad (16)$$

Chemical Kinetics, Thermodynamic, and Transport Properties

The details of the five-species air chemical kinetics, thermodynamic, and transport properties used in this study are given in Refs. 10 and 14. Since the assumption of constant Prandtl number results in large error in mixture thermal conductivity for high temperature nonequilibrium flow, this assumption is removed in this study. The mixture thermal conductivity is calculated using the approximate formula derived by Mason and Saxena[18] which has been shown[18-19] to be accurate to within 10%.

Method of Solution

The method used in this study for solving the nonequilibrium VSL equations with slip boundary conditions is essentially the same as described in Ref. 14 for no-slip conditions. However, there are some unavoidable changes made. One of these involves an extra iterative routine for shock quantities at every station to satisfy the modified Rankine-Hugoniot relations. Within each of these iterations, similar steps as for the no-slip case are taken to obtain converged shock-layer profiles. These iterations cause the computing time to increase by about a factor of 5.

Another significant change made involves the technique Moss[14] has used to obtain stagnation solutions. Rather than assuming flow quantities at the shock to be symmetric or asymmetric with respect to the stagnation streamline as other researchers did,[20] Moss used a truncated series similar to that used by Kuo. The resulting series expansion for tangential velocity at the shock [Eq. C(46) of Ref. 14] has been reformulated in this study for slip conditions.

Table 1 Trajectory conditions and effective body parameters for STS-2 entry

Altitude, km	U^*_∞, km/s	ρ^*_∞, kg/m^3	T^*_∞, K	Mach No.	Angle of attack	θ, deg	R^*_N, m	λ_∞/R_N	ε	$N_{Re,\infty}$	T^*_w, K
92.35	7.50	2.184×10^{-6}	180	27.90	40.4	41.15	1.295	0.028	0.13	1752	1040-650
96.58	7.50	1.001×10^{-6}	184	27.60	39.9	40.65	1.271	0.061	0.19	772	890-540
99.49	7.50	5.906×10^{-7}	190	27.14	41.8	42.50	1.362	0.096	0.24	474	780-480
105.31	7.49	2.212×10^{-7}	213	25.59	40.9	41.60	1.317	0.259	0.39	156	500[a]
109.75	7.49	1.049×10^{-7}	242	24.02	40.7	41.40	1.309	0.539	0.57	66	420[a]

[a]Constant temperature is assumed for the windward centerline.

Table 2 Freestream compositions[a]

Altitude, km	C_O	C_{O_2}	C_{N_2}	$\overline{M}^*$
92.35	0.0	0.240	0.760	28.96
95.58	0.011	0.217	0.772	28.65
99.49	0.022	0.200	0.778	28.36
105.31	0.042	0.171	0.787	27.80
109.75	0.065	0.146	0.789	27.23

[a]From Ref. 24, based on exospheric temperature of 1000 K.

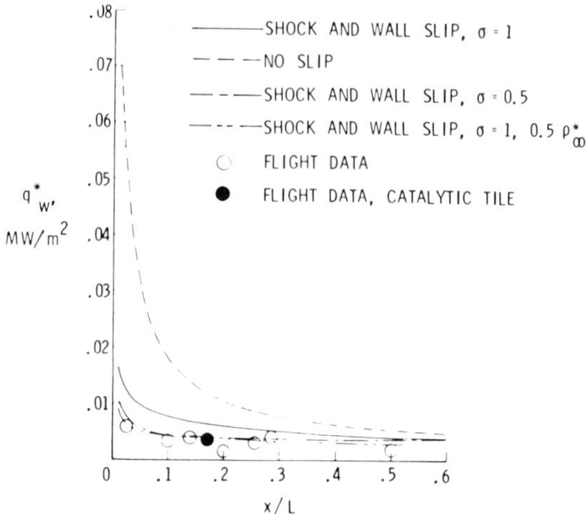

Fig. 2 Comparison of measured and calculated heating distribution at an altitude of 110 km.

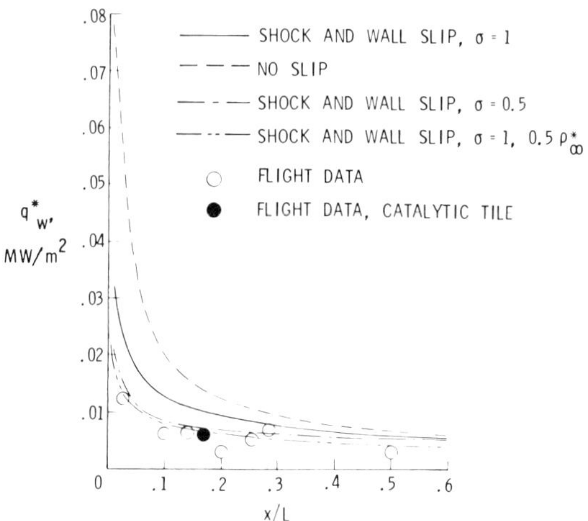

Fig. 3 Comparison of measured and calculated heating distribution at an altitude of 105 km.

Results and Discussion

Laminar convective heating rates and typical flowfield results are presented for the windward centerline of the Space Shuttle at several of the selected STS-2 flight conditions above 90 km in altitude where the difference in heating due to slip phenomena is signifi-

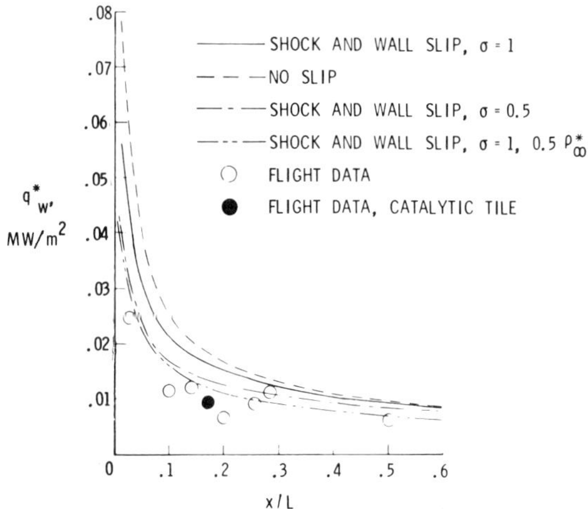

Fig. 4 Comparison of measured and calculated heating distribution at an altitude of 100 km.

cant. The VSL results calculated with and without slip boundary conditions are compared to the flight heating rate values as deduced[21] from surface temperature measurements. The results obtained by varying surface accommodation coefficient and freestream density are also presented.

Freestream Conditions

The freestream conditions for the STS-2 flight and the equivalent body parameters (Ref. 10) θ and R_N for which solutions have been obtained are listed in Table 1, where the altitude ranges from 92 to 110 km. The calculated freestream Reynolds number based on the nose radius ranges from 1752 to 66, while the Reynolds number parameter ε and Knudsen number γ_∞/R_N ranges from 0.13 to 0.57 and 0.028 to 0.54, respectively. The trajectory parameters, such as velocity and angle of attack, were determined from a trajectory reconstruction process as described in Ref. 22. Freestream density and temperature were determined by a process[23] that combined atmospheric modeling and meteorological data taken close to the time of Shuttle flight. Since measured meteorological data are not available above 90 km, the Jacchia-Roberts model[24] was used in this region, with the boundary values chosen to match the meteorological profiles below 90 km.[23] Therefore, the uncertainty in freestream density above 90 km obtained from this process is expected to

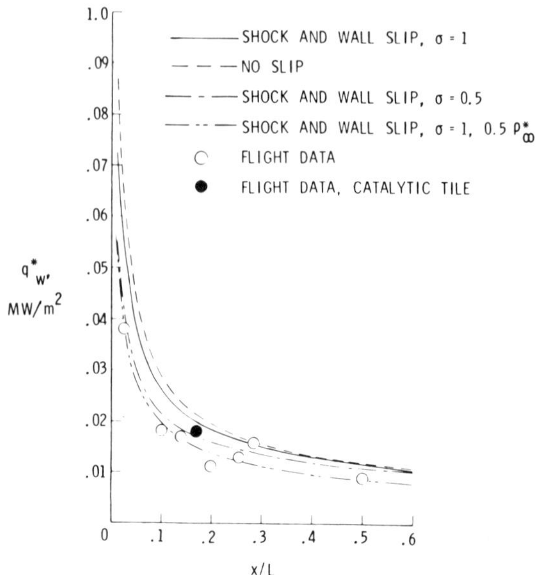

Fig. 5 Comparison of measured and calculated heating distribution at an altitude of 96 km.

increase with altitude, but its magnitude cannot be determined. The corresponding freestream compositions that contain a significant amount of dissociated species were obtained from Ref. 24 and are listed in Table 2.

Heating Data and Present Results

The deduced heating data and the calculated finite catalytic wall results obtained with and without both shock and wall slip conditions are shown as a function of axial distance along the windward centerline in Figs. 2-5 (96 km and above), separate values of accommodation coefficient (σ = 1 and 0.5) are assumed. Since the uncertainty in freestream density can be very large at this altitude range, calculation is also made with half of the value of estimated[23] freestream density and full accommodation for slip conditions. The results shown in Figs. 2-5 indicate that the calculated heating distributions with slip boundary conditions agree better with the flight data than those without slip conditions. The agreement improves when the accommodation coefficient or the freestream density is decreased to half of its full value. It is conceivable that a combination of a more realistic value for accommodation coefficient (say, 0.7

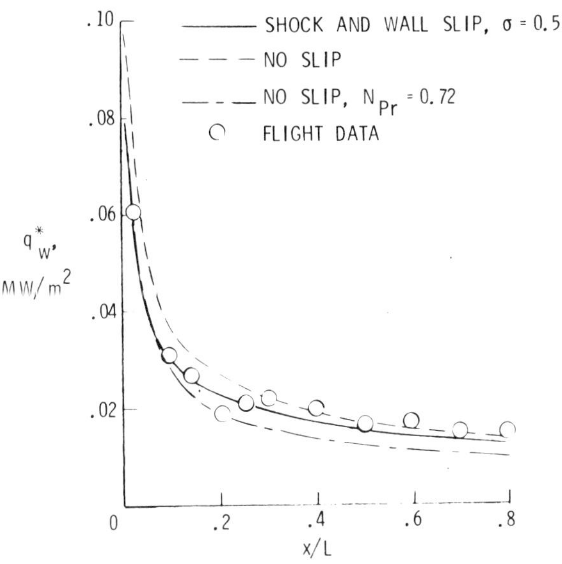

Fig. 6 Comparison of measured and calculated heating distribution at an altitude of 92 km.

or 0.8) and a more moderate reduction in estimated free-stream density (say 30%) might have been the true situation. Since the previous study[10] showed good agreement between flight data and calculated results, the same chemical reaction rates used here are not considered to be accountable for such large discrepancy. For the lowest altitude case (Fig. 6) considered, the difference between the calculated result for no-slip and slip (with 50% accommodation) case appears to be small when compared to the range of scatter in the data for the aft region of the vehicle. However, for the forward region, a superior prediction is obtained with the slip conditions. Also shown in Fig. 6 is the calculated result for another no-slip case, with the Prandtl number assumed to be 0.72. It is presented here to demonstrate that such an assumption of constant Prandtl number can result in a quite different answer.

As mentioned earlier, the results obtained in this study are for finite catalytic wall conditions. Since all the input wall temperatures are below 1040 K as tabulated in Table 1, the catalytic efficiencies γ_i used are based on the Vycor data[25] shown in Fig. 14 of Ref. 25. However, the accuracy of γ_i has very little effect on heating results in the low-density region;[10] and, therefore, it will not be a subject of concern for

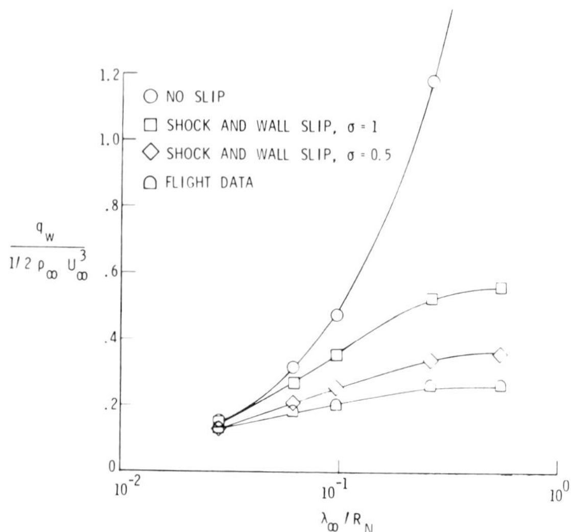

Fig. 7 Heat-transfer coefficient at x/L = 0.025 as a function of Knudsen number.

this study. It is interesting to note that at 96 km and above the measured heating rate for the catalytic tiles shown by the shaded circles fall within the scatter of those for the noncatalytic tiles.

The altitude range chosen for the present study is based on the extent of validity for VSL method at one end and the significance of slip effect on heating at the other end. It was shown in Ref. 4 that at a Reynolds number of 90 the perfect-gas VSL solutions for a drag coefficient of a 45-deg hyperboloid were in agreement with the Navier-Stokes solution within 2% in downstream region and 15% in the stagnation region. Therefore, for the highest altitude case considered in this study (with $N_{Re,\infty}$= 66 and ε = 0.57), the results obtained are expected to be quite reliable.

The effect of slip phenomena on surface heating across the regions of various degrees of rarefaction can be seen from Figs. 7-9, where the heat-transfer coefficient is plotted as a function of freestream Knudsen number at the body location of x/L = 0.025, 0.14, and 0.255, respectively. The results are obtained from the no-slip case, from slip with full and half accommodation, and also from flight data. The no-slip result is seen to diverge very quickly with increasing Knudsen number. And the effect of slip on heating reduction is more significant for a larger Knudsen number, and strongest for the stagnation region. It also shows here that when the

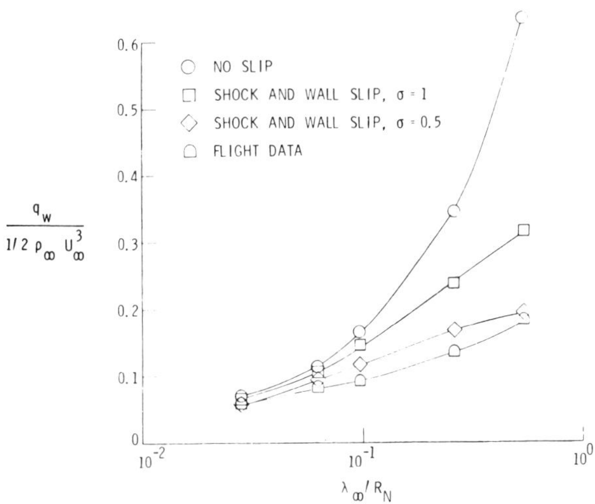

Fig. 8 Heat-transfer coefficient at x/L = 0.14 as a function of Knudsen number.

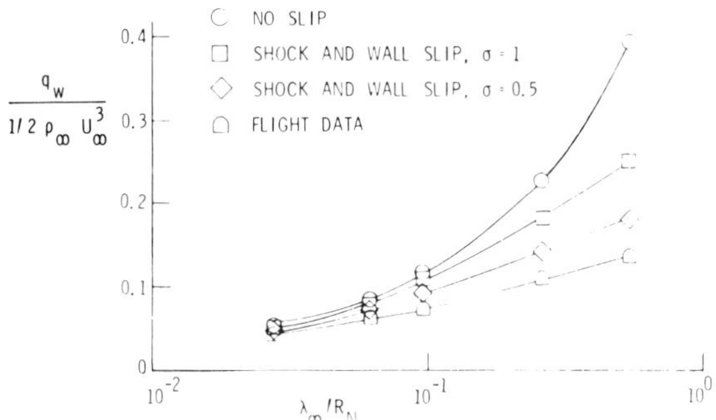

Fig. 9 Heat-transfer coefficient at x/L = 0.255 as a function of Knudsen number.

accommodation is reduced to 0.5, the agreement with the flight data becomes closer but still overpredicts the experiment. A similar finding was also obtained in Ref. 26, with Monte Carlo results that indicate a closer agreement with the flight data but overpredict the flight data at an altitude of 110 km when 50% accommodation was assumed. It should be mentioned here that the definition of accommodation coefficient in this study is identical to that in Ref. 26.

Flowfield Structure

Typical shock-layer profiles obtained with and without shock slip and wall slip conditions are presented

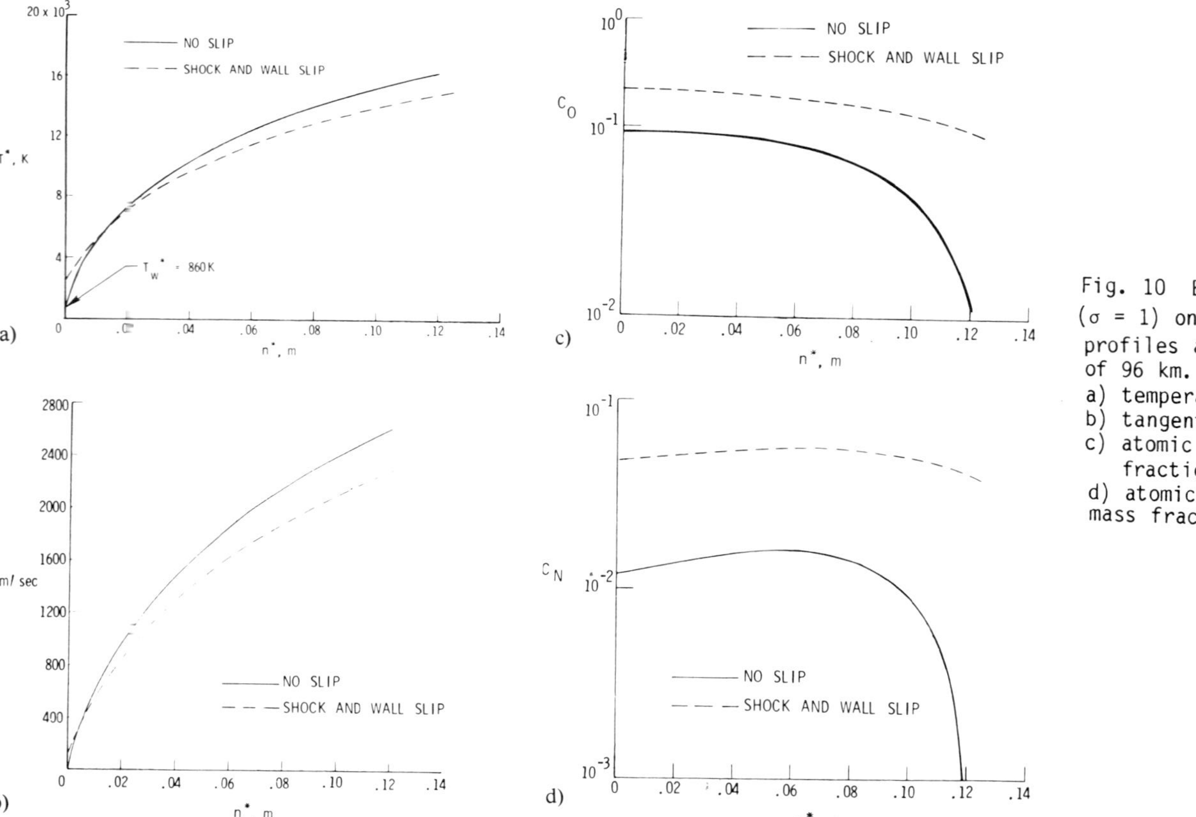

Fig. 10 Effect of slip ($\sigma = 1$) on shock-layer profiles at an altitude of 96 km. $x/L = 0.02$:
a) temperature;
b) tangential velocity;
c) atomic oxygen mass fraction;
d) atomic nitrogen mass fraction.

in Fig. 10 at a body location of x/L = 0.02 for the case of 96 km altitude. The different temperature profiles shown in Fig. 10a indicate that while the actual wall temperature is about 860 K, the gas temperature near the wall jumps to a value of 2300 K owing to the slip phenomenon. The temperature at the shock location is lower for the slip case than for the no-slip case; however, in reality there is a shock structure of finite thickness that extends beyond the calculated shock location. Similar results for tangential velocity are also depicted in Fig. 10b with a velocity slip occurring near the wall for the slip result. The atomic oxygen and nitrogen mass fraction profiles plotted in Figs. 10c and 10d indicate a higher degree of dissociation for the slip case than for the no slip case throughout the shock layer. Similar findings are also obtained for the Monte-Carlo results[24] when compared to the VSL results with no-slip condition. The nonzero value of atomic mass fraction at the shock for the slip case indicates that the chemical reaction has already taken place in the shock structure that exists beyond the current shock location.

Conclusions

The present study demonstrates the usefulness of a nonequilibrium viscous shock-layer solution including shock slip and wall slip conditions to predict Shuttle entry heating at an altitude range between 92 to 110 km. Although no precise agreement can be mentioned owing to the unknowns in surface accommodation or freestream density in high altitude, the qualitative agreement reached between the present result and flight data and also the Monte Carlo results of Moss and Bird is very encouraging. For the comparison made between the STS-2 flight data and viscous shock-layer results calculated with and without slip conditions, the following conclusions can be drawn:

1) The calculated heating distributions with slip boundary conditions agree better with the flight data than the no-slip results. The agreement improves when the accommodation coefficient or freestream density is decreased to half of its full value, suggesting the possibility of less than full accommodation for the tile surface and (or) an overestimate of freestream density. It is noted that no actual measurements of density were made for the Shuttle flights above 90 km, and the values assumed were based on atmospheric models.

2) The heat-transfer coefficient for no-slip results diverges very severely from the flight data, as the flow becomes rarefied especially near the nose region.

3) The effect of slip on heating reduction is more pronounced for larger Knudsen numbers and more significant for the stagnation region than for the aft region of the vehicle.

4) A higher degree of dissociation is predicted for slip results than for no-slip results. However, in this flight regime, the effect of catalicity is found to be negligible, as expected.

References

[1]Walberg, G. D., "A Review of Aeroassisted Orbit Transfer," AIAA Paper 82-1378, AIAA 9th Atmospheric Flight Mechanics Conference, San Diego, CA, Aug. 1982.

[2]Shinn, J. L. and Jones, J. J., "Chemical Nonequilibrium Effects on Flowfields for Aeroassisted Orbital Transfer Vehicles," AIAA Paper 83-0214, AIAA 21st Aerospace Sciences Meeting, Reno, NV, Jan. 1983.

[3]Davis, R. T., "Numerical Solution of the Hypersonic Viscous Shock-Layer Equations," AIAA Journal, Vol. 8, May 1970, pp. 843-851.

[4]Anderson, E. C. and Moss, J. N., "Numerical Solution of the Steady-State Navier-Stokes Equations for Hypersonic Flow About Blunt Axisymmetric Bodies," NASA TM X-71977, June 1974.

[5]Hendricks, W. L., "A Similarity Solution of the Navier-Stokes Equations with Wall Catalysis and Slip for Hypersonic, Low Reynolds Number Flow Over Spheres," AIAA Paper 75-675, AIAA 10th Thermophysics Conference, Denver, CO, May 1975.

[6]Li, C. P., "Hypersonic Nonequilibrium Flow Past a Sphere at Low Reynolds Number," AIAA Paper 74-173, AIAA 12th Aerospace Sciences Meeting, Washington, D.C., Jan. 1974.

[7]Scott, C. D., "Viscous Reacting Flows with Wall Slip and Catalysis Applied to Spheres in Arc Jets and Flight," AIAA Paper 74-174, AIAA 12th Aerospace Sciences Meeting, Washington, D.C., Jan. 1974.

[8]Little, H. R., "An Experimental Investigation of Surface Conditions on Hyperboloids and Paraboloids at a Mach Number of 10," AEDC-TR-69-225, U.S. Air Force, Jan. 1970.

[9]Chrusciel, E. T., Lewis, C. H., and Sugimura, T., "Slip Effects in Hypersonic Rarefied Flow," Rarefied Gas Dynamics: AIAA Progress in Astronautics and Aeronautics, Vol. 74, edited by S. S. Fisher, AIAA, New York, 1980, pp. 1040-1054.

[10]Shinn, J. L., Moss, J. N., and Simmonds, A. L., "Viscous Shock-Layer Heating Analysis for the Shuttle Windward-Symmetry Plane with Surface Finite Catalytic Recombination Rates," AIAA Paper 82-0842, AIAA Thermophysics Conference, St. Louis, MO, June 1982.

[11]Davis, R. T., "Hypersonic Flow of a Chemically Reacting Binary Mixture Past a Blunt Body," AIAA Paper 70-805, AIAA 3rd Fluid and Plasma Dynamics Conference, Los Angeles, CA, June 1970.

[12]Scott, C. D., "Wall Boundary Equations with Slip and Catalysis for Multicomponent, Nonequilibrium Gas Flows," NASA TM X-58111, 1974.

[13]Hendricks, W. L., "Slip Conditions with Wall Catalysis and Radiation for Multicomponent, Nonequilibrium Gas Flow," NASA TM X-64942, June 1974.

[14]Moss, J. N., "Reacting Viscous-Shock-Layer Solutions with Multicomponent Diffusion and Mass Injection," NASA TR-411, June 1974.

[15]Patterson, G. N., Molecular Flow of Gases, John Wiley and Sons, New York, 1956.

[16]Shidlovskiy, V. P., Introduction to the Dynamics of Rarefied Gases, J. A. Laurmann, translation editor, American Elsevier Publishing Company, New York, 1967.

[17]Maslen, S. H., "On Heat Transfer in Slip Flow," Journal of the Aeronautical Sciences, Vol. 25, June 1958, pp. 400-401.

[18]Mason, E. A. and Saxena, S. C., "Approximate Formula for the Thermal Conductivity of Gas Mixtures," The Physics of Fluids, Vol. 1, 1958, pp. 361-369.

[19]Biolsi, L. (private communication, 1983).

[20]Miner, E. W. and Lewis, C. H., "Computer Users' Guide for a Chemically Reacting Viscous Shock-Layer Program," NASA CR-2551, May 1975.

[21]Throckmorton, D. A., "Benchmark Aeroheating Data From the First Flights of the Space Shuttle Orbiter," AIAA Paper 82-0003, AIAA 20th Aerospace Sciences Meeting, Orlando, FL, Jan. 1982.

[22]Compton, H. R., Findlay, J. T., Kelly, G. M., and Heck, M. L., "Shuttle (STS-1) Entry Trajectory Reconstruction," AIAA Paper 81-2459, AIAA Flight Testing Conference, Las Vegas, NV, Nov. 1981.

[23]Price, J. M. and Blanchard, R. C., "Determination of Atmospheric Properties for STS-1 Aerothermodynamic Investigations," AIAA Paper 81-2430, AIAA Flight Testing Conference, Las Vegas, NV, Nov. 1981.

[24]Jacchia, L. G., "Thermospheric Temperature, Density, and Composition: New Models," Research in Space Science, Smithsonian Institution Astrophysical Observatory, SAO Special Report 375, March 1977, p. 79.

[25]Steward, D. A. and Rakich, J. V., "Catalytic Surface Effects Experiment on Space Shuttle," AIAA Paper 81-1143, AIAA 16th Thermophysics Conference, Palo Alto, CA, June 1981.

[26]Moss, J. N. and Bird, G. A., "Direct Simulation of Transitional Flow for Hypersonic Reentry Conditions," AIAA Paper 84-0223, AIAA 22nd Aerospace Sciences Meeting, Reno, NV, Jan. 1984.

Problems of Rate Chemistry in the Flight Regimes of Aeroassisted Orbital Transfer Vehicles

Chul Park*
NASA Ames Research Center, Moffett Field, California

Abstract

The dissociating and ionizing nonequilibrium flows behind a normal shock wave are calculated for the density and vehicle regimes appropriate for aeroassisted orbital transfer vehicles; the departure of vibrational and electron temperatures from the gas temperature as well as viscous transport phenomena are accounted for. From the thermodynamic properties so determined, radiative power emission is calculated using an existing code. The resulting radiation characteristics are compared with the available experimental data. Chemical parameters are varied to investigate their effect on the radiation characteristics. It is concluded that the current knowledge of rate chemistry leads to a factor-of-4 uncertainty in nonequilibrium radiation intensities. The chemical parameters that must be studied to improve the accuracy are identified.

Nomenclature

AOTV = aeroassisted orbital transfer vehicle
C = chemical reaction rate coefficient
$\bar{C}$ = average molecular velocity = $[8\,kT/(\pi m)]^{1/2}$

Presented as Paper 84-1730 at AIAA 19th Thermophysics Conference, Snowmass, Colorado, June 25-28, 1984.

*Research Scientist.

D_i = coefficient of diffusion of species i
E = energy per unit volume
h_i = enthalpy of species i
I_i = ionization potential for electron-impact ionization process i
k = Boltzmann constant
m = mass of a molecule
n = number density
NEQAIR = nonequilibrium air radiation
p = pressure
Q_r = radiative power loss
q_r = nonequilibrium radiative heat flux, Eq. (13)
s = temperature exponent in rate coefficient expression
T = heavy-particle translational-rotational temperature
T_d = characteristic reaction temperature in rate coefficient
T_v = vibrational-electron temperature
t = time
u = velocity
w_i = rate of chemical production of species i
x = distance from shock wave
x_E = distance to the point where radiation intensity is 1.1 times the equilibrium value
x_p = distance to the point where radiation intensity is the peak
γ_i = species concentration, mole/g
ε = radiative power emission per unit volume, W/cm^3
λ = thermal conductivity
μ = viscosity
ν = collision frequency
ρ = density
σ = vibrational excitation cross section, Eq. (6)
τ = vibrational relaxation time

Subscripts

E = equilibrium
e = electron
h = heavy particle
i = species i or reaction i
m = molecule

s = shock wave (postshock condition)
t = total
v = vibrational
∞ = preshock condition

Introduction

Reusable aeroassisted orbital transfer vehicles (AOTVs)[1-4] can substantially lower the cost of delivering payloads to various orbits in the Earth-moon space. The benefit is partly a result of using the aerodynamic forces produced over the heat-shield surfaces of the AOTVs during atmospheric flight to replace rocket propulsion for maneuvering.[4] One of the problems in designing such vehicles is the uncertainty in the magnitude of radiative heat-transfer rates.[5] As is well known, convective heat-transfer rates are reduced by flying at high altitudes. To fly at high altitudes and still produce the required aerodynamic forces, however, requires that the heat shield be large. The combination of large size and high altitude causes the convective heat-transfer rates to the body to be fairly small.[4]

The combination of large size and high altitude does not reduce the radiative heat-transfer rates to the same extent as the convective heat-transfer rates; however, the phenomenon known as nonequilibrium radiation enhancement tends to maintain the radiative heat fluxes to the heat-shield surface at a nearly fixed value.[5-7] This is believed to be caused by a region of high temperature and high concentration of excited atoms and molecules that is created during the process of thermal and chemical relaxation behind a shock wave. This nonequilibrium enhancement phenomenon tends to be offset by two phenomena known as collision limiting and as truncation.[5,6] The collision-limiting phenomenon refers to the decrease, as density decreases, in the ratio between the frequency of the molecular collisions causing the excitation of atoms and molecules and the frequency of radiative depopulation of these excited atoms and molecules. As a result, radiative emission becomes comparatively weaker. Truncation refers to the fact that the boundary layer over the heat shield becomes thicker as density decreases and eventually engulfs the region of enhanced radiation. These phenomena are poorly understood, and existing laboratory data regarding them show an uncertainty of a factor of about 4 at the flight velocities anticipated for the AOTV.[5] The prototype flight data taken from projects Fire, Apollo, and PAET are seemingly contradictory: Fire and Apollo data indicate a substantial collision-limiting or truncation phenomenon or both, whereas PAET data do not.[5,7]

One way to help reduce the uncertainty is to use a theoretical approach. By this approach, one endeavors to develop a computer code that numerically reproduces the existing ground-based laboratory data. If one succeeds in doing so, one would gain some confidence in using the same code for flight conditions. Toward this end, a computer code named NEQAIR (nonequilibrium air radiation) has been developed recently.[8] Assuming that the gas under consideration has a sufficiently high density to establish a Maxwell/Boltzmann distribution within the translational, rotational, and vibrational modes of atoms, molecules, and electrons (but not for the electronic excitation), the code computes the nonequilibrium electronic excitation of atoms and molecules for the given nonequilibrium species concentration and for three temperatures: the translational-rotational temperature of heavy particles, the vibrational temperature, and the electron translational temperature. From the number densities of the excited electronic states so computed, the code determines the radiative properties.

To provide the input data required by the NEQAIR code, one must determine the nonequilibrium thermodynamic states existing in the shock layer over the heat shield of an AOTV. The state variables can be determined by integrating chemical rate equations simultaneously with the flow-conservation equations. The chemical reactions for air have been studied for many years, and there is a wealth of information on the rate-coefficient values dictating the reactions. It has never been shown, however, whether the exiswting rate-chemistry models and rate-coefficient values are sufficiently accurate in theoretically reproducing the experimentally observed radiation characteristics in a nonequilibrium flow.

The present paper attempts to examine the accuracy of the existing rate-chemistry data base. This is done by studying the radiation phenomenon in a one-dimensional flow behind a normal shock wave. Chemical reactions are computed simultaneously with the flow-conservation equations following the flow. From the computed thermodynamic state variables, the radiative characteristics are calculated using the NEQAIR code. The computed radiative characteristics are then compared with the experimental data obtained in shock tubes and ballistic ranges. Since most of the experiments were carried out with a test gas that was a mixture of 79% nitrogen and 21% oxygen, the calculations are made also for this gas mixture. By varying the thermochemical parameters of the problem, the effect of the uncertainties in the parameters on the radiative characteristics is shown. The discrepancies between the theory and experiment are pointed out, and the possible causes for the discrepancies are hypothesized. Through this study, the present paper identifies the chemical parameters that need to be studied further in future.

Method of Calculation

The existing standard method of computing chemical reactions in a gas flow is applicable for density regimes sufficiently high for a single temperature to exist. In this regime, all particles (atoms, molecules, and electrons) have the same translational temperature and all internal modes of these particles are excited to a Boltzmann distribution corresponding to the temperature.[9] For a monatomic ionizing flow, however, it has been shown that electron-translational temperature can deviate significantly from the heavy-particle translational temperature even at moderately high densities.[10-12] Deviation of vibrational temperature from the translational temperature has also been known to occur at fairly high densities.[13,14] In a shock layer over an AOTV, the pressure is of the order of 0.01 atm, which is sufficiently high to cause a Maxwellian/Boltzmann distribution in translational, rotational, and vibrational modes, but not necessarily to bring these temperatures to the same value. Only the rotational mode of the molecules tends to equilibrate rapidly with the translational mode of the heavy particles (atoms and molecules) and, therefore, can be assumed to be at the same temperature as the heavy-particle translational mode. The vibrational temperature and the translational temperature of the electrons, however, must be regarded in general as different from the heavy-particle translational temperature.

A general three-temperature (heavy-particle translational-rotational, vibrational, and electron-translational) gas model is presently being developed for use in the study of the AOTV problems.[15] The three-temperature gas model requires knowledge of chemical-reaction-rate coefficients as a function of the three temperatures, as well as the interactions among them. Since these are mostly unknown, such a model is not likely to be useful in the near future.

In order to satisfy the immediate need, a two-temperature gas model is introduced in the present work. In this model, vibrational and electron temperatures are assumed to be the same but different from the translational-rotational temperature of the heavy particles. This assumption is based on the knowledge that the energy exchange between the electron-translational mode and the vibrational mode is very rapid.[15] The assumption is believed to be valid everywhere except in the viscous boundary layer adjacent to the wall. The boundary conditions on the vibrational and electron-translational modes at the wall can be radically different and could cause separation between the two temperatures there.[16] Another region of concern is immediately behind the shock wave where a similar phenomenon is possible. Here, however, the two-temperature model is likely to be valid: the vibrational and electron temperatures immediately behind the shock wave are linked closely to those ahead of the shock wave. (Electrons exist ahead of a shock wave as a result of precursor ionization.) The electron-vibrational interaction in the precursor region will tend to equalize the two temperatures. The present analysis is applied only to the one-dimensional flow of infinite extent behind a normal shock wave free of boundary layer, and hence the two-temperature model is appropriate.

Even with this simplification, one needs an assumption regarding the dependence of the rate coefficients on temperatures. Here, the rates of collisional dissociation reactions such as $N_2 + N \rightarrow N + N + N$ are assumed to be determined by the vibrational temperature, and those of exchange reactions such as $O + N_2 \rightarrow NO + N$ are dictated by the translational temperature. The first assumption follows the generally accepted belief that dissociation occurs as a result of ladder-climbing of vibrational energy levels.[17] The second assumption arises from the fact that the rates of exchange reactions are limited mostly by the frequency of collisions.

For the one-dimensional flow under consideration, overall mass and momentum obey the following differential equations:

$$\frac{\partial \rho}{\partial t} + \frac{\partial}{\partial x}(\rho u) = 0 \tag{1}$$

$$\frac{\partial}{\partial t}(\rho u) + \frac{\partial}{\partial x}(\rho u^2 + p) = \frac{\partial}{\partial x}\left(\frac{4}{3}\mu\frac{\partial u}{\partial x}\right) \tag{2}$$

Here, t is the time; x is the distance from shock wave; ρ is the density; u is the velocity; p is the pressure; and μ is the bulk viscosity.[15] In deriving Eq. (2), the second coefficient of viscousity is assumed to be equal to $-(2/3)\mu$. In deriving the overall energy equation, the contribution of the vibrational and electron kinetic modes in conduction must be accounted for independently of that of the heavy-particle translational-rotational mode because of the difference in the temperatures. Denoting the translational-rotational temperature by T, vibrational-electron temperature by T_v, and the conductivities associated with the translational-rotational mode, electron-translational mode, and vibrational mode by λ_h, λ_e, and λ_v, respectively, the energy equation becomes[15]

$$\frac{\partial E}{\partial t} + \frac{\partial}{\partial x}[(E + p)u] = \frac{\partial}{\partial x}\left(\frac{4}{3}\mu u\frac{\partial u}{\partial x} + \frac{\gamma_h}{\gamma_t}\lambda\frac{\partial T}{\partial x} + \frac{\gamma_e}{\gamma_t}\lambda_e + \frac{\gamma_m}{\gamma_t}\lambda_v\frac{\partial T_v}{\partial x} + \sum_i \rho D_i h_i\frac{\partial \gamma_i}{\partial x}\right) - Q_r \tag{3}$$

Here, γ_i is the species concentration (in moles per gram); D_i is the diffusion coefficient; and h_i is the enthalpy of species i. The quantity γ_h denotes the concentrations of all heavy particles, γ_e of electrons, γ_m of molecules, and γ_t of the total concentration. The quantity E is the internal energy plus kinetic energy per unit volume (see Ref. 15), and Q_r is the radiative power loss per unit volume; Q_r is taken to be zero in the present work.

The species concentration γ_i satisfies the equation

$$\frac{\partial}{\partial t}(\rho\gamma_i) + \frac{\partial}{\partial x}(\rho u\gamma_i) = \frac{\partial}{\partial x}\left(\rho D_i \frac{\partial\gamma_i}{\partial x}\right) + w_i \tag{4}$$

where w_i is the rate of chemical production of the species.[15] The expression for w_i is well known and involves rate coefficients and equilibrium constants. The equilibrium constants are determined using the most recent spectroscopic data.[18] As stated earlier, the rate coefficients are expressed as a function of either T or T_v for the heavy particles. Whenever an electron is involved, the rate coefficient is evaluated using electron temperature because the collision frequency between a heavy particle and an electron is dictated by electron temperature alone.[19]

The equation describing the conservation of vibrational energy and the equation describing the conservation of electron kinetic energy are both well known.[13-15,20] By adding these two equations, one obtains the equation of conservation of the sum of vibrational and electron energies in the form

$$\begin{aligned}
&\frac{\partial}{\partial t}[(1.5\gamma_e + \gamma_m)\rho T_v] + \frac{\partial}{\partial x}[(2.5\gamma_e + \gamma_m)\rho u T_v] \\
&= \frac{\partial}{\partial x}\left[\frac{1.2022 \times 10^{-8}}{\gamma_t}(\gamma_e\lambda_e + \gamma_m\lambda_v)\frac{\partial T_v}{\partial x}\right] \\
&\quad + \left[3\gamma_e \sum_i \frac{m_e}{m_i}\nu_i + \frac{\gamma_m}{\tau}\right]\rho(T - T_v) \\
&+ u\frac{\partial}{\partial x}(\rho\gamma_e T_v) - 1.2022 \times 10^{-8}\rho \times \sum_i I_i w_i - Q_r
\end{aligned} \tag{5}$$

The equation is derived assuming that vibrational energy for one molecule is expressible as kT_v where k is the Boltzmann constant. This assumption is valid when T_v is much larger than the characteristic temperature of vibration (the vibrational energy gap divided by k = 2240 K for O_2 and 3390 K for N_2). As will be shown later, T_V is always greater than the characteristic temperatures in the present problem; hence, the assumption is justified. The quantities m_e and m_i are electron mass and the mass of (heavy particle) species i, respectively. The quantity ν_i is the collision frequency for transfer of kinetic energy between the ith heavy particle and electron, and τ is the vibrational relaxation time. The quantity I_i is the ionization potential for the ith electron-impact ionization reaction. The constant 1.2022×10^{-8} results from the conversion of mole-unit to the cgs unit. The five terms on the right-hand side of Eq. (5) denote the effects of conduction, elastic energy exchange between the heavy-particle translational mode and the combined vibration-electron energy mode, ohmic heating caused by the charge separation, inelastic energy exchange caused by electron-impact ionization, and radiative power loss, respectively.

The conservation equations (1) through (5) involve three categories of rate-controlling parameters: the collision frequencies ν_i's, the vibrational relaxation time τ, and the chemical-reaction-rate coefficients. The present work adopts the work of Yos[21] for determining collision frequencies, as well as transport properties μ, Di, and λ (D_i is calculated individually for each i). The work of Millikan and White[22] is used for vibrational relaxation times. Figure 1 shows the variation of vibrational relaxation times for N_2 and O_2 with the heavy-particle temperature T. The existing experimental data[22] extend to a temperature of about 8000 K. The empirical equation of Millikan and White describes the experimental data correctly below about 5000 K. At temperatures between 5000 and 8000 K, the equation tends to underestimate the experimental data for O_2. Nothing is known of the relaxation time at the temperatures above 8000 K that are expected in the shock layer over an AOTV.

Since the kinetic energy of a gas molecule is much higher than the characteristic vibrational temperature, the vibrational relaxation rate (which is the reciprocal of the relaxation time) can be expressed in terms of an effective excitation cross section σ in the form

$$\text{vibrational relaxation rate} = 1/\tau = \sigma\bar{C}n \qquad (6)$$

where $\bar{C}$ is the average molecular velocity

$$\bar{C} = [(8kT/(\pi m)]^{1/2}$$

and n is the number density of the molecule under consideration. The cross section σ must be smaller than the elastic cross section, which is of the order of 10^{-16} cm^2 in this high-temperature range. In Fig. 1, the relaxation times calculated using Eq. (6) are shown for $\sigma = 10^{-16}$ and 10^{-17} cm^2. As seen in the figure, the times are much larger than those predicted by the formula of Millikan and White at high temperatures. An expression that asymptotically approaches the formula of Millikan and White at low temperatures and Eq. (6) at high temperatures is

$$\tau \text{ (total)} = \tau \text{ (Millikan and White)} + \tau \text{ [Eq. (6)]} \quad (7)$$

This is shown in fig. 1 also. Most calculations are car-

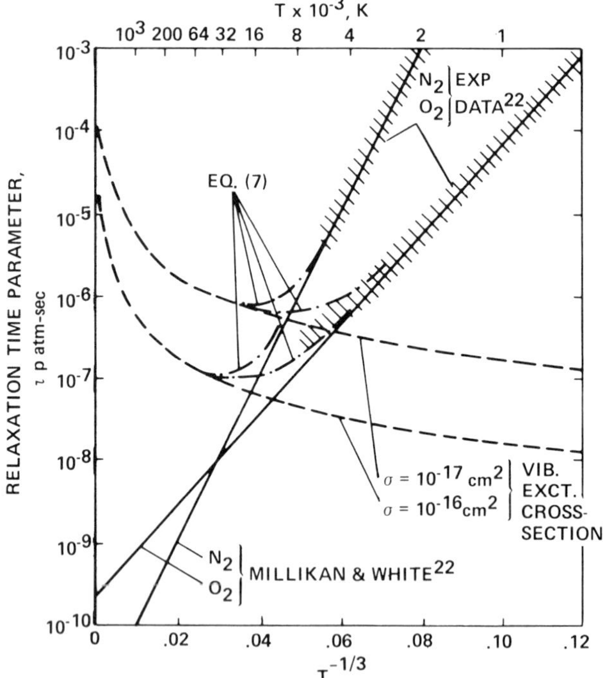

Fig. 1 Vibrational relaxation times as a function of heavy-particle translational-rotational temperature T.

ried out in the present work using a cross-section value of 10^{-16} cm^2; for purposes of comparison, a few cases are calculated using a value of 10^{-17} cm^2. As is apparent in Fig. 1, the relaxation time τ is different for N_2, O_2, and NO (for simplicity, NO is not shown in Fig. 1). The quantity τ in Eq. (5) is the average value

$$1/\tau = X_{N_2}/\tau(N_2) + X_{O_2}/\tau(O_2) + X_{NO}/\tau(NO)$$

where X is the species mole fraction.

It is customary to express the chemical-reaction-rate coefficient in the form

$$\text{rate coefficient} = CT^S \exp(-T_d/T)$$

For the cases in which vibrational-electron temperature T_v is the controlling temperature, this is amended as

$$\text{rate coefficient} = CT_v^S \exp(-T_d/T_v)$$

The values of C, S, and T_d have been compiled for air reactions by several authors. The most recent such compilations, done by Kang and Dunn[23] and Park and Menees[24] are reproduced in Table 1; they are the values used in the present work. From the table, one sees that the two sets of rate-coefficient values differ by a factor of about 3 in the temperature range of interest.

For the purpose of the present study, the steady-state solutions for Eqs. (1-5) must be obtained. In the steady-state limit, the equations constitute a boundary-value problem in which boundary conditions are required at both ends of the spatial domain. The most general method of solving a boundary-value problem is to treat it as a time-dependent problem. For this purpose, the temporal derivative terms must be retained.

In the present work, three different versions of Eqs. (1-5) are solved: 1) a one-temperature inviscid flow model in which $T_v = T$ and all transport terms are zero; 2) a two-temperature inviscid flow model in which the transport terms are zero; and 3) a two-temperature viscous flow model in which all terms are retained. In cases (1) and (2), the problem can be converted into an initial-

Table 1 List of reactions and rate coefficients

No.	Reaction	M	Temperature	Kang and Dunn[23] C, $cm^3/(mole \cdot s)$	Kang and Dunn[23] s	Kang and Dunn[23] T_d, K	Park and Menees[24] C, $cm^3/(mole \cdot s)$	Park and Menees[24] s	Park and Menees[24] T_d, K
1	$O_2 + M \rightarrow O + O + M$	N	T_v	3.6^{18}	-1	59,500	8.25^{19}	-1.0	59,500
		O	T_v	2.0^{19}	-1	59,500	8.25^{19}	-1.0	59,500
		N_2	T_v	7.2^{18}	-1	59,500	2.75^{19}	-1.0	59,500
		O_2	T_v	3.2^{19}	-1	59,500	2.75^{19}	-1.0	59,500
		NO	T_v	3.6^{18}	-1	59,500	2.75^{19}	-1.0	59,500
2	$N_2 + M \rightarrow N + N + M$	N	T_v	4.1^{22}	-1.5	113,200	1.11^{22}	-1.6	113,200
		O	T_v	1.9^{17}	-0.5	113,200	1.11^{22}	-1.6	113,200
		N_2	T_v	4.7^{17}	-0.5	113,200	3.7^{21}	-1.6	113,200
		O_2	T_v	1.9^{17}	-0.5	113,200	3.7^{21}	-1.6	113,200
		NO	T_v	1.9^{17}	-0.5	113,200	3.7^{21}	-1.6	113,200
3	$NO + M \rightarrow N + O + M$	N	T_v	7.8^{20}	-1.5	75,500	4.6^{17}	-0.5	75,500
		O	T_v	7.8^{20}	-1.5	75,500	4.6^{17}	-0.5	75,500
		N_2	T_v	3.9^{20}	-1.5	75,500	2.3^{17}	-0.5	75,500
		O_2	T_v	3.9^{20}	-1.5	75,500	2.3^{17}	-0.5	75,500
		NO	T_v	7.8^{20}	-1.5	75,500	2.3^{17}	-0.5	75,500
4	$NO + O \rightarrow N + O_2$		T	3.2^{9}	1.0	19,700	2.16^{8}	1.29	19,220
5	$O + N_2 \rightarrow N + NO$		T	7.3^{13}	0	38,000	3.18^{13}	0.10	37,700
6	$N + O \rightarrow NO^+ + e$		T	1.4^{6}	1.5	31,900	1.53^{10}	0.37	32,000
7	$O + O \rightarrow O_2^+ + e$		T	1.6^{17}	-0.96	80,800	3.85^{10}	0.49	80,600
8	$O + O_2^+ \rightarrow O_2 + O^+$		T	2.9^{18}	-1.11	28,000	6.85^{13}	-0.52	18,600
9	$N_2 + N^+ \rightarrow N + N_2^+$		T	2.0^{11}	0.81	13,000	9.85^{12}	-0.18	12,100
10	$N + N \rightarrow N_2^+ + e$		T	1.4^{13}	0	67,800	1.79^{10}	0.77	67,500
11	$O + NO^+ \rightarrow NO + O^+$		T	3.6^{15}	-0.6	50,800	2.75^{13}	0.01	51,000
12	$N_2 + O^+ \rightarrow O + N_2^+$		T	3.4^{19}	-2.0	23,000	6.33^{13}	-0.21	22,200
13	$N + NO^+ \rightarrow NO + N^+$		T	1.0^{19}	-0.93	61,000	2.21^{15}	-0.02	61,100
14	$O_2 + NO^+ \rightarrow NO + O_2^+$		T	1.8^{15}	0.17	33,000	1.03^{16}	-0.17	32,400
15	$O + NO^+ \rightarrow O_2 + N^+$		T	1.3^{13}	0.31	77,300			
16	$NO^+ + N \rightarrow N^+ + N_2^+$						1.70^{13}	0.40	35,500
17	$O + e \rightarrow O^+ + e + e$		T_v	3.6^{31}	-2.91	158,000	1.95^{34}	-3.78	158,500
18	$N + e \rightarrow N^+ + e + e$		T_v	1.1^{32}	-3.14	168,000	1.25^{35}	-3.82	168,600

value problem by eliminating the temporal derivative terms. Solutions can be obtained in this case by marching in x, starting from the shock wave; the initial values are obtained from the Rankine-Hugoniot relations for cases (1) and (2).

For case (3), the Rankine-Hugoniot relations must be modified to account for the viscous transport effects; that is, the shock-slip effect must be included to provide the upstream boundary conditions. The viscous terms must be retained in principle in both the preshock and the postshock flows. For purposes of simplification, however, the terms are eliminated for the preshock flow by assuming that the spatial derivatives of all dependent variables are zero there. The preshock quantities ρ_∞, u_∞, p_∞, $Y_{i\infty}$, and T_∞ are taken to be the ambient atmospheric values. However, the $T_{v\infty}$ (and the corresponding E_∞) value is varied between the atmospheric value and 8000 K in the present calculation in order to study the effect of precursor phenomena; vibrational temperature of the flow immediately ahead of the shock wave could be significantly higher than the atmospheric value because of the precursor phenomena. The resulting equations are

$$\rho_s u_s = \rho_\infty u_\infty \tag{8}$$

$$\rho_s u_s^2 + p_s = \rho_\infty u_\infty^2 + p_\infty + \frac{4}{3}\left(\mu \frac{\partial u}{\partial x}\right)_s \tag{9}$$

$$(E_s + p_s)u_s = (E_\infty + p_\infty)u_\infty + \left(\frac{4}{3}\mu u \frac{\partial u}{\partial x} + \frac{Y_h}{Y_t}\lambda_h \frac{\partial T}{\partial x} + \frac{Y_e}{Y_t}\lambda_e + \frac{Y_m}{Y_t}\lambda_v \frac{\partial T_v}{\partial x} + \sum_i \rho D_i h_i \frac{\partial Y_i}{\partial x}\right)_s \tag{10}$$

$$\rho_s u_s Y_{is} = \rho_\infty u_\infty Y_{i\infty} + \left(\rho D_i \frac{\partial Y_i}{\partial x}\right)_s \tag{11}$$

and

$$(2.5Y_e + Y_m)_s \rho_s u_s T_{v_s} = (2.5Y_e + Y_m)_\infty \rho_\infty u_\infty T_{v_\infty} + \left[\frac{1.2022^{-8}}{Y_t}(Y_e\lambda_e + Y_m\lambda_v)\frac{\partial T_v}{\partial x}\right]_s \tag{12}$$

where the subscript s denotes the postshock conditions. The downstream boundary conditions for case (3) are obtained by setting the second derivatives of the dependent variables to zero. Solutions are obtained by alternating the timewise integration between the flow set, Eqs. (1-3), and the chemistry set, Eqs. (4) and (5). The implicit time-marching technique of Beam and Warming is employed for both sets.[25]

For cases (1) and (2), the spatial integration was carried out to very large x-values in order to ensure that equilibrium is reached. The necessary total number of integrating steps varied from approximately 1000 to 10,000, depending on the initial values. Computing times up to about 300 s were necessary on the Cray-XMP computer for the calculation. For case (3), the maximum x-value was chosen to be the equilibration distance (the point where radiation intensity reaches 1.1 times the equilibrium value) determined experimentally in a shock tube. The spatial distance was divided into 128 grid points. Approximately 13,000 time-steps were necessary to reach a steady-state solution, each step requiring about 1 s on Cray-XMP.

The nonequilibrium thermodynamic state variables calculated by the methods of (1)-(3) above are fed to the NEQAIR code to determine the radiative properties.[8] The NEQAIR calculation was performed at 80 to 128 selected spatial points. Only the optically thin emission power integrated over wavelengths longer than 250 nm is computed. Approximately 1 s was necessary to execute one NEQAIR calculation.

Results

In Fig. 2, the calculated species mole fractions are shown for the case in which $p_\infty = 0.1$ Torr, $T_\infty = 300$ K, and $u_\infty = 10$ km/s; there is a wealth of shock-tube data for this case. The case shown is for $T_{v\infty} = 5000$ K, $\sigma = 10^{-16}$ cm^2, and the rate-coefficient set of Park in Table 1. In the figure, the mole fractions at $x = \infty$ denote the equilibrium values. The figure shows that although the mole fractions of N and O are lower, those of all other species are higher than in equilibrium, forming an overshoot for some of the species. This phenomenon can be attributed to the difference in the reaction rates with which the two groups of species are associated; most reactions are faster than those of the dissociation of N_2 and O_2 which form N and O. An incomplete dissociation of N_2 and O_2 implies that there is a greater thermal energy to which the fast-forming species tend to equilibrate temporarily; a near-equilibrium is reached among the fast-forming species with N_2 and O_2 only partly dissociated. Noteworthy is the overshoot in electron concentration--by a factor of almost 5 in this case. Electron density is usually associated with radiation, and hence its overshoot foreshadows an overshoot in radiation intensity.

One also sees in Fig. 2 that the energetic species such as N, O, N^+, and O^+ are present in finite concentrations immediately behind the shock wave, x = 0. These species are not present in the ambient atmosphere, and their presence at x = 0 is caused by the backward (against flow) diffusion represented by the second term in

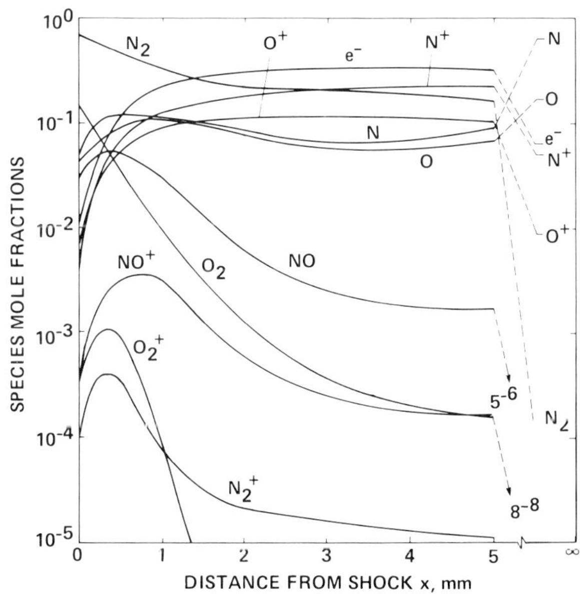

Fig. 2 Species mole fractions calculated for p_∞ = 0.1 Torr, t_∞ = 300 K, u_∞ = 10 km/s, $T_{v\infty}$ = 5000 K, $\sigma = 10^{-16}$ cm^2, rate coefficients of Park and Menees,[24] and two-temperature viscous model.

the right-hand side of Eq. (11). Reactions involving such energetic species are usually fast, and hence their early presence is likely to affect the overall rate chemistry significantly. One sees, therefore, that the viscous processes, and the accompanying shock-slip phenomenon, are significant in this environment.

In Figs. 3a and 3b, the calculated temperatures T and T_v and radiative power emission per unit volume ε

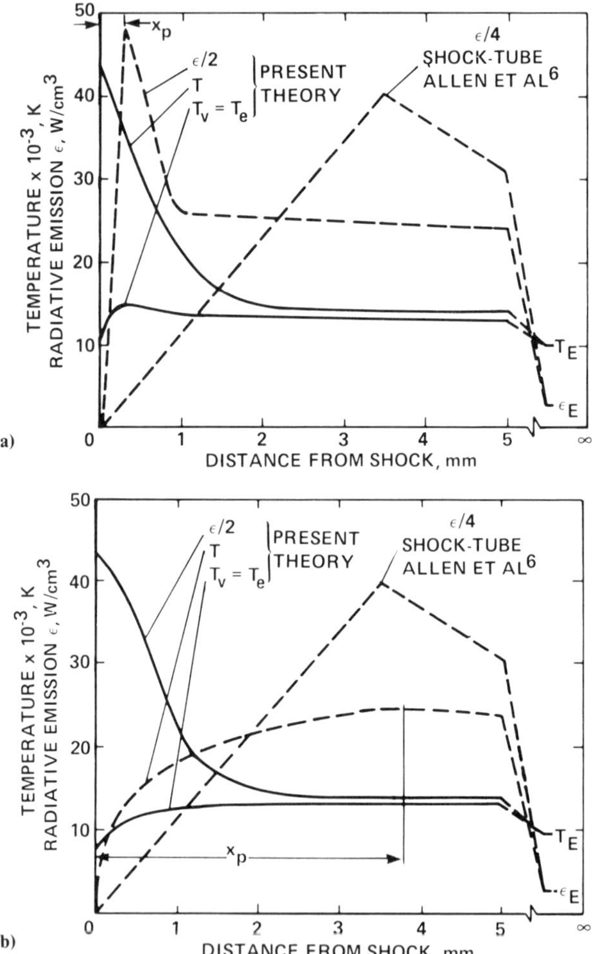

Fig. 3 Heavy-particle translational-rotational temperature T, vibrational-electron temperature T_v, and radiation power emission ε calculated for $p_\infty = 0.1$ Torr, $T_\infty = 300$ K, $u_\infty = 10$ km/s, $T_{v\infty} = 5000$ K, rate coefficients of Park and Menees,[24] and two-temperature viscous model: a) $\sigma = 10^{-16}$ cm^2; b) $\sigma = 10^{-17}$ cm^2.

are plotted. The preshock conditions and the method used are the same as in Fig. 2, except that two different σ values, 10^{-16} cm^2 (Fig. 3a) and 10^{-17} cm^2 (Fig. 3b), are used. For comparison, the experimentally determined radiative powers are shown also. As the figure shows, the two temperatures are significantly different in the region immediately behind the shock wave. The heavy-particle translational-rotational temperature T is in excess of 40,000 K at $x = 0$, while the vibrational-electron temperature T_V stays below 15,000 K. The large σ value (Fig. 3a) produces an early rise in T_V, leading to a peak at $x \approx 0.4$ mm, whereas the small σ value (Fig. 3b) leads to a gradual rise to a peak at $x = 3.7$ mm. A large σ value produces a smaller relaxation time τ (see Fig. 1) and is velieved to lead to a faster equilibration between T and T_V. The behavior of the radiation power ε is closely associated with that of T_V; the rise and fall of ε roughly coincides with that of T_V. This is because the internal degrees of freedom of atoms and molecules tend to be excited mainly according to vibrational and electron temperatures, both of which are represented by T_V. Though not shown here, electron concentration also is associated closely with T_V; the most significant electron-producing reactions are the electron-impact ionization, reactions (17) and (18) in Table 1, which are dictated by T_V. It should be noted also that both T and T_V are significantly higher in the nonequilibrium domain than the equilibrium temperature T_E. The temperatures T and T_V reach their equilibrium values only when N_2 and O_2 have dissociated to their equilibrium values.

In Fig. 3, the radiative overshoot phenomenon observed experimentally is reproduced qualitatively by the calculation. The location of the peak radiation point x_p calculated with $\sigma = 10^{-17}$ cm^2 agrees approximately with the experimental value. Referring to Fig. 1, this means that the relaxation time τ at the high temperatures of interest, $20{,}000 < T < 40{,}000$ K, is an order of magnitude larger than that predicted by the extrapolation of the formula of Millikan and White.[22] The calculated radiative power ε is, however, only about half of the experimental value.

One notices in Figs. 3a and 3b that the postshock vibrational-electron temperature T_{VS} is significantly higher than the assumed preshock value $T_{v\infty}$ of 5000 K. This is caused by the backward conduction of vibrational-electron energy represented by the second term in the right-hand side of Eq. (12). Because of the close association of T_v with radiation, the high initial value should be considered an important phenomenon.

Though not shown here, the effect of varying the preshock vibrational-electron temperature $T_{v\infty}$ was studied. At high stream densities, that is, for ρ_∞ between approximately 3×10^{-8} and 2×10^{-7} g/cm^3, the postshock T_{vs} remained relatively unchanged; for example, a $T_{v\infty}$ value of 300 K produced T_{vs} of 6950 K for the conditions of Fig. 3a. However, at lower preshock densities, the postshock T_{vs} was closer to $T_{v\infty}$. A complex relationship, one beyond the scope of this work, seems to exist between $T_{v\infty}$, T_{vs}, and the freestream density. One must recognize, however, that $T_{v\infty}$ affects the overall rate chemistry significantly in certain flight regimes, and therefore, that the precursor phenomena, which determine $T_{v\infty}$, remain significant parameters of the problem.

The peak radiation distance x_p and the equilibration distance x_E, defined as the distance where radiation intensity equals 1.1 times the equilibrium value, are calculated and plotted in Fig. 4 along with the experimental data.[6] The calculations are made with the previously discussed method (1), designated 1-T, INV in Fig. 4; method (2), designated 2-T, INV; and method (3), designated 2-T, VIS with $\sigma = 10^{-16}$ or 10^{-17} cm^2. In addition to the reaction-rate coefficient sets of Kang and Park shown in Table 1, a specially fast rate-coefficient set, denoted FAST in Fig. 4, is considered. The fast set is a variation of the set by Park and is obtained by increasing the coefficient C of reactions (4) and (5) in Table 1 by a factor of 100, and by assuming that the charged species N^+, O^+, N_2^+, O_2^+, and NO^+ participate in reactions (1)-(3) with C values 10 times greater than those of their neutral counterparts. The fast reaction set increases the net rate of dissociation of N_2 and O_2 by at least an order of magnitude in the near-equilibrium region.

From Fig. 4, one can make three observations:

1) The chemical-reaction-rate coefficients selected by Kang and Park give substantially different results from each other, implying that the characteristic distances x_p

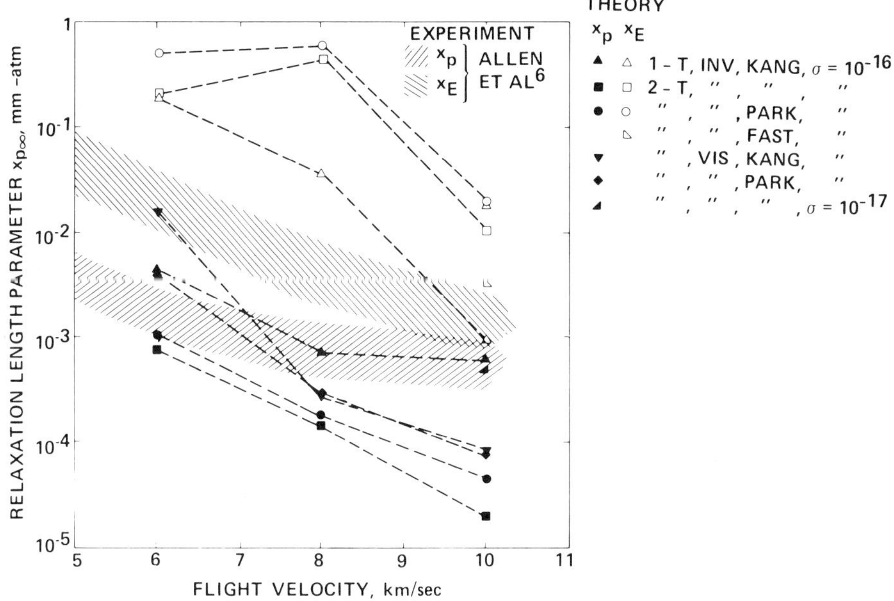

Fig. 4 Peak radiation point x_p and equilibration point (where radiation is 1.1 times the equilibrium value) x_E, for p_∞ = 0.1 Torr, T_∞ = 300 K, and $T_{v\infty}$ = 5000 K.

and x_E are relatively sensitive functions of rate coefficients. Since the differences in rate-coefficient values between the Park and Kang sets represent the extent of uncertainty in the rate-coefficient values, one must conclude that the current knowledge of rate-coefficient values is inadequate in predicting nonequilibrium radiation accurately.

2) The peak point x_{p_2} is best reproduced by assuming σ to be 10^{-17} cm^2 or by using the one-temperature gas model. Considering that the two temperatures are most likely significantly different in the present environment, the good agreement with the one-temperature model should be considered fortuitous. A small σ, and hence large τ (see Fig. 1 and the preceding paragraphs), seems to be a more plausible explanation.

3) All methods considerably overestimate the equilibration distance x_E, except the one-temperature model and the fast-reaction set. One can understand why the one-

temperature model results in a small x_E; all rates are evaluated in this case by T which is larger than T_v. Again, the good agreement with the one-temperature model should be considered fortuitous. The fact that the fast-reaction set yields a better result suggests that the existing rate-chemistry model underestimates the overall reaction rate when a gas is nearly in equilibrium.

The effect of the fast-reaction rates on equilibration distance is shown again in Fig. 5, in which the mole fraction of N_2 and radiation emission ε are shown for the region away from the shock wave where equilibrium is approached. Concentration of N_2 is chosen here because, as was mentioned earlier, its dissociation is one of the slowest processes. The experimental radiation data are shown also for comparison.[6] As seen in the figure, the fast-reaction set causes the equilibrium to be reached within a distance that is an order of magnitude shorted than that in Park's set.[24] The decay of radiation intensity to the equilibrium value is related closely to the decay in N_2 concentration. However, even the fast-reaction set fails to reproduce correctly the experimentally observed pattern of radiation decay.[6] The rate-coefficient set of Park and Menees[24] predicts the nonequi-

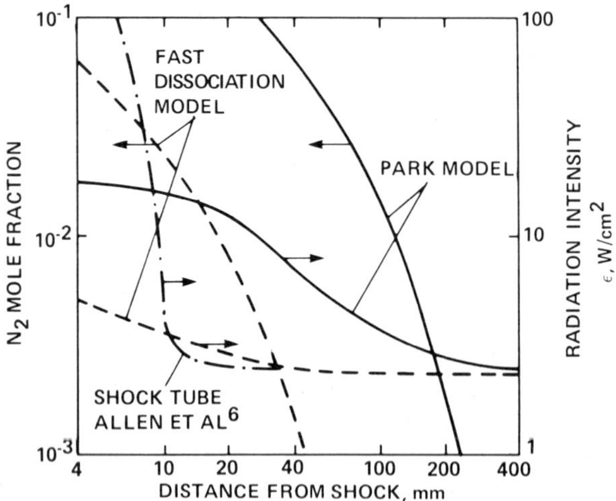

Fig. 5 Mole fraction of N_2 and radiative power emission in the region approaching equilibrium, $p_\infty = 0.1$ Torr, $T_\infty = 300$ K, $T_{v\infty} = 5000$ K, two-temperature inviscid model, and $\sigma = 10^{-16}$ cm^2.

librium radiation to persist over a distance that is an order of magnitude longer than that observed experimentally.

In the stagnation region of a blunt body, the shock layer is relatively thin, and hence, radiation from the layer can be approximated by that from an infinite slab. If the radiation is optically thin, as is assumed in this work, then the radiative heat flux at any point x equals half the integration of the emission ε over x from zero to the point. The integral taken from the shock wave to the equilibration point $x = x_E$ is known as nonequilibrium radiative heat flux, denoted here by q_r

$$q_r = \frac{1}{2} \int_0^{x_E} \varepsilon \, dx \qquad (13)$$

A problem arises here regarding the numerical value of x_E. As seen in Figs. 4 and 5, the calculated x_E values are considerably larger than those determined experimentally. The q_r values are calculated, therefore, using both sets of x_E values, that is, the experimental and the theoretical values. The results are shown in Fig. 6 for $u_\infty = 10$ km/s. All parameters are the same here as in Fig. 2, except for the freestream density ρ_∞ which was varied from 10^{-9} to 1.7×10^{-7} g/cm^3. Figure 6 includes the experimental q_r value[6] and the available energy flow $(1/2)\rho_\infty u_\infty^3$ for comparison.

As expected, Fig. 6 shows that the experimentally determined x_E values lead to smaller heat fluxes than the theoretical x_E values. The former is smaller and the latter is larger than the experimental q_r. The q_r values decrease as ρ_∞ decreases. At the density corresponding to the altitude of 80 km, a typical anticipated flight altitude of an AOTV, the two q_r values are 9 and 30 W/cm^2, respectively. The decline in q_r with ρ_∞ is a result of radiative depopulation of the excited atoms and molecules, and can be interpreted to be due to a slow onset of collision-limiting phenomena. Thus, even though the present theory fails to determine the nonequilibrium radiative heat-flux values accurately, it describes the onset of collision-limiting phenomena qualitatively. The present result demonstrates also the advantage of flying

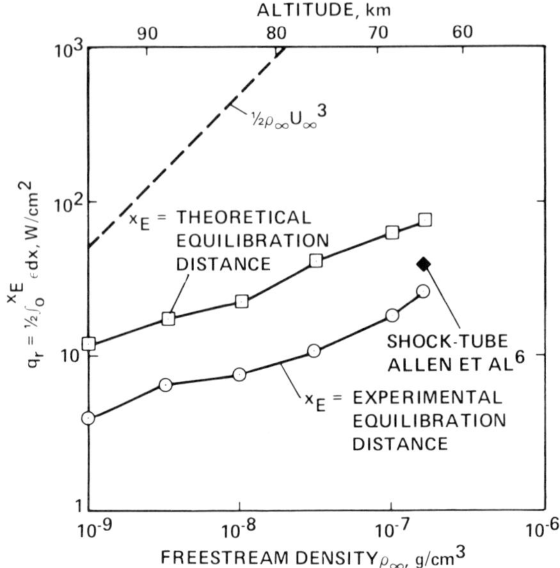

Fig. 6 Nonequilibrium radiative heat fluxes calculated for u_∞ = 10 km/s, T_∞ = 300 K, $T_{v\infty}$ = 5000 K, $\sigma = 10^{-16}$ cm^2, and rate coefficients of Park and Menees.[24]

at high altitudes. Compared with the available energy flow $(1/2)\rho_\infty u_\infty^3$, the computed q_r values are fairly small. This implies that the energy removal by radiation is small, and that the reduction in energy caused by radiation can be neglected as a first approximation, as was done in the present work.

The nonequilibrium radiative heat flux q_r corresponding to the experimentally determined equilibration distance x_E is calculated at flow velocities u_∞ of 6 to 10 km/s using the various different calculation methods. The assumed parameters are p_∞ = 0.1 Torr, $\sigma = 10^{-16}$ cm^2, T_∞ = 300 K, and $T_{v\infty}$ = 5000 K. The results are shown in Fig. 7, where they are compared with the existing shock-tube[5,6] and ballistic-range data.[26] As the figure shows, the different methods yield approximately the same results at 10 km/s. At this velocity, the calculated q_r values are centered at about 23 W/cm^2, which falls between the 40-W/cm^2 value obtained by a shock tube[6] and the 10-W/cm^2 value obtained at a ballistic range.[26] At lower velocities, however, agreement among the various methods and between the theory and experiment becomes poorer; some computed points lie outside the experimental bounds. The discrepancy can be attributed partly to the uncertainty in the measured values; the total number of independent measurements is too few to make a valid assessment. Though not shown, calculations were repeated with σ values changed to 10^{-17} cm^2, with $T_{v\infty}$ changed from 300 K to 8000 K, and with the fast-reaction-rate-coefficient set. These parameters had little effect on the calculated q_r values corresponding to the experimentally determined x_E's, though they had a significant effect on those corresponding to the theoretically determined x_E's.

Discussion

The foregoing results indicate that the current understanding of rate chemistry in air in the ionizing regime is inadequate for the purpose of predicting nonequilibrium radiation. The existing theory and an improved (two-temperature) model considerably overestimate the equilibration distance and underestimate the peak radiation distance at 10 km/s. At $p_\infty = 0.1$ Torr and

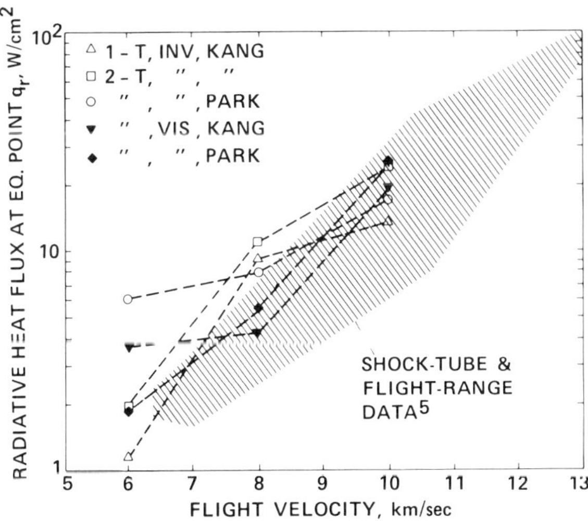

Fig. 7 Nonequilibrium radiative heat flux corresponding to the experimentally determined equilibration distance x_E, for $p_\infty = 0.1$ Torr, $T_\infty = 300$ K, $T_{v\infty} = 5000$ K, $\sigma = 10^{-16}$ cm^2, and rate coefficients of Park and Menees.[24]

u_∞ = 10 km/s, for which most accurate experimental data are available, the calculated radiative fluxes are about half the experimental value obtained in a shock tube and twice those obtained in ballistic ranges. Specifically, the following observations can be made.

1) The existing formula of Millikan and White[22] used customarily for the calculation of vibrational relaxation times is inadequate at high temperatures. The vibrational relaxation time dictates the relaxation behavior in the early part of the flow, that is, in the region immediately following the shock wave. According to the present analysis, the true relaxation times seem to be much longer at the high temperatures prevailing in this region than predicted by the formula. The erroneous relaxation-time value leads to erroneous peak-intensity points and, consequently, to erroneous radiation characteristics.

2) The conventional expression for chemical-reaction-rate coefficients employing only one temperature is inadequate. The conventional concept of molecular dissociation based on vibrational ladder-climbing processes would imply that the vibrational temperature is the temperature controlling the dissociation rates. The present results show that this model has a relatively small effect on the behavior of chemical reactions during the early period (because this period is dictated mostly by the vibrational relaxation time), but it has a great effect toward the end of the nonequilibrium region; the model considerably overestimates the equilibration distance. To be precise, dissociation can occur through the rotational ladder-climbing in addition to the vibrational ladder-climbing process. Since rotational temperature is higher than the vibrational temperature in the nonequilibrium regime, the rotational process will be contributing significantly to the overall dissociation rate. Only an expression that correctly accounts for this effect can be expected to predict the dissociation rates correctly.

It is possible that dissociation processes are accelerated by the excitation of internal degrees of freedom or by ions and electrons. For example, an electronically excited oxygen atom could cause reactions (4) and (5) in Table 1 to proceed at rates faster than could a ground-state atom because of its large reaction cross sections. Alternatively, electrons may participate in the collisional dissociation processes, reactions (1) through (3) in Table 1, as a third body, that is, the third body M = e. If the reaction cross sections of these processes were the same as those for heavy particles (M = heavy particle), then the resulting rates will be about 200 times faster than with the heavy particles because of the faster average speeds of electrons.

3) The existing experimental data base on nonequilibrium radiation is inadequate for verifying the accuracy of the theories. At the flow velocity of 10 km/s for which the most accurate set of data exists, the experimental data are scattered by a factor of about 4. At other velocities, there is an insufficient number of independent measurements to enable valid assessment of their accuracy.

Conclusions

Vibrational relaxation times calculated using the formula of Millikan and White[22] considerably underestimate the relaxation times at high temperatures, leading to an underestimation of the distance at which radiation reaches the maximum intensity. The conventional reaction-rate model considerably overestimates the distance at which the flow reaches equilibrium, leading to an overestimation of the nonequilibrium radiation heat flux. Further research into these two areas is needed. A nonequilibrium radiative heat flux of about 20 W/cm^2 is predicted by the present two-temperature theory at $p_\infty = 0.1$ Torr and $u_\infty = 10$ km/s with an uncertainty of a factor of ±2. The heat flux decreases steadily with decreasing velocity and density.

References

[1]Walberg, G. D., "A Review of Aeroassisted Orbit Transfer," AIAA Paper 82-1378, AIAA 9th Atmospheric Flight Mechanics Conference, San Diego, Calif., Aug. 1981.

[2]Menees, G. P., "Trajectory Analysis of Radiative Heating for Planetary Missions with Aerobraking of Spacecraft," AIAA Paper 83-0407, AIAA 21st Aerospace Sciences Meeting, Reno, Nev., Jan. 1983.

[3]Menees, G. P., "Thermal-Protection Requirements for Near-Earth Aeroassisted Orbital-Transfer Vehicle Missions," AIAA

Paper 83-1513, AIAA 18th Thermophysics Conference, Montreal, Canada, June 1983; published elsewhere in this volume.

[4]Menees, G. P., Park, C., and Wilson, J. F., "Design and Performance Analysis of a Conical-Aerobrake Orbital-Transfer Vehicle Concept," AIAA Paper 84-0410, AIAA 22nd Aerospace Sciences Meeting, Reno, Nev., Jan. 1984; published elsewhere in this volume.

[5]Park, C., "Radiation Enhancement by Nonequilibrium in Earth's Atmosphere," AIAA Paper 83-0410, AIAA 21st Aerospace Sciences Meeting, Reno, Nev., Jan. 1983.

[6]Allen, R. A., Rose, P. H., and Camm, J. C., "Nonequilibrium and Equilibrium Radiation at Super-Satellite Reentry Velocities," Research Rept. 156, AVCO-Everett Research Laboratory, Everett, Mass., Sept. 1962.

[7]Whiting, E. E., Arnold, J. O., Page, W. A., and Reynolds, R. M., "Composition of the Earth's Atmosphere by Shock-Layer Radiometry during the PAET Entry Probe Experiment," Journal of Quantitative Spectroscopy and Radiative Transfer, Vol. 13, No. 9, Sept. 1973, pp. 837-859.

[8]Park, C., "Calculation of Nonequilibrium Radiation in AOTV Flight Regimes," AIAA Paper 84-0306, AIAA 22nd Aerospace Sciences Meeting, Reno, Nev., Jan. 1984; published elsewhere in this volume.

[9]Bittker, D. A. and Scullin, V. J., "General Chemical Kinetics Computer Program for Static and Flow Reactions with Application to Combustion and Shock-Tube Kinetics," NASA TN D-6586, Jan. 1972.

[10]Bowen, S. W. and Park, C., "Computer Study of Nonequilibrium Excitation in Recombining Nitrogen Plasma Nozzle Flows," AIAA Journal, Vol. 9, No. 3, Mar. 1971, pp. 493-499.

[11]Park, C., "Comparison of Electron and Electronic Temperatures in Recombining Nozzle Flow of Ionized Nitrogen-Hydrogen Mixture. Part 1, Theory," Journal of Plasma Physics, Vol. 9, Pt. 2, July 1973, pp. 187-215.

[12]Park, C., "Comparison of Electron and Electronic Temperatures in Recombining Nozzle Flow of Ionized Nitrogen-Hydrogen Mixture. Part 2, Experiment," Journal of Plasma Physics, Vol. 9, Pt. 2, July 1973, pp. 217-234.

[13]Stollery, J. L., Smith, J. E., and Park, C., "The Effects of Vibrational Relaxation on Hypersonic Nozzle Flows," The High Temperature Aspects of Hypersonic Flow, edited by W. C. Nelson, Pergamon Press, New York, 1964, pp. 49-66.

[14]Stollery, J. L. and Park, C., "Computer Solutions to the Problem of Vibrational Relaxation in Hypersonic Nozzle Flows," Journal of Fluid Mechanics, Vol. 19, No. 1, July 1964, pp. 113-123.

[15]Lee, J. H., "Basic Governing Equations for AOTV Flight Regimes," AIAA Paper 84-1729, AIAA 19th Thermophysics Conference, Snowmass, Colo., June 1984; published elsewhere in this volume.

[16]Okuno, A. F. and Park, C., "Stagnation-Point Heat Transfer Rate in Nitrogen Plasma Flows: Theory and Experiment," Transaction of the American Society of Mechanical Engineers, Series C, Journal of Heat Transfer, Vol. 92, No. 3, Aug. 1970, pp. 372-384.

[17]Keck, J. C., "Diffusion Theory of Nonequilibrium Dissociation and Recombination," Journal of Chemical Physics, Vol. 43, No. 7, Oct. 1963, pp. 2284-2298.

[18]Bourcier, S., Tables of Constants and Numerical Data: Spectroscopic Data Related to Diatomic Molecules, Pergamon Press, Oxford, 1970.

[19]Park, C., "Collisional Ionization and Recombination Rates of Atomic Nitrogen," AIAA Journal, Vol. 7, No. 8, Aug. 1969, pp. 1653-1654.

[20]Appleton, J. P. and Bray, K. N. C., "The Conservation Equations for a Nonequilibrium Plasma," Journal of Fluid Mechanics, Vol. 20, Pt. 4, Dec. 1964, pp. 659-672.

[21]Yos, J. M., "Transport Properties of Nitrogen, Hydrogen, Oxygen, and Air to 30,000 K," Technical Memorandum RAD TM-63-7, AVCO-RAD, Wilmington, Mass., Mar. 1963.

[22]Millikan, R. C. and White, D. R., "Systematics of Vibrational Relaxation," Journal of Chemical Physics, Vol. 39, No. 12, Dec. 1963, pp. 3209-3213.

[23]Kang, S. W. and Dunn, M. G., "Theoretical and Measured Electron-Density Distributions for the RAM Vehicle at High Altitudes," AIAA Paper 72-689, AIAA 5th Fluid and Plasma Dynamics Conference, Boston, Mass., June 1972.

[24]Park, C. and Menees, G. P., "Odd Nitrogen Production by Meteoroids," Journal of Geophysical Research, Series C, Vol. 83, No. 8, Aug. 1978, pp. 4029-4035.

[25]Beam, R. M. and Warming, R. F., "an Implicit Finite-Difference Algorithm for Hyperbolic Systems in Conservation-Law Form," Journal of Computational Physics, Vol. 22, No. 1, Sept. 1976, pp. 87-110.

[26]Page, W. A. and Arnold, J. O., "Shock-Layer Radiation of Blunt Bodies at Entry Velocities," NASA TR R-193, Apr. 1964.

Progress in Noncatalytic Surfaces for Metallic Heat Shields

R. T. Swann,* G. M. Wood, Jr.,† and R. D. Brown,‡
NASA Langley Research Center, Hampton, Virginia
B. T. Upchurch,§ and G. J. Allen¶
Old Dominion University, Norfolk, Virginia

Abstract

The magnitude of the atom recombination coefficients needed for metal heat-shield surfaces on Shuttle-type vehicles is analyzed and discussed. Prior work which identifies surfaces having low catalytic activity is reviewed. Arc tunnel tests to evaluate catalytic activity are described, and the difficulties of such tests are discussed. Results of laboratory tests are presented which show major differences between atom recombination and atom exchange from molecules. Results of surface analysis show that bulk and surface composition of a coating are different.

Introduction

Shuttle heating experiments have shown clearly that aerodynamic heating of the reusable surface insulation (RSI) tiles is reduced because the tile surfaces are partially noncatalytic to atom recombination.[1,2,3] Although the ceramic tiles used on the current shuttle provide reliable and thermally efficient protection for the structure, operating factors including cost and turn-around time for future transportation vehicles might favor

---

Presented as Paper 84-1734 at AIAA 19th Thermophysics Conference, Snowmass, Colorado, June 25-28, 1984.

*Chief Scientist, Materials Division.

†Research Scientist, Instrument Research Division.

‡Aerothermodynamist, Loads and Aerolasticity Division.

§Assistant Professor of Chemistry; currently, Consulting Chemist, Chemicon Corporation, Virginia Beach, Virginia.

¶Graduate Student; currently, Analytical Chemist, Quality Assurance Materials Test and Evaluation Laboratory, Naval Weapons Station, Yorktown, Virginia.

the use of a metallic heat shield. If metal heat shields are used, they should also have noncatalytic surfaces to reduce the temperature. Lower temperatures lead to lower weight, longer lifetime, and in many cases substantially reduced cost.

This paper reports on an investigation of requirements and possibilities for reduced catalytic activity on metallic heat shields. The effects of catalytic activity coefficients on surface temperatures for a particular shuttle trajectory point are examined analytically and discussed in terms of heat-shield material requirements. Available data on catalytic activity of surfaces to atom recombination are reviewed to identify materials having low activity as well as to assess the effect of factors that are believed to reduce activity.

This project includes both arc tunnel and laboratory tests. The arc tunnel tests were intended to show the relative catalytic activity of pure oxides under realistic entry heating conditions. Results of these tests are presented and difficulties are discussed. The laboratory tests were intended to clarify basic recombination mechanisms and to provide reliable values for recombination rates at elevated temperatures. Progress in this area is discussed. Since a basic understanding of recombination mechanisms will require knowledge of surface composition, a brief review of preliminary spectroscopic results is also presented.

Background on Catalytic Activity

The efficiency of various surfaces in catalyzing the recombination of oxygen and nitrogen atoms has drawn considerable study over an extended time. The purpose of this review is to assemble the factors from those studies which may relate to selection of materials having low catalytic activity. First, the general features of past results will be summarized as they relate to the goals of this study. Then, the range of reported values for some of the less active surfaces will be reviewed to identify the levels of activity that might be achieved on heat-shield surfaces. Since heat-shield surfaces will generally be oxides, attention here centers on oxides.

Research on surface catalytic activity, relative to recombination of oxygen and nitrogen atoms, has concentrated on oxygen atoms on silica (SiO_2) or silica-based materials. Silica has probably been widely selected because it is chemically stable and easy to use. This selection of silica is helpful to the present study for two reasons. First, catalytic activity is generally low on silica-based materials. The second important factor is that "results taken all together imply that, for the oxide surfaces, the relative activities are the same for an extraordinary variety of processes.[1] This factor suggests that trends established with oxygen on silica, such as temperature dependence, apply to nitrogen on silica or to oxygen and nitrogen on other oxide surfaces. For example, the data of Greaves and Linnett[4] generally, but not uniformly, confirm the similar temperature dependence of oxygen recombination rates on various oxides.

The similarity of relative catalytic activities of various oxide surfaces for a broad range of processes was used in the present investigation. In the absence of other information, a result obtained with one gas was assumed to apply to the other. Upon this hypothesis, data with chlorine or hydrogen atoms, or with oxygen and nitrogen molecules, can be used to infer catalytic activity.[6] However, as will be shown, usefulness of data on molecules is limited. Nevertheless, at a more fundamental level, this similarity of relative activities implies that the same kinds of processes occur at the same sites, even though the gases or oxide surfaces are different. Thus, a surface treatment which reduces oxygen recombination rates should also reduce nitrogen recombination rates. In addition, if the processes do occur at the same sites, then, in a mixture of nitrogen and oxygen, the oxygen and nitrogen atoms compete for the same "recombination sites." Therefore, the specific oxygen recombination rate in the mixture should be reduced by the fraction $(1-f_N)$ relative to the rate with oxygen atoms only, where f_N is the average fraction of active sites occupied by nitrogen atoms. Thus, the recombination rate for each component in a mixture of oxygen and nitrogen atoms should be slower than for either gas alone. This effect has not been reported.

Qualitative Features of Atom Recombination at Surfaces

A brief summary of the qualitative features of reported results follows, with some discussion for clarity where needed.

1) Atomic recombination at oxide surfaces is of first order.[4,5,7] Linnett and Marsden[8] suggest, on the basis of all evidence available to them, that for oxygen atoms the primary recombination mechanism is collision and combination of gaseous atoms with surface oxygen atoms.

2) Oxygen molecules are dissociatively chemisorbed on oxygen,[5] most likely as O^-. Isotope exchange or

"scrambling" ($^{16}O_2$ + $^{18}O_2$ → 2 $^{16}O^{18}O$)shows that adsorbed oxygen atoms can migrate and combine with atoms from other molecules on the surface. However, Dickens and Sutcliffe[5] suggest that this process has negligible influence on recombination coefficients. Reference 6 shows that nitrogen molecules are not chemisorbed, or are not mobile, on silica at temperatures of interest for metallic heat shields because isotope exchange does not occur. Reference 6 also shows that gas-phase oxygen molecules exchange atoms both with other gas-phase atoms and with the oxide surface. Therefore, to the extent that chemisorbed molecules or surface oxygen atoms enter directly into the recombination process, the mechanisms for nitrogen and for oxygen recombination must differ.

3) Oxygen recombination on silica occurs at a small number of fairly active sites. Linnett and Marsden[8] suggest that approximately one surface oxygen site in 30 is active. If data obtained at low temperature[5] is extrapolated to $1/T = 0$, then a recombination probabablity of 0.063 is obtained for oxygen atom recombination on silica.

4) The more exothermic oxides are the least active catalytically.[7]

5) The least active substances tend to be glass formers.[4] These materials may have a smaller effective surface area. That is, they are smoother even on an atomic scale.

6) Catalytic activity can be very sensitive to slight changes in material or test conditions. For example, Breen[9] reports substantial variation in results with time. This sensitivity to slight changes offers the possibility that treatments can be developed which will greatly reduce the number of active sites.

Room Temperature Recombination of Oxygen Atoms

Table 1 contains representative values of the room temperature recombination probability for oxygen atoms on selected oxide surfaces. These were selected as the ones having lowest catalytic activity. As mentioned earlier, many of these are glass formers or network formers.[10] The others are generally intermediates or network modifiers.

The data show generally good agreement of rank-order between different materials. The results in the first two columns were obtained by the same investigators. The numbers in the first column were obtained later, using better techniques, and are presumably their best values for the recombination rates.

Table 1 Reported values of recombination rates of oxygen atoms on oxide surfaces at room temperature

Oxide	Recombination probability x 10^5					Properties
Pyrex	3.1-4.5		12			a
Boric acid	6.3		28			a
Arsenic	8.1		46			a
Phosphoric acid	12					a
Germanium	13					a
Gallium	13					b
Silica	16	71	35	12-31	12-107	a
Chromic oxide	25		20			
Antimony	27					
Zinc oxide	44		59			a,b,c
Vanadium pentoxide	48					
Lead oxide	63		580			c,b
Tin dioxide	100					b
Molybdenum	100					
Calcium	160					b
Aluminum	210		180-340			a,c

[a]f = network former; [b]m = network modifier; [c]i = intermediate.

Examination of the data in Table 1 leads to two tentative but important conclusions. First, network formers, intermediates, and modifiers are the materials which have low catalytic activity. Second, the catalytic activity is sensitive to small changes in or at the surface. This is supported by the fact that large day-to-day changes in activity are reported in reference 9 for silica, and by the large difference found in the activity of Pyrex compared to the basic oxide or silica.

Elevated Temperature Recombination of Oxygen Atoms

Relatively little elevated temperature data is available on the recombination of atoms at surfaces. The most data is available for oxygen recombination on silica surfaces. The results of Dickens and Sutcliffe[5] are in qualitative agreement with those of Greaves and Linnett[11] for temperatures up to 350°C; those of Breen[9] also tend to support these values in this temperature range, although his temperature measurements were not satisfactory.

At higher temperature, results for oxygen recombination on coated RSI are available from Scott[12] as well as Breen.[9] Breen's recombination probabilities are considerably higher than those given by or extrapolated from Scott's[12] data above 800°C. In any case, all observers agree that recombination probability increases with increasing temperature.

In view of the uncertainties inherent in the various measurements, the procedure used by Shinn et al.[13] is certainly warranted. In this approach, recombination probabilities in the range of the measured values are selected on the basis of the fit they provide between calculations and heating data measured in flight experiments. With regard to the present investigation where the specific goal is to find materials with low catalytic activity, these difficulties with elevated temperature experiments suggest that tests be performed at room temperature when possible.

Recombination of Atoms Other than Oxygen

Reference 14 reports room temperature recombination probabilities for nitrogen atoms on silica surfaces of less than 10^{-5}. Reference 15 found recombination probabilities for chlorine atoms on silica at room temperature to be about 10^{-4}. For chlorine, the overall activation energy of the recombination process is small (in the order of the heat of absorption of Cl), and the low number value of γ_{Cl} suggests a small number of active sites.[16]

This brief review of the literature on atom recombination at surfaces suggests that to get low recombination rates, attention should center on so-called network or glass formers. The effects of some chemical treatments on catalytic activity have also been reported.[16] However, the available information does not provide real insight into materials or treatments which can produce surfaces that are effectively noncatalytic to atom recombination during atmospheric entry.

Required Catalytic Activity

In this basic study of noncatalytic surfaces for metallic heat shields, an early need was to establish the payoff in reduced heating which is available with various levels of catalytic activity. Reference 6 suggests that a recombination probability of 0.002 is desirable, and that relatively little benefit is obtained if the probability is 0.05 or greater. However, those results were obtained from a static analysis intended to provide plausibility to researchers with materials backgrounds. The present paper contains a more definitive examination of the quantitative effects of oxygen and nitrogen atom recombination probabilities on aerodynamic heating.

The results presented here use the same calculation methods, computer program, entry conditions, and thermodynamic and transport data as those used by Shinn, Moss, and Simmonds.[13] The data are examined at one entry time near peak heating for a Shuttle entry; the altitude is 74.98 km and the velocity is 7.2 km/s.

Results are presented in terms of surface radiation equlibrium temperature for three locations on the bottom surface centerline. The temperature is calculated from the heating rate, assuming the emittance is 0.8. In these calculations, the catalytic activities (recombination probabilities) γ_O and γ_N are constant over the entire vehicle; they have the values indicated in the figures.

The effect of catalytic activity on surface temperature for a location near the nose of the vehicle (2 1/2 % chord, X/L = 0.025) is shown in Fig. 1. Values of catalytic activity in the range that may be easily achievable in practice (about 0.01) give surface temperatures significantly below the catalytic wall temperature (1340°C). A further point can be made relative to materials requirements. At 1000°C, candidate superalloy heat-shield metals have much lower strength and shorter lifetime than at room temperature, and currently 1100°C can be taken as the upper limit of use temperature for superalloys. If the entry condition analyzed here were the maximum temperature point on the design trajectory, then the recombination rates must be less than $\gamma_N = 0.001$ and $\gamma_O = 0.01$ for superalloys to be used

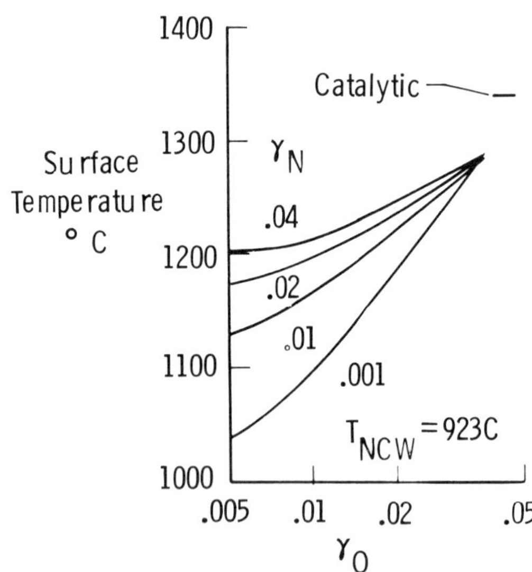

Fig. 1 Effect of atom recombination probability near Shuttle noise: altitude = 74.98 km; velocity = 7.2 km/s.

at this vehicle location. Therefore, with surfaces similar to those on the current Shuttle, the transition from ceramics to a metallic heat shield would be made downstream of this body point. However, for a much lower recombination rate ($\gamma_N = \gamma_O = 1.0 \times 10^{-5}$), the temperature can be reduced to 923°C, well within the capability of superalloys. These surface activities are in the reported range of room temperature values for nitrogen atom recombination. On the other hand, if improved alloys having 1200°C use temperature can be developed, they could be used at this vehicle location if their recombination probabilities were 0.02 or less.

Surface temperature increases rapidly with increasing oxygen recombination rate γ_O, particularly when γ_N is small; increasing γ_N leads to substantially higher temperatures. However, if γ_O is high enough to yield significant concentrations of molecular oxygen at the surface, the following gas phase reactions become important.

$$N + O_2 \rightarrow NO + O$$

$$N + NO \rightarrow N_2 + O$$

The overall effect of these reactions is to recombine N atoms in the gas phase so that the surface recombination probability of nitrogen atoms (γ_N) has negligible effect on surface temperature. This is the case if γ_O reaches a value as high as 0.04.

This effect of oxygen atom recombination on nitrogen atom concentration is apparent in Fig. 2. With a small value of γ_O, the concentration of NO is very small near

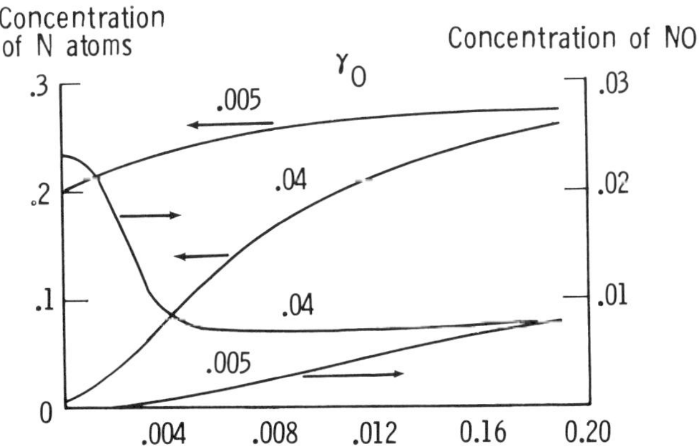

Fig. 2 Effect of oxygen atom recombination probability on boundary-layer nitrogen atom concentration profile.

the surface. This occurs because little molecular oxygen is available to generate NO, while the reaction $N + NO \rightarrow N_2 + O$ depletes the small supply of NO generated further out in the boundary layer. On the other hand, when γ_O is large, molecular oxygen is available in sufficient quantities to recombine virtually all of the atomic nitrogen in the boundary layer. This leads to the practical conclusion that if the oxygen recombination rate at a surface is high (about 0.04), then the nitrogen recombination rate is not important.

Whereas the temperature reductions shown in Fig. 1 seem significant, their practical importance can be considered in Fig. 3. This figure shows the temperature distribution along the Shuttle bottom centerline for five different catalytic conditions. First, the difference between the temperature of the catalytic wall and the surface having a catalytic activity of 0.04 corresponds to a slight shift upstream of the point where a superalloy heat shield can be used. With a 1000°C temperature limit for superalloy heat shields, a catalytic surface will require ceramic heat shields for 10 % of its length, whereas with $\gamma = 0.01$, ceramic is required only for 5 % of the length. Furthermore, the catalytic activities that might be achieved with incremental improvements in the Shuttle-type coatings do not result in any temperature reduction near the rear of the vehicle. However, a highly noncatalytic coating, that is, a coating having a recombination probability close to that of nitrogen atoms on silica at room temperature ($\gamma_N = 10^{-5}$) , would permit

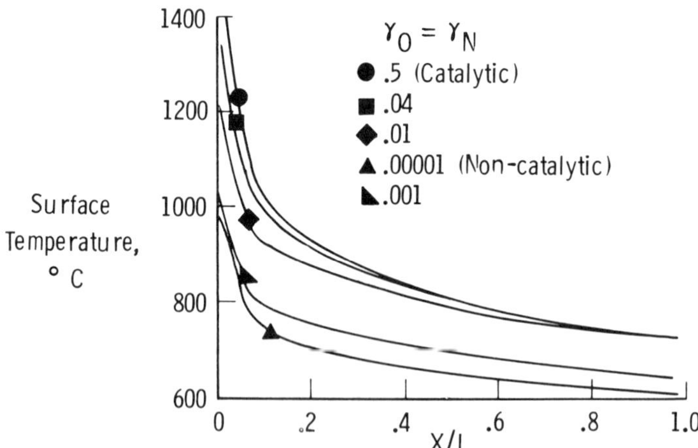

Fig. 3 Effect of atom recombination probability on Shuttle temperature distribution.

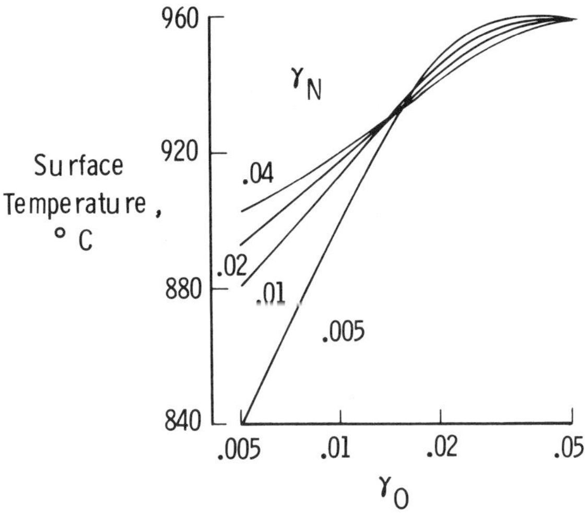

Fig. 4 Effect of atom recombination probability on Shuttle surface temperature at X/L = 0.155.

use of superalloys all the way from X/L = 0.0005 to the rear of the vehicle. In addition, the surface would remain some 200°C cooler than a catalytic surface for its entire length (see Fig. 3). A recombination probability as high as 0.001 retains most of this temperature reduction.

Figure 4 shows the effect of catalytic activity on surface temperature at the 15 % chord (X/L = 0.15). The effect is much less than for the 2.5 % chord (Fig. 2). For the 97 % chord (Fig. 5), reversals from the expected patterns emerge. That is, lower nitrogen recombination probabilities result in higher temperatures. The surface temperature reaches a peak for γ_O about 0.01, and decreases with further increases in γ_O. The magnitude of the effects is small, but the trends are unmistakable. This result is caused by the fact that the concentration of atoms is higher by a factor sufficient to maintain the heating rate despite the lower recombination probability.

The calculations presented here show that recombination probabilities substantially less than 0.01 are needed to get maximum benefit of a noncatalytic surface. On the other hand, available materials and heating data suggest that values attainable at entry temperatures will be 0.01 and higher. However, the fact that lower recombination probabilities lead to higher surface temperatures at the rear of the Shuttle surface

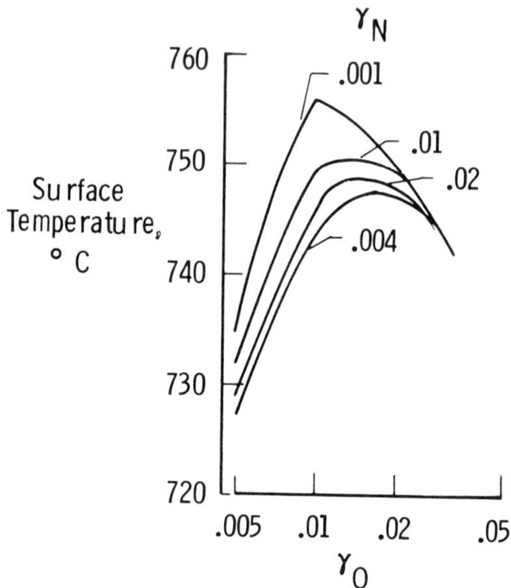

Fig. 5 Effect of atom recombination probability on Shuttle surface temperature at X/L = 0.97.

suggests that a section of high catalytic activity near the nose might remove much of the dissociation from the flow. This would cause higher temperatures at the front of the vehicle, but these temperatures could be borne by a ceramic heat shield. With dissociation removed from the flow, temperatures would be lower on the remainder of the vehicle, making the use of metallic heat shields more attractive. Calculations show the expected effect, but its magnitude is too small to have much effect on metal heat-shield design.

The calculated results are shown in Fig. 6. The surface is assumed catalytic for $X/L < 0.021$; for $X/L \geqslant 0.021$, the recombination probability was 0.01. For comparison, results with a recombination probability of 0.01 for the entire length of the vehicle are shown. The temperature reduction achieved reaches a maximum of about 35°C, and most of the reduction occurs near the point where the recombination probability decreased. Similar results were obtained when a longer span of catalytic surface was assumed. These calculations show that substantial surface temperature reduction cannot be achieved this way, that is, by removing dissociation from the flow with a section of catalytic surface.

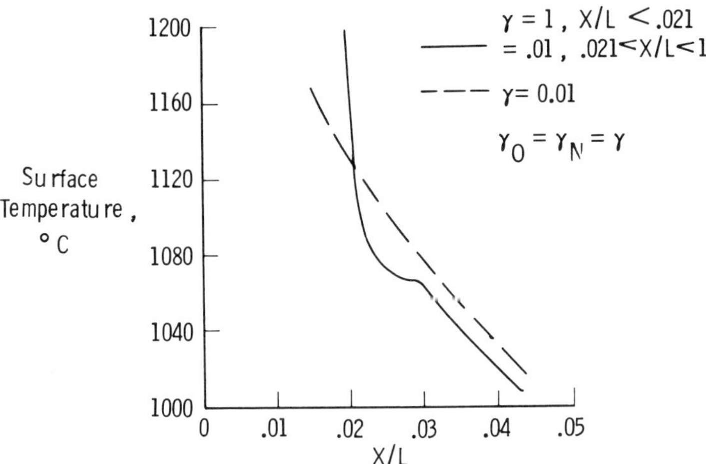

Fig. 6 Downstream effect of catalytic surface.

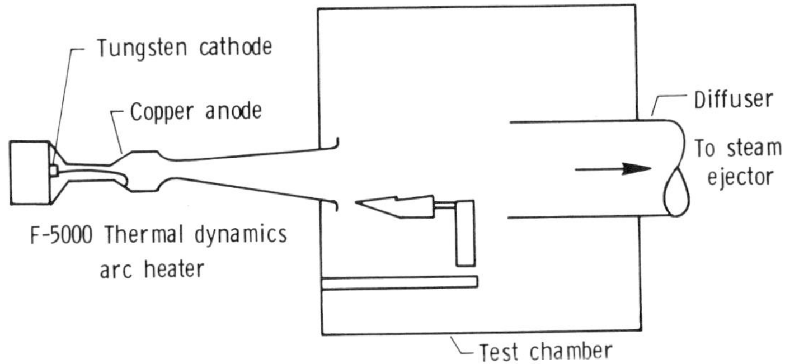

Fig. 7 Schematic of 1-MW arc tunnel.

The results and discussion presented here have been for a Shuttle-type configuration and a particular trajectory point, therefore, they may not apply to other vehicles or entry conditions. For example, Shinn and Jones[17] and Scott[18] show that for an aerobraking orbital transfer vehicle, which is a blunted configuration, the reduction in heating as a result of a noncatalytic surface is substantial, and the reduction persists to the edge of the vehicles windward surface.

Arc Tunnel Tests

The objective of this part of the project was to obtain relative catalytic activity values for a wide range of pure oxides under the most realistic ground-test conditions available. The tests were conducted in the Langley Research Center 1-MW Aerothermal Arc Tunnel.[19] Figure 7 presents a simplified diagram of the tunnel. The test environments used in this investigation were as follows: stream enthalpy 5800 Btu/lb; heating rate (Inconel calorimeter) 7 Btu/ft^2/s; surface pressure 0.0015 atm; Mach no. 4.1. The apparatus and test setup were similar to that of reference 23, and enthalpy was determined by the methods of reference 21. The specimen was mounted on a wedge, and was placed behind a rearward facing step to provide more uniform heating.[22] Specimens were heated until they reached an equlibrium temperature.

Specimens

The specimens were 5 cm x 5 cm Inconel calorimeters with an appropriate coating. A reference catalytic calorimeter was prepared by oxidation of Inconel in air at 1800°F for two hours before arc tunnel heating. The test coatings were plasma sprayed on an Inconel surface which had been sand blasted and cleaned, then covered with a thin coating of NiAl. Finally, the test coating was applied to the NiAl surface.

Because the available literature suggests that SiO_2 has the lowest recombination efficiency for oxygen and nitrogen atoms, tests of such a coating were considered particularly important for this study. However, efforts to apply such a coating by plasma spraying were unsuccessful. Therefore, a silicon coating was applied to the surface. This coating oxidized fairly early in the arc tunnel test to yield a SiO_2 surface. Application of the other coatings by plasma spraying was straightforward, although the pure oxide coatings did not prove to be adherent under thermal cycling because of thermal expansion mismatch with the substrate calorimeter. For testing, the calorimeter was placed in a foamed ceramic holder which was mounted in a wedge specimen holder as in reference 20.

Emittance

The surface emittance of each specimen was measured after testing, using the technique described in reference 23. In this technique, spectral emittance is measured at temperature and then integrated to yield total normal emittance. The emittance values obtained at 1573°C are shown in Table 2. The emittance of the SiO_2 surface seems high, but this is because only a small thickness of

Table 2 Summary of arc tunnel test results

Material	Surface temperature, °F	Surface emittance	Heating rate, Btu/ft^2/s	Relative heating rate
Oxidized Inconel	1590	0.812	7.0	1.0
Silicon	1420	0.772	4.53	0.65
Glass ($SiO_2/Na_2O/CaO$)	1410	0.743	4.61	0.66
Zirconia	1760	0.500	5.92	0.85
Alumina	1730	0.545	6.12	0.87

silicon oxidized and the emittance comes primarily from the silicon beneath the surface.

Discussion of Arc Tunnel Tests

The intent of these tests was to obtain, for a relatively complete set of candidate pure materials, relative catalytic activity at high temperatures in dissociated air. Results, to date, are very limited primarily because of difficulties in preparing the specimens (calorimeters). The specimens tested were prepared by plasma spraying. Some of the materials would not adhere to the surface during plasma spraying and others separated from the surface during arc tunnel testing.

The surfaces tested in these arc tunnel tests were selected to study materials with low catalytic activity. Silicon dioxide has the lowest reported activity, so its inclusion in the tests was essential. Alumina and zirconia were included because they have long been identified as constituents which increase the activity of silica.[16] Efforts to get other pure oxide surfaces were unsuccessful, because of coating difficulties and safety considerations.

The results of the arc tunnel tests are summarized in Table 2. The relative catalytic activities, that is, the relative heating rates in Table 2 are in the expected order. The oxidized Inconel surface experiences the highest heating rate; alumina and zirconia surfaces are the next highest,although heating was substantially less than for the Inconel; the silicon surface had the lowest heating rate. The soda-lime glass, which was tested to provide a complex surface for further analysis, experienced about the same heating as the silicon surface. The heating to the silicon surface was higher than in reference 20 for a proprietary coating tested under similar conditions. No composition analysis of the silicon surface was performed after arc tunnel testing, so conversion of the silicon surface to SiO_2 may have been incomplete.

Because of the limited number of materials tested, no definite conclusions can be drawn from the arc tunnel tests. Many of the materials which need to be evaluated, such as As_2O_3 and BeF_2, have such health hazards or uncertainties that they could not be flame sprayed under available conditions. Arc tunnel testing of these materials in pure form would also present serious contamination problems. Of course, these hazards might also impede application to entry vehicles. However, for a thorough understanding of the catalytic activity of possible heat-shield surfaces, evaluation of these materials is needed. In the future, for as many of these materials as the hazards can be overcome, either arc tunnel or laboratory evaluation will be conducted. In view of the health hazards, or at least uncertainties involved, arc tunnels do not provide a suitable method to screen pure oxides for catalytic activity. However, after basic understanding of the factors controlling catalytic activity has been acquired, arc tunnel evaluation will be necessary. Presumably, small amounts of specific materials will be necessary to control the number of active sites on a surface along with their activity. In general, if a coating is too hazardous for arc tunnel evaluation, then it is probably too hazardous for vehicle application.

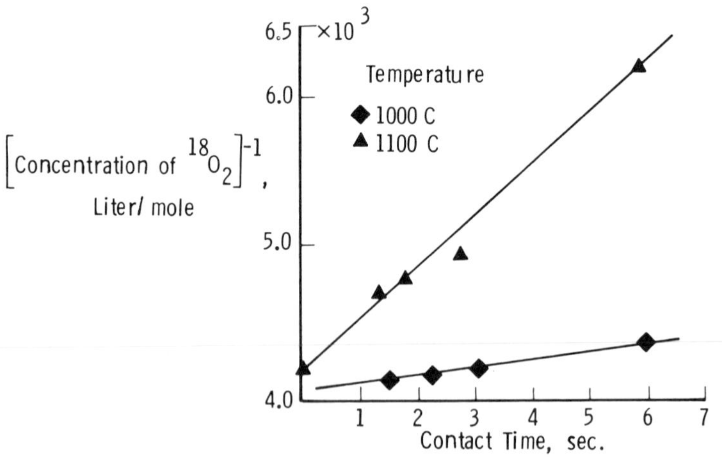

Fig. 8 Atom exchange between molecular oxygen and α-quartz indicated activation energy = 65.2 kcal/mole.

Laboratory Tests

The plan for this project was to evaluate surface catalytic activity by generating atoms in a microwave discharge, flowing them past the test surface, and measuring recombination with a mass spectrometer. A mass spectrometer was chosen over thermal methods for three reasons. First, the mass spectrometer eliminates questions about the energy accommodation coefficient. Second, it permits evaluation of catalytic activity of high temperature surfaces. Finally, it is compatible with composition sampling at selected locations within the test chamber. Unfortunately, the first apparatus constructed for these tests did not deliver atoms to the mass spectrometer because of recombination in a capillary tube. Therefore, no new spectrometric data are available.

Early tests on this project were conducted with molecules rather than with atoms.[6] This was a convenient way to check some aspects of the experimental apparatus. Reference 4 suggests enough commonality between atomic and molecular processes at oxide surfaces to make such tests useful. These tests showed that atoms in gas-phase molecules exchange with atoms from SiO_2 on the surface, and that at the surface, gas-phase molecules exchange atoms with each other. The efficiency of gas-solid atom exchange was of the order of 10^{-8} exchanges/collision on α-quartz; this process is second order (Fig. 8). For oxygen atoms on silica, the recombination probability is greater than 10^{-4} recombinations/collision and is first order. These differences in magnitude and reaction order show that major differences exist between the two processes. However, common features may exist as well.

From reference 5, molecular oxygen is generally chemisorbed on oxides, as is atomic oxygen. Then recombination of oxygen atoms, which is first order, occurs by direct recombination of adsorbed and incident atoms. On the other hand, exchange between gas-phase molecules and solid-phase atoms, which is second order,

Table 3 Composition of glass coating from ESCA

		Composition, atomic %		
Element	Initial composition	Run 1	Run 2	Run 3
Si	76	42	47	52
Ca	12	7	5	9
Na	12	51	48	39

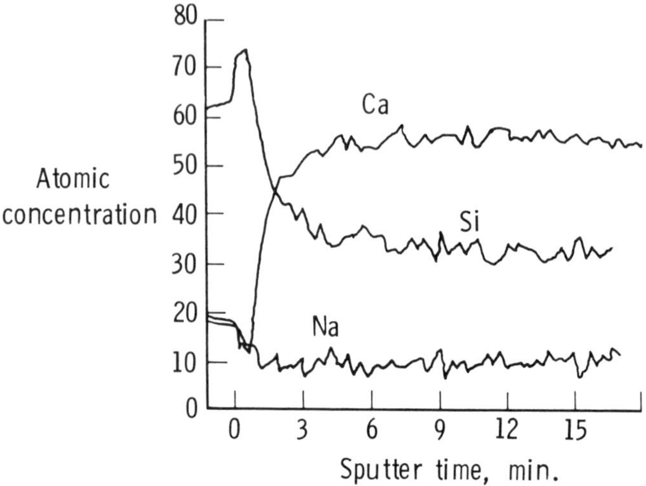

Fig. 9 Atom concentration profile in glass coating after arc tunnel test.

must be controlled by a surface migratory mechanism. Although the rate-controlling step is different, the rate in each case is proportional to the concentration of active sites on the surface. Therefore, studies using molecular oxygen may be useful in research designed to reduce the concentration of active sites. However, studies to obtain a basic understanding of the recombination process must be performed primarily with atoms.

Surface Analysis

An entry vehicle surface must be mechanically sound and have high emittance as well as low catalytic activity. Since all these requirements are unlikely to be met by a pure substance, a complex coating will be needed. For catalytic activity, more than for other properties of interest, performance depends on the outermost atomic layers at the surface. Furthermore, this surface composition, in general is not the same as the bulk composition of the phase. Therefore, a noncatalytic surface must maintain the necessary composition at the surface, while the bulk of the coating retains the composition necessary for emittance and thermal expansion control.

In view of these requirements, detailed knowledge is needed of the surface composition and structure for various bulk formulations. In this study, a preliminary investigation was undertaken of the composition of the glass coating that was tested in the arc tunnel. First, the composition near the surface (~50 A) was determined by electron spectroscopy for chemical analysis (ESCA), with the result shown in the Table 3.

These results show that compared to the initial bulk composition of the coating, the surface after arc tunnel testing is considerably depleted in silicon and calcium atoms while the concentration of sodium is substantially increased. The different compositions for various ESCA runs may result from changes in composition from point to point on the surface, since differences in color were apparent. The increased sodium concentration is the opposite of that found in reference 16, which found less sodium near the surface. The reason for this difference is not known.

To study the composition profile near the surface, Auger electron spectroscopy (AES) was used while the surface was sputtered away with high-energy ions. The surface was initially covered with a gold coating 200 to 400 A thick. Concentrations of silicon, calcium, and sodium were monitored by AES as the surface was sputtered away at approximately 500 A/min. After about one min, the gold coating is removed and the spectroscopic results should accurately represent the concentration profile. Since no standard was available, absolute levels cannot be determined. Figure 9 shows that the surface is depleted in calcium and enriched in silicon. No definite conclusion can be reached from the AES data about the concentration of sodium in this coating.

These analyses show that the surface composition of a coating, which controls catalytic activity of the surface, need not be the same as the bulk composition. Therefore, some measure of independent contol of catalytic activity, emittance, and thermal expansion is clearly possible. Extensive experiments of the kind described here will be necessary to develop an optimum coating for a metallic TPS. However, the key problem remains the identification of a surface material with extremely low catalytic efficiency for both oxygen and nitrogen atom recombination.

Concluding Remarks

This study has not progressed far enough to recommend a "best" catalytic surface for metal heat shields. However, reasonably detailed aerothermodynamic calculations suggest that a recombination coefficient of about 0.001 to 0.002 is needed for an effectively noncatalytic surface under atmospheric entry conditions. This agrees with an earlier analysis based upon static considerations. Furthermore, these detailed calculations show that whenever significant concentrations of molecular oxygen are present, nitrogen atoms recombine in the gas phase through a series of reactions involving nitric oxide.

A review of the literature shows that of the materials studied, a silicon dioxide surface has the lowest elevated temperature catalytic activity, and surfaces generally considered to be noncatalytic to atom recombination usually have a large silica content. Available evidence suggests that recombination occurs at relatively few sites, but that these sites are fairly active. If catalytic activity is to be reduced below that for silica, these sites must be eliminated or deactivated. Further insight into the basic recombination process will be necessary to achieve the goal.

The arc tunnel proved to be a poor device for screening the relative catalytic activity of pure oxides. Many of the materials of interest were too hazardous to test in an arc tunnel, and most of the others proved to be poorly adherent to a metal substrate.

Spectroscopic analysis of a coating tested in the arc tunnel showed that surface composition varied substantially from bulk composition of the coating. This shows that a coating might be developed which has surface properties optimized for minimum catalytic activity, while bulk properties are tailored for emittance and thermal expansions. As a final word, practical coatings for metals have already been shown to have catalytic activity in the general range of that available on shuttle TPS surfaces. Lower activities are needed, however, so far, neither theory nor experience show how lower values can be achieved.

Acknowledgments

The authors are grateful to Dr. J. L. Shinn for furnishing the computer program to calculate the heating of noncatalytic surfaces. Special appreciation is also extended to Rise′ M. Williams for preparation of the manuscript.

References

[1]Stewart, D. A., Rakich, J. V. and Lanfranco, M. J., "Catalytic Surface Effects Experiment on the Space Shuttle," "Thermophysics of Atmospheric Entry," Progress in Astronautics and Aeronautics, Vol. 82, 1982, pp. 248-272.

[2]Rakich, J. V., Stewart, D. A. and Lanfranco, M. J., "Results of a Flight Experiment on the Catalytic Efficiency of the Space Shuttle

Heat Shield," Entry Vehicle Heating and Thermal Protection Systems: Space Shuttle, Solar Starprobe, Jupiter Galileo Probe, Progress in Astronautics and Aeronautics, Vol. 83, 1983, pp. 97-122.

[3]Stewart, D. A., Rakich, J. V. and Lanfranco, M. J., "Catalytic Surface Effects on Space Shuttle Thermal Protection System During Earth Entry of Flights STS-2 through STS-5," Shuttle Performance: Lessons Learned, NASA CP-2283, Pt. 2, 1983.

[4]Greaves, J. C. and Linnet, J. W., "Recombination of Atoms at Surfaces. Pt. 5 Oxygen Atoms at Oxide Surfaces," Transactions of the Faraday Society, Vol. 55, 1959, p. 1346.

[5]Dickens, P. G. and Sutcliffe, M. S., "Recombination of Oxygen Atoms on Oxide Surfaces. Pt. 1 Activation Energies of Recombination," Transactions of Faraday Society, Vol. 60, 1964, p. 1272.

[6]Swann, R. T., Wood, G. M., Brown, K. D., Upchurch, B. T., Miller, I. M. and Allen, G., "Catalytic Activity of Sufaces for Metallic TPS," NASA CP-2315, June 1984.

[7]Greaves, J. C. and Linnett, J. W., "The Recombination of Oxygen Atoms at Sufaces," Transactions Faraday Society, Vol. 54, 1958, p. 1323.

[8]Linnett, J. W., F. R. S. and Marsden, D. G. H., "The Recombination of Oxygen Atoms at Salt and Oxide Surfaces," Proc. Royal Soc. A, Vol. 234, 1956, p. 504.

[9]Breen, J., et al., "Catalysis Study for Space Shuttle Vehicle Thermal Protection Systems," NASA CR 134124, 1974.

[10]Von Vlec, Llf., "Physical Ceramics for Engineers," Addison-Wesley Publishing Co., Reading, Mass., 1964.

[11]Greaves, J. C. and Linnett, J. W., "Recombination of Atoms at Surfaces. Pt. 6 Recombination of Oxygen Atoms on Silica from 20°C to 600°C," Transactions of Faraday Society, Vol. 55, 1959, p. 1955.

[12]Scott C. S., "Catalytic Recombination of Nitrogen and Oxygen on High Temperature Reusable Surface Insulation," AIAA Paper 80-1477, AIAA 15th Thermophysics Conference, Snowmass, Colorado, July 1980.

[13]Shinn, J. L., Moss, J. N., and Simmonds, A. L., "Viscous-Shock Layer Heating Analysis for the Shuttle Windward Plane with Surface Finite Catalytic Recombination Rates," AIAA Paper 82-0842, AIAA/ASME 3rd Joint Thermophysics, Fluids, Plasma, and Heat Transfer Conference, June 1982.

[14]Rahman, M. L. and Linnett, J. W., "Recombination of Atoms at Surfaces. Pt. 10 Nitrogen Atoms at Pyrex Surfaces," Transactions Faraday Society, Vol. 67, 1971, p. 170.

[15]Ashrae, P. G., Parker, A. J., and Stearn, P. E., "Recombination of Chlorine Atoms on Surfaces, Pt. 1, ESR Studies with Pyrex,

Silica, Metal and Salt Covered Surfaces," Transactions of the Faraday Society, Vol. 67, 1971, p. 3081.

[16]Little, L. H., "Infrared Spectra of Absorbed Species," Academic Press, New York, 1966.

[17]Shinn, J. L. and Jones, J. J., "Chemical Non-Equilibrium Effects on Flowfields for Aeroassist Orbital Transfer Vehicles," AIAA Paper 83-0214, AIAA 21st Aerospace Sciences Meeting, Jan. 1983.

[18]Scott, C. D., Reid, R. C., Maria, R. J., Li, C. P. and Derry, S. M., "The Aerothermodynamics and Thermal Protection Challenges for an Aerobraking Orbital Transfer Vehicle," NASA CP Pending.

[19]Chapman, A. J., "Thermal Performance of 625-kg/m^3 Elastromeric Ablative Materials and Spherically Blunted 0.44-Radian Cones," NASA TMX-2612, Nov. 1972.

[20]Pittman, C. M., Brown, R. D., and Shideler, J. L., "Evaluation of a Noncatalytic Coating for Metallic TPS," NASA TM 85745, Jan. 1984.

[21]Fay, J. A., and Riddell, F. R., "Theory of Stagnation Point Heat Transfer in Dissociated Air," Journal Aeronautical Science, Vol. 25, No. 2, Feb. 1958, pp. 73-85, 121.

[22]Brown, R. D. and Jakubowski, A. K., "Heat-Transfer and Pressure Distributions for Laminar Separated Flows Downstream of Rearward-Facing Steps With and Without Mass Suction," NASA TN D-7430, 1974.

[23]Edwards, S. F., Kantsios, A. G., Voros, John P., and Stewart, W. F., "Apparatus Description and Data Analysis of a Radiometric Technique for Measurements of Spectral and Total Normal Emittance," NASA TN D-7798, 1975.

Author Index for Volume 96

PROGRESS IN ASTRONAUTICS AND AERONAUTICS SERIES VOLUMES

VOLUME TITLE/EDITORS

***1. Solid Propellant Rocket Research** (1960)
Martin Summerfield
Princeton University

***2. Liquid Rockets and Propellants** (1960)
Loren E. Bollinger
The Ohio State University
Martin Goldsmith
The Rand Corporation
Alexis W. Lemmon Jr.
Battelle Memorial Institute

***3. Energy Conversion for Space Power** (1961)
Nathan W. Snyder
Institute for Defense Analyses

***4. Space Power Systems** (1961)
Nathan W. Snyder
Institute for Defense Analyses

***5. Electrostatic Propulsion** (1961)
David B. Langmuir
Space Technology Laboratories, Inc.
Ernst Stuhlinger
NASA George C. Marshall Space Flight Center
J.M. Sellen Jr.
Space Technology Laboratories, Inc.

***6. Detonation and Two-Phase Flow** (1962)
S.S. Penner
California Institute of Technology
F.A. Williams
Harvard University

***7. Hypersonic Flow Research** (1962)
Frederick R. Riddell
AVCO Corporation

***8. Guidance and Control** (1962)
Robert E. Roberson
Consultant
James S. Farrior
Lockheed Missiles and Space Company

***9. Electric Propulsion Development** (1963)
Ernst Stuhlinger
NASA George C. Marshall Space Flight Center

***10. Technology of Lunar Exploration** (1963)
Clifford I. Cummings and Harold R. Lawrence
Jet Propulsion Laboratory

***11. Power Systems for Space Flight** (1963)
Morris A. Zipkin and Russell N. Edwards
General Electric Company

***12. Ionization in High-Temperature Gases** (1963)
Kurt E. Shuler, Editor
National Bureau of Standards
John B. Fenn, Associate Editor
Princeton University

***13. Guidance and Control—II** (1964)
Robert C. Langford
General Precision Inc.
Charles J. Mundo
Institute of Naval Studies

***14. Celestial Mechanics and Astrodynamics** (1964)
Victor G. Szebehely
Yale University Observatory

***15. Heterogeneous Combustion** (1964)
Hans G. Wolfhard
Institute for Defense Analyses
Irvin Glassman
Princeton University
Leon Green Jr.
Air Force Systems Command

***16. Space Power Systems Engineering** (1966)
George C. Szego
Institute for Defense Analyses
J. Edward Taylor
TRW Inc.

***17. Methods in Astrodynamics and Celestial Mechanics** (1966)
Raynor L. Duncombe
U.S. Naval Observatory
Victor G. Szebehely
Yale University Observatory

***18. Thermophysics and Temperature Control of Spacecraft and Entry Vehicles** (1966)
Gerhard B. Heller
NASA George C. Marshall Space Flight Center

***19. Communication Satellite Systems Technology** (1966)
Richard B. Marsten
Radio Corporation of America

*Out of print.

***20. Thermophysics of Spacecraft and Planetary Bodies: Radiation Properties of Solids and the Electromagnetic Radiation Environment in Space** (1967)
Gerhard B. Heller
NASA George C. Marshall Space Flight Center

***21. Thermal Design Principles of Spacecraft and Entry Bodies** (1969)
Jerry T. Bevans
TRW Systems

***22. Stratospheric Circulation** (1969)
Willis L. Webb
Atmospheric Sciences Laboratory, White Sands, and University of Texas at El Paso

***23. Thermophysics: Applications to Thermal Design of Spacecraft** (1970)
Jerry T. Bevans
TRW Systems

24. Heat Transfer and Spacecraft Thermal Control (1971)
John W. Lucas
Jet Propulsion Laboratory

25. Communication Satellites for the 70's: Technology (1971)
Nathaniel E. Feldman
The Rand Corporation
Charles M. Kelly
The Aerospace Corporation

26. Communication Satellites for the 70's: Systems (1971)
Nathaniel E. Feldman
The Rand Corporation
Charles M. Kelly
The Aerospace Corporation

27. Thermospheric Circulation (1972)
Willis L. Webb
Atmospheric Sciences Laboratory, White Sands, and University of Texas at El Paso

28. Thermal Characteristics of the Moon (1972)
John W. Lucas
Jet Propulsion Laboratory

29. Fundamentals of Spacecraft Thermal Design (1972)
John W. Lucas
Jet Propulsion Laboratory

30. Solar Activity Observations and Predictions (1972)
Patrick S. McIntosh and Murray Dryer
Environmental Research Laboratories, National Oceanic and Atmospheric Administration

31. Thermal Control and Radiation (1973)
Chang-Lin Tien
University of California at Berkeley

32. Communications Satellite Systems (1974)
P.L. Bargellini
COMSAT Laboratories

33. Communications Satellite Technology (1974)
P.L. Bargellini
COMSAT Laboratories

34. Instrumentation for Airbreathing Propulsion (1974)
Allen E. Fuhs
Naval Postgraduate School
Marshall Kingery
Arnold Engineering Development Center

35. Thermophysics and Spacecraft Thermal Control (1974)
Robert G. Hering
University of Iowa

36. Thermal Pollution Analysis (1975)
Joseph A. Schetz
Virginia Polytechnic Institute

37. Aeroacoustics: Jet and Combustion Noise; Duct Acoustics (1975)
Henry T. Nagamatsu, Editor
General Electric Research and Development Center
Jack V. O'Keefe, Associate Editor
The Boeing Company
Ira R. Schwartz, Associate Editor
NASA Ames Research Center

38. Aeroacoustics: Fan, STOL, and Boundary Layer Noise; Sonic Boom; Aeroacoustic Instrumentation (1975)
Henry T. Nagamatsu, Editor
General Electric Research and Development Center
Jack V. O'Keefe, Associate Editor
The Boeing Company
Ira R. Schwartz, Associate Editor
NASA Ames Research Center

39. Heat Transfer with Thermal Control Applications (1975)
M. Michael Yovanovich
University of Waterloo

40. Aerodynamics of Base Combustion (1976)
S.N.B. Murthy, Editor
Purdue University
J.R. Osborn, Associate Editor
Purdue University
A.W. Barrows and J.R. Ward, Associate Editors
Ballistics Research Laboratories

41. Communications Satellite Developments: Systems (1976)
Gilbert E. LaVean
Defense Communications Agency
William G. Schmidt
CML Satellite Corporation

42. Communications Satellite Developments: Technology (1976)
William G. Schmidt
CML Satellite Corporation
Gilbert E. LaVean
Defense Communications Agency

43. Aeroacoustics: Jet Noise, Combustion and Core Engine Noise (1976)
Ira R. Schwartz, Editor
NASA Ames Research Center
Henry T. Nagamatsu, Associate Editor
General Electric Research and Development Center
Warren C. Strahle, Associate Editor
Georgia Institute of Technology

44. Aeroacoustics: Fan Noise and Control; Duct Acoustics; Rotor Noise (1976)
Ira R. Schwartz, Editor
NASA Ames Research Center
Henry T. Nagamatsu, Associate Editor
General Electric Research and Development Center
Warren C. Strahle, Associate Editor
Georgia Institute of Technology

45. Aeroacoustics: STOL Noise; Airframe and Airfoil Noise (1976)
Ira R. Schwartz, Editor
NASA Ames Research Center
Henry T. Nagamatsu, Associate Editor
General Electric Research and Development Center
Warren C. Strahle, Associate Editor
Georgia Institute of Technology

46. Aeroacoustics: Acoustic Wave Propagation; Aircraft Noise Prediction; Aeroacoustic Instrumentation (1976)
Ira R. Schwartz, Editor
NASA Ames Research Center
Henry T. Nagamatsu, Associate Editor
General Electric Research and Development Center
Warren C. Strahle, Associate Editor
Georgia Institute of Technology

47. Spacecraft Charging by Magnetospheric Plasmas (1976)
Alan Rosen
TRW Inc.

48. Scientific Investigations on the Skylab Satellite (1976)
Marion I. Kent and Ernst Stuhlinger
NASA George C. Marshall Space Flight Center
Shi-Tsan Wu
The University of Alabama

49. Radiative Transfer and Thermal Control (1976)
Allie M. Smith
ARO Inc.

50. Exploration of the Outer Solar System (1976)
Eugene W. Greenstadt
TRW Inc.
Murray Dryer
National Oceanic and Atmospheric Administration
Devrie S. Intriligator
University of Southern California

51. Rarefied Gas Dynamics, Parts I and II (two volumes) (1977)
J. Leith Potter
ARO Inc.

52. Materials Sciences in Space with Application to Space Processing (1977)
Leo Steg
General Electric Company

53. Experimental Diagnostics in Gas Phase Combustion Systems (1977)
Ben T. Zinn, Editor
Georgia Institute of Technology
Craig T. Bowman, Associate Editor
Stanford University
Daniel L. Hartley, Associate Editor
Sandia Laboratories
Edward W. Price, Associate Editor
Georgia Institute of Technology
James G. Skifstad, Associate Editor
Purdue University

54. Satellite Communications: Future Systems (1977)
David Jarett
TRW Inc.

55. Satellite Communications: Advanced Technologies (1977)
David Jarett
TRW Inc.

56. Thermophysics of Spacecraft and Outer Planet Entry Probes (1977)
Allie M. Smith
ARO Inc.

57. Space-Based Manufacturing from Nonterrestrial Materials (1977)
Gerard K. O'Neill, Editor
Princeton University
Brian O'Leary, Assistant Editor
Princeton University

58. Turbulent Combustion (1978)
Lawrence A. Kennedy
State University of New York at Buffalo

59. Aerodynamic Heating and Thermal Protection Systems (1978)
Leroy S. Fletcher
University of Virginia

60. Heat Transfer and Thermal Control Systems (1978)
Leroy S. Fletcher
University of Virginia

61. Radiation Energy Conversion in Space (1978)
Kenneth W. Billman
NASA Ames Research Center

62. Alternative Hydrocarbon Fuels: Combustion and Chemical Kinetics (1978)
Craig T. Bowman
Stanford University
Jorgen Birkeland
Department of Energy

63. Experimental Diagnostics in Combustion of Solids (1978)
Thomas L. Boggs
Naval Weapons Center
Ben T. Zinn
Georgia Institute of Technology

64. Outer Planet Entry Heating and Thermal Protection (1979)
Raymond Viskanta
Purdue University

65. Thermophysics and Thermal Control (1979)
Raymond Viskanta
Purdue University

66. Interior Ballistics of Guns (1979)
Herman Krier
University of Illinois at Urbana-Champaign
Martin Summerfield
New York University

67. Remote Sensing of Earth from Space: Role of "Smart Sensors" (1979)
Roger A. Breckenridge
NASA Langley Research Center

68. Injection and Mixing in Turbulent Flow (1980)
Joseph A. Schetz
Virginia Polytechnic Institute and State University

69. Entry Heating and Thermal Protection (1980)
Walter B. Olstad
NASA Headquarters

70. Heat Transfer, Thermal Control, and Heat Pipes (1980)
Walter B. Olstad
NASA Headquarters

71. Space Systems and Their Interactions with Earth's Space Environment (1980)
Henry B. Garrett and Charles P. Pike
Hanscom Air Force Base

72. Viscous Flow Drag Reduction (1980)
Gary R. Hough
Vought Advanced Technology Center

73. Combustion Experiments in a Zero-Gravity Laboratory (1981)
Thomas H. Cochran
NASA Lewis Research Center

74. Rarefied Gas Dynamics, Parts I and II (two volumes) (1981)
Sam S. Fisher
University of Virginia at Charlottesville

75. Gasdynamics of Detonations and Explosions (1981)
J.R. Bowen
University of Wisconsin at Madison
N. Manson
Université de Poitiers
A.K. Oppenheim
University of California at Berkeley
R.I. Soloukhin
Institute of Heat and Mass Transfer, BSSR Academy of Sciences

76. Combustion in Reactive Systems (1981)
J.R. Bowen
University of Wisconsin at Madison
N. Manson
Université de Poitiers
A.K. Oppenheim
University of California at Berkeley
R.I. Soloukhin
Institute of Heat and Mass Transfer, BSSR Academy of Sciences

77. Aerothermodynamics and Planetary Entry (1981)
A.L. Crosbie
University of Missouri-Rolla

78. Heat Transfer and Thermal Control (1981)
A.L. Crosbie
University of Missouri-Rolla

79. Electric Propulsion and Its Applications to Space Missions (1981)
Robert C. Finke
NASA Lewis Research Center

80. Aero-Optical Phenomena (1982)
Keith G. Gilbert and Leonard J. Otten
Air Force Weapons Laboratory

81. Transonic Aerodynamics (1982)
David Nixon
Nielsen Engineering & Research, Inc.

82. Thermophysics of Atmospheric Entry (1982)
T.E. Horton
The University of Mississippi

83. Spacecraft Radiative Transfer and Temperature Control (1982)
T.E. Horton
The University of Mississippi

84. Liquid-Metal Flows and Magnetohydrodynamics (1983)
H. Branover
Ben-Gurion University of the Negev
P.S. Lykoudis
Purdue University
A. Yakhot
Ben-Gurion University of the Negev

85. Entry Vehicle Heating and Thermal Protection Systems: Space Shuttle, Solar Starprobe, Jupiter Galileo Probe (1983)
Paul F. Bauer
McDonnell Douglas Astronautics Company
Howard E. Collicott
The Boeing Company

86. Spacecraft Thermal Control, Design, and Operation (1983)
Howard E. Collicott
The Boeing Company
Paul E. Bauer
McDonnell Douglas Astronautics Company

87. Shock Waves, Explosions, and Detonations (1983)
J.R. Bowen
University of Washington
N. Manson
Université de Poitiers
A.K. Oppenheim
University of California at Berkeley
R.I. Soloukhin
Institute of Heat and Mass Transfer, BSSR Academy of Sciences

88. Flames, Lasers, and Reactive Systems (1983)
J.R. Bowen
University of Washington
N. Manson
Université de Poitiers
A.K. Oppenheim
University of California at Berkeley
R.I. Soloukhin
Institute of Heat and Mass Transfer, BSSR Academy of Sciences

89. Orbit-Raising and Maneuvering Propulsion: Research Status and Needs (1984)
Leonard H. Caveny
Air Force Office of Scientific Research

90. Fundamentals of Solid-Propellant Combustion (1984)
Kenneth K. Kuo
The Pennsylvania State University
Martin Summerfield
Princeton Combustion Research Laboratories, Inc.

91. Spacecraft Contamination: Sources and Prevention (1984)
J.A. Roux
The University of Mississippi
T.D. McCay
NASA Marshall Space Flight Center

92. Combustion Diagnostics by Nonintrusive Methods (1984)
T.D. McCay
NASA Marshall Space Flight Center
J.A. Roux
The University of Mississippi

93. The INTELSAT Global Satellite System (1984)
Joel Alper
COMSAT Corporation
Joseph Pelton
INTELSAT

94. Dynamics of Shock Waves, Explosions, and Detonations (1984)
J.R. Bowen
University of Washington
N. Manson
Universite de Poitiers
A.K. Oppenheim
University of California
R.I. Soloukhin
Institute of Heat and Mass Transfer, BSSR Academy of Sciences

95. Dynamics of Flames and Reactive Systems (1984)
J.R. Bowen
University of Washington
N. Manson
Universite de Poitiers
A.K. Oppenheim
University of California
R.I. Soloukhin
Institute of Heat and Mass Transfer, BSSR Academy of Sciences

96. Thermal Design of Aeroassisted Orbital Transfer Vehicles (1985)
H.F. Nelson
University of Missouri-Rolla

(Other Volumes are planned.)